Modern Organic Chemistry

Modern Organic Chemistry

Edited by **Saul Rowen**

NY RESEARCH PRESS

New York

Published by NY Research Press,
23 West, 55th Street, Suite 816,
New York, NY 10019, USA
www.nyresearchpress.com

Modern Organic Chemistry
Edited by Saul Rowen

International Standard Book Number: 978-1-63238-484-3 (Hardback)

The publisher's policy is to use permanent paper from mills that operate a sustainable forestry policy. Furthermore, the publisher ensures that the text paper and cover boards used have met acceptable environmental accreditation standards.

Trademark Notice: Registered trademark of products or corporate names are used only for explanation and identification without intent to infringe.

Printed in the United States of America.

Contents

Preface

Organic chemistry refers to the scientific study of the compounds which have carbon bonds. Organic compounds also have oxygen, nitrogen, chlorine, bromine or sulphur. Their study incorporates examining their structure, composition, properties, bonding and reactions. Modern organic chemistry uses many different techniques to study organic compounds like nuclear magnetic resonance (NMR) spectroscopy which deals with atom connectivity, elemental analysis which refers to deduction of elemental composition of a molecule, mass spectrometry which is the study of molecular weight and pattern of its structure, crystallography which deals with finding molecular geometry, etc. This book will trace the progress made in this field and its sub-fields and also highlight some of the key theories and their applications. It will unfold the innovative aspects of this area. Those with an interest in this subject will find this book helpful. It will serve as a valuable guide for students and researchers alike. It will also help new researchers by foregrounding their knowledge in this field.

The researches compiled throughout the book are authentic and of high quality, combining several disciplines and from very diverse regions from around the world. Drawing on the contributions of many researchers from diverse countries, the book's objective is to provide the readers with the latest achievements in the area of research. This book will surely be a source of knowledge to all interested and researching the field.

In the end, I would like to express my deep sense of gratitude to all the authors for meeting the set deadlines in completing and submitting their research chapters. I would also like to thank the publisher for the support offered to us throughout the course of the book. Finally, I extend my sincere thanks to my family for being a constant source of inspiration and encouragement.

Editor

Diastereoselective Catalytic Hydrogenation of Schiff Bases of *N*-Pyruvoyl-(*S*)-Proline Esters

Toratane Munegumi[1]*, Shokichi Ohuchi[2], Kaoru Harada[2]

[1]Department Science Education, Naruto University of Education, Naruto, Japan
[2]Department of Chemistry, University of Tsukuba, Tsukuba, Japan
Email: *tmunegumi@naruto-u.ac.jp

Abstract

Diastereoselective catalytic hydrogenation of pyruvic acid esters, amides, and their Schiff bases has been well studied over a long period to show that proline is one of the most effective chiral auxiliaries. Proline derivatives have been used as auxiliaries in the diastereoselective catalytic hydrogenation of pyruvamide Schiff bases. The diastereoselective hydrogenation resulted in up to a 78% enantiomeric excess of the amino acid derived from the hydrolysis of the dipeptide products. The chelation hypothesis explains the stereochemistry of the catalytic hydrogenation using (*S*)-proline esters in the amide moiety and the two chiral centers in the amide and Schiff base moieties.

Keywords

Diastereoselective; Catalytic Hydrogenation; Proline; Pyruvamide; Schiff Base

1. Introduction

Pyruvic acid is an important precursor for the biosynthesis of amino acids [1] [2]. In the pathway of amino acid biosynthesis, pyruvic acid possessing no chiral centers can be derived via a pyridoxal at the active center of the enzymes involved [1] [2]. This chiral induction may be explained by the role of the chiral active center of enzymes composed of a homochiral protein. The pyruvic acid intermediates and transition states always form a chiral complex at the active center of the enzyme proteins. The newly formed chiral amino acid and the enzyme active center including pyridoxal form a type of diastereomeric intermediate, specifically an L-amino acid plus

---
*Corresponding author.

L-protein and a D-amino acid plus L-protein. The former intermediate is often more stable than the latter and yields L-amino acids. The diastereomeric intermediate formation from pyruvic acid and L-protein is recognized as a diastereoselective reaction. The diastereoselectivity [3] is almost exclusive.

On the other hand, the diastereoselective hydrogenation of biosynthesis has been imitated by organic synthesis [4]. Model reactions using pyridoxal *in vitro* and catalytic hydrogenation exemplify this synthesis. Diastereoselective catalytic hydrogenation of pyruvic acid and related compounds has long been investigated and demonstrates that the chelation of a vicinal C=O plus C=N group or two vicinal C=O groups to palladium catalysts (**Figure 1**) affects higher diastereoselectivity [5]-[20]. The chiral center of pyruvic acid amide moiety [11] [15]-[20] can influence the configuration of the newly formed chiral center. In particular, proline and its derivatives as chiral auxiliaries have demonstrated their effectiveness [18] [20] in diastereoselective catalytic hydrogena- tion.

In this study, the diastereoselective hydrogenation process follows the formation of Schiff bases having a chiral center in the pyruvic acid amide moiety. We used (*S*)-proline derivatives as chiral auxiliaries in the pyruvic acid amide moiety and benzylamine in the Schiff base moiety. We also examined diastereoselective hydrogenation using chiral 1-phenylethylamines in the Schiff base moiety to examine the effect of a double chiral auxiliary. (*S*)-Proline is the same as L-proline, and we use *R*, *S*-nomenclature in preference to D, L-nomenclature to indicate the stereochemistry of chiral amines used in this study. Preparation and hydrogenation of substrates were conducted as shown in **Figures 2** and **3**.

2. Results and Discussion

2.1. Diastereoselective Catalytic Hydrogenation Using Achiral Benzylamine

Results of the catalytic hydrogenation of Schiff bases of *N*-pyruvoyl-(*S*)-proline esters using two methods: A (Pd(OH)₂-C) and B (Pd-C and then Pd(OH)₂-C) are shown in **Table 1**.

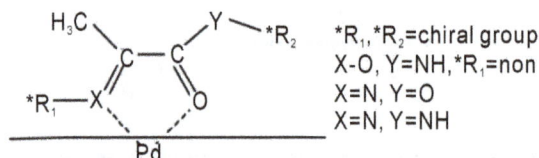

Figure 1. The chelation hypothesis of the diastereoselective catalytic hydrogenation.

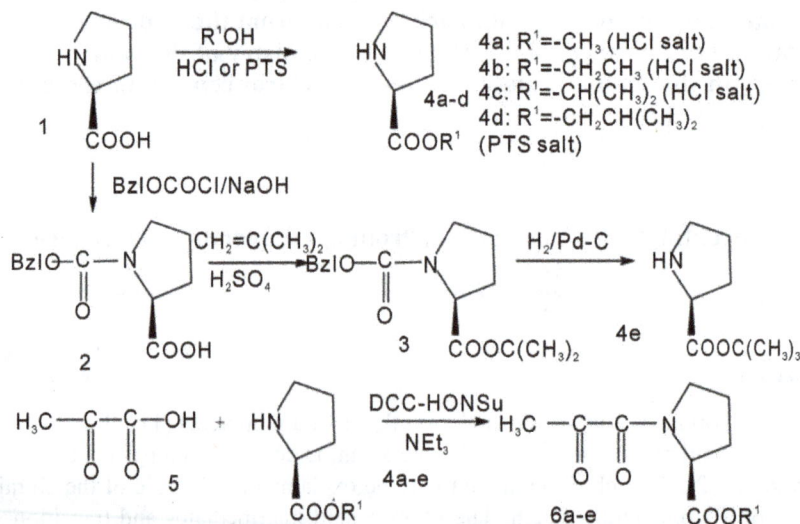

Figure 2. Preparation of *N*-pyruvoyl-(*S*)-proline esters. PTS: *p*-toluenesulfonic acid; Bzl: benzyl; DCC: *N*, *N*′-dicyclohexylcarbodiimide; HOSu: N-hydroxy-succinimide; NEt₃: triethylamine.

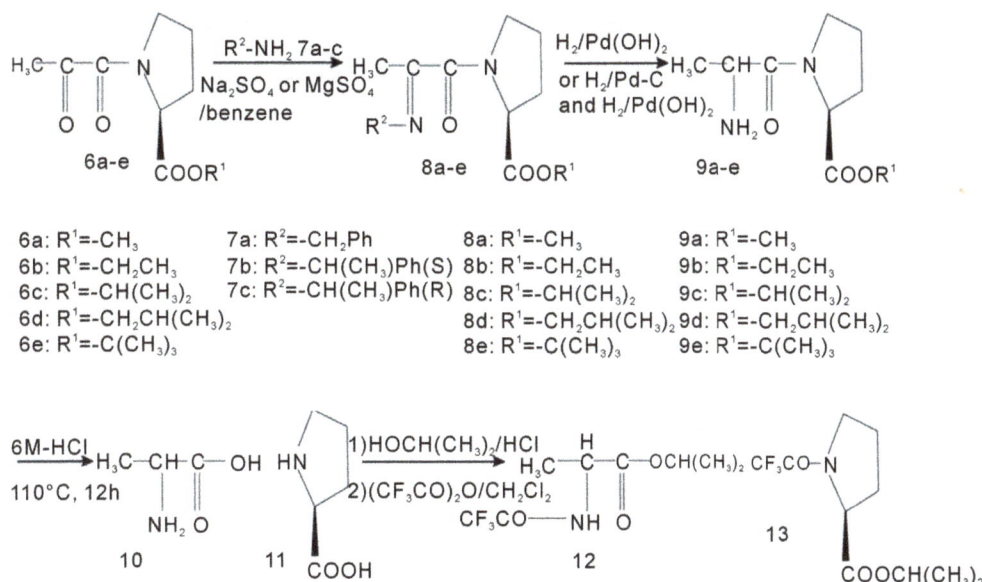

Figure 3. Preparation of Schiff bases of N-pyruvoyl-(S)-proline esters and their reaction process after hydrogenation.

The configuration of enantiomeric excess (e.e.) of alanine was (S)-configuration (20% to 69%). With the exception of the 2-methyl-2-propyl ester (No. 34 to 36), the e.e. of alanine did not cause a big difference (44% - 69%), because of the size of the ester moiety. The reactions gave a lower e.e. as follows: 22% in 2-propanol, 23% tetrahydrofuran, and 28% in ethyl acetate.

Although the e.e. obtained using method A was similar to that obtained using method B, the yield of the reaction using method A was slightly smaller than that using method B. Comparing entry No. 1 with 2, and entry No. 5 with 6, and so on provides typical examples. Yield of lactic acid is shown (10% to 40%) for several reactions. Lactic acid is formed by the hydrogenation of the remaining N-pyruvoyl-(S)-proline esters without the formation of Schiff bases.

2.2. Diastereoselective Catalytic Hydrogenation Using Achiral Benzylamine at Different Temperatures

Table 2 shows the results of asymmetric hydrogenation of Schiff bases of N-pyruvoyl-(S)-proline esters by method A (Pd(OH)$_2$-C) at different temperatures.

The configuration of alanine in enantiomeric excess (e.e.) was (S)-configuration (23% to 78%). With the exception of 2-methyl-2-propyl ester (No. 10, 12), the e.e. of alanine did not cause a big difference (40% to 78%) because of the size of the ester moiety. The highest e.e.s from **8a** and **8c** were 75% in 2-propanol at −10°C (No. 6) and 78% in ethanol at −10°C (No. 13), respectively. The yield of the reactions was lower at lower temperatures and in 2-propanol, tetrahydrofuran, and ethyl acetate.

2.3. Plausible Stereochemistry of Diastereoselective Catalytic Hydrogenation of Schiff Bases of *N*-Pyruvoyl-(*S*)-Proline Esters

From the results shown in **Tables 1** and **2**, a plausible steric pathway is shown in **Figure 4**.

Substrates **8a-e** have two main rotatable bonds (C–C bond and C–N bond) as shown in **Figure 4**. There are four important conformers, including s-cis [21] and s-trans forms. These conformers may form intermediates (a-1), (b), (c), or (d) on the palladium catalyst. There is the equilibrium between the stable chelation intermediate (a-1) and the unstable intermediate (b), and between the stable intermediates (c) and the unstable intermediate (d). Intermediates (a-1) and (c) will be in a higher proportion than intermediates (b) and (d). The lower proportion of intermediates (b) and (d) results from their larger steric repulsion. Important conformers (a-1) and (c) can be hydrogenated to give (S)-alanyl-(S)-proline esters and (R)-alanyl-(S)-proline esters, respectively. The products

Table 1. Diastereoselective catalytic hydrogenation of Schiff bases (R^1=Bzl) of N-pyruvoyl-(S)-proline esters at 30°C.

No.	Schiff base	Solvent[a] (method)[b]	Yield of Ala (Lac)[c] (%)	S (e.e. %)
1	**8a**	MeOH (A)	50 (23)	64
2		MeOH (B)	82	58
3		EtOH (A)	58 (26)	59
4		EtOH (B)	58	53
5		2-PrOH (A)	34 (40)	55
6		2-PrOH (B)	57	55
7		2-Me-1-PrOH (B)	20	67
8		THF (A)	38 (40)	47
9		AcOEt (A)	45 (22)	52
10		AcOEt (B)	25	44
11	**8b**	MeOH (A)	58 (25)	63
12		MeOH (B)	65	62
13		EtOH (A)	55 (26)	66
14		EtOH (B)	67	60
15		2-PrOH (A)	60 (25)	48
16		2-PrOH (B)	70	58
17		2-Me-2-PrOH (B)	67	69
18		THF (A)	59 (20)	45
19		AcOEt (A)	62 (27)	52
20	**8c**	MeOH (A)	49 (40)	65
21		MeOH (B)	48	55
22		EtOH (A)	38 (33)	63
23		EtOH (B)	46	55
24		2-PrOH (A)	27 (28)	58
25		2-PrOH (B)	50	53
26		2-Me-2-PrOH (B)	50	49
27		THF (A)	23 (20)	47
28		AcOEt (A)	28 (22)	55
29		AcOEt (B)	34	51
30	**8d**	MeOH (B)	30	62
31		EtOH (B)	17	63
32	**8e**	MeOH (A)	20 (17)	55
33		EtOH (A)	18 (21)	51
34		2-PrOH (A)	22 (17)	20
35		THF (A)	16 (10)	23
36		AcOEt (A)	19 (15)	28

[a]2-PrOH, 2-propanol; THF, tetrahydrofuran; AcOEt, ethyl acetate; [b]method A, H_2/Pd(OH)$_2$; method B, H_2/Pd-C and then H_2/Pd(OH)$_2$-C; [c]Yield of lactic acid in the bracket.

formed after passing through intermediate (a-1) can explain the actual results in **Tables 1** and **2** better than that after passing through intermediate (c). Five-membered ring formation on palladium catalyst had been observed to make the chelation intermediate reasonable [22].

2.4. Diastereoselective Catalytic Hydrogenation Using Chiral 1-Phenylethylamines

Table 3 shows the results of diastereoselective catalytic hydrogenation of chiral Schiff bases of N-pyruvoyl-(S)-proline esters in methanol.

Table 2. Diastereoselective catalytic hydrogenation of Schiff bases (R^1=Bzl) of N-pyruvoyl-(S)-proline esters at different temperatures by method A.

No.	Schiff base	Solvent	Temp (°C)	Yield of Ala (%)	S (e.e.%)
1	8a	MeOH	30	50	64
2		MeOH	10	29	62
3		MeOH	−10	32	67
4		MeOH	−30	10	50
5		EtOH	−10	33	46
6		2-PrOH	−10	20	75
7		THF	−10	24	45
8		AcOEt	−10	28	71
9	8c	MeOH	30	38	63
10		MeOH	10	23	29
11		MeOH	−10	39	77
12		MeOH	−30	8	23
13		EtOH	−10	40	78
14		2-PrOH	−10	22	42
15		THF	−10	15	40
16		AcOEt	−10	25	58

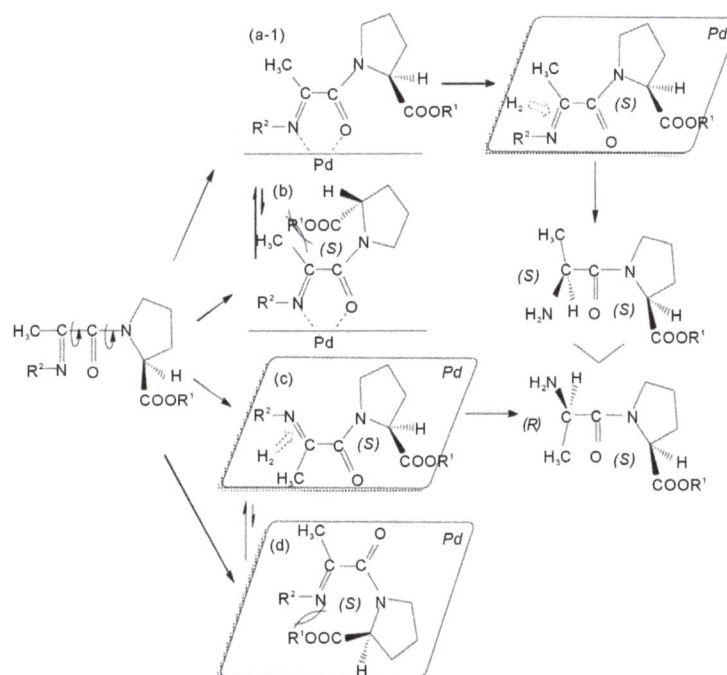

Figure 4. Plausible stereochemical course in the diastereoselective hydrogenation of Schiff bases of N-pyruvoyl-(S)-proline esters and benzylamine.

The configuration of alanine in enantiomeric excess (e.e.) was (S)-configuration. The e.e. was always larger when the configuration of 1-phenylethylamine was (R), rather than when the configuration of 1-phenylethylamine was (S). For example, No. 3 and No. 4 gave 67% and 54%, respectively. The results suggest that (R)-1-phenylethylamine produces more stable intermediates, which yield (S)-alanyl-(S)-proline esters, than (S)-1-phenylethylamine.

Figure 5 shows a plausible stereochemical pathway for the diastereoselective catalytic hydrogenation of Schiff bases between chiral amines and N-pyruvoyl-(S)-proline esters.

Table 3. Diastereoselective catalytic hydrogenation of chiral Schiff bases of *N*-pyruvoyl-(*S*)-proline esters in methanol.

No.	Schiff base	method	Temp (°C)	Config. in R2	Yield of Ala (%)	S (e.e. %)
1	8a	A	30	R	49	62
2		A	30	S	29	56
3		A	10	R	47	67
4		A	10	S	33	55
5		A	−10	R	32	42
6		A	−10	S	20	28
7	8b	B	30	R	33	71
8		B	30	S	31	39
9	8c	A	30	R	34	64
10		A	30	S	31	25
11		B	30	R		62
12		B	30	S	30	47
13		A	10	R	30	66
14		A	10	S	36	52
15		A	−10	R	22	57
16		A	−10	S	25	55
17	8d	B	30	R	22	74
18		B	30	S	19	53

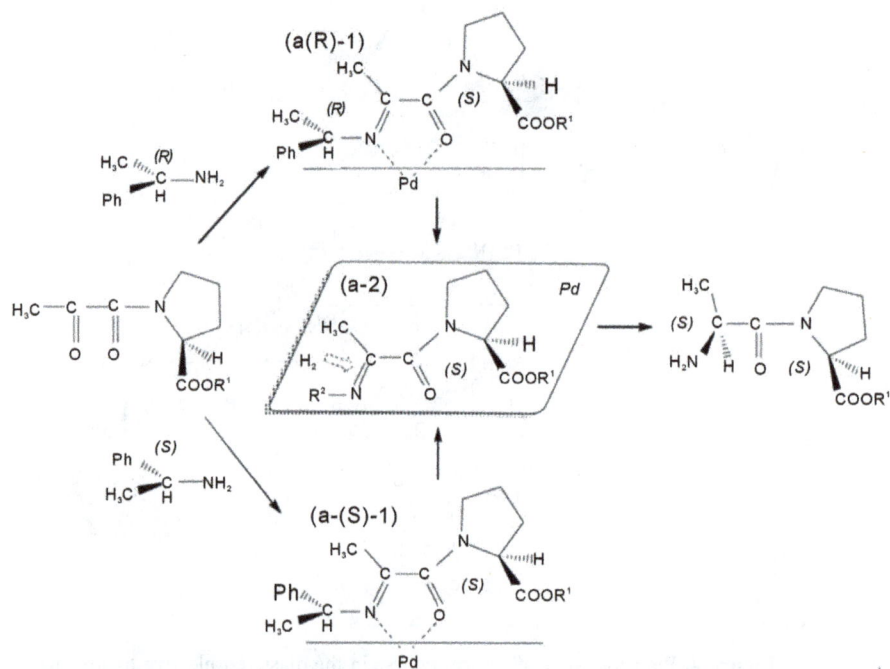

Figure 5. Plausible stereochemical course in the diastereoselective hydrogenation of Schiff bases of *N*-pyruvoyl-(*S*)-proline esters and benzylamine.

Figure 5 modifies **Figure 4** to extend the chelation hypothesis to explain the double stereo-induction using two chiral centers.

N-Pyruvoyl-(*S*)-proline esters react with (*R*)-1-phenyl-ethylamine or (*S*)-1-phenylethylamine to afford their Schiff bases. These Schiff bases form chelation intermediates (a-(*R*)-1) and (a-(*S*)-1) during the hydrogenation. Schiff bases of *N*-pyruvoyl-(*S*)-proline esters with achiral amine and chiral amines preferentially yielded (*S*)-alanyl as shown in **Tables 1-3**. These results suggest that the stereochemistry of the catalytic hydrogenation is mainly controlled by the chirality of the proline ester moiety as distinct from the chiral amine moiety in Schiff bases. However, the e.e. of alanine moiety depends on the configuration of the amine moiety as shown in **Table 3**. (R)-1-Phenylethylamine yielded a higher e.e. of (S)-alanine than (S)-1-phenylethylamine did. The results may be caused by the difference in the orientation of the substituted group in the chelation intermediates. The chelation hypothesis suggests that the substituted groups of the amine moiety were in different positions in the intermediates (a-(R)-1) and (a-(S)-1).

Intermediate (a-(R)-1) directs the methyl group of the amine moiety to the opposite side, the phenyl group to the front side, and the hydrogen to the down side. The intermediate (a-(S)-1) directs the methyl group of the amine moiety to the front side, the phenyl group to the opposite side, and the hydrogen to the down side. Because phenyl is relatively more bulky than methyl, and carboxyl ester is relatively more bulky than hydrogen, we can compare the total bulkiness between the front and the opposite sides in the chelation intermediates.

Intermediate (a-(R)-1) directs methyl plus hydrogen groups to the opposite side and phenyl plus carboxyl ester groups to the front side. Therefore, the opposite side is the less bulky side and is more likely to be adsorbed as an intermediate (a-2) and hydrogenated on the catalyst. The intermediate (a-(S)-1) directs phenyl plus hydrogen groups to the opposite side and methyl and carboxyl ester groups to the front side. In that case, although the chirality of proline ester moiety controls the stereochemistry of the chiral induction, the adsorption from both sides competes to decrease the e.e. of alanine.

The conformation of the amine moiety in this chelation intermediate is different from the conformation proposed for Schiff bases of pyruvic esters [6]. It may be because the amine moieties in the chelation intermediates (a-(R)-1 and a-(S)-1) have narrower space from the catalyst than the amine moiety of Schiff bases of pyruvic acid esters have.

3. Experimental

3.1. Instrumentation

Gas chromatographic separation of enantiomeric alanine obtained by hydrolysis of hydrogenation products was achieved using a Hitachi 163 gas chromatograph equipped with a chiral capillary column (Chirasil-Val, 25 m × 0.3 mm I.D.). Nitrogen was used as the carrier gas at 30 mL/min. The column oven temperature was programmed to increase from 80°C to 170°C at 4°C/min. A flame ionization detector was used.

Yield of the newly formed chiral amino acid was determined by using a high-performance liquid chromatography (HPLC) system composed of a Jasco UVdec-100-V UV spectrophotometer and Jasco TRI Rotor-V flow pump, equipped with a reversed phase C-18 column (TSK Inertsil ODS, TOSO, Tokyo, Japan) (4.6 mm × 250 mm). Water alone was used as the eluent. The flow rate was controlled at 0.5 ml/min. The spectrophotometer was set to detect absorption at 210 nm. The peaks on the chromatograms were integrated using a SIC Chromatocorder.

A Hitachi model 260-50 infrared spectrophotometer (Hitachi, Tokyo, Japan) was used to record the infrared spectra. A Jasco DIP-181 Digital Polarimeter was used to determine the optical rotation of the compounds synthesized from L-aspartic acid. An EX-270 NMR system (JEOL, Tokyo, Japan) was used to determine ^{1}H NMR spectra.

3.2. Chemicals

Solvents, (S)-proline (**1**), pyruvic acid (**5**), and benzylamine (**7a**) were purchased from Wako Pure Chemical Industry (Osaka, Japan). Optically active amines, (S)-1-phenylethylamine (**7b**, $[\alpha]_D^{25}$ = +39.0 (neat), 96% e.e.) and (R)-1-phenylethylamine (**7b**, $[\alpha]_D^{25}$ = -39.0 (neat), 96% e.e.) were purchased from Aldrich (Milwaukee, USA). N,N'-Dicyclohexylcarbodiimide and N-hydroxysuccinimide were purchased from Watanabe Chemical Industries (Hiroshima, Japan). 5% Palladium on charcoal was purchased from Nippon Engelhard. 10% Palladium hydroxide on charcoal was used.

3.2.1. Preparation of N-Benzyloxycarbonyl-(S)-Proline (2)

(S)-Proline (**1**) (11.5 g, 100 mmol) was dissolved in 0.5 M-sodium hydrogen carbonate solution (300 ml). To the cooled solution, benzyloxycarbonyl chloride (18.8 g, 110 mmol) was added dropwise. After the resulting suspension was stirred for 24 h at room temperature, the reaction mixture was extracted with ether. The remaining aqueous layer was acidified with concentrated hydrochloric acid to a pH of 2. The acidified aqueous solution was extracted with ethyl acetate. The obtained ethyl acetate layer was extracted with brine, and the remaining organic layer was dried with anhydrous magnesium sulfate. After removal of the magnesium sulfate by filtration, the filtrate was evaporated in vacuo, and the resulting precipitate was recrystallized with ethyl acetate to give 22.5 g (90%). M.p. 74°C - 75°C. $[\alpha]_D^{20}$ –38.0 (c 1.09, methanol).

3.2.2. (S)-Proline 2-Methyl-2-Propyl Ester (3)

Sulfuric acid (2 ml) was added to a solution containing N-benzyloxycarbonyl-(S)-proline (2) (6.00 g, 24.0 mmol) and dichloromethane (50 ml) in a pressure bottle. To the resulting solution cooled in a dry ice-acetone mixture was added 2-butene liquid that had been cooled below boiling point until the weight of the reaction mixture increased by 5 g. The reaction mixture in the sealed pressure bottle was stirred for 3 days at room temperature. The reaction mixture was neutralized with sodium hydrogen carbonate and extracted with water. The resulting organic layer was dried with anhydrous magnesium sulfate and was evaporated in vacuo to give an oily N-benzyloxycarbonyl-(S)-proline 2-methyl-2-propyl ester (6.66 g, 91.0%). $[\alpha]_D^{20}$ –45.3 (c 1.00, ethyl acetate).

3.2.3 Preparation of N-Pyruvoyl-(S)-Proline Methyl Ester (6a)

A suspension comprising (S)-Proline (1) (8.00 g, 69.4 mmol) and methanol (120 ml) was saturated with hydrogen chloride in an ice bath. The resulting suspension was stirred for 2 days, and then solvents were evaporated in vacuo to give (S)-proline methyl ester hydrogen chloride (4a) (11.80 g, 100%). $[\alpha]_D^{20}$ –45.1 (c 1.00, methanol). (S)-Proline methyl ester hydrogen chloride (4a) (8.03 g, 48.4 mmol) was neutralized with triethylamine (4.98 g, 48.4 mmol) in ethyl acetate (40 ml) in an ice bath. The solution was added to a mixture of pyruvic acid (5) (4.26 g, 48.4 mmol), dicyclohexylcarbodiimide (11.0 g, 53.3 mmol) and N-hydroxysuccinimide (5.57 g, 48.4 mmol) in an ice bath. The reaction was continued for 3 h in the ice bath and for a further 18h at room temperature. After removal of N,N'-dicyclohexylurea, the reaction mixture was evaporated in vacuo to give an oily product. The oily product was loaded onto a silica gel column and chromatography conducted using a developing agent composed of benzene-ethyl acetate (12:1 (v/v)) to afford an oily pure product 3.17 g (33%). ¹H-NMR (CDCl₃, δ): 1.50 - 2.50 (4H, m), 2.35 (3H, d), 3.64 (3H, s), 3.50 (2H, t), 4.75 (1H, m). IR (liquid, cm⁻¹): 1740, 1710, 1640 (C=O). $[\alpha]_D^{22}$ –75.1 (c 0.86, ethyl acetate). Elemental analysis: Calcd. for C₉H₁₃NO₄: C, 54.65%; H, 6.83%; N, 6.76%. Found: C, 54.26%; H, 6.57%; N, 6.73%.

3.2.4. Preparation of N-Pyruvoyl-(S)-Proline Ethyl Ester (6b)

(S)-Proline ethyl ester hydrochloride (4b) was prepared from (S)-proline (1) (8.00 g, 69.4 mmol) in ethanol (120 ml) in a manner similar to the preparation of methyl ester as an oily product (12.5 g, 100%). $[\alpha]_D^{20}$ –46.1 (c 0.41, methanol). In a manner similar to that for the preparation of the methyl ester, N-pyruvoyl-(S)-proline ethyl ester was prepared (5.70 g, 55%) from corresponding (S)-proline ethyl ester (4b) (8.69 g, 48.4 mmol). ¹H-NMR (CDCl₃, δ): 1.25 (3H, t), 1.50 - 2.50 (4H, m), 2.30 (3H, d), 3.60 (2H, m), 4.05 (2H, m), 4.70 (1H, m). IR (liquid, cm⁻¹): 1740, 1710, 1640 (C=O). $[\alpha]_D^{22}$ –75.9 (c 0.99, ethyl acetate). Elemental analysis: Calcd. for C₁₀H₁₅NO₄: C, 56.32%; H, 7.09%; N, 6.56%. Found: C, 56.02%; H, 7.13%; N, 6.47%.

3.2.5. Preparation of N-Pyruvoyl-(S)-Proline 2-Propyl Ester (6c)

(S)-Proline-2-propyl ester hydrochloride (4c) was prepared from (S)-proline (1) (8.00 g, 69.4 mmol) in 2-propanol (120 ml) in a manner similar to that used for the preparation of the oily methyl ester (13.4 g, 100%). $[\alpha]_D^{20}$ –47.8 (c 0.91, methanol). In a manner similar to that used for the preparation of the methyl ester, N-pyruvoyl-(S)-proline isopropyl ester (6c) was prepared (6.07 g, 55%) from corresponding (S)-proline 2-propyl ester (9.37 g, 48.4 mmol) in 2-propanol (120 mL). ¹H-NMR (CDCl₃, δ): 1.15 - 1.30 (6H, dd), 1.80 - 2.25 (4H, br), 2.32 - 2.38 (3H, d), 3.40 - 3.90 (2H, m), 4.20 - 4.50 (1H, m). IR (liquid, cm⁻¹): 1740, 1710, 1630 (C=O). $[\alpha]_D^{22}$ –68.1 (c 1.09, ethyl acetate). Elemental analysis: Calcd. for C₁₁H₁₇NO₄: C, 58.13%; H, 7.54%; N, 6.16%. Found: C, 57.88%; H, 7.60%; N, 6.22%.

3.2.6. Preparation of N-Pyruvoyl-(S)-Proline 2-Methyl-1-Propyl Ester (6d)

(S)-Proline (1) (5.76 g, 50.0 mmol), p-toluenesulfonic acid monohydrate (10.45 g, 55.0 mmol) and 2-methyl-1-propanol (37.4 g, 500 mmol) were refluxed together in benzene (180 ml) for 12h with the azeotropic removal of water using a Dean-Stark apparatus. After refluxing, proline spot (R_f = 0.21) disappeared, but a product spot (R_f = 0.52) clearly appeared after thin layer chromatography (developing solvent: 1-butanol-acetic acid-water (volumetric ratio: 4-1-2) of the reaction mixture, which was evaporated in vacuo to give an oily product (**4d**).

A benzene solution (100 ml) of the product (**4d**) was washed three times with saturated sodium hydrogen carbonate. The resulting benzene layer was dried with anhydrous magnesium sulfate and evaporated to give an oily product. A part of the product (1.01 g, 5.9 mmol: free ester) was dissolved with pyruvic acid (0.52 g, 5.9 mmol) in 10 ml ethyl acetate. To the cooled solution was added N-hydroxysuccinimide (0.78 g, 6.8 mmol) and dicyclohexylcarbodiimide (1.34 g, 6.5 mmol) in ethyl acetate (10 ml). After stirring at 0°C for 2 h and at room temperature for 11 h, the reaction mixture was filtered to give a pale yellow solution, which was washed with 0.5 M HCl (80 ml × 3), saturated sodium hydrogen carbonate (80 ml × 3), and brine. The resulted ethyl acetate layer was dried with anhydrous magnesium sulfate and evaporated in vacuo to give a brownish oily product, which was purified using flash chromatography (eluate: ethyl acetate-1-hexane (1 - 5)) to give an oil (0.61 g, 43%). ^{1}H-NMR (CCl$_4$, δ): 0.87 - 0.99 (6H, d), 1.40 - 2.4 (4H, br), 2.29 - 2.35 (3H, d), 3.30 - 3.70 (1H, m), 3.73 - 3.84 (2H, d), 4.50 - 5.0 (1H, m). $[\alpha]_D^{20}$ −51.8 (c 1.06, ethanol). Elemental analysis: Calcd. for C$_{12}$H$_{19}$NO$_4$: C, 59.73%; H, 7.93%; N, 5.80%. Found: C, 59.33%; H, 7.93%; N, 6.18%.

3.2.7. Preparation of N-Pyruvoyl-(S)-proline 2-Methyl-2-Propyl Ester (6e)

N-Benzyloxycarbonyl-(S)-proline 2-methyl-2-propyl ester (**3**) (6.10 g, 20.0 mmol) dissolved in ethyl acetate (80 ml) was hydrogenolyzed over 5% palladium on charcoal under hydrogen. After removal of the catalyst by filtration, the filtrate was evaporated in vacuo to give an oily product, (S)-proline t-butyl ester of 2.15 g (63%). This compound was all used for the synthesis of N-pyruvoyl-(S)-proline t-butyl ester (**6e**) in the same manner as for the synthesis of others (**6a-c**). ^{1}H-NMR (CDCl$_3$, δ): 1.44 (9H, s), 2.07 (4H, m), 2.40 (3H, d), 3.50 - 3.83 (2H, m), 4.60 - 4.83 (1H, m). IR (liquid, cm^{-1}): 1740, 1720, 1640 (C = O). $[\alpha]_D^{22}$ −70.8 (c 0.84, methanol). Elemental analysis: Calcd. for C$_{12}$H$_{19}$NO$_4$: C, 59.73%; H, 7.93%; N, 5.80%. Found: C, 59.81%; H, 8.11%; N, 6.08%.

3.3. Hydrogenolysis of Schiff Bases

N-Pyruvoyl-(S)-proline methyl ester (**6a**) (43.8 mg, 0.22 mmol) was dissolved in benzene (**7a**) (1 ml), (S)-1-phenylethylamine (**7b**), or (R)-1-phenylethylamine (**7c**). To the solution was added anhydrous sodium sulfate (500 mg) for method A or magnesium sulfate (300 mg) for method B under an atmosphere of nitrogen. A dehydration reaction was conducted at room temperature for 24 h. After removal of sodium sulfate or magnesium sulfate by filtration, the filtrate was evaporated in vacuo, and the resulting yellow oil was redissolved in methanol. The hydrogenation and hydrogenolysis were conducted using alternative methods, A and B.

Method A is as follows: the hydrogenation and subsequent hydrogenolysis were conducted over 15 mg of 10% hydroxyl palladium on charcoal under a hydrogen atmosphere at 30°C for 4 days. Method B is as follows: the hydrogenation was conducted over 30 mg of 5% palladium on charcoal at 30°C for 24 h, and then the filtrate was subsequently hydrogenolyzed over 30 mg of 10% palladium hydroxide on charcoal at 30°C for 3 h. The work-up for the reaction mixtures resulting from methods A and B was same. The filtrate obtained by the removal of catalyst was evaporated in vacuo to give a precipitate, which was hydrolyzed in 5 ml 6 M HCl at 110°C for 12 h.

3.4. Sample Preparation for Analysis of Hydrolysates of Hydrogenolysis Mixture

The resulting hydrolysate of hydrogenolysis mixture was diluted with pure water to 10 ml. An amount of 2 ml of the solution was analyzed by HPLC to determine yield. The remaining aqueous solution was evaporated in vacuo to dryness leaving a residue, to which was added 5 ml of 1.5 M HCl/2-propanol, and the solution was refluxed for 3 h. The 2-propanol solution was evaporated in vacuo, and the obtained residue was redissolved in a mixture of dichloromethane (2 ml) and trifluoroacetic anhydride (1 ml). The solution was refluxed for 1 h and was evaporated in vacuo to give an oily product. The oily product was dissolved in ethyl acetate. After washing with 1 M HCl, 4% sodium hydrogen sulfate and brine, the ethyl acetate layer was dried with anhydrous magnesium sulfate, later removed, and evaporated to an oily product for enantiomeric separation by gas chromatography.

4. Conclusions

1) Diastereoselective catalytic hydrogenation of Schiff bases of N-pyruvoyl-(S)-proline esters was conducted to afford the highest e.e. of 78% of (S)-alanine from 8c in ethanol at −10°C.

2) Preference of (S)-alanine-(S)-proline ester was explained by the chelation hypothesis based on the two-step

adsorption process.

3) Double stereo-induction by the chiral centers in the (*S*)-proline ester and the chiral amine in the Schiff moiety was conducted, and the stereochemistry can be explained by the chelation hypothesis [5].

References

[1]   Voet, D. and Voet, J.G. (2010) Biochemistry. 4th Edition, Wiley, Hoboken.

[2]   McMurry, J.E. and Begley, T.P. (2005) The Organic Chemistry of Biological Pathways. Roberts and Company Publishers, Englewood.

[3]   Zimmerman, S.C. and Breslow, R. (1984) Asymmetric Synthesis of Amino Acids by Pyridoxamine Enzyme Analogs Utilizing General Base-Acid Catalysis. *Journal of the American Chemical Society*, **106**, 1490-1491. http://dx.doi.org/10.1021/ja00317a054

[4]   Michael, B.S. and March, J. (2007) March's Advanced Organic Chemistry. 6th Edition, John Wiley & Son, Inc., Hoboken.

[5]   Harada, K. and Munegumi, T. (1991) Reduction of C=X to CHXH by Catalytic Hydrogenation. In: B. M. Trost and I. Fleming, Eds., *Comprehensive Organic Synthesis*, Elsevier Science Ltd., Oxford, 139-158. http://dx.doi.org/10.1016/B978-0-08-052349-1.00222-5

[6]   Matsumoto, K. and Harada, K. (1966) Stereoselective Syntheses of Optically Active Amino Acids from Menthyl Esters of α-Keto Acids. *Journal of Organic Chemistry*, **31**, 1956-1958. http://dx.doi.org/10.1021/jo01344a064

[7]   Harada, K. and Matsumoto, K. (1967) Sterically Controlled Syntheses of Optically Active Organic Compounds. V. Sterically Controlled Synthesis of Optically Active α-Amino Acids from α-Keto Acids by Reductive Amination. *Journal of Organic Chemistry*, **32**, 1794-1800. http://dx.doi.org/10.1021/jo01281a020

[8]   Harada, K. and Matsumoto, K. (1968) Sterically Controlled Syntheses of Optically Active Organic Compounds. VI. Solvent Effect in a Sterically Controlled Synthesis of Optically Active α-Amino Acids from α-Keto Acids by Hydrogenolytic Asymmetric Transamination. *Journal of Organic Chemistry*, **33**, 4467-4470. http://dx.doi.org/10.1021/jo01276a036

[9]   Harada, K. and Yoshida, T. (1970) A Temperature Effect in an Asymmetric Synthesis by Hydrogenolytic Asymmetric Transamination. *Journal of the Chemical Society D: Chemical Communications*, **1970**, 1071-1072. http://dx.doi.org/10.1039/c29700001071

[10] Harada, K. and Yoshida, T. (1970) Syntheses of Optically Active α-Amino Acids from Esters of α-Keto Acids by Hydrogenolytic Asymmetric Transamination, a Solvent Effect. *Bulletin of the Chemical Society of Japan*, **43**, 921-925. http://dx.doi.org/10.1246/bcsj.43.921

[11] Harada, K. and Matsumoto, K. (1971) Sterically Controlled Syntheses of Optically Active Organic Compounds. XIV. Syntheses of Dipeptides from N-(α-Ketoacyl)-α-Amino Acid Esters. *Bulletin of the Chemical Society of Japan*, **44**, 1068-1071. http://dx.doi.org/10.1246/bcsj.44.1068

[12] Harada, K. and Kataoka, Y. (1978) Asymmetric Synthesis of Alanine by Hydrogenolytic Asymmetric Transamination. *Tetrahedron Letters*, **19**, 2103-2106. http://dx.doi.org/10.1016/S0040-4039(01)94761-6

[13] Harada, K. and Kataoka, Y. (1978) The Temperature Dependence of Hydrogenolytic Asymmetric Transamination between Esters of Optically Active Phenylglycine and Pyruvic Acid. *Chemistry Letters*, **7**, 791-794. http://dx.doi.org/10.1246/cl.1978.791

[14] Harada, K. and Tamura, M. (1979) Asymmetric Synthesis of Alanine by Hydrogenolytic Asymmetric Transamination between (*R*)-2-Amino-2-Phenylethanol and Ethyl Pyruvate. *Bulletin of the Chemical Society of Japan*, **52**, 1227-1228. http://dx.doi.org/10.1246/bcsj.52.1227

[15] Harada, K., Munegumi, T. and Nomoto, S. (1981) Asymmetric Hydrogenation of Chiral Pyruvamides. *Tetrahedron Letters*, **22**, 111-114. http://dx.doi.org/10.1016/0040-4039(81)80162-1

[16] Harada, K. and Munegumi, T. (1983) Asymmetric Hydrogenations of N-Pyruvoyl-(*S*)-Amino Acids Esters. *Bulletin of the Chemical Society of Japan*, **56**, 2774-2777. http://dx.doi.org/10.1246/bcsj.56.2774

[17] Harada, K. and Munegumi, T. (1984) Asymmetric Catalytic Hydrogenations of Chiral α-Keto Amides. *Bulletin of the Chemical Society of Japan*, **57**, 3203-3232. http://dx.doi.org/10.1246/bcsj.57.3203

[18] Munegumi, T., Fujita, M., Maruyama, T., Shiono, S., Takasaki, M. and Harada, K. (1987) Asymmetric Catalytic Hydrogenation of N-Pyruvoyl-(*S*)-Proline Esters. *Bulletin of the Chemical Society of Japan*, **60**, 249-253. http://dx.doi.org/10.1246/bcsj.60.249

[19] Munegumi, T. and Harada, K. (1988) Asymmetric Catalytic Hydrogenations of Oximes and Benzylimino Derivatives

of Chiral Pyruvamides. *Bulletin of the Chemical Society of Japan*, **61**, 1425-1427.
http://dx.doi.org/10.1246/bcsj.61.1425

[20] Munegumi, T., Maruyama, T., Takasaki, M. and Harada, K. (1990) Diastereoselective Catalytic Hydrogenations of
N^{α}-Pyruvoyl-(S)-Prolinamide. *Bulletin of the Chemical Society of Japan*, **63**, 1832-1834.
http://dx.doi.org/10.1246/bcsj.63.1832

[21] Dorman, D.E. and Bovery, F.A. (1973) Carbon-13 Magnetic Resonance Spectroscopy. Spectrum of Proline in Oligo-
peptides. *Journal of Organic Chemistry*, **38**, 2379-2383. http://dx.doi.org/10.1021/jo00953a021

[22] Osawa, M., Hatta, A., Harada, K. and Suetaka, W. (1976) Infrared Dichroism and Molecular Orientation of Ethyl
2-Hydroxyimino-3-Oxo-3-Phenylpropinate in Thin Films on Pd Metal Surfaces. *Bulletin of the Chemical Society of
Japan*, **49**, 1512-1516. http://dx.doi.org/10.1246/bcsj.49.1512

Visible Light Induced Knoevenagel Condensation Catalyzed by Starfruit Juice of *Averrhoa carambola*

Rammohan Pal*, **Taradas Sarkar**

Department of Chemistry, Acharya Jagadish Chandra Bose College, Kolkata, India
Email: *pal_rammohan@yahoo.com

Abstract

Aqueous starfruit juice catalyzed a simple and efficient Knoevenagel condensation of aromatic aldehydes with malononitrile has been developed under visible light. Products were obtained in yields up to 98% after short reaction times and they were isolated by simple filtration in pure crystallization states. The method is green and economically viable. A plausible mechanism for photochemical Knoevenagel condensation reaction catalyzed by starfruit juice was also predicted.

Keywords

Knoevenagel Condensation, Aqueous Starfruit Juice, Aldehydes, Malononitrile, Visible Light

1. Introduction

Knoevenagel condensation, first demonstrated by Emil Knoevenagel in 1894 [1], is one of the most important and widely employed methods for carbon-carbon double bond formation in synthetic chemistry [2]-[4]. It has been used for the preparation of a wide range of substituted electrophilic alkenes, and for the synthesis of intermediate such as coumarin derivatives which are useful in perfumes, cosmetics and bioactive compounds [5]-[8]. In addition, Knoevenagel condensation products exhibit inhibition of antiphosphorylation of EGF-receptor and antiproliferative activity [9]. As a result of their importance from a pharmacological, industrial and synthetic point of view a large number of methods for the Knoevenagel condensation have been reported using various Lewis bases/acids [10]-[14]. The use of Green Chemistry protocol based on microwave assisted reaction [15]-[20], ultrasound irradiation [21] [22], biotechnology-based approach [23] [24], solid phase [25] [26], green solvent like ionic liquid [27]-[33] or water [34]-[36], and grindstone method under solvent-free condition [37]-[39]

---

*Corresponding author.

have also been developed. It is noteworthy to observe that all these protocols have some drawbacks, such as use of expensive catalyst, high thermal conditions, disposal of toxic solvents and catalyst, long reaction time often pose a problem.

In the past two decades, classical organic chemistry had been rewritten around new approaches that search for products and processes in the chemical industry that are environmentally acceptable [40]. Therefore, to address depletion of natural resources and preservation of ecosystem is just urgent to develop so called "greener technologies" to make chemical agents for well being of human health [41].

An attractive area in organic synthesis involves photochemical reactions particularly using visible light in environment-friendly solvent like water or aqueous ethanol and is generally considered as a clean and green procedure. This type of photo-activation of substrate very often minimizing the formation of by-products and for this reason, photochemical reactions occupy an interesting position and excellent reviews/paper have been published [42]-[47]. The use of water as a reaction medium is not only inexpensive and environmentally benign but also provides completely different reactivity [48]. It has been suggested that the effect of water on organic reaction may be due to the high internal pressure exerted by a water solution which results from the high cohesive energy of water [49].

A number of organic reactions using natural catalysts such as clay [50]-[53], natural phosphate [54] [56], animal bone [57], and also various fruit juice's are reported in literature. Due to acidic nature, aqueous fruit juice like lemon [58]-[64], pineapple [65] [66], coconut [67], *Acacia concinna* [68], *Sapindus trifolistus* [69] and *Tamarindus indica* [70] [71] fruit has been found to be a suitable replacement for various homogeneous acid catalysis. In accordance with this, we report the Knoevenagel condensation of aromatic aldehydes with malononitrile in presence of aqueous starfruit juice, a natural, green and biocatalyst system stimulated by visible light.

Starfruit (*Averrhoa carambola*) (**Figure 1**) is grown extensively in Philippines, Indonesia, Malaysia, India, Bangladesh, Latin America and Sri Lanka. It has long been one of the most popular of the citrus tropical and subtropical fruits, largely because of its attractive flavor and refreshing sugar-acid balance. Starfruit juice of *Averrhoa carambola* shows antioxidant properties due to scavenging of nitric oxide (NO) and antimicrobial activities against *E. coli*, *Klebsiella spp.* and *Staphylococcus aureus* [72].

The main ingredients [73] [74] of 100 gm of unripe starfruit contain water (89 - 91 g), protein (0.38 g), fat (0.08 g), carbohydrates (9.38 g), sugars (3.98 g), edible fiber (0.8 - 0.9 g), calcium (4.4 - 6.0 mg), phosphorous (15.5 - 21 mg), sodium (2 mg) and potassium (133 mg). Fresh mature unripe fruit were found to have a total acid content of 12.51 mg consisting of 5 mg oxalic acid, 4.37 mg tartaric acid, 1.32 mg citric acid, 1.21 mg malic acid, 0.22 mg succinic acid, 0.26 - 0.53 mg ascorbic acid and 0.39 mg pantothenic acid. The composition of the starfruit juice varies with geographical, cultural and seasonal harvesting and processing. An aqueous extract of starfruit juice is acidic due to presence of edible organic acids and hence it will be work as an acid catalyst for acid catalyzed reactions.

2. Results and Discussion

In continuation of our research interest concerning the investigation of new natural catalyst and development of

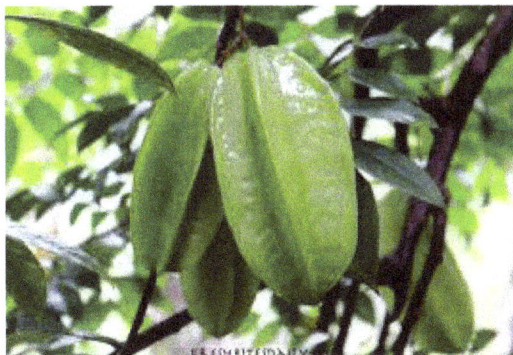

Figure 1. Photoghrapy of starfruit of *Averrhoa carambola*.

new methodologies (62 - 64, 70, 71) herein is reported in this paper, for the first time a highly efficient, eco-friendly and economic method for Knoevenagel condensation of aldehydes with malononitrile using aqueous starfruit juice stimulated by visible light affording 2-(substituted phenylidene) malononitrile (**Scheme 1**).

The photochemical reactions were found to be very clean and the products were obtained in extremely pure crystalline states with an average yield of 75% - 98% and the reaction time varied on an average 2 - 7 min. The products were isolated from the reaction mixture in pure crystalline form by cooling in an ice-bath and need no further crystallization for aromatic aldehydes and the results are given in **Table 1**.

The scope of application of the presented method is demonstrated by using various substituted aromatic aldehydes to react with malononitrile. The procedure was successfully applied for heteroaromatic aldehydes (entries 18, 19), and the ether (entries 3, 9, 15, 16), esters (entries 14, 15) linkages in the aromatic aldehydes were unaffected under photochemical conditions. The reaction was further explored for the synthesis of *p*-bis-2-(phenylidene) malononitrile (**3t**) in 95% yield by the condensation of terephthalaldehyde (entry 20) with two mole of malononitrile under similar reaction conditions.

When the same reactions were performed at room temperature for 1 h, only 30% - 35% of the corresponding products were isolated. On the hand, under refluxing conditions for 10 min only 30% - 40% of **3** were isolated and the yield of the products increased to 60% - 65% after 3 - 4 h. The microwave irradiation reaction, accomplished in an average time periods 2 - 3 min. In all the above cases products were isolated by column chromatography or required further crystallization from appropriate solvents. Thus, the present method in comparison with room temperature, thermal and microwave irradiation one, is encouragingly effectual and smoothly for aromatic aldehydes free from any adhering by byproduct or side products.

In the present instance, we speculate that the reaction may plausibly be initiated by homolytic C-H bond cleavage of malononitrile (**2**) in the presence of light to produce a radical **I** and hydrogen radical which is immediately converted to a transient anion radical **II** due to a weak interaction with water molecules. Aldehyde (**1**) becomes activated by protonation from starfruit juice to produce a protonated species **III**. One electron transfer from **II** to **IV** produced a radical **V**, which couples with radical **I** to form **VI**. Protonation of **VI** followed by dehydration from **VII** to form the title compound **3** as depicted in **Scheme 2**.

All the products (**Table 1**) are known compounds and their structures are confirmed by comparison of melting

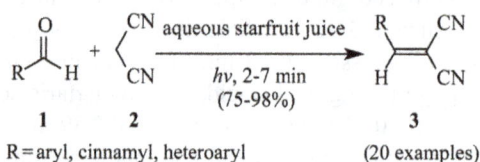

Scheme 1. Visible light induced Knoevenagel condensation reaction catalyzed by starfruit juice.

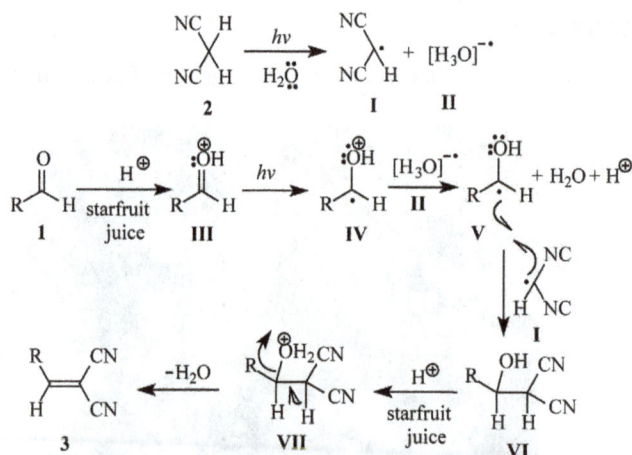

Scheme 2. Plausible mechanistic pathway for the photochemical Knoevenagel condensation of aldehydes and malononitrile catalyzed by starfruit juice.

Table 1. Results of photochemical Knoevenagel condensation of aldehydes with malononitrile catalyzed by aqueous starfruit juice.

Entry	Substrate (1)	Product (3)[a]	Yield (%)[b]	Time (min)	M.p.(°C) [Lit. Value][c]
1	**1a**	**3a**	92	3	81 - 82 [80]-[83][37]
2	**1b**	**3b**	94	4	133 - 134 [134] [135][27]
3	**1c**	**3c**	85	5	113 - 115 [116][66]
4	**1d**	**3d**	75	4	160 [159][66]
5	**1e**	**3e**	95	2	165 [164][71]
6	**1f**	**3f**	82	4	182 - 184 [185] [186][20]
7	**1g**	**3g**	93	4	164 - 166 [162]-[164][19]
8	**1h**	**3h**	80	.7	158 [156][66]
9	**1i**	**3i**	92	6	132 - 134 [135] [136][20]
10	**1j**	**3j**	90	5	182 [180][22b]

Continued

Entry	Reactant	Product	Yield (%)	Time	m.p. (°C)
11	**1k** (2-nitrobenzaldehyde)	**3k**	95	3	134 [136]-[138][27]
12	**1l** (3-nitrobenzaldehyde)	**3l**	88	4	104 - 106 [107] [108][20]
13	**1m** (4-nitrobenzaldehyde)	**3m**	90	2	161 - 163 [160]-[162][19]
14	**1n** (C₆H₅COO-C₆H₄-CHO)	**3n**	94	7	152 - 154 [152][71]
15	**1o** (C₆H₅COO, MeO substituted benzaldehyde)	**3o**	92	7	140 - 141 [140] [141][71]
16	**1p** (piperonal)	**3p**	98	4	200 - 202 [198][71]
17	**1q** (cinnamaldehyde)	**3q**	78	5	125 - 127 [127]-[129][37]
18	**3r** (pyridine derivative)	**3r**	90	3	80 - 82 [84] [85][27]
19	**1s** (bisindolyl aldehyde)	**3s**	78	7	320 - 322 [320]-[322][71]
20	**1t** (terephthalaldehyde, OHC-C₆H₄-CHO)	**3t**	95	3	300 [298]-[300][71]

[a]All products were identified by their physical and spectral data; [b]Isolated yields; [c]References for literature melting point.

points and spectral data with their literature data.

3. Conclusion

We have described a potentially efficient, absolutely clean, and high yielding eco-friendly methodology, for the photochemical Knoevenagel condensation of various aromatic aldehydes with malononitrile catalyzed by aqueous starfruit juice. The present protocol devoid of any toxic catalysts, solvents or solid supports and may be considered as an excellent improvement over the existing methods.

4. Experimental Section

All reactions were run in dried glassware. Reagents were purchased (Spectrochem or SRL or LOBA) and used without further purification. Melting points were determined on a Kofler block and uncorrected. Reactions were irradiated in a 200 W tungsten lamp (Philips India Ltd). ^{1}H and ^{13}C NMR and spectra were obtained in CDCl$_3$ or DMSO-d$_6$ on a Bruker AV-300 (300 MHz) spectrometers using TMS as an internal standard. Analytical samples were dried *in vacuo* at room temperature. The carbon, hydrogen and nitrogen percentages in synthesized products were analyzed by Perkin-Elmer 2400 series II C, H, N analyzers. Thin layer chromatography was carried out on silica gel.

4.1. Preparation of Aqueous Extract of Starfruit Juice

The mature green starfruit were purchased from the local market. The starfruit were cut into pieces with the help of knife. The hard green material (20 g) was boiled with water (50 ml), cooled and it was centrifuged using micro centrifuge (REMI RM-12C). The clear portion of the aqueous extract (pH = 3.5) of the starfruit was used as catalyst for the reactions.

4.2. General Procedure for Photochemical Knoevenagel Condensation Reaction

Different aromatic aldehydes (**1a-s**) (10 mmol) or (**1t**) (5 mmol), malononitrile (10 mmol), and aqueous starfruit juice (5 ml, pH = 3.5) were taken in a round bottomed flask and irradiated with a 200 W tungsten lamp (Philips India Ltd). The reaction time varied from 2 - 7 min monitored by TLC. Upon completion of the reaction, the reaction mixture was cooled and the crystalline products (**3a-t**) so obtained was filtered, washed with water and dried in vacuo. The Knoevenagel condensation products were isolated in excellent yields in essentially pure form.

4.3. Spectral Data for Some Selected Compounds

2-(3-*Hydroxyphenylmethylene*)*malononitrile* (**3e**): Yellow crystal, Yield: 95%, mp. 165°C; ^{1}H NMR (300 MHz, DMSO-d$_6$): δ 7.08 (d, 7.5 Hz, 1H), 7.35 - 7.44 (m, 3H), 8.44 (s, 1H, H-C=C), 10.12 (s, 1H, OH); Anal. Calcd. for C$_{10}$H$_6$N$_2$O, C, 70.58; H, 3.55; N, 16.46%, found C, 70.22; H, 3.87; N, 16.21%.

2-(4-*Benzoyloxyphenylmethylene*)*malononitrile* (**3n**): Colorless crystal, Yield: 94%, mp. 152°C - 154°C; ^{1}H NMR (300 MHz, CDCl$_3$): δ 7.43 (d, 8.7 Hz, 2H), 7.54 (t, 7.5 Hz, 2H), 7.66 - 7.78 (m, 1H), 7.78 (s, 1H, H-C=C), 8.01 (d, 8.7 Hz, 2H), 8.20 (d, 7.8 Hz, 2H); ^{13}C NMR (75 MHz, CDCl$_3$): δ 82.54 (=C<), 112.49 (CN), 113.61 (CN), 123.12, 128.41, 128.53, 128.75, 130.29 (-CH=), 132.37, 134.20, 155.56, 158.56, 164.24 (ester carbonyl); DEPT - 90 (75 MHz, CDCl$_3$): 123.11, 128.74, 130.28, 132.35, 134.18, 158.52; DEPT - 135 (75 MHz, CDCl$_3$): 123.11, 128.74, 130.28, 132.35, 134.18, 158.51; Anal. Calcd. for C$_{17}$H$_{10}$N$_2$O$_2$, C, 74.45; H, 3.67; N, 10.21%, found C, 74.11; H, 3.81; N, 10.43%.

2-(4-*Benzoyloxy-3-methoxyphenylmethylene*)*malono-nitrile* (**3o**): Colorless crystal, Yield: 92%, mp. 140°C - 141°C; ^{1}H NMR (300 MHz, CDCl$_3$): δ 3.89 (s, 3H, OMe), 7.34 (d, 8.4 Hz, 1H), 7.43 (dd, 8.7 and 1.8 Hz, 1H), 7.53 (t, 7.5 Hz, 2H), 7.64 - 7.69 (m, 1H), 7.74 (d, 1.8 Hz, 1H), 7.76 (s, 1H, H-C=C), 8.20 (d, 8.7 Hz, 2H); Anal. Calcd. for C$_{18}$H$_{12}$N$_2$O$_3$, C, 71.05; H, 3.97; N, 9.21%, found C, 70.85; H, 4.02; N, 9.52%.

2-(3,4-*Methylenedioxyphenylmethylene*)*malononitrile* (**3p**): Yellow crystal, Yield: 98%, mp. 200°C - 202°C; ^{1}H NMR (300 MHz, CDCl$_3$): δ 6.12 (s, 2H, -O-CH$_2$-O-), 6.93 (d, 8.1 Hz, 1H), 7.32 (dd, 8.1 and 1.5 Hz, 1H), 7.59 (s, 1H, H-C=C), 7.60 (s, 1H); Anal. Calcd. for C$_{11}$H$_6$N$_2$O$_2$, C, 66.67; H, 3.05; N, 14.14%, found C, 67.01; H, 3.21; N, 14.32%.

2-[{*p*-3, 3'-*Bis*(2-*methylindolyl*)*methyl*}*phenyl-methylene*]*malononitrile* (**3s**): Pale-yellow crystal, Yield: 78%, mp. 320°C - 322°C; ^{1}H NMR (300 MHz, CDCl$_3$): δ 2.09 (s, 6H, Me), 6.04 (s, 1H, Ar-CH), 6.84 - 6.93 (m, 4H), 7.06 (t, 6.9 Hz, 2H), 7.28 (d, 9.0 Hz, 2H), 7.44 (d, 8.1 Hz, 2H), 7.72 (s, 1H, H-C=C), 7.80 (d, 8.7 Hz, 2H), 7.80 (br. s, 2H, NH); Anal. Calcd. for C$_{29}$H$_{22}$N$_4$, C, 81.67; H, 5.20; N, 13.14%, found C, 81.33; H, 5.40; N, 13.25%.

p-*Bis*-2-(*phenylmethylene*)*malononitrile* (**3t**): White crystal, Yield: 95%, mp. 300°C; ^{1}H NMR (300 MHz, DMSO-d$_6$): δ 8.09 (s, 4H), 8.63 (s, 2H, H-C=C); ^{13}C NMR (75 MHz, DMSO-d$_6$): δ 84.71 (=C<), 112.14 (CN), 113.80 (CN), 130.83 (-CH=), 135.32 (aromatic quarternary), 159.80 (aromatic -CH=); DEPT - 90 (75 MHz, DMSO-d$_6$): 130.83, 159.81; DEPT - 135 (75 MHz, DMSO-d$_6$): 130.84, 159.81; Anal. Calcd. for C$_{14}$H$_6$N$_4$, C, 73.04; H, 2.63; N, 24.34%, found C, 72.95, H, 2.76; N, 24.45%.

Acknowledgements

Financial assistance from the UGC Minor Research Project No. PSW-130/11-12 (ERO), New Delhi, India is gratefully acknowledged.

References

[1]   Knoevenagel, E. (1894) Uebereine Darstellungsweise der Glutarsaure. *Berichte der Deutschen Chemischen Gesell-schaft*, **2**, 2345-2346. http://dx.doi.org/10.1002/cber.189402702229

[2]   Knoevenagel, E. (1898) Condensationen zwisschen malonester und aldehyden unter dem einfluss von ammoniak und organischen aminen. *Chemische Berichte*, **31**, 2585-2595. http://dx.doi.org/10.1002/cber.18980310307

[3]   Jones, G. (1967) The Knoevenagel Condensation. *Organic Reactions*, **15**, 204-599.

[4]   Tietze, L.F. and Beifuss, U. (1991) The Knoevenagel Reaction. In: Trost, B.M., Ed., *Comprehensive Organic Synthesis*, Pergamon Press, Oxford, 341-394. http://dx.doi.org/10.1016/B978-0-08-052349-1.00033-0

[5]   Yu, N., Aramini, J.M., Germann, M.W. and Huang, Z. (2000) Reactions of Salicylaldehydes with Alkyl Cyanoacetates on the Surface of Solid Catalysts: Synthesis of 4H-Chromene Derivatives. *Tetrahedron Letters*, **41**, 6993-6996. http://dx.doi.org/10.1016/S0040-4039(00)01195-3

[6]   Gallos, J., Discordia, R.P., Crispino, G.A., Li, J., Grosso, J.A., Polniaszek, V. and True, V.C. (2003) A Mild and Efficient Synthesis of 4-Aryl-Quinolin-2(1H)-Ones via a Tandem Amidation/Knoevenagel Condensation of 2-Amino-Benzophenones with Esters or Lactones. *Tetrahedron Letters*, **44**, 4271-4273. http://dx.doi.org/10.1016/S0040-4039(03)00889-X

[7]   Xing, C. and Zhu, S. (2004) Unexpected Formation of Tetrasubstituted 2,3-Dihydrofurans from the Reactions of β-Keto Polyfluoroalkanesulfones with Aldehydes. *Journal of Organic Chemistry*, **69**, 6486-6488. http://dx.doi.org/10.1021/jo049317y

[8]   Tietze, L.F. and Rackelmann, N. (2004) Domino Reactions in the Synthesis of Heterocyclic Natural Products and Analogs. *Pure and Applied Chemistry*, **76**, 1967-1983. http://dx.doi.org/10.1351/pac200476111967

[9]   Vijender, M., Kishor, P. and Satyanarayana, B. (2008) Zirconium Tetrachloride-SiO$_2$ Catalyzed Knoevenagel Condensation: A Simple and Efficient Protocol for the Synthesis of Substituted Electrophilic Alkenes. Arkivoc, **2008**, 122-128. http://dx.doi.org/10.3998/ark.5550190.0009.d14

[10] Rao, P.S. and Venkataratnam, R.V. (1991) Zinc Chloride as a New Catalyst for Knoevenagel Condensation. *Tetrahedron Letters*, **32**, 5821-5822. http://dx.doi.org/10.1016/S0040-4039(00)93564-0

[11] Prajapati, D. and Sandhu, J.S. (1993) Cadmium Iodide as a New Catalyst for Knoevenagel Condensation. *Journal of the Chemical Society, Perkin Transactions* 1, **1**, 739-740. http://dx.doi.org/10.1039/p19930000739

[12] Lehnert, W. (1970) Verbesserte Variante der Knoevenagel-Kondensation Mit TiCl$_4$/THF/Pyridine (I). Alkyliden-und Arylidenmalonester bei 0 - 25°C. *Tetrahedron Letters*, **11**, 4723-4724. http://dx.doi.org/10.1016/S0040-4039(00)89377-6

[13] Dai, G., Shi, D., Zhou, L. and Huaxue, Y. (1995) Knoevenagel Condensation Catalysed by Potassium Fluoride/Alumina. *Chinese Journal of Applied Chemistry*, **12**, 104-108.

[14] Gill, C., Pandhare, G., Raut, R., Gore, V. and Gholap, S. (2008) Knoevenagel Condensation: A Simple and Efficient Protocol of Electrophilic Alkenes Catalyzed by Anhydrous Ferric Sulphate with Remarkable Reusability. *Bulletin of the Catalysis Society of India*, **7**, 153-157.

[15] Bogdal, D. (1998) Coumarins: Fast Synthesis by Knoevenagel Condensation under Microwave Irradiation. *Journal of Chemical Research* (*Synopsis*), **8**, 468-469. http://dx.doi.org/10.1039/a801724g

[16] Kumar, H.M.S., Reddy, B.V.S., Anjaneyulu, S. and Yadav, J.S. (1998) Non Solvent Reaction: Ammonium Acetate

Catalyzed Highly Convenient Preparation of Trans-Cinnamic Acid. *Synthetic Communications*, **28**, 3811-3815. http://dx.doi.org/10.1080/00397919808004934

[17] Kumar, H.M.S., Reddy, B.V.S., Reddy, P.T., Srinivas, D. and Yadav, J.S. (2000) Silica Gel Catalyzed Preparation of Cinnamic Acid under Microwave Irradiation. *Organic Preparations Procedures International*, **32**, 81-83. http://dx.doi.org/10.1080/00304940009356750

[18] Mogilaiah, K. and Reddy, C.S. (2003) An Efficient Friedlander Condensation Using Sodium Fluoride as Catalyst in the Solid State. *Synthetic Communications*, **33**, 3131-3134. http://dx.doi.org/10.1081/SCC-120023427

[19] Mallouk, S., Bougrin, K., Laghzizil, A. and Benhida, R. (2010) Microwave-Assisted and Efficient Solvent-Free Knoevenagel Condensation. A Sustainable Protocol Using Calcium Hydroxyapatite as Catalyst. *Molecules*, **15**, 813-823.

[20] Bhuiyan, M.M.H., Hossain, M.I., Alam, M.A. and Mahmud, M.M. (2012) Microwave Assisted Knoevenagel Condensation: Synthesis and Antimicrobial Activities of Some Arylidene-Malononitriles. *Chemistry Journal*, **2**, 30-36.

[21] McNulty, J., Steere, J.A. and Wolf, S. (1998) The Ultrasound Promoted Knoevenagel Condensation of Aromatic Aldehydes. *Tetrahedron Letters*, **39**, 8013-8016. http://dx.doi.org/10.1016/S0040-4039(98)01789-4

[22] Palmisano, G., Tibiletti, F., Penoni, A., Colombo, F., Tollari, S., Garella, D., Tagliapietra, S. and Cravotto, G. (2011) Ultrasound-Enhanced One-Pot Synthesis of 3-(Het)arylmethyl-4-hydroxycoumarins in Water. *Ultrasonics Sonochemistry*, **18**, 652-660. http://dx.doi.org/10.1016/j.ultsonch.2010.08.009

[23] Pratap, U.R., Jawale, D.V., Waghmare, R.A., Lingampalle, D.L. and Mane, R.A. (2011) Synthesis of 5-arylidene-2, 4-thiazolidinediones by Knoevenagel Condensation Catalyzed by Baker's Yeast. *New Journal of Chemistry*, **35**, 49-51. http://dx.doi.org/10.1039/c0nj00691b

[24] Wang, C.H., Guan, Z. and He, Y.H. (2011) Biocatalytic Domino Reaction: Synthesis of 2*H*-1-benzopyran-2-one Derivatives Using Alkaline Protease from *Bacillus licheniformis*. *Green Chemistry*, **13**, 2048-2054. http://dx.doi.org/10.1039/c0gc00799d

[25] Xia, Y., Yang, Z.Y., Brossi, A. and Lee, K.H. (1999) Asymmetric 'Solid-Phase Synthesis of (3'*R*,4'*R*)-Di-*O*-cis-acyl 3-Carboxyl Khellactones. *Organic Letters*, **1**, 2113-2115. http://dx.doi.org/10.1021/ol991168w

[26] Guo, G., Arvanitis, E.A., Pottorf, R.S. and Player, M.P. (2003) Solid-Phase Synthesis of a Tyrphostin Ether Library. *Journal of Combinatorial Chemistry*, **5**, 408-413. http://dx.doi.org/10.1021/cc030003i

[27] Ying, A.G., Liu, L., Wu, G.F., Chen, X.Z., Ye, W.D., Chen, J.H. and Zhang, K.Y. (2009) Knoevenagel Condensation Catalyzed by DBU Brönsted Ionic Liquid without Solvent. *Chemical Research in Chinese Universities*, **25**, 876-881.

[28] Khan, F.A., Dash, F.J., Satapathy, R. and Upadhyay, S.K. (2004) Hydrotalcite Catalysis in Ionic Medium: A Recyclable Reaction System for Heterogeneous Knoevenagel and Nitroaldol Condensation. *Tetrahedron Letters*, **45**, 3055-3058. http://dx.doi.org/10.1016/j.tetlet.2004.02.103

[29] Verdia, P., Santamarta, F. and Tojo, E. (2011) Knoevenagel Reaction in [MMIm][MSO$_4$]: Synthesis of Coumarins. *Molecules*, **16**, 4379-4388. http://dx.doi.org/10.3390/molecules16064379

[30] Hu, Y., Chen, J., Le, Z.G. and Zheng, Q.G. (2005) Organic Reactions in Ionic Liquids: Ionic Liquids Ethylammonium Nitrate Promoted Knoevenagel Condensation of Aromatic Aldehydes with Active Methylene Compounds. *Synthetic Communications*, **35**, 739-744. http://dx.doi.org/10.1081/SCC-200050380

[31] Xin, X., Guo, X., Duan, H., Lin, Y. and Sun, H. (2007) Efficient Knoevenagel Condensation Catalyzed by Cyclic Guanidinium Lactate Ionic Liquid as Medium. *Catalysis Communications*, **8**, 115-117. http://dx.doi.org/10.1016/j.catcom.2006.05.034

[32] Santamarta, F., Verdía, P. and Tojo, E. (2008) A Simple, Efficient and Green Procedure for Knoevenagel Reaction in [MMIm][MSO$_4$] Ionic Liquid. *Catalysis Communications*, **9**, 1779-1781.

[33] Yue, C., Mao, A., Wei, Y. and Lü, M. (2008) Knoevenagel Condensation Reaction Catalyzed by Task-Specific Ionic Liquid under Solvent-Free Conditions. *Catalysis Communications*, **9**, 1571-1574. http://dx.doi.org/10.1016/j.catcom.2008.01.002

[34] Bigi, F., Conforti, M.L., Maggi, R., Piccinno, A. and Sartori, G. (2000) Clean Synthesis in Water: Uncatalysed Preparation of Ylidenemalonitriles. *Green Chemistry*, **2**, 101-103. http://dx.doi.org/10.1039/b001246g

[35] Wang, S., Ren, Z., Cao, W. and Tong, W. (2001) The Knoevenagel Condensation of Aromatic Aldehydes with Malononitrile or Ethyl Cyanoacetate in the Presence of CTMAB in Water. *Synthetic Communications*, **31**, 673-677. http://dx.doi.org/10.1081/SCC-100103255

[36] Oskooie, H.A., Heravi, M.M., Derikvand, F., Khorasani, M. and Bamoharram, F.F. (2006) On Water: An Efficient Knoevenagel Condensation Using 12-Tungstophoric Acid as a Reusable Green Catalyst. *Synthetic Communications*, **36**, 2819-2823. http://dx.doi.org/10.1080/00397910600770631

[37] Pasha, M.A. and Manjula, K. (2011) Lithium Hydroxide: A Simple and an Efficient Catalyst for Knoevenagel Condensation under Solvent-Free Grindstone Method. *Journal of Saudi Chemical Society*, **15**, 283-286.

http://dx.doi.org/10.1016/j.jscs.2010.10.010

[38] Rong, L., Li, X., Wang, H., Shi, D., Tu, S. and Zhuang, Q. (2006) Efficient Green Procedure for the Knoevenagel Condensation under Solvent-Free Conditions. *Synthetic Communications*, **36**, 2407-2412. http://dx.doi.org/10.1080/00397910600640289

[39] Ren, Z., Cao, W. and Tong, W. (2002) The Knoevenagel Condensation Reaction of Aromatic Aldehydes with Malononitrile by Grinding in the Absence of Solvents and Catalysts. *Synthetic Communications*, **32**, 3475-3479. http://dx.doi.org/10.1081/SCC-120014780

[40] Okkerse, C. and van Bekkun, H. (1999) From Fossil to Green. *Green Chemistry*, **1**, 107-114. http://dx.doi.org/10.1039/a809539f

[41] Anastas, P.T. and Warner, J.C. (2000) Green Chemistry: Theory and Practice. Oxford University Press, New York.

[42] Hoffmann, N. (2008) Photochemical Reactions as Key Steps in Organic Synthesis. *Chemical Reviews*, **108**, 1052-1103. http://dx.doi.org/10.1021/cr0680336

[43] Fagnoni, M., Dondi, D., Ravelli, D. and Albini, A. (2007) Photocatalysis for the Formation of the C-C Bond. *Chemical Reviews*, **107**, 2725-2756. http://dx.doi.org/10.1021/cr068352x

[44] Ghosh, S. and Das, J. (2011) A Novel Photochemical Wittig Reaction for the Synthesis of 2-Aryl/Alkylbenzofurans. *Tetrahedron Letters*, **52**, 1112-1116. http://dx.doi.org/10.1016/j.tetlet.2010.12.104

[45] Ghosh, S., Das, J. and Chattopadhyay, S. (2011) A Novel Light Induced Knoevenagel Condensation of Meldrum's Acid with Aromatic Aldehydes in Aqueous Ethanol. *Tetrahedron Letters*, **52**, 2869-2872. http://dx.doi.org/10.1016/j.tetlet.2011.03.123

[46] Ghosh, S., Das, J. and Saikh, F. (2012) A New Synthesis of 2-Aryl/Alkylbenzofurans by Visible Light Stimulated Intermolecular Sonogashira Coupling and Cyclization Reaction in Water. *Tetrahedron Letters*, **53**, 5883-5886. http://dx.doi.org/10.1016/j.tetlet.2012.08.078

[47] Ghosh, S., Saikh, F., Das, J. and Pramanik, A.K. (2013) Hantzsch 1,4-Dihydropyridine Synthesis in Aqueous Ethanol by Visible Light. *Tetrahedron Letters*, **54**, 58-62. http://dx.doi.org/10.1016/j.tetlet.2012.10.079

[48] Li, C. and Chen, L. (2006) Organic Chemistry in Water. *Chemical Society Reviews*, **35**, 68-82. http://dx.doi.org/10.1039/b507207g

[49] Breslow, R. (1991) Hydrophobic Effects on Simple Organic Reactions in Water. *Accounts of Chemical Research*, **24**, 159-164. http://dx.doi.org/10.1021/ar00006a001

[50] Ramesh, E. and Raghunathan, R. (2009) Microwave-Assisted K-10 Montmorillonite Clay-Mediated Knoevenagel Hetero-Diels-Alder Reactions: A Novel Protocol for the Synthesis of Polycyclic Pyrano[2,3,4-*kl*]xanthenes Derivatives. *Synthetic Communications*, **39**, 613-625. http://dx.doi.org/10.1080/00397910802417825

[51] Habibi, D. and Marvi, O. (2006) Montmorillonite KSF and Montmorillonite K-10 Clays as Efficient Catalysts for the Solventless Synthesis of Bismaleimides and Bisphthalimides Using Microwave Irradiation. *Arkivoc: Online Journal of Organic Chemistry*, **2006**, 8-15. http://dx.doi.org/10.3998/ark.5550190.0007.d02

[52] Chakrabarty, M., Mukherjee, A., Arima, S., Harigaya, Y. and Pilet, G. (2009) Expeditious Reaction of Ninhydrin with Active Methylene Compounds on Montmorillonite K10 Clay. *Monatshefte Für Chemie—Chemical Monthly*, **140**, 189-197. http://dx.doi.org/10.1007/s00706-008-0066-6

[53] Wada, S. and Suzuki, H. (2003) Calcite and Fluorite as Catalyst for the Knoevenagel Condensation of Malononitrile and Methyl Cyanoacetate under Solvent-Free Conditions. *Tetrahedron Letters*, **44**, 399-401. http://dx.doi.org/10.1016/S0040-4039(02)02431-0

[54] Zahouily, M., Mounir, B., Charki, H., Mezdar, A., Bahlaouan, B. and Ouammou, M. (2006) Investigation of the Basis Catalytic Activity of Natural Phosphates in the Michael Condensation. *Arkivoc: Online Journal of Organic Chemistry*, **2006**, 178-186. http://dx.doi.org/10.3998/ark.5550190.0007.d19

[55] Zahouily, M., Bahlaouan, B., Rayadh, A. and Sebti, S. (2004) Natural Phosphates and Potassium Fluoride Doped Natural Phosphate: Efficient Catalyst for the Construction of a Carbon-Nitrogen Bond. *Tetrahedron Letters*, **45**, 4135-4138. http://dx.doi.org/10.1016/j.tetlet.2004.03.164

[56] Sebti, S., Smahi, A. and Solly, A. (2002) Natural Phosphate Doped with Potasiuum Fluoride and Modified with Sodium Nitrate: Efficient Catalysts for the Knoevenagel Condensation. *Tetrahedron Letters*, **43**, 1813-1815. http://dx.doi.org/10.1016/S0040-4039(02)00092-8

[57] Riadi, Y., Mamouni, R., Azzalou, R., Boulahjar, R., Abrouki, Y., El Haddad, M., Routier, S., Guillaumet, G. and Lazar, S. (2010) Animal Bone Meal as an Efficient Catalyst for Crossed-Aldol Condensation. *Tetrahedron Letters*, **51**, 6715-6717. http://dx.doi.org/10.1016/j.tetlet.2010.10.056

[58] Deshmukh, M.B., Patil, S.S., Jadhav, S.D. and Pawar, P.B. (2012) Green Approach for Knoevenagel Condensation of Aromatic Aldehydes with Active Methylene Group. *Synthetic Communications*, **42**, 1177-1183.

http://dx.doi.org/10.1080/00397911.2010.537423

[59] Patil, S., Jadhav, S.D. and Deshmuk, M.B. (2011) Natural Acid Catalyzed Multi-Component Reaction as a Green Approach. *Archives of Applied Science Research*, **3**, 203-208.

[60] Sachdeva, H., Saroj, R., Khaturia, S. and Dwivedi, D. (2013) Environ-Economic Synthesis and Characterization of Some New 1,2,4-Triazole Derivative as Organic Fluorescent Materials and Potent Fungicidal Agents. *Organic Chemistry International*, **2013**, Article ID 659107.

[61] Patil, S., Jhadav, S.D. and Patil, U.P. (2012) Natural Acid Catalyzed Synthesis of Schiff Base under Solvent-Free Condition: As a Green Approach. *Archives of Applied Science Research*, **4**, 1074-1078.

[62] Pal, R., Khasnobis, S. and Sarkar, T. (2013) First Application of Fruit Juice of *Citrus limon* for Facile and Green Synthesis of Bis- and Tris(indolyl)methanes in Water. *Chemistry Journal*, **3**, 7-12.

[63] Pal, R. (2013) Microwave-Assisted Eco-Friendly Synthesis of Bis-, Tris(indolyl)methanes and Synthesis of Di-bis(indolyl)methanes Catalyzed by Fruit Juice of *Citrus limon* under Solvent-Free Conditions. *IOSR Journal of Applied Chemistry*, **3**, 1-8. http://dx.doi.org/10.9790/5736-0340108

[64] Pal, R. (2013) New Greener Alternative for Biocondensation of Aldehydes and Indoles Using Lemon Juice: Formation of Bis-, Tris-, and Tetraindoles. *International Journal of Organic Chemistry*, **3**, 136-142. http://dx.doi.org/10.4236/ijoc.2013.32015

[65] Patil, S., Jadhav, S.D. and Mane, S. (2011) Pineapple Juice as a Natural Catalyst: An Excellent Catalyst for Biginelli Reaction. *Journal of Organic Chemistry*, **1**, 125-131.

[66] Patil, S., Jadhav, S.D. and Deshmukh, M.B. (2013) Eco-Friendly and Economic Method for Knoevenagel Condensation by Employing Natural Catalyst. *Indian Journal of Chemistry*, **52B**, 1172-1175.

[67] Fonseca, A.M., Monte, F.J., de Oliveira, M.C.F., de Mattos, M.C.M., Cordell, G.A., Braz-Filho, R. and Lemos, T.L.G. (2009) Coconut Water (*Cocos nucifera* L.)—A New Biocatalyst System for Organic Synthesis. *Journal of Molecular Catalysis B: Enzymatic*, **57**, 78-82. http://dx.doi.org/10.1016/j.molcatb.2008.06.022

[68] Mote, K., Pore, S., Rashinkar, G., Kambale, S., Kumbhar, A. and Salunkhe, R. (2010) Acacia Concinna Pods: As a Green Catalyst for Highly Efficient Synthesis of Acylation of Amines. *Archives of Applied Science Research*, **2**, 74-80.

[69] Pore, S., Rashimkar, G., Mote, K. and Salunkhe, R. (2010) Aqueous Extract of the Pericarp of *Sapindus trifoliatus* Fruits: A Novel 'Green' Catalyst for the Aldimine Synthesis. *Chemistry and Biodiversity*, **7**, 1796-1800. http://dx.doi.org/10.1002/cbdv.200900272

[70] Pal, R. (2013) A Convenient, Rapid and Eco-Friendly Synthesis of Bis-, Tris(indolyl)methanes and Synthesis of Tetraindolyl Compounds Catalyzed by Tamarind Juice under Microwave Irradiation. *International Journal of Chemtech Applications*, **2**, 26-40.

[71] Pal, R. (2014) Visible Light Induced Knoevenagel Condensation: A Clean and Efficient Protocol Using Aqueous Fruit Extract of *Tamarindus indica* as Catalyst. *International Journal of Advanced Chemistry*, **2**, 27-33.

[72] Sripanidkulchai, B., Tattawasart, U., Laupattarakasem, P. and Wongpanich, V. (2002) Anti-Inflammatory and Bactericidal Properties of Selected Indigeneous Medicinal Plants Used for Dysuria. *Thai Journal of Pharmaceutical Sciences*, **26**, 33-38.

[73] Morton, J.F. (1987) Carambola. In: Dowling, C.F., Ed., *Fruits of Warm Climates*, Flair Books, Miami, 125-128.

[74] Dasgupta, P., Chakraborty, P. and Bala, N.N. (2013) *Averrhoa carambola*: An Updated Review. *International Journal of Pharma Research & Review*, **2**, 54-63.

3

In Vitro Antimicrobial Activity of Some Novel 3-(Substituted Phenyl) Isocoumarins, 1(2*H*)-Isoquinolones and Isocoumarin-1-Thiones

Zaman Ashraf[1]*, Aamer Saeed[2]

[1]Department of Chemistry, Allama Iqbal Open University, Islamabad, Pakistan
[2]Department of Chemistry, Quaid-I-Azam University, Islamabad, Pakistan
Email: *mzchem@yahoo.com

Abstract

The work reports antibacterial and antifungal activity of some 3-(substituted phenyl) isocoumarins (1H-2-benzopyran-1-ones), isocarbostyrils 1(2H)-isoquinolones, the nitrogen analogues of isocoumarins and isocoumarin-1-thiones, the thio derivatives of isocoumarins. The antimicrobial activity was determined against ten different Gram positive and Gram negative bacterial strains and three fungal strains. The bacterial strains were *Klebsiella pneumonae* (ATCC 6633), *Staphylococcus aureus* (ATCC 29213), *Micrococcus luteus* (ATCC 9341), *Pseudomonas aeruginosa* (ATCC 33347), *Escherichia coli* (ATCC 25922), *Salmonella typhi* (ATCC 19430), *Lactobacillus bulgaricus*, (ATCC 25929), *Pasteurella multocida* A (ATCC 9150), *Staphylococcus epidermidis* (ATCC 29232) and *Proteus vulgaris* (ATCC 49565) and fungal strains were *Aspergillus flavus, Aspergillus nigar* and *Aspergillus pterus*. Agar well diffusion method was followed for antibacterial activity and poison plate method was adopted for antifungal assay. Chloramphenicol and fluconazole used as standard drugs for antibacterial and antifungal activity respectively. In general, these compounds exhibited high antibacterial potential than antifungal. Comparative study reveals that the 1-thio derivatives are more active than parent isocoumarins but 1(2H)-isoquinolones, are less active. Most of these compounds showed poor activity but some of these compounds exhibited moderate to good activity against *Staphylococcus epidermidis, Klebsiella pneumonae, Escherichia coli* and *Proteus vulgaris*, compared with the standard drug.

Keywords

Antimicrobial Activity; Isocoumarins; Isocarbostyrils; 1-Thioisocoumarins

---

*Corresponding author.

1. Introduction

Isocoumarins are the secondary metabolites of fungi, bacteria, plants and insect venoms and pheromones. A huge number of them have been isolated from fungi, lichens and bacteria. Some higher plants, insect and marine organisms are also the rich source of these secondary metabolites [1] [2]. They exhibit a broad range of pharmacological activities including antimalarial, antimicrobial, immunomodulatory, antifungal, anti-inflammatory, cytotoxic, antiangiogenic and antiallergic [3]-[9]. Isocoumarins are also used as a lead compound for the identification of insecticides which selectively bind at the insect GABA receptor [10].

3-Substituted isocoumarins also show anti HIV activity *in vitro*, diuretic, antihypertensive, antiarrythmics, β-sympatholytics, anticorrosive, laxatives, asthmolytic, phytotoxic, and are useful in the treatment of emphysema [11]. Isocoumarin derivatives are potently inhibits endothelial cell proliferation, migration, sprouting, tube formation in vitro, and tumor growth *in vivo* [12].

1(2H)-isoquinolones are the nitrogen analogues of isocoumarins. 1(2H)-isoquinolone derivatives are found in several bioactive natural product such as thalifolin, doryphorine [13], ruprechstyril [14], narciclasine [15], lycoricidine [16] and the alkaloids coryaldine and thalflavine. Substitutd isoquinolones exhibit antidepressant, anti-inflamatory, analgesic, hypolipidimic, and analeptic activities have also been reported.

Though a number of new antibiotics have been produced in the last three decades, yet resistance to these drugs by microorganisms has developed. Some antibiotics have become almost obsolete because of drug resistance [17]. In general, bacteria have the genetic ability to transmit and acquire resistance to drugs utilized as therapeutic agents [18]. Consequently new drugs must be synthesized and assayed against these pathogenic resistant microorganisms for the sake of life.

The present study has been designed to determine the *in vitro* antibacterial and antifungal activity of 3-(substituted phenyl)isocoumarins, 1-thioisocoumarins and 1(2H)-isoquinolones against some pathogenic and nonpathogenic strains. A comparison of the efficacy of these classes of compounds and then establishes a structure activity relationship between differently substituted analogues of isocoumarins, 1-thioisocoumarins and isoquinolones is also discussed.

2. Materials and Method

In this *in vitro* antimicrobial assay, we used three series of the compounds *i.e.* 3-(substituted phenyl) isocoumarins (1H-2-benzopyran-1-ones) (1 - 10), (isocarbostyrils 1(2H)-isoquinolones, the nitrogen analogues of isocoumarins (1H-2-enzopyran-1-ones) (11 - 20) and 1-thioisocoumarins (1H-isochromenes-1-thiones) (21 - 30). Each series contains ten compounds having different functionality at position 3. All of the compounds used in this study have been synthesized, purified and characterized by the author [19] [20]. All chemical used are of analytical grade. The purified samples were dissolved in DMSO 5 mg/ml which is the negative control in this bioassay. The antibiotic chloramphenicol and fluconazole were used as standard drugs for antibacterial and antifungal activity respectively.

2.1. Antibacterial Activity

The antibacterial assay was performed by agar well diffusion method against ten different Gram positive and Gram negative bacterial strains [21]. The bacterial strains *Escherichia coli, Klebsiella pneumonae* (ATCC 6633), *Staphylococcus aureus* (ATCC 29213), *Micrococcus luteus* (ATCC 9341), *Pseudomonas aeruginosa* (ATCC 33347), *Escherichia coli* (ATCC 25922), *Salmonella typhi* (ATCC 19430), *Lactobacillus bulgaricus*, (ATCC 25929), *Pasteurella multocida* A (ATCC 9150), *Staphylococcus epidermidis* (ATCC 29232) and *Proteus vulgaris* (ATCC 49565) were selected in this study. *Micrococcus luteus, Staphylococcus aureus* and *Staphylococcus epidermidis* are the Gram positive whilst the remaining seven are Gram negative bacteria. All of the tested microorganisms were maintained on nutrient agar at 4°C and sub-cultured before use.

The bacteria studied are clinically important ones causing several infections and it is essential to overcome them through some active therapeutic agents. Each tested bacterium was sub-cultured in nutrient broth at 37°C for 24 h. One hundred microliters of each bacterium was spread with the help of sterile spreader on to a sterile Muller-Hinton agar plate so as to achieve a confluent growth. The plates were allowed to dry and wells (6mm diameter) were punched in the agar with the help of cork borer. 0.1 mL of the each compound solution (5 mg/mL) in DMSO was introduced in to the well and the plates were incubated overnight at 37°C.

The antimicrobial spectrum of the compounds was determined for the bacterial species in terms of size of the zones around each well and results are presented in **Tables 1-3**. The diameters of the zone of inhibition produced by the compounds were compared with those produced by the commercial antibiotic chloramphenicol (5 mg/mL). This is the common antibiotic used for the treatment of infections caused by gram positive and gram negative bacteria. The control activity was deducted from the test and the results obtained were plotted. The experiment was performed three times to minimize the error and the mean values are presented.

2.2. Antifungal Activity

Antifungal activity of these three series of compounds was determined by using three fungal strain; *Aspergillus flavus*, *Aspergillus nigar* and *Aspergillus pterus* using poison plate method [22]. Potato dextrose agar (PDA) plates were equipped by using pour plate technique for each compound. A 2% concentration of the synthesized compounds in DMSO as a solvent was used. A 2% solution of fluconazole was used as standard. A drug free control was included and plates were observed for growth after 48 h of static incubation at 30°C and results are presented in **Tables 4-6**. All of the synthesized compounds showed poor antifungal activity against the selected fungal strains. The 1-thioisocoumarins showed good growth inhibition than the parent isocoumarins and isoquinolones. The *Aspergillus niger* is the most resistant strain against these isocoumarins derivatives.

3. Results and Discussion

The antibacterial activity of the 3-phenylsubstituted isocoumarins, 3-phenylsubstituted isoquinolin-1(2H)-ones and 3-phenylsubstituted 1H-isochromenes-1-thiones was determined against ten bacterial strains and reported in **Tables 1-3** respectively. The results of the antibacterial assay of these three series of compound reflect that the 3-phenyl substituted isocoumarins are more active as compared to their nitrogen analogues but less active as compared to their thio analogues. The structures of the compounds of these three series are shown in **Figure 1**.

Table 1. *In vitro* Antibcterial activity of 3-substituted isocoumarins (1 - 10).

Compds.	E. c.	K. p.	L. b.	M. l.	P. m.	P. v.	P. a.	S. t.	S. a.	S. e.
1	-	-	-	2	0.5	-	-	-	-	-
2	1.5	-	-	-	-	-	-	-	-	-
3	1	-	6.5	1	-	-	7	-	-	-
4	-	-	1	1.5	-	-	-	-	-	-
5	8	4	4	4	3.5	5.5	4	4	6	7
6	-	8	-	-	4	-	2	-	-	7
7	-	-	2	1	-	-	-	-	-	-
8	1	-	-	1	-	-	-	-	-	-
9	1.5	-	-	-	-	-	-	-	-	0.5
10	-	-	-	-	-	-	-	-	-	-
Standard	18	10	13	13	12	13	13	14	13	13

Table 2. *In vitro* Antibcterial activity of 3-substituted isoquinolones (11 - 20).

Compds.	E. c.	K. p.	L. b.	M. l.	P. m.	P. v.	P. a.	S. t.	S. a.	S. e.
11	1	-	-	-	-	-	-	-	-	-
12	-	-	-	-	0.5	-	-	-	-	1
13	-	-	-	-	-	-	2	-	-	-
14	-	-	-	-	-	-	-	-	-	-
15	2	3	1.5	2	2.5	1.5	-	3	3.5	-
16	3	7	4.5	2.5	5	1.5	2	-	2.5	-
17	-	-	-	-	-	-	3	-	-	-
18	1	-	-	-	-	-	-	-	-	-
19	0	-	-	-	-	-	-	-	0.5	-
20	1.5	-	1	-	-	-	-	-	-	-
Standard	18	-	13	13	12	13	13	14	13	13

Table 3. *In vitro* Antibcterial activity of 3-substituted-1-thioisocoumarins (21 - 30).

Compds.	E. c.	K. p.	L. b.	M. l.	P. m.	P. v.	P. a.	S. t.	S. a.	S. e.
21	15	-	1	1	-	-	1	3	1	-
22	4	-	-	0.5	-	-	-	-	1	-
23	15	-	-	9	0.5	7.5	5.5	1	1	9
24	1	-	-	-	0.5	-	3	1	0.5	-
25	-	1.5	-	-	-	-	0.5	0.5	-	-
26	-	5	-	0.5	-	-	1	1	-	-
27	3	8.5	3	1	1	-	1	3	2	7.5
28	1.5	5	-	0.5	-	-	1.5	0.5	0.5	9
29	1	-	-	1	-	-	-	-	1	-
30	0.5	-	-	-	1	-	-	-	1	-
Standard	18	10	13	13	12	13	13	14	13	13

*Activity of each sample is measured by subtracting the activity of DMSO, (-) No activity. *Escherichia coli* (**E. c.**), *Klebsiella pneumonae* (**K. p.**), *Lactobacillus bulgaricus* (**L. b.**), *Micrococcus luteus* (**M. l.**), *Pasteurella multocida* (**P. m.**), *Proteus vulgaris* (**P. v.**), *Pseudomonas aeruginosa* (**P. a.**), *Salmonella typhi* (**S. t.**), *Staphylococcus aureus* (**S. a.**) and *Staphylococcus epidermidis* (**S. e.**).

Table 4. *In vitro* Antifungal activity of 3-substituted isocoumarins (1 - 10).

Strains	1	2	3	4	5	6	7	8	9	10	Standard
Aspergillus flavus	5	-	11	6	22	3	7	-	9	25	37
Aspergillus niger	-	3	-	-	11	-	-	6	-	14	23
Aspergillus pterus	4	6	-	7	26	-	9	-	4	28	36

Table 5. *In vitro* Antifungal activity of 3-substituted isoquinolones (11 - 20).

Strains	11	12	13	14	15	16	17	18	19	20	Standard
Aspergillus flavus	7	-	9	2	16	-	-	4	-	19	37
Aspergillus niger	-	-	-	-	5	-	-	6	-	8	23
Aspergillus pterus	4	-	-	5	18	-	5	-	-	21	36

Table 6. *In vitro* Antifungal activity of 3-substituted-1-thioisocoumarins (21 - 30).

Strains	21	22	23	24	25	26	27	28	29	30	Standard
Aspergillus flavus	6	-	15	9	24	-	10	-	27	32	37
Aspergillus niger	7	10	-	-	13	-	-	9	15	17	23
Aspergillus pterus	5	8	-	10	27	-	12	-	29	34	36

(-) No activity.

Among all the ten differently 3-phenylsubstituted isoquinolin-1(2*H*)-ones only the 3-(2,4-dichlorophenyl)-1(2*H*) isoquinolone **(18)** shows moderate to potent activity against these tested microorganisms. It shows potent activity against *K. pneumonae* and have moderate efficacy against *L. bulgaricus* and *P. multocida*. These results indicate that in case of 1(2*H*) isoquinolones presence of two electronegative halogen (chlorine) functionality is important in showing antibacterial activity. It is inactive against gram negative bacteria and all of the remaining 1(2*H*) isoquinolones are inactive against both selected gram positive and gram negative bacterial strains.

Most of the 3-phenylsubstituted isocoumarins are inactive against these tested gram positive and gram negative bacterial strains. The 3-(4-nitrophenyl) isocoumarin **(7)** exhibits moderate activity against all the selected gram positive and gram negative bacterial strains. The compound 3-(3-flourophenyl)isocoumarin **(9)** shows moderate activity against *L. bulgaricus* and *P. auriginosa* which are gram positive bacterial strains but is inactive against all the remaining gram positive and gram negative bacterial strains.

It was found that 3-phenylsubstituted 1*H*-isochromenes-1-thiones show potent activity against gram positive bacteria and three derivatives also exhibit activity against gram negative bacteria. 3-(3-Iodophenyl)-1*H*-isochromenes-1-thiones is most active against *E. coli* but inactive towards all other tested microorganisms. Similarly

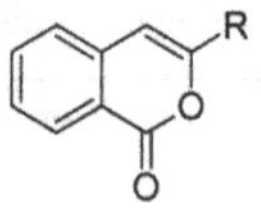

(1-10) R=

(1) = C$_{15}$H$_{31}$,   (2) = 3-I-C$_6$H$_4$,   (3) = 2-Br-C$_6$H$_4$,

(4) = 2-Cl-C$_6$H$_4$CH$_2$,   (5) = 4-F-C$_6$H$_4$,   (6) = 2-Cl-C$_6$H$_3$N,

(7) = 4-NO$_2$-C$_6$H$_4$,   (8) = 2,4-DiCl-C$_6$H$_3$, (9) = 3-F-C$_6$H$_4$,

(10) = 4-F-2-Cl-C$_6$H$_3$

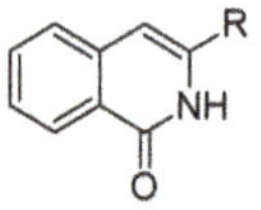

(11-20) R=

(11) = C$_{15}$H$_{31}$,   (12) = 3-I-C$_6$H$_4$,   (13) = 2-Br-C$_6$H$_4$,

(14) = 2-Cl-C$_6$H$_4$CH$_2$, (15) = 4-F-C$_6$H$_4$,   (16) = 2-Cl-C$_6$H$_3$N,

(17) = 4-NO$_2$-C$_6$H$_4$,   (18) = 2,4-DiCl-C$_6$H$_3$, (19) = 3-F-C$_6$H$_4$,

(20) = 4-F-2-Cl-C$_6$H$_3$

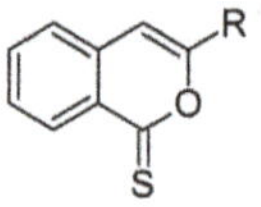

(21-30) R=

(21) = C$_{15}$H$_{31}$,   (22) = 3-I-C$_6$H$_4$,   (23) = 2-Br-C$_6$H$_4$,

(24) = 4-F-C$_6$H$_4$CH$_2$,   (25) = 4-Cl-C$_6$H$_4$, (26) = 4-OCH$_3$-C$_6$H$_4$,

(27) = 4-NO$_2$-C$_6$H$_4$,   (28) = 2,4-DiCl-C$_6$H$_3$, (29) = 3-F-C$_6$H$_4$,

(30) = 4-F-2-Cl-C$_6$H$_3$

Figure 1. Structures of the compounds.

3-pentadecyl-1*H*-isochromenes-1-thiones **(21)** and 3-(2-chloro-4-flourophenyl)-1*H*-isochromenes-1-thiones **(30)** shows activity against *K. pneumonae* and *S. epidermidis* but are inactive against all other bacterial strains. 3-(2,4-dichlorophenyl)-1*H*-isochromenes-1-thiones **(28)** has maximum potential in inhibiting the growth of *K. pneumonae* but possess moderate activity against the *S. epidermidis*. 3-(3-flourophenyl)-1*H*-isochromenes-1-thiones **(29)** is the member of this series which shows maximum effectiveness against both gram positive (*E. coli, L. bulgaricus, P. vulgaricus*) and gram negative (*M. luteus, S. epidermidis*) bacteria. Some other members of this series also possess moderate activity against these studied microorganisms.

   The antifungal activity results showed that most of the isaocoumarin analogues exhibited poor to moderate activity but some of the compounds displayed excellent growth inhibition. 3-(4-fluorophenyl)isocoumarin **(5)** and 3-(2-chloro-4-fluorophenyl)isocoumarin **(10)** showed good growth inhibition against *Aspergillus flavus and Aspergillus pterus* but moderate activity against *Aspergillus nigar*. Most of the 3-substituted isoquinolones **(11 - 20)** depict poor or no antifungal activity against the selected fungal strains. 3-(4-fluorophenyl)-1(2*H*) isoquino-lone **(15)** and 3-(2-chloro-4-fluorophenyl)-1(2*H*)-isoquinolone **(20)** possess moderate activity against *Aspergillus flavus and Aspergillus pterus* but are not effective against *Aspergillus nigar*.

   Some of the isocoumarin-1-thiones portrayed good to excellent activity against tested fungal stains. 3-(3-flourophenyl)-1*H*-isochromenes-1-thiones **(29)** and 3-(2-chloro-4-flourophenyl)-1*H*-isochromenes-1-thiones **(30)** represented excellent growth inhibition against *Aspergillus flavus* and *Aspergillus pterus* and good activity against *Aspergillus nigar*. 3-(4-chlorophenyl)-1*H*-isochromenes-1-thiones **(25)** having good antifungal potential against tested fungal strains. The thio derivatives of isocoumarins possess greater antimicrobial activity than ni-trogen analogues and parent isocoumarins **Figure 2**. The comparison of the antifungal activity among the three classes also presented in **Figure 3**.

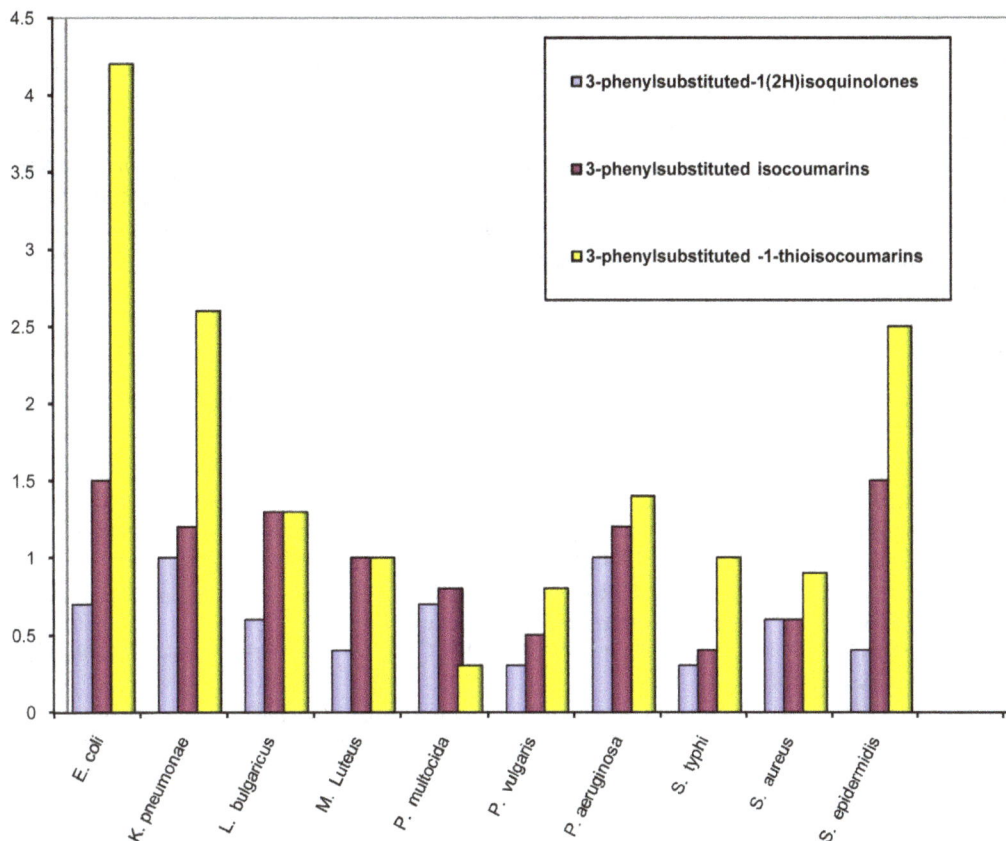

Figure 2. Comparison of the antibacterial activity of 3-phenylsubstituted isocoumarins (1 - 10), 3-phenylsubstituted isoquinolin-1(*2H*)-ones (11-20) and 3-phenylsubstituted 1*H*-isochromenes-1-thiones (21 - 30).

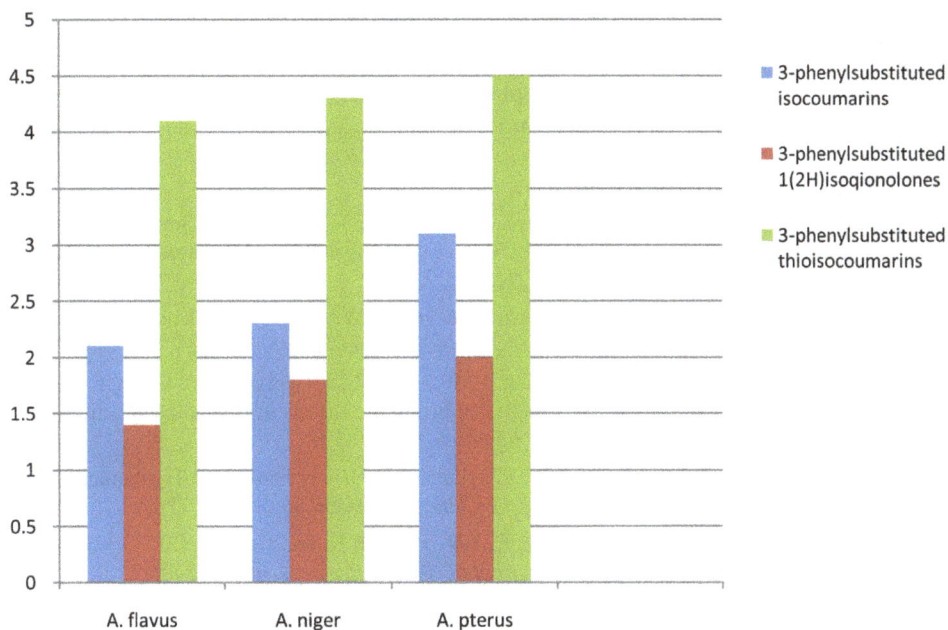

Figure 3. Comparison of the antifungal activity of 3-phenylsubstituted isocoumarins (1 - 10), 3-phenylsubstituted isoquinolin-1(2*H*)-ones (11 - 20) and 3-phenylsubstituted 1*H*-isochromenes-1-thiones (21 - 30).

4. Conclusion

We have concluded from this antibacterial assay that when the isocoumarins are converted into 1-thiones the biological activity of the resulting derivatives is increased and nitrogen analogues are less active as compared to parent isocoumarins. Most probably this is due to the high hydrophobicity of the sulphur analogues. In the nitrogen derivatives polarity is increased but hydrophobicity is decreased and as a result of decrease in the lipophilicity activity is decreased. Hydrophobic functionalities are the necessities for these compounds in exhibiting biological activity. The electronegativity of the substituents present at 3-phenyl ring also play an important role in biological activity. The fluorine and chlorine substituted derivatives are more active than the bromine and iodine ones.

References

[1]   Napolitano, E. (1997) The Synthesis of Isocoumarins over the Last Decade. *Organic Preparations and Procedures International*, **29**, 631-634. http://dx.doi.org/10.1080/00304949709355245

[2]   Sankawa, U. (1999) Comprehensive Natural Products Chemistry. Elsevier Science Publishers Ltd., Amsterdam.

[3]   Hisashi, M., Hiroshi, S. and Masayuki, Y. (1999) Structure-Requirement of Isocoumarins, Phthalides and Stilbenes from Hydrangeae Dulcis Folium for Inhibitory Activity on Histamine Release from Rat Peritoneal Mast Cells. *Bioorganic & Medicinal Chemistry*, **7**, 1445-1450. http://dx.doi.org/10.1016/S0968-0896(99)00058-9

[4]   Masayuki, Y., Emiko, H., Yoshikazu, N., Kimiyo, I., Hisashi, M., Hiroshi, S., Johji, Y. and Nobutoshi, M. (1994) Development of Bioactive Functions in Hydrangeae Dulcis Folium. III. On the Antiallergic and Antimicrobial Principles of Hydrangeae Dulcis Folium (1). Thunberginol A, B, and F. *Chemical and Pharmaceutical Bulletin*, **42**, 2225-2230. http://dx.doi.org/10.1248/cpb.42.2225

[5]   (a) Hisashi, M., Hiroshi, S., Johji, Y. and Masayuki, Y. (1998) Immunomodulatory Activity of Thunberginol A and Related Compounds Isolated from Hydrangeae Dulcis Folium on Splenocyte Proliferation Activated by Mitogens. *Bioorganic & Medicinal Chemistry Letters*, **8**, 215-220. http://dx.doi.org/10.1016/S0960-894X(97)10221-9

[6]   Authrine, C.W., James, B.G., James, A.S. and David, M. (1996) Cercophorins A-C: Novel Anti Fungal and Cytotoxic Metabolites from the Coprophilous Fungus Cercophora Areolata. *Journal of Natural Products*, **59**, 765-769. http://dx.doi.org/10.1021/np9603232

[7]   Koohei, N., Mikikio, Y., Yoshiko, T., Kenichi, K. and Shoichi, N. (1981) Antifungal Activity of Oosponol, Oospolactone, Phylloducin, Hydrangenol, and Some Other Related Compounds. *Chemical and Pharmaceutical Bulletin*, **29**, 2689-2691. http://dx.doi.org/10.1248/cpb.29.2689

[8]   Takuya, F., Yoshiyasu, F. and Yoshinori, A. (1986) Polygonolide, an Isocumarin from Polygonum Hydropiper Possessing Anti-Inflammatory Activity. *Phytochemistry*, **25**, 517-520. http://dx.doi.org/10.1016/S0031-9422(00)85513-2

[9]   Jeong, H.L., Yun, J.P., Hang, S.K., Young, S.H., Kyu-Won, K. and Jung, J.L. (2001) Anti-Angiogenic Activities of Novel Isocoumarins, AGI-7 and Sescandelin. *The Journal of Antibiotics*, **54**, 463-466. http://dx.doi.org/10.7164/antibiotics.54.463

[10]   Ozoe, Y., Kuriyama, T., Tachibana, Y., Harimaya, K., Takahashi, N., Yaguchi, T., Suzuki, E., Imamura, K. and Oyama, K. (2004) Isocumarin Derivatives as a Novel GABA Receptor Ligand from Neosartorya Quadricincta. *Journal of Pest Science*, **29**, 328-331. http://dx.doi.org/10.1584/jpestics.29.328

[11]   Hudson, J.B., Graham, E.A., Harris, L. and Ashwood-Smith, M.J. (1993) The Unusual UVA-Dependent Antiviral Properties of the Furoisocoumarin, Coriandrin. *Phytochemistry and Photobiology*, **57**, 491-496.

[12]   Corinne, L.R., Naoki, A., Jennifer, G.T., Michael, B., William, M.D., George, D.K., Susan, L.R., Michael, M., Robert, F., Raghu, K., Donald, K. and Surender, K. (2002) Antineoplastic Effects of Chemotherapeutic Agents Are Potentiated by NM-3, an Inhibitor of Angiogenesis. *Cancer Research*, **62**, 789-795.

[13]   Chen, C.Y., Chang, F.R. and Teng, C.M. (1999) Cheritamina, a New N-Fatty Acyl Tryptamine and Other Constituents from the Stem of Annona Cherimola. *Journal of the Chinese Chemical Society*, **46**, 77-86.

[14]   Pettit, G.R., Meng, Y.H., Herald, D.L., Graham, K.A.N., Pettit, R.K. and Doubek, D.L. (2003) Isolation and Structure of Ruprechstyril from *Ruprechtia tangarana*. *Journal of Natural Products*, **66**, 1065-1069. http://dx.doi.org/10.1021/np0300986

[15]   David, G., Theodore, M. and Tomas, H. (1999) A Short Chemoenzymatic Synthesis of (+)-Narciclasine. *Tetrahedron Letters*, **40**, 3077-3283.

[16]   Richard, C.T. and James, K. (1990) An Annulative, Carbohydrate-Based Approach to Pancratistatin and Structurally-Related Phenanthridone Alkaloids. Synthesis of (+)-Tetrabenzyllycoricidine. *The Journal of Organic Chemistry*, **55**,

6076-6078. http://dx.doi.org/10.1021/jo00312a007

[17] Ekpendu, T.O., Akshomeju, A.A. and Okogun, J.I. (1994) Antiinflamatory, Antimicrobial Activity. *Letters in Applied Microbiology*, **30**, 379-384.

[18] Cohen, M.L. (1992) Epidemiology of Drug Resistance, Implications for a Post-Antimicrobial Era. *Science*, **257**, 1050-1057. http://dx.doi.org/10.1126/science.257.5073.1050

[19] Aamer, S. and Zaman, A. (2008) Synthesis of Some 3-Aryl-1*H*-Isochromene-1-Thiones. *Journal of Heterocyclic Chemistry*, **45**, 679-682. http://dx.doi.org/10.1002/jhet.5570450307

[20] Aamer, S. and Zaman, A. (2008) An Efficient Synthesis of Some 3-Aryl-Isoquinolin-1(2H)-Ones; *Chemistry of Heterocyclic Compounds*, **8**, 1203-1208.

[21] Zaman, A., Aun, M., Muhammad, I. and Ahmed, H. (2011) *In Vitro* Antibacterial and Antifungal Activity of Methanol, Chloroform and Aqueous Extracts of *Origanum vulgare* and Their Comparative Analysis. *International Journal of Organic Chemistry*, **1**, 257-261. http://dx.doi.org/10.4236/ijoc.2011.14037

[22] Shastri, R.V. and Varudkar, J.S. (2009) Synthesis and Antimicrobial Activity of 3-Propen-1,2-Benzisoxazole Derivatives. *Indian Journal of Chemistry*, **48B**, 1156-1160.

4

A Modified Method for the Synthesis of Tetradentate Ligand Involving Peptide Bond

Pulimamidi Rabindra Reddy*, Ravula Chandrashekar, Hussain Shaik, Battu Satyanarayana

Department of Chemistry, Osmania University, Hyderabad, India
Email: *profprreddy@gmail.com

Abstract

In view of the importance of picolinic acid (Pa) in preventing cell growth and arresting cell cycle, attempts were made to design, synthesize and characterize two new Pa based tetradentate ligands (DPPTR and DPPTY) with a modified procedure. The procedure reported here avoids by-products and provides better yield and purity.

Keywords

Amide Bond Formation, Tetradentate Ligand, Peptide Bond, Coupling Reagent

1. Introduction

Synthesis of ligands with peptide bond has attracted lot of attention due to their importance in biological systems. Many reagents were used to get the desired peptides [1]-[5]. Among them DCC (N,N'-dicyclohexylcarbodiimide), EDCI (3-(Ethyliminomethyleneamino)-N,N-dimethylpropan-1-amine), HOBT (1-Hydroxybenzotriazole) and HATU (1-[Bis(dimethylamino)methylene]-1H-1,2,3-triazolo[4,5-b]pyridinium 3-oxid hexafluorophosphate) were extensively used [6]-[11]. However, with DCC and EDCI-HOBT, there is a possibility of by-product formation, reduction in the yield and makes purification difficult. With DCC, multiple by-products like anhydride, urea and N-acylurea were formed [12] [13]. To avoid this HATU was used earlier for the synthesis of simple peptides [14] [15]. In view of this, We adopted a modified procedure employing HATU and DIEA (N,N-Diisopropylethylamine) for the synthesis of tetradentate ligand involving peptide bond. The base (DIEA) has an advantage due to the presence of an isopropyl group which helps in increasing the basicity on N-atom and facili-

---

*Corresponding author.

tates the proton abstraction from the acid. Since Pa and their derivatives are known to prevent the cell growth and arrest the cell cycle [16], it was thought important to design, synthesize and characterize two new Picolinic acid based tetradentate ligands, N-(3-(1H-indol-3-yl)-1-oxo-1-(pyridine-2-yl methylamino)propan-2-yl) picolinamide (DPPTR) and N-3-(4-hydroxyphenyl)-1-oxo-1-(pyridine-2-ylmethylamino) propan-2-yl) picolinamide (DPPTY) using HATU and DIEA. The synthesis of ligands involves three steps (**Scheme 1**). The intermediates and final ligands were isolated and characterized. The results show that the yields for the intermediates are >90% and for final ligands > 80% with good purity.

2. Results and Discussion

2.1. DPTR-I/DPTY-I

The ESI-Mass spectra of DPTR-I (**Figure 1**) shows m/z peak at 324 indicating that the DPTR-I molecular ion species appeared as $[M + H]^+$ and for DPTY-I (**Figure 2**, (Suppl Material, **SM**)) the mass spectra shows a peak at 301 specifying that the DPTY-I molecular ion species also appeared as $[M + H]^+$. The M.ps 130°C - 133°C for DPTR-I and 128°C - 131°C for DPTY-I.

^{1}H-NMR spectra of DPTR-I/DPTY-I

DPTR-I (**Figure 3**): ^{1}H-NMR (400 MHz, CD$_3$OH): δ = 10.02 (s, 1H) 7.94 (d, 1H), 7.74 (d, 1H), 7.14 - 7.12 (m, 2H), 6.75 - 6.72 (m, 1H), 6.61 (d, 1H), 6.48 - 6.44 (m, 1H), 6.29 (d, 1H), 6.21 - 6.16 (m, 1H), 6.13 - 6.05 (m, 1H), 2.77 (s, 3H), 2.72 (brs, 1H), 1.65 - 1.62 (m, 2H) [17] [18].

DPTY-I (**Figure 4**, SM): ^{1}H-NMR (400 MHz, DMSO-d6): δ = 9.24 (s, 1H), 8.80 (d, 1H), 8.65 (d, 1H), 7.99 (t, 1H), 7.98 (d, 1H), 7.63 - 7.60 (m, 1H), 7.00 (d, 2H), 6.63 (d, 2H), 4.73 - 4.68 (m, 1H), 3.64 (S, 3H), 3.09 (d, 2H).

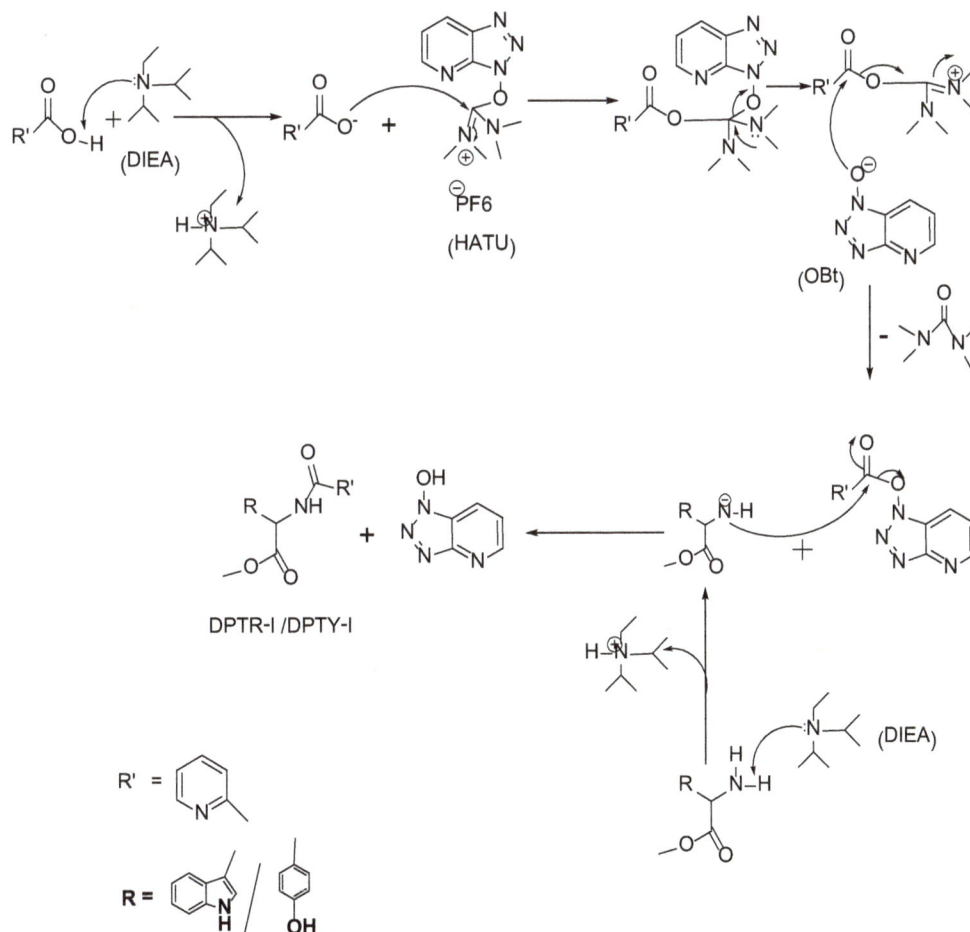

The proposed mechanism showing the formation of DPTR-I/DPTY-I.

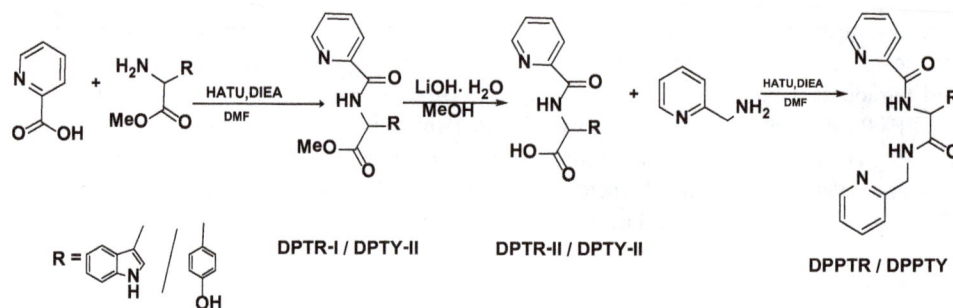

Scheme 1. Synthesis of ligands (DPPTR/DPPTY).

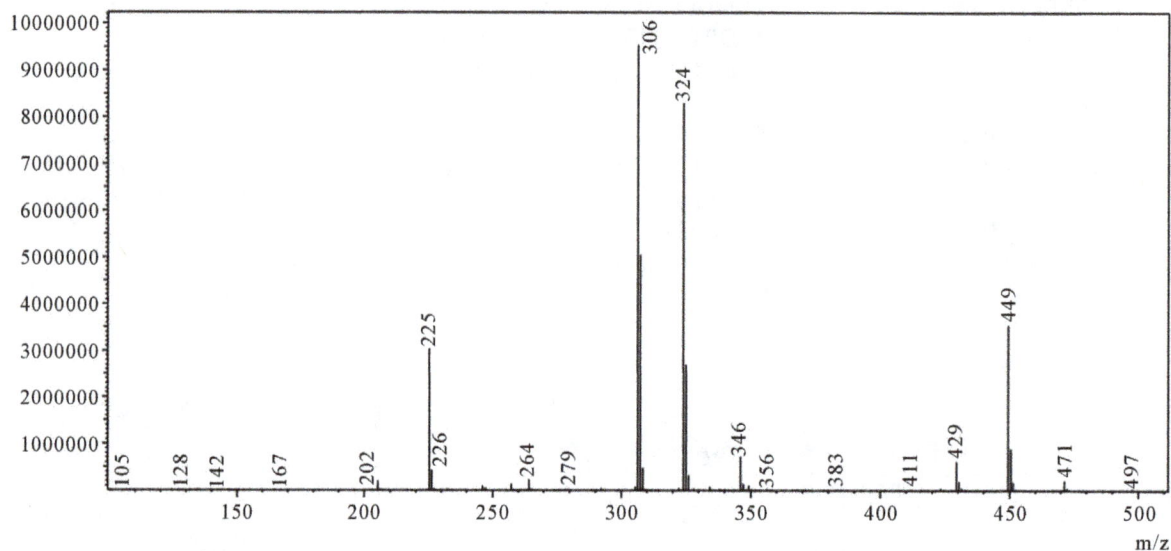

Figure 1. ESI-Mass spectra of DPTR-I.

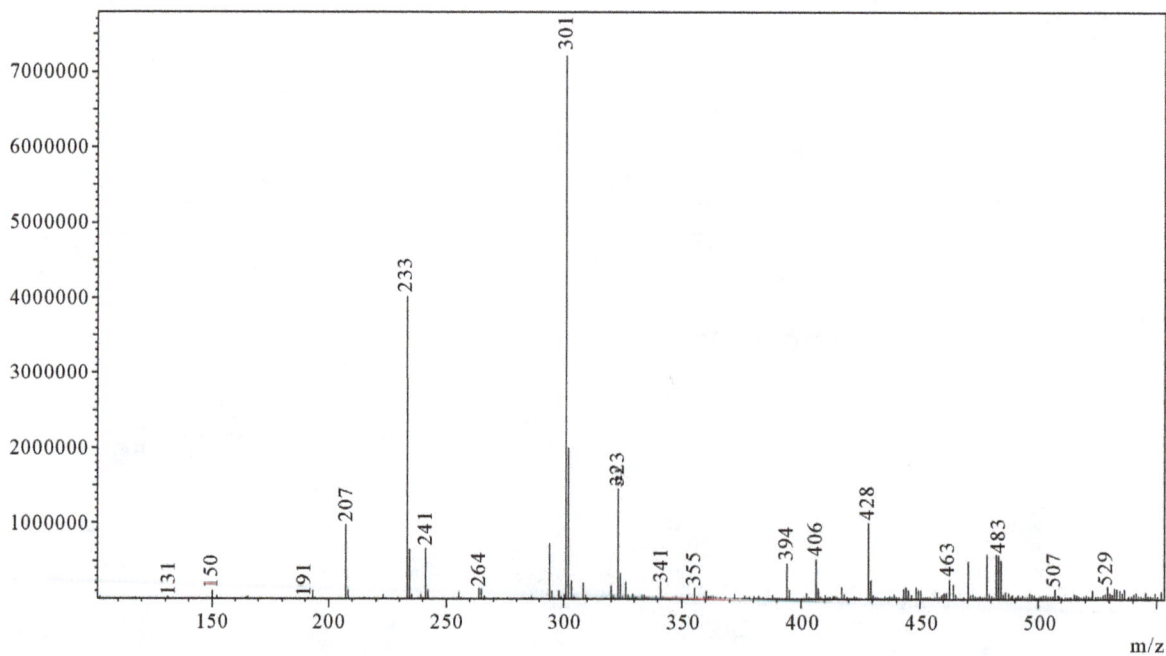

Figure 2. ESI-Mass spectra of DPTY-I.

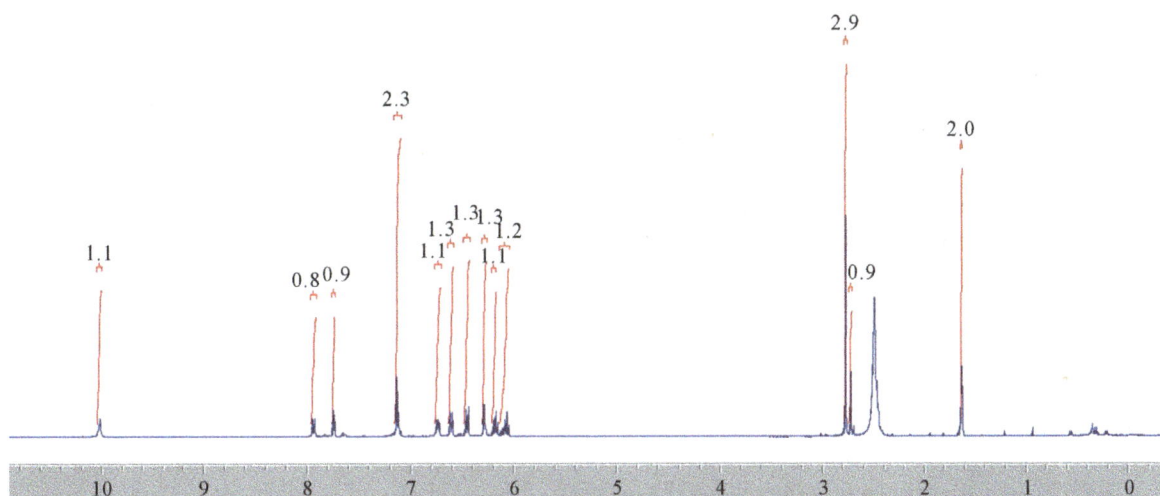

Figure 3. ^{1}H-NMR spectra of DPTR-I.

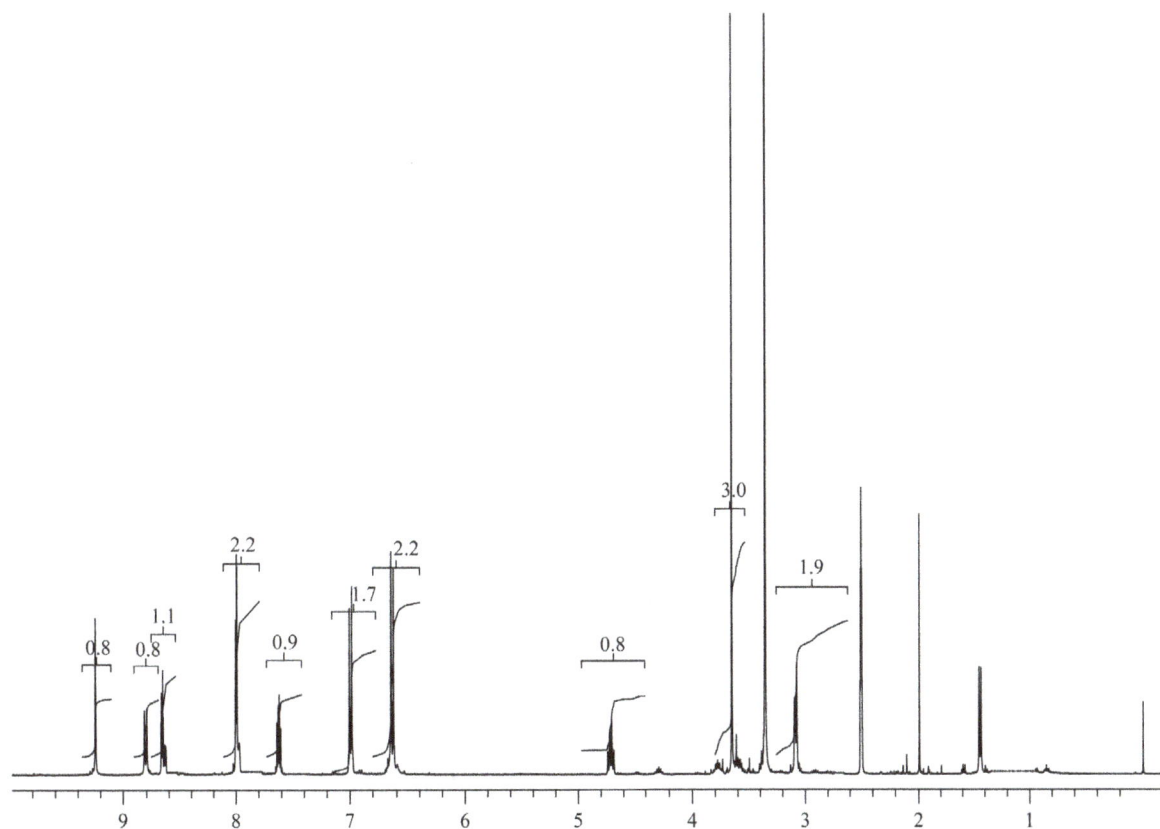

Figure 4. ^{1}H-NMR spectra of DPTY-I.

The mechanism involves,

1) Proton abstraction occurs from Pa followed by the addition of carboxylate anion (-Coo$^-$) to the electron deficient carbon atom of HATU. This results in the formation of new C-O bond.

2) The resulting anion reacts with the newly formed activated carboxylic acid derived from intermediate to form an OBt activated ester.

3) The amine reacts with the OBt activated ester to form the amide product.

Similar mechanism was proposed earlier for the formation of amide bond [19]-[21].

2.2. DPTR-II/DPTY-II

The ESI-Mass spectra of DPTR-II (**Figure 5**) shows m/z peak at 315 suggesting that the DPTR-II molecular ion species appeared as $[M + Li]^+$ and for DPTY-II (**Figure 6**, SM) the mass spectra shows a peak at 287 indicating that the DPTY-II molecular ion species appeared as $[M + H]^+$. The M.ps 131°C - 134°C for DPTR-II and 129°C - 132°C for DPTY-II.

^{1}H-NMR spectra of DPTR-II/DPTY-II

DPTR-II (**Figure 7**): ^{1}HNMR (400 MHz, DMSO-d$_6$): δ = 9.88 (s, 1H), 7.67 (d, 1H), 7.19 - 7.08 (m, 3H), 6.69 - 6.66 (m, 1H), 6.58 (d, 1H), 6.37 (d, 1H), 6.19 (d, 1H), 6.09 - 6.05 (m, 1H), 5.93 - 5.89 (m, 1H), 3.40 - 3.38 (m, 1H), 2.37 - 2.32 (m, 2H).

DPTY-II (**Figure 8**, SM): 1H-NMR (400 MHz, DMSO-d6): δ = 9.22 (s, 1H), 8.65 - 8.60 (m, 2H), 8.01 - 7.98 (m, 2H), 7.62 (d, 1H), 6.99 (d, 2H), 6.22 (d, 2H), 4.67 - 4.62 (m, 1H), 3.08 (d, 2H).

IR spectra of DPTR-II/DPTY-II

The IR spectra for the above intermediates were recorded to confirm the presence of acidic proton (COOH functional group).

DPTR-II (Figure 9): IR v_{max} (MeOH): 3323 (CONH), 3098 (COOH), 1625 (C=O), 1516 (C=C) 1437 (C=N), 1245 (C-N) cm^{-1} [22].

DPTY-II (Figure 10, SM); IR v_{max} (MeOH): 3330 (CONH), 2956 (COOH), 1739 (C=O), 1439 (C=C), 1367 (C=N), 1218 (C-N) cm^{-1}.

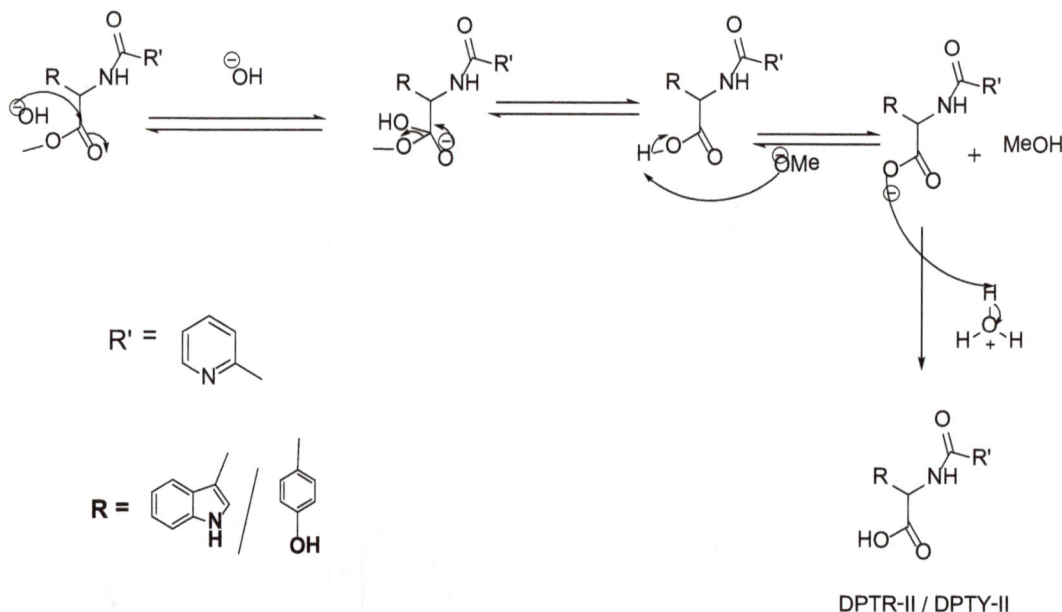

The proposed mechanism showing the formation of DPTR-II/DPTY-II.

The general mechanism for the de-esterification by base involves a series of equilibria. The hydroxide anion adds to the carbonyl group of the ester. The direct products are a carboxylic acid salt and an alcohol. To convert the salt to the corresponding carboxylic acid, acidic workup of the product mixture was performed [23].

2.3. Final Ligands: DPPTR/DPPTY

The ESI-Mass spectra of DPPTR-(**Figure 11**) shows m/z peak at 400 indicating that the DPPTR molecular ion species appeared as $[M + H]^+$ and for DPPTY (**Figure 12**, SM) the mass spectra shows a peak at 377 suggesting that the DPPTY molecular ion species appeared as $[M + H]^+$. The M.ps 136°C - 138°C for DPPTR and 133°C - 135°C for DPPTY.

^{1}H-NMR spectra of DPPTR/DPPTY

DPPTR (**Figure 13**); 1H-NMR (400 MHz, DMSO-d6): δ = 10.68 (s, 1H), 9.82 (d, 2H), 8.59 - 8.50 (m, 1H),

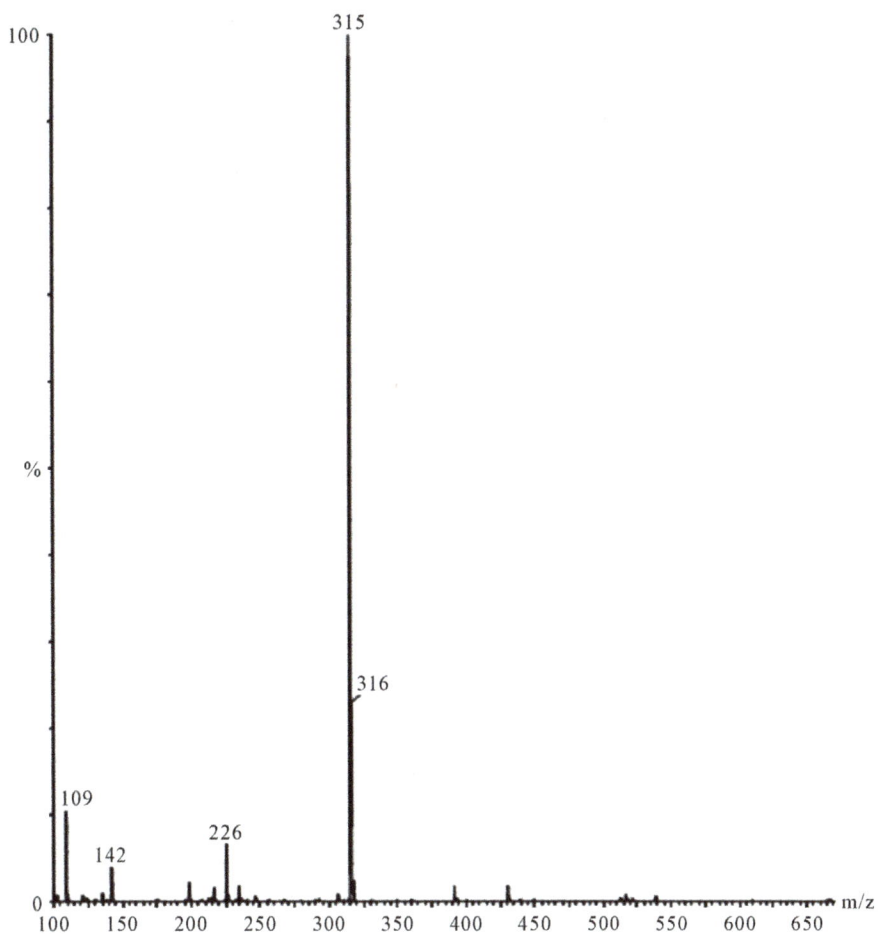

Figure 5. ESI-Mass spectra of DPTR-II.

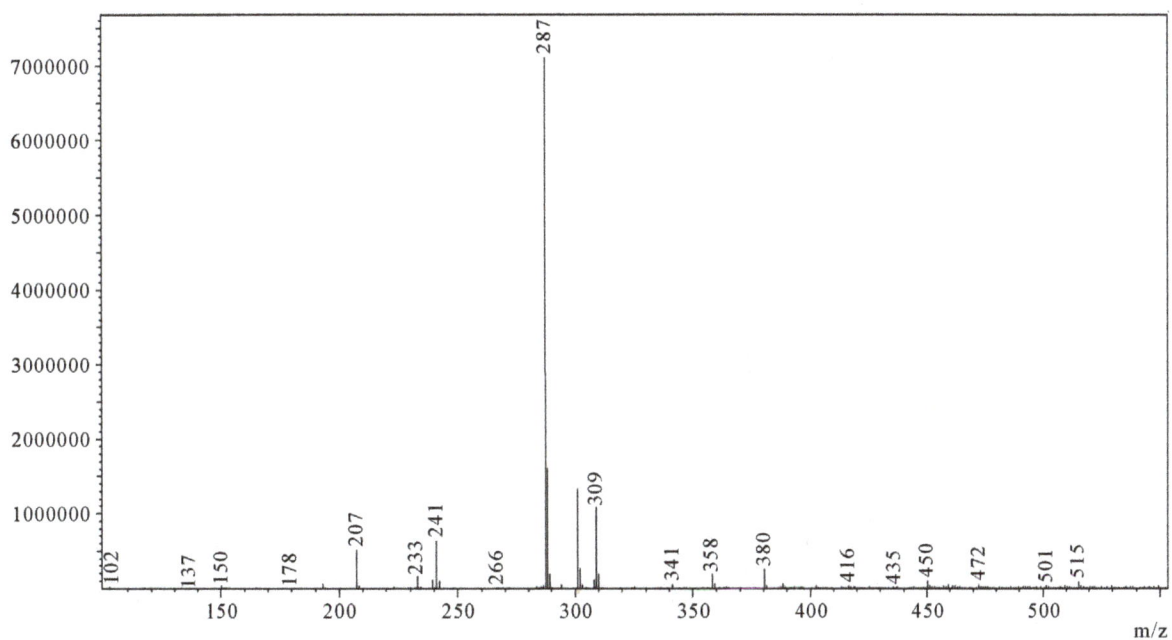

Figure 6. ESI-Mass spectra of DPTY-II.

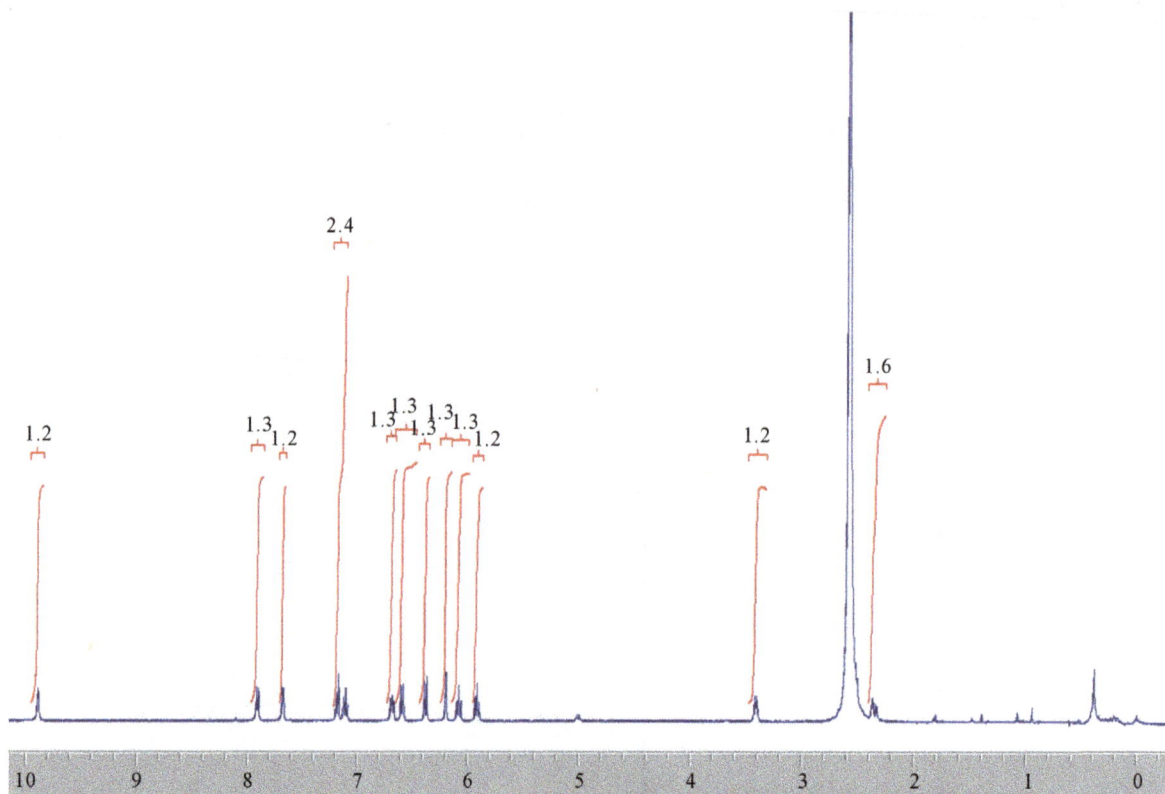

Figure 7. ^{1}H-NMR spectra of DPTR-II.

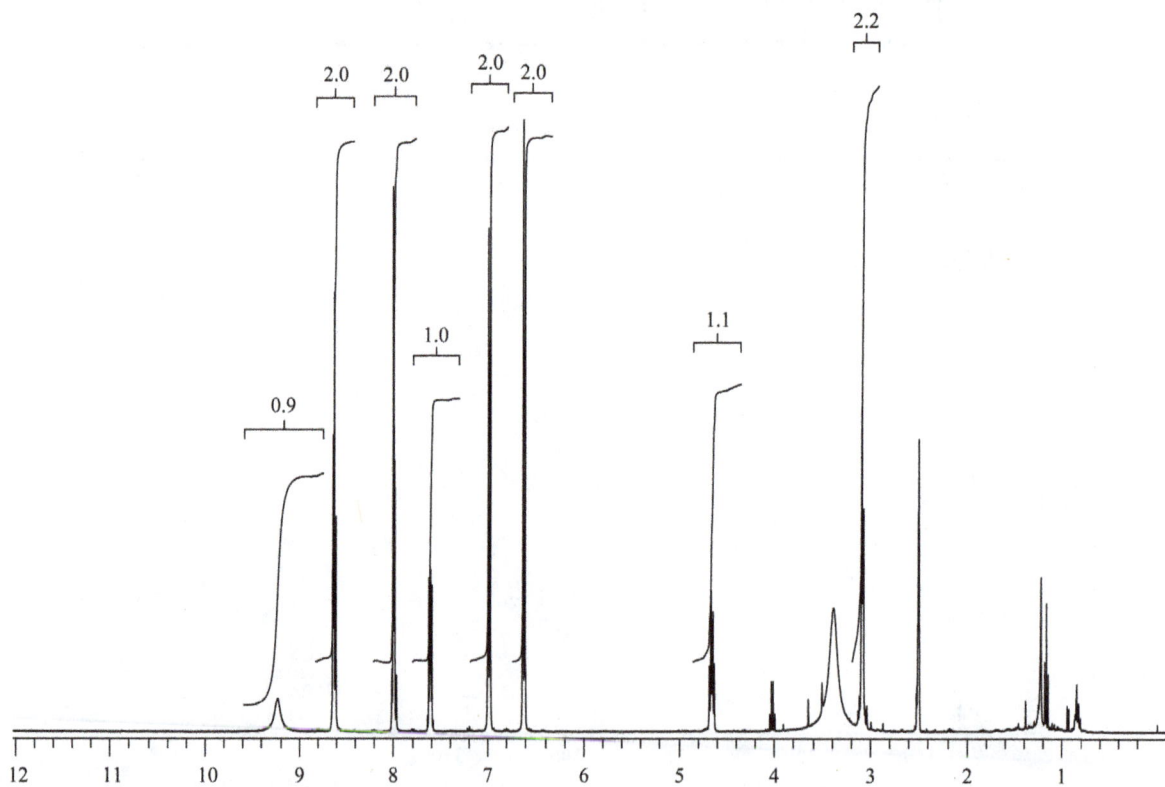

Figure 8. ^{1}H-NMR spectra of DPTY-II.

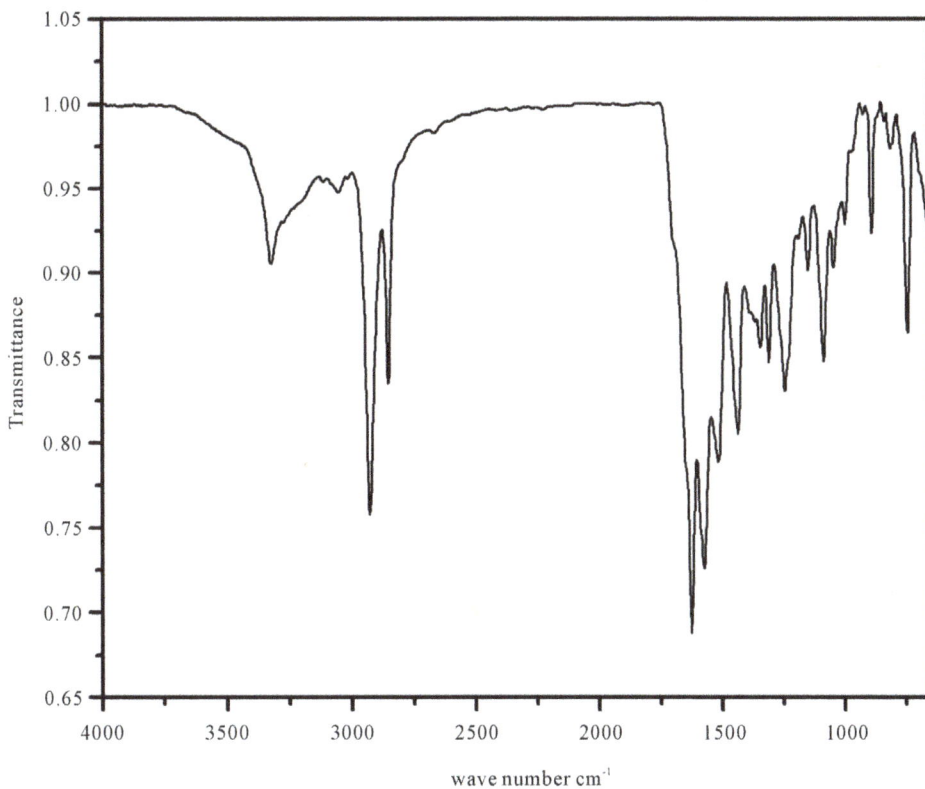

Figure 9. Infrared spectra of DPTR-II.

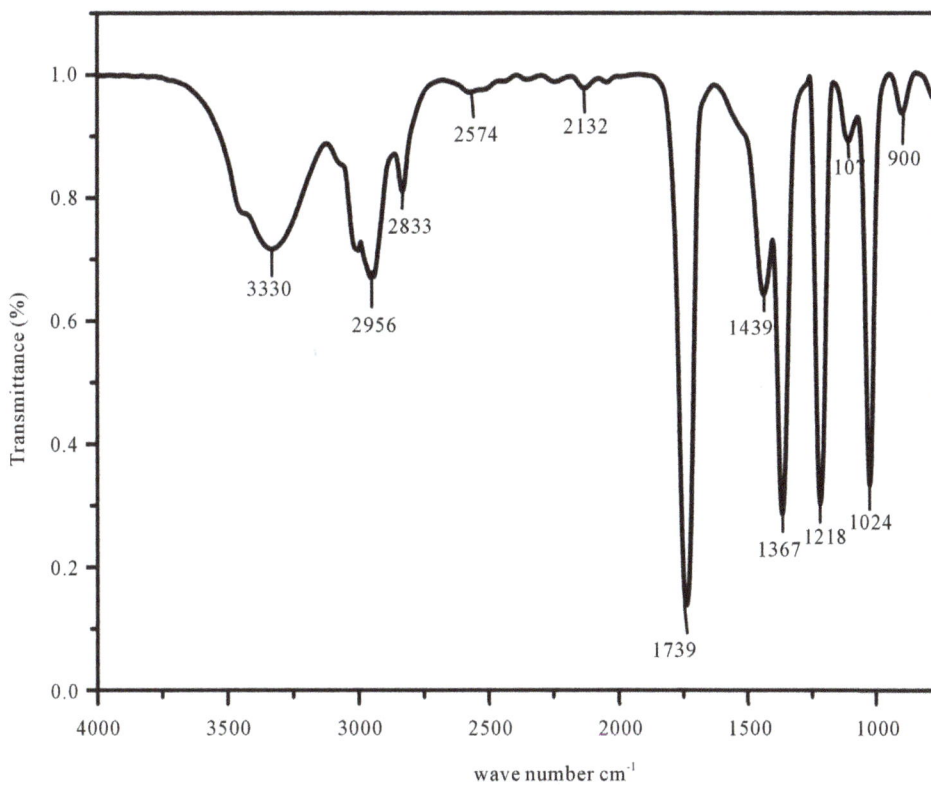

Figure 10. Infrared spectra of DPTY-II.

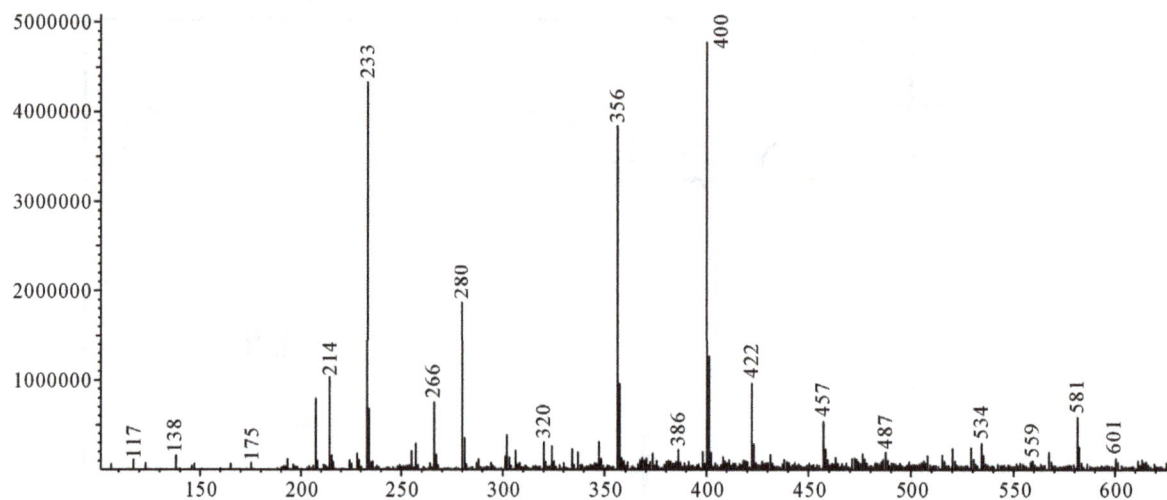

Figure 11. ESI-Mass spectra of DPPTR.

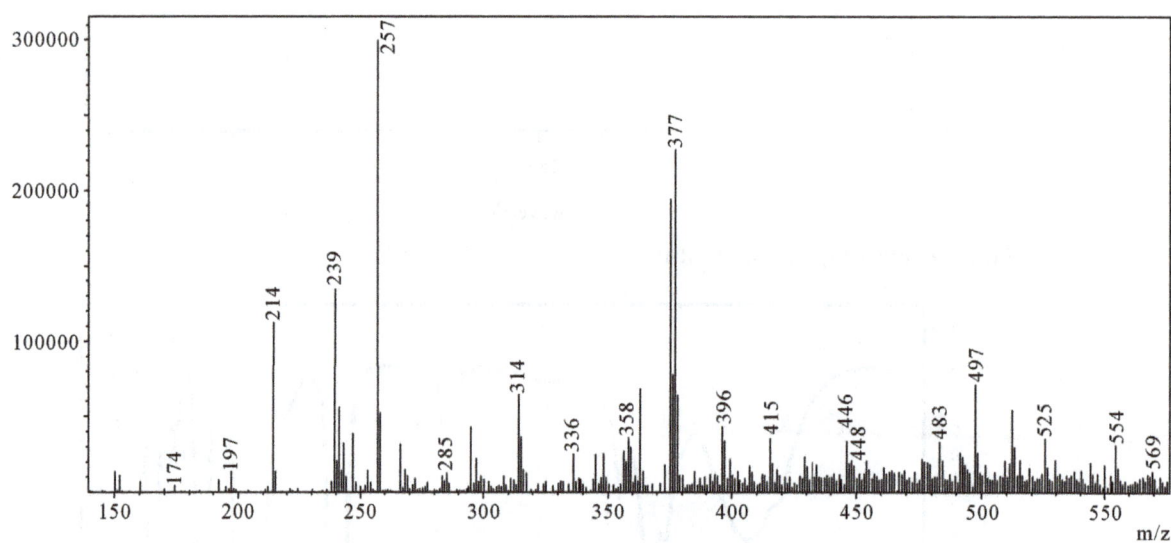

Figure 12. ESI-Mass spectra of DPPTY.

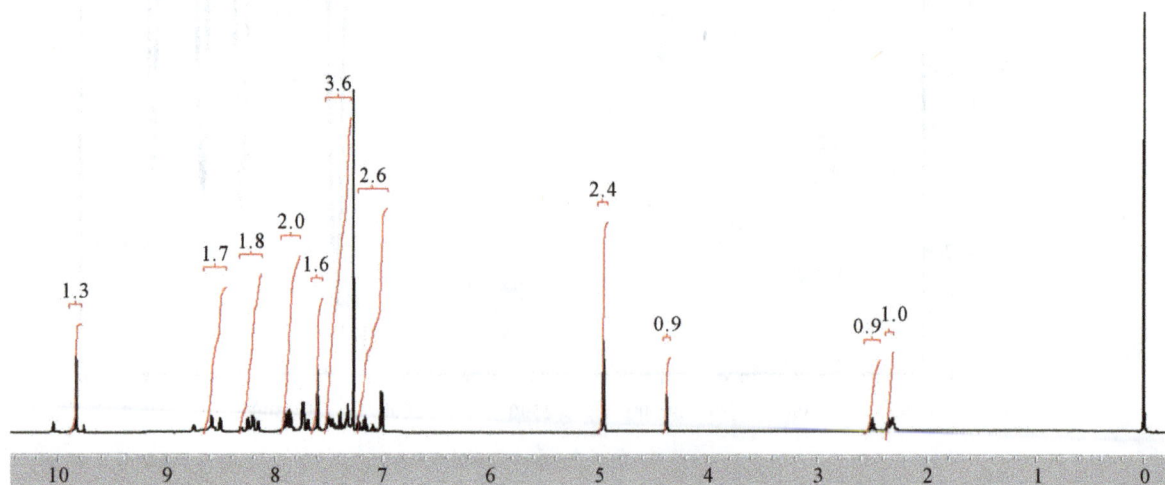

Figure 13. ^{1}H-NMR spectra of DPPTR.

8.26 - 8.14 (m, 2H), 7.91 - 7.85 (m, 2H), 7.74 - 7.67 (m, 2H), 7.59 (s, 1H), 7.50 - 7.31 (m, 2H), 7.23 - 7.08 (m, 1H), 7.01 - 6.99 (m, 1H), 4.95 (d, 3H), 4.37 (d, 1H), 2.50 - 2.47 (m, 1H), 2.36 - 2.31 (m, 1H)

DPPTY (**Figure 14**, SM); 1-HNMR (400 MHz, DMSO-d_6): δ = 8.82 (t, 1H), 8.65 (t, 2H), 8.48 (d, 1H), 8.02 (d, 2H), 7.75 (t, 1H), 7.62 - 7.59 (m, 1H), 7.25 (t, 1H), 7.10 (d, 1H), 7.00 (d, 2H), 6.60 (d, 2H), 4.79 - 4.74 (m, 1H), 4.43 - 4.34 (m, 2H), 3.06 - 2.99 (m, 2H).

IR spectra of DPPTR/DPPTY

DPPTR (**Figure 15**): IR υ_{max} (MeOH): 3307 (CONH), 1660 (C=O), 1516 cm^{-1} (C=C), 1478 (C=N), 1232 (C-N) cm^{-1}.

DPPTY (**Figure 16**, SM): IR υ_{max} (MeOH): 3323 (CONH), 1741 (C=O), 1371 (C=C), 1443 (C=N), 1224 (C-N) cm^{-1}.

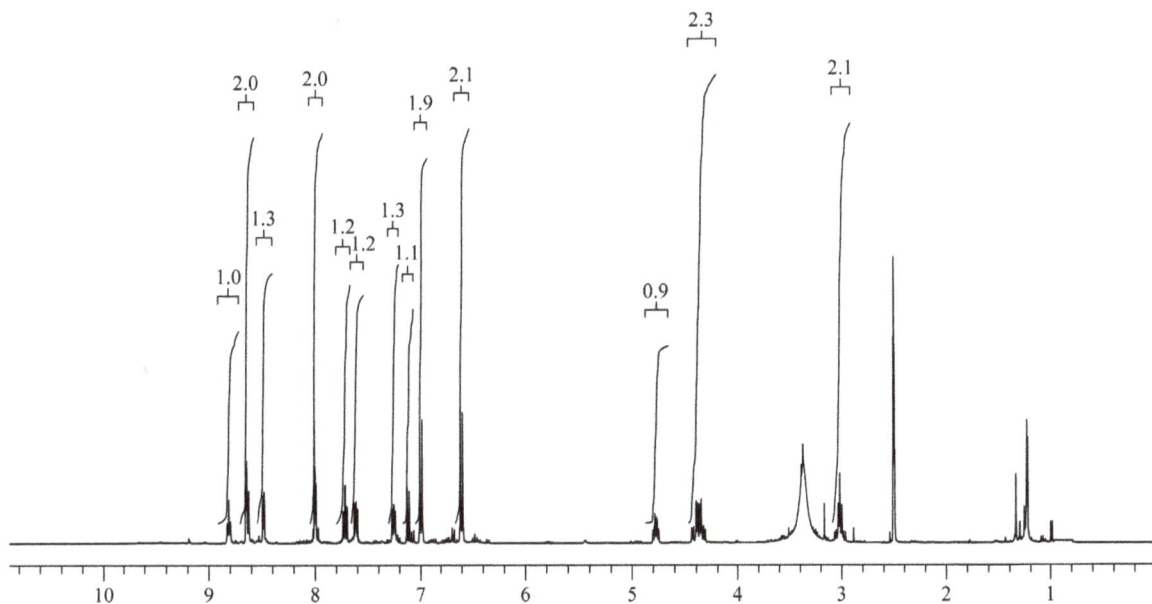

Figure 14. ^{1}H-NMR spectra of DPPTY.

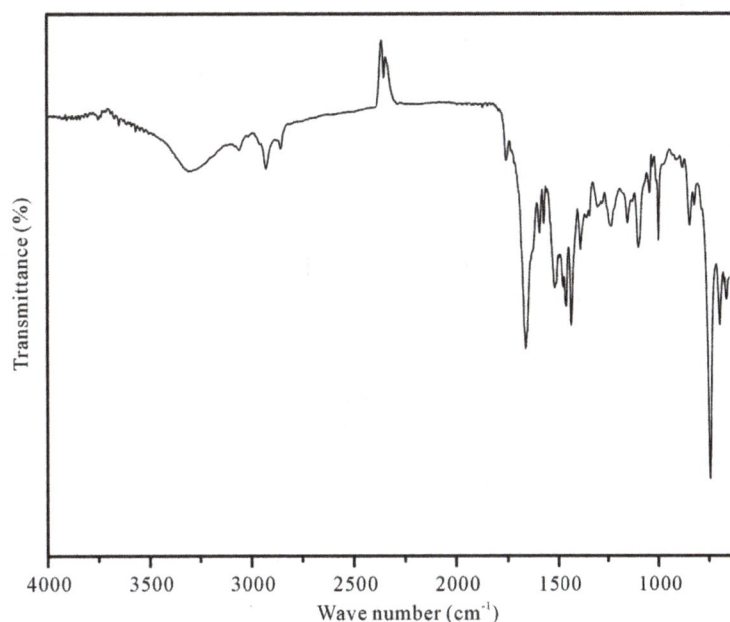

Figure 15. Infrared spectra of DPPTR.

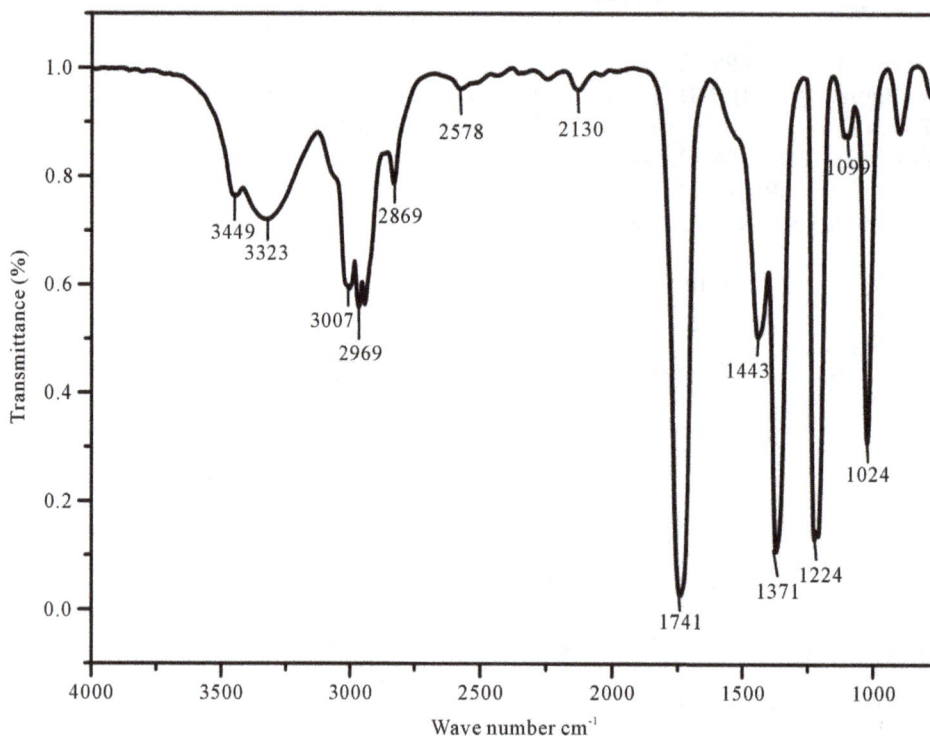

Figure 16. Infrared spectra of DPPTY.

The mechanism for the formation of final ligands is similar to that described for the formation of DPTR-I/ DPTY-I except that the starting materials are different.

3. Conclusion

Two new tetradentate ligands involving peptide bond were synthesized with a modified procedure and characterized. The procedure is simple and avoids by-products and results in better yields. Since small molecular bio-ligands containing peptide bond are known to play an important role as biomimetics, construction of such mimics can lead to a better understanding of the biological complexity at a molecular level. Therefore, the procedure described here will provide an opportunity to synthesize new small molecules.

4. Experimental Section

4.1. Material and Methods

Picolinic acid, Tryptophan-methyl ester, Tyrosine-methyl ester, picolylamine and $LiOH \cdot H_2O$ are obtained from sigma chemical company (99% purity), USA. HATU and solvents (DIEA, methanol, Ethylacetate, n-Hexane and dimethylformamide) were purchased from Merck, India and were of analar grade. The chemicals were used as supplied. The TLC silica gel plates (60 F_{254}) were obtained from Merck. Infrared spectra were recorded on a Perkin-Elmer FT-IR spectrometer in the range of 4000 - 750 cm^{-1} using Methanol as solvent. ESI mass spectra for the ligands were recorded on a Quattro Lc (Micro mass, Manchester, UK) triple quadruple mass spectrometer with Mass Lynx software and Shimadzu, model LC-MS; 8030. The ^{1}H-NMR spectra were recorded on a Bruker Biospin and Avance-III 400 MHz Fourier Transform Digital NMR Spectrometer, Switzerland using DMSO as solvent and TMS as the internal standard. The melting points were recorded on a cintex melting point instrument and are uncorrected. All reactions were carried under N_2 atmosphere.

4.2. Synthesis of Peptides

The synthesis of peptides involves three steps (**Scheme 1**). The following procedure was adopted for the synthesis.

DPPTR:

For the synthesis of DPPTR, 2-picolinic acid (0.2 g, 1.62 mmol) was dissolved in dry DMF (10 mL) and HATU (0.74 g, 1.95 mmol) and DIEA (0.62 g, 4.86 mmol) were added. The solution was cooled to 0°C. The solution was stirred for 30 min followed by the addition of Tryptophan-methyl ester (0.62 g, 2.43 mmol). The mixture was warmed to room temperature and the stirring continued for another 12 h. After the workup, the solvent was removed under reduced pressure and the remaining solid was washed with petroleum ether to afford the compound, **DPTR-I** (yield: 0.481 g, 93%). In the second step, the protected methyl ester (OMe) was removed by saponification using LiOH in MeOH to get **DPTR-II**. It was purified by column chromatography (yield: 0.419 g, 91%). Finally, **DPTR-II** (0.42 g, 1.35 mmol) was dissolved in dry DMF (10 ml) and HATU (0.61 g, 1.62 mmol) and DIEA (0.52 g, 4.05 mmol) were added and the mixture was cooled to 0°C. The solution was stirred for 30 min and the picolylamine (0.22 g, 2.03 mmol) was added. The mixture was warmed to room temperature and stirred for another 12 h. After the workup, the solvent was removed under reduced pressure. The residue was purified by column chromatography on silica gel (eluent: hexane/ethyl acetate) to afford the compound **DPPTR** (yield: 0.459 g, 85%). The **DPPTY** was synthesized as per the procedure described in SM.

Acknowledgements

The financial support from the Council of Scientific and Industrial Research (01/2569/12-EMR-II) and University Grants Commission (41-286/2012-SR), Govt. of India is gratefully acknowledged.

References

[1] Han, S.-Y. and Kim, Y.-A. (2004) Recent Development of Peptide Coupling Reagents in Organic Synthesis. *Tetrahedron*, **60**, 2447-2467. http://dx.doi.org/10.1016/j.tet.2004.01.020

[2] Al berico, F. (2004) Developments in Peptide and Amide Synthesis. *Current Opinion in Chemical Biology*, **8**, 211-221. http://dx.doi.org/10.1016/j.cbpa.2004.03.002

[3] Bodanszky, M. and Bodansky, A. (1984) The Practice of Peptide Synthesis. Springer, New York.

[4] Sheehan, J.C. and Hess, G.P. (1955) A New Method of Forming Peptide Bonds. *Journal of the American Chemical Society*, **17**, 1067-1068. http://dx.doi.org/10.1021/ja01609a099

[5] Valuer, E. andBradley, M. (2009) Amide Bond Formation: Beyond the Myth of Coupling Reagents. *The Royal Society of Chemistry*, **38**, 606-631.

[6] Sheehan, J.C., Cruickshank, P.A. and Boshart, G.L. (1961) Convenient Synthesis of Water-Soluble Carbodiimides. *The Journal of Organic Chemistry*, **26**, 2525-2528. http://dx.doi.org/10.1021/jo01351a600

[7] Himaja, M., Prathap, K.J., Mali, S.V. and Ramana, M.V. (2011) Synthesis Antibacterial and Insecticidal of a New Series of 4-(5-Methyl-1,2,4-triazolo[1,5-a]pyridine-7-ylamino)-N-(aryl)benzamides. *Indian Journal of Hetero Cyclic Chemistry*, **20**, 317-320.

[8] Brogly, L., Dappen, M., Thorsett, E. and Tucker, J. (2001) Amide Bond Formation Using HATU in High throughput Synthesis. *Proceedings of the 222nd ACS Nation Meeting*, Chicago, 26-30 August 2001, MEDI-112.

[9] Xiao, Z., Yang, M.G., Li, P. and Percy, H. (2009) Synthesis of 3-Subbstited-4(3H)-Quinazolinones via HATU-Mediated Coupling of 4-Hydroxyquinazolines with Amines. *Organic Letters*, **11**, 1421-1424. http://dx.doi.org/10.1021/ol802946p

[10] Prasad, B., Sambarkar, A. and Patil, C. (2012) Synthesis of Amides from Acids and Using Coupling Reagents. *Journal of Current Pharmaceutical Research*, **10**, 22-24.

[11] Glunz, P.W. (2004) Preparation of Pyrrolo[1,2-a]Pyrimidinones as Coagulation Cascade Inhibitors. From PCT Int. Appl., WO2004002406 A2 20040108.

[12] Joullie, M.M. and Lassen, K.M. (2010) Evolution of Amide Bond Formation. *ARKIVOC*, **viii**, 189-250.

[13] Londregan, A.T., Storer, G., Wooten, C., Yang, X.J. and Warmus, J. (2009) An Improved Amide Coupling Procedure for the Synthesis of N-(Pyridin-2-yl)Amines. *Tetrahedron Letters*, **50**, 1986-1988. http://dx.doi.org/10.1016/j.tetlet.2009.02.071

[14] Tran, T.P., Mullins, P.B., am Ende, C.W. and Pettersson, M. (2013) Synthesis of Pyridopyrazine-1,6-Diones from 6-Hydroxypicolinic Acids via One-Pot Coupling/Cyclization Reaction. *Organic Letters*, **15**, 642-645. http://dx.doi.org/10.1021/ol303463e

[15] ChemBark. News, Analysis, and Commentary for the World of Chemistry & Chemical Research, Amide Bond Formation Using HATU. http://block.chembark.com/2007/04/01

[16] Zhang, H.K., Zhang, X., Mao, B.Z., Li, Q. and He, Z.H. (2004) Alpha-Picolinic Acid, a Fungal Toxin and Mammal Apoptosis-Inducing Agent, Elicits Hypersensitive-Like Response and Enhances Disease Resistance in Rice. *Cell Research*, **14**, 27-33. http://dx.doi.org/10.1038/sj.cr.7290199

[17] Fulmer, G.R., Miller, A.J.M., Sherden, N.H., Gottlieb, H.E., Nundelman, A., Stoltz, B.M., Bercaw, J.E. and Goldberg, K.I. (2010) NMR Chemical Schifts of Trace Impurities: Common Laboratory Solvents, Organics, and Gases in Deuterated Solvents Relevant to the Organometallic Chemist. *Organometallics*, **29**, 2176-2179.

[18] Gottlieb, H.E., Kotlyar, V. and Nundelman, A. (1997) NMR Chemical Shifts of Common Laboratory Solvents as Trace Impurities. *The Journal of Organic Chemistry*, **62**, 7512-7515. http://dx.doi.org/10.1021/jo971176v

[19] Carpino, L.A. (1993) 1-Hydroxy-7-Azabenzotriazole. An Efficient Peptide Coupling Additive. *Journal of the American Chemical Society*, **115**, 4397-4398. http://dx.doi.org/10.1021/ja00063a082

[20] Carpino, L.A., Imazuumi, H., El-Faham, A., Ferrer, F.J., Zhamg, C., Lee, Y., Foxmam, B.M., Heanklein, P., Hanay, C., Mugge, C., Wensahuh, H., Klose, J., Beyermanm, M. and Bienert, M. (2002) The Uronium/Guanidinium Peptide Coupling Reagents: Finally the True Uronium Salts. *Angewandte Chemie International Edition*, **41**, 441-445. http://dx.doi.org/10.1002/1521-3773(20020201)41:3<441::AID-ANIE441>3.0.CO;2-N

[21] Carpino, L.A., Imazumi, H., Foxman, B.M., Vela, M.J., Henklein, P., El-Faham, A., Klose, J. and Bienert, M. (2000) Comparison of the Effects of 5- and 6-HOAt on Model Peptide Coupling Reactions Relative to the Cases for the 4- and 7-Isomers. *Organic Letters*, **2**, 2253-2256. http://dx.doi.org/10.1021/ol006013z

[22] Nakamoto, K. (1978) Infrared and Raman Spectra of Inorganic and Coordination Compounds. 3rd Editon, John Wiley & Sons, New York, 228.

[23] Mc Murray, J. (1992) Organic Chemistry. 3rd Edition, Cole Publishing Company, California, 814.

5

Synthesis of New Fluorine/Phosphorus Substituted 6-(2'-Amino Phenyl)-3-Thioxo-1,2,4-Triazin-5(2H, 4H)One and Their Related Alkylated Systems as Molluscicidal Agent as against the Snails Responsible for Bilharziasis Diseases

Abeer N. Al-Romaizan*, Mohammed S. T. Makki, Reda M. Abdel-Rahman

Department of Chemistry, Faculty of Science, King Abdul Aziz University,
Jeddah, KSA
Email: *ar-orkied@hotmail.com

Abstract

New fluorine substituted 6-(5'-fluoro-2'-triphenylphosphiniminophenyl) 3-thioxo-1,2,4-triazin-5 (2H, 4H) one (2) was obtained via Wittig's reaction of the corresponding 6-(5'-fluoro-2'-amino-phenyl)-3-thioxo-1,2,4-triazinone (1). Behavior of compound 2 towards alkylating agents and/or oxidizing agents was studied were, N-hydroxyl (3), Mannich base (4,5), S-alkyl (6,7,8) and thiazolo [3,2-b][1,2,4] triazinones (10-14) and or 3-disulfide (18), 3-sulfonic acid 19 and 1,2,4-triazin-3,5-Dionne (20) derivatives obtained. Structures of the new products are established by elemental and spectral data. The new targets obtained screened as Molluscicidalagents against Biomophlaria Alexandrina snails responsible for Bilharziasis diseases, in compare with Baylucide as standard drug.

Keywords

Fluorine, Phosphorus, Sulfur-1,2,4-Triazine, Characteristic Properties, Molluscicidal
Activity

---

*Corresponding author.

1. Introduction

The incorporation of fluorine atoms into a heterocyclic nitrogen molecule frequently provides properties of pharmacological interest as compared to their non-fluorinated analogs [1]-[5]. Also, bonded phosphorus atoms with S, O, N and C-atoms of heterocyclic systems enhance their important properties as herbicides, pesticides and insecticides [6]-[11]. On the other hand, 3-thioxo-1,2,4-triazin-5-one derivatives and their N- and S-alkyl derivatives have gained considerable attention due to their well as medicinal utility such as anti-HIV, anti AIDS and anticancer agents [12]-[16]. Literature reveals that no reports of a molecular scaffold containing these important cores. With this based upon these observations. The present work aims to synthesis and chemical reactivity of 1,2,4-triazinone bearing, fluorine, phosphorus and sulfur atoms through alkylation reactions and the new systems as Molluscicidal agents against Biomophalaria Alexandrina snails by removal from the wastewater (Clean water).

2. Experimental

Melting points were determined with an electro-thermal Bibbly Stuart Scientific Melting point SMPI (UK). A Perkin Elmer (Lambda EZ-210) double beam spectrophotometer (190 - 1100 nm) used for recording the electronic spectra. A Perkin Elmer model RXI-FT-IR 55,529 cm^{-1} used for recording the IR spectra (EtOH as solvents). A Brucker advance DPX 400 MHz using TMS as an internal standard for recording the $^1H/^{13}C$ NMR spectra in deuterated DMSO (δ in ppm). AGC-MS-QP 1000 Ex model is used for recording the mass spectra. Hexafluorobenzene was used as external standard for ^{19}F NMR at 8425 MHz and ^{31}P (in $CDCl_3$, 101.25 MHz) [17]. Elemental analysis was performed on Micro Analytical Center of National Reaches Center-Dokki, Cairo, Egypt. Compound 1 was prepared according the reported method [14] and compound 15 as procedure published [18].

6-(5'-Fluor phenyl)2'amino-3-thioxo-1,2,4-triazine-5(2H, 4H)one (1)

Equimolar mixture of 5-fluoroisatin (in 100 ml NaOH, 5%) and thiosemicarbazide (in 10 ml H_2O) reflux for 2 h, then cold and poured onto ice-HCl. The solid result was filtered off and crystallized from EtOH as yellow crystals to give 1. Yield (80%), m.p. 263°C - 265°C. Analytical data, Found C = 44.91, H = 2.90, F = 7.58, N = 23.40, S = 13.29%; Calculated for $C_9H_7FN_4OS$ (238) C = 45.37, H = 2.94, F = 7.98 N = 23.52, S = 13.44%, M/Z (256, M + H_2O, 5%), base peak (68, 100%), 148 (21), 136 (18), 110 (30), 96 (50), 82 (58), 70 (78); UV: (λ_{max} EtOH) 280 nm. IR vcm^{-1} = 3424 (NH_2) 3258, 3169 (NH, NH), 1685 (C=O), 1618 (NH_2), 1545 (C=N), 1263 (C-F): 858, 818 (aryl CH) 685 (C-F) ^1H NMR (DMSO) = 14.66, 16.66, 10.90 (each 1H, s, 3NH), 8.68 - 8.06, 7.69 - 7.64, 7.39, 7.39 - 7.30 (3H, aryl protons) ^{13}C NMR = δ 179.47 (C=S), 162 (C=O), 159 - 157 (spin coupling C-F), 138.54 (C=N), 131.82, 121.8, 121.51 (aromatic carbons), 78.14, 77.71 (C5-C6 1,2,4-triazine).

6-[5'-Fluoro-2'(triphenylphosphinimino)phenyl]-3-thioxo-1,2,4-triazine-5(2H, 4H)one (2)

A mixture of 1 (0.01 mol) and triphenyl phosphine (0.01 mol) in acetonitrile (20 ml), THF (20 ml) reflux for 2 h then cold. The solid produced and crystallized from EtOH to give 2 as deep yellowish crystals. Yield (70%) m.p. 249°C - 250°C. Analytical date Found C = 64.60, H = 3.96, F = 3.70, N = 11.01, S = 6.33%. Calculated for $C_{27}H_{20}FN_4OPS$ (498): C = 65.06, H = 4.01, F = 3.81, N = 11.24, S = 6.42%, M/Z (498.00) 370 (4), 290 (10), 171 (60), 159 (100), 128 (20), 118 (40), 102 (65), 92 (58), 76 (58), 65 (40); UV: (λ_{max} EtOH) 310 nm. IR vcm^{-1} = 3335 (NH), 1658 (C=O), 1380 (P=N), 1250 (C-F), 1087 (C-S) 1045, (P-N) 879 (aryl CH), 650 (C-F). ^1H NMR (DMSO): 14.56, 12.78 (each 1H, s, 2NH), 8.21, 7.76, 7.66, 7.65, 7.65, 7.64, 7.484, 7.840, 7.47, 7.464, 7.460, 7.45, 7.398, 7.391, 7.38, 7.37, 7.29, 7.28, 7.28, 7.27, 7.02, 7.01, 7.009, 7.005, 6.994, 6.990, 6.866, 6.859, 6.852, 6.845, (18H of aromatic protons). ^{13}C NMR (DMSO) = δ 179.74 (C=S), 163.0 (C=O), 138.62 (C-F), 131 (C=N), 118.94 - 107.94 (aromatic carbons), 7.76, 77.21 (C$_5$-C$_6$ of 1,2,4-triazine).

2,4-Di(hydroxymethyl)-6-(5'-fluoro-2'-triphenylphosphiniminophenyl)-3-thioxo-1,2,4-triazine-5-one (3)

A mixture of 2 (0.01 mol) and formaldehyde (0.02 mol), in methanol (50 ml) reflux for 2 h, cold. The solid obtained filtered off and crystallized from MeOH to give 2 as faint yellow crystals. Yield = (65%) m.p. 280°C - 282°C. Analytical date: Found C = 61.92, H = 4.21, F = 5.23, N = 9.93, S = 5.43% Calculated for $C_{29}H_{24}FN_4PSO_3$ (558): C = 62.36, H = 4.30, F = 3.40, N = 10.03, S = 5.73%. UV: (λ_{max} EtOH) = 363 nm. IR vcm^{-1} = 3346 (b, 2OH) 2974, 2889 (2CH$_2$), 1646 (C=O), 1382 (P=N), 1240 (C-F), 1086 (C-S), 1046 (P-N), 879 (aryl CH), 755 (C-F). ^1H NMR (DMSO) δ 8.34 - 6.84 (18 aromatic protons), 4.8, 4.4 (each s, 2H, alcoholic 2OH) 2.62, 2.58 (each s, 4H, 2CH$_2$). ^{13}C NMR (DMSO) = δ 179.86 - 179.68 (C=S), 163.07 (C=O), 159.62, 158.03 (C-F), (C-N), 138.60 (C=N), 132.121 - 107.94 (aromatic carbons), 77.75 77.32 (C5-C6 of 1,2,4-triazine), 40.57 -

40.46, 40.32 - 4.14 (2 CH$_2$).

2,4-Di(Piperidinomethyl)-6-(5'-fluoro-2'-triphenylphosphin-iminophenyl)-3-thioxo-1,2,-4-triazine-5-one (4)

A mixture of **2** (0.01 mol), piperidine (0.02 mol) and formaldehyde (0.02 mol) in methanol (50 ml) reflux for 2 h, cold. The solid produced filtered off and crystallized from MeOH to give **4** as yellow crystals. Yield = (60%) m.p. 179°C - 180°C. Analytical date found C = 67.41, H = 5.83, F = 2.55, N = 11.97, S = 4.33%. Calculated for C$_{39}$H$_{42}$FN$_6$OPS (692): C = 67.63, H = 12.13, F = 2.74, N = 12.13, S = 4.62%. IR vcm^{-1} = 3062 (aromatic CH), 2936, 2840 (aliphatic CH$_2$), 1721 (C=O), 1538 (C=N), 1468 (deform CH$_2$), 1389) (P=N), 1248 (C-F), 1184 (C=S), 1049 (P-N), 885, 815, 754 (aryl CH), 709 (C-F). ^{1}H NMR (DMSO): δ 8.23 - 6.80 (18H, aromatic), 2.95, 2.92, 2.89 and 2.58 (CH$_2$ of piperdine, N-CH$_2$-N). ^{13}C NMR (DMSO) = δ 172 (C=S), 154 (C=O), 147.25 (C-F), 137.54 (C=N), 116.10 - 108 (aromatic carbons), 77.80, 77.39 (C$_5$-C$_6$ of 1,2,4-triazine), 40 - 58, 40.47, 40.33, 40.19, 40.05 (CH$_2$ of piperidine) 39.91 - 39.63 (N-CH$_2$-N).

2,4-Di(4'-arylaminomethyl)-6-(5'-fluoro-2'-triphenylphosphiniminophenyl)-3-thioxo-1,2,4-triazine-5-ones (5a & 5b)

A mixture of **2** (0.01 mol) and formaldehyde (0.02 mol), 4-fluoroaniline and/or 4-aminoantipyrine (0.02 mol) in methanol (50 ml) warm under reflux for 2 h, then cold. The solid thus obtained, filtered off and crystallized from EtOH to give **5a & 5b** as yellow crystals.

Compound **5a**, yield (75%) m.p. 210°C - 212°C. Analytical data: Found C = 65.80, H = 4.11, F = 7.47, N = 10.89, S = 4.01%. Calculated for C$_{41}$H$_{32}$F$_3$N$_6$OPS (744), C = 66.12, H = 4.30; F = 7.66; N = 11.29, S = 4.30%. M/Z = 744 (M, 0, 0%), 370 (1), 367 (25), 290 (60), 272 (30), 248 (20), 218 (42), 169 (100), 128 (85), 102 (100), 65 (100). UV: (λ_{max} EtOH) 364 nm; IR vcm^{-1} = 3343 (aryl-NH), 2974, 2889 (CH$_2$), 1650 (C = O), 1382 (P = N), 1250 (C-F), 1086 (C-S), 1045 (P-N) 879 (aryl CH). ^{1}H NMR (DMSO) = δ 12.71 (s, 1H, NH), 12.55 (s, 1H, NH), 7.41, 7.39, 7.32, 7.06, 7.01, 6.99, 6.98, 6.97, 6.91, 6.90 & 6.89, 6.88, 6.878, 6.873, 6.86 & 6.85, 6.84, 6.83, 6.77 & 6.768, 6.761, 6.754, 6.742, 6.735 (aromatic CH), 5.18 - 5.15 & 5.14 - 5.13 (4H, CH$_2$ of N CH$_2$-NH). ^{13}C NMR = (DMSO) = δ 178.0 (C=S), 161.02 (C=O) 144.99 (C-F), 138.37 (C=N), 130.65 - 114.11 (aromatic carbons) 77.57, 77.14 (C$_5$-C$_6$ of 1,2,4-triazine), 40.61 - 40.35, 39.94 - 39.66 (2N-CH$_2$-NH).

Compound **5b**, yield (60%); m.p. 200°C - 202°C. Analytical data: Found C = 65.89, H = 4.91, F = 2.04, N = 14.83, S = 3.35%; Calculated for C$_{51}$H$_{46}$FN$_{10}$O$_3$PS (928), C = 65.94, H = 9.95, F = 2.04, N = 15.08, S = 3.44%. IR vcm^{-1}: 3277 (b, NH), 2974 (aliphatic CH$_3$) 2840 (CH$_2$), 1697, 1660 (2C=O), 1604 (C=N), 1542 (C=N), 1482 (deform, CH$_2$), 1416 (P=N), 1370 (NCSN), 1274 (C-F), 1152 (C-S), 1045 (P-N), 878, 810, 750 (aromatic CH), 650 (C-F). ^{1}H NMR (DMSO) δ 7.48 - 7.26 (aromatic CH) 5.07 (s, 1H, NH), 3.1 - 3.0 (s, 1H, NH), 2.99 - 2.88, 2.84 - 2.82 (2 CH$_2$), 2.62 - 2.41, 2.34 - 2.23, 2.229 - 2.2221, 2.16 - 2.07 (4 Me). ^{13}C NMR = (DMSO) δ 179.81 (C=S) 159.11 (C=O), 142 (C-F), 138.0 (C=N), 129.38 - 123.31 (aromatic carbons), 77.54, 77.11 (C$_5$-C$_6$ of 1,2,4-triazin), 40.61 - 40.08 & 39.94 - 39.66 & 36.75 - 35.49 (N-CH$_2$), 18.42, 15.15 (N-Me, C-Me).

[6-(5'-Fluoro-2'-triphenylphosphiniminophenyl)-5-oxo-1,2,4-triazine-3-yl]thioacetic acid (6)

Equimolar mixture of **2** and monochloroacetic acid in DMF (20 ml) warm for 30/min, then poured onto ice. The solid yielded filtered off and crystallized form EtOH to give **6** as faint yellow crystals. Yield (80%), m.p. 187°C - 188°C. Analytical data: Found: C = 62.42, H = 3.81, F = 3.20, N = 9.85, S = 5.57%; Calculated for C$_{29}$H$_{22}$FN$_4$O$_3$PS (556). C = 62.58, H = 3.95, F = 3.41, N = 10.07, S = 5.75. IR vcm^{-1} = 3327 (b, OH, NH), 2973, 2884 (CH$_2$), 1659 (b, 2 C=O), 1440 (deform CH$_2$), 1380 (P=N), 1250 (C-F), 1087 (C-S), 1045 (P-N) 880 (Ar CH), 810 (Ar CH). ^{1}H NMR (DMSO) δ = 10.31 (s, 1H, NH), 8.06, 8.0, 7.98 - 7.97, 79.86 - 7.84, 7.73, & 7.726, 7.721, 7.14, 7.08, 7.67 & 7.66, 7.65, 7.63, 7.53 - 7.51, 7.48 - 7.35 & 7.34, 6.997, 6.992, 6.838 - 6.824 (18 CH, aromatic) & 4.74 (s, 1H, OH of COOH), 3.86 - 3.24 (2H, CH$_2$). ^{13}C NMR: (DMSO): δ 168.29 (C=S), 165.40 (C=O), 157.16 (C=O), 142.13 (C-F) 131.35 (C=N), 130.04 - 102.25 (aromatic carbons), 72.43, 72.22 (C$_5$-C$_6$ of 1,2,4-traizine), 34.95 - 34.81 (CH$_2$ carbon).

1,1-Di[6-(5'-Fluoro-2'-triphenylphosphiniminophenyl)-5-oxo-1,2,-4-triazine-3'yl]dimercaptoacetic acid (7)

A mixture of **2** (0.02 mol) and 1,1-dichloloracetic acid (0.01 mol) in DMF (20 ml) reflux for 30 min, cold then poured into ice. The resulted solid filtered off and crystallized from dioxin to give **7** as faint yellow crystals, yield (60%) m.p. 238°C - 240°C. Analytical data: Found C = 63.45, H = 3.49, F = 3.39, N = 10.39, S = 5.88%, Calculated for C$_{56}$H$_{40}$F$_2$N$_8$O$_4$P$_2$S$_2$ (1052) C = 63.87, H = 3.80, F = 3.61, N = 10.64, S = 6.08%. IR vcm^{-1} = 3425, 3259, 3170 (OH, NH, NH), 1865, 1680 (C=O), 1618 (C=N), 1476, 1452 (aliphatic CH), 1360 (P=N), 1252 (C-F), 1193 (C-S), 1045 (P-N), 903, 859, 818, 758 (aryl CH) 685 (C-F). ^{1}H NMR (DMSO): δ 12.79, 12.78 (each s, 2NH), 10.75 (s, 1H, OH), 8.21 - 6.84 (18 CH, aromatic), 2.82 - 2.59 (s, 1H, CH) ^{13}C NMR: δ 179.72 (C=S),

163 (C=O), 159 (C-S), 158.0 (C-S), 138.61 (C-F), 132.18 (C=N), 121.12 - 107.94 (aromatic carbons), 77.66, 77.45 (C$_5$-C$_6$) of 1,2,4-triazine), 40.57 - 40.46 (-CH-).

Tri[6-(5'-fluoro-2'-triphenxylphosphin-iminophenyl)-5-oxo-1,2,-4-triazine-3'yl]trimercaptoacetic acid (8)

A mixture of **2** (0.03 mol) and 1,1,1-trichloroacetic acid (0.01 mol) in DMF (20 ml) warm for 30 min then cold and poured on to ice. The produced solid filtered off and crystallized from Et OH to give **8** as reddish crystals. Yield (60%); m.p. 189°C - 190°C. Analytical data: Found C = 63.89, H = 3.45, F = 3.55 N, 10.67, S = 5.83%. Calculated for C$_{83}$H$_{58}$F$_3$N$_{12}$O$_5$P$_3$S$_3$ (1548); C = 64.34, H = 3.74, F = 3.68, N = 10.85, S = 6.2%. UV (λ_{max} EtOH) 359 nm. IR vcm^{-1} = 3500 - 3100 (b, 3NH, OH) 1716 (C=O), 1624 (NH = OH of 1,2,4-triazinone) 1537 (C=N) 1471 (aliphatic CH). 1390 (P=N), 1300 (NCSN), 1260 (C-F), 1200 (C-S), 1645 (P-N), 920, 850, 780 (aryl CH), 650 (C-F). ^{1}H NMR (DMSO) = δ 12.95, 12.72, 12.33 (each s, 3H, NH), 10.84 (s, 1H, OH), 8.51, 8.23, 8.01, 7.92, 7.89, 7.71, 7.7, 7.69, 7.65, 7.63, 7.59, 7.58, 7.57 - 7.54, 7.50 - 7.47, 7.40 - 7.37, 7.33 - 7.31, 7.02 - 6.98, 6.86 - 6.83 (aromatic CH). ^{13}C NMR = (DMSO) δ 179.61 (C=S), 163 (C=O), 159.5 (C-S), 157 (C-F) 138.58 (C=N), 137.79 (C=N), 132.91 (C=N), 131.99 - 107.93 (aromatic CH), 77.92, 77.49 (C$_5$-C$_6$ of 1,2,4-triazine).

6(5'-Fluoro-2'-triphenylphosphin-iminophenyl)2,-4-dihydro-thiazolo[3,2-b][1,2,4]triazine-3,7-dione (9)

Equimolar mixture of **2** and monochloroactic acid in DMF (20 ml) reflux for 2 h then cold and poured onto ice. The solid obtained filtered off and crystallized from dioxan to give **9** as brown ppt, Yield (60%) m.p. 224°C. Compound **6** (0.50 mg) heat above its melting point (60°C higher) for 10 min, cold then treat with MeOH. The solid produced filtered off and crystallized from dioxan to give **9** as brown ppt. Yield (58%), m.p. 225°C - 227°C. Analytical data Found C = 64.40, H = 3.51, F = 3.35, N = 9.93, S = 5.48% Calculated for C$_{29}$H$_{20}$FN$_4$O$_2$PS (538); C = 64.68, H = 3.71 F = 3.53, N = 10.40. S = 5.94%. UV: (λ_{max} EtOH) 352 nm. IR vcm^{-1} = 3204 (b-OH), 1694 (C=O), 1623 (C=N), 15,636, 1475 (CH$_2$), 1380 (P=N), 12,999 (C-F), 1148 (C-S) 816, (aromatic CH), 711 (C-F). ^{1}H NMR (DMSO) = δ 10.79 (s, 1H, Phenolic OH), 8.23 (s, 1H, CH of thazole), 7.95 - 7.47, 7.35, 7.35, 6.98, 6.80 (aromatic CH). ^{13}C NMR = (DMSO): δ 167.21 (C=S), 147.47, (C=O), 136.64 (C-F), 132.97 (C=N), 131.92 - 128.54, 118.80 - 118.14, 113.79 - 113.73 (aromatic carbons), 111.04 - 110.99 (-CH=), 77.80, 77.38 (C$_5$-C$_6$ of 1,2,4-triazine).

6(5'-Fluoro-2'-triphenylphosphiniminophenyl)-5-oxo-3-(cyanomethylthia)-2H-1,2,4-triazine (10)

A mixture of **2** (0.01 mol) and chloroacetinitrile (0.01 mol) in DMF (20 ml) warm (10 min) then cold and poured onto ice. The result solid filtered off and crystallized from dioxan to give **10** as faint Yellow crystals. Yield (70%); m.p. 214°C - 215°C. Analytical data: Found C = 64.39, H = 3.58, F = 3.11, N, 12.85, S = 5.75%. Calculated for C$_{29}$H$_{21}$FN$_5$OPS (537); C = 64.80, H = 3.91, F = 3.53, N = 13.03, S = 5.95%. M/Z = 537 (5%) 281 (20), 207 (60), 149 (20), 113 (30), 85 (100), 58 (100). UV: (λ_{max} EtOH) 321 nm. IR vcm^{-1} = 3424, 3167 (NH, S-CH=C=NH) 2100 - 2085 (C≡N), 1646 (C=O), 1595 (C=N), 1481 (CH$_2$), 1370 (P-N), 967, 839, 762 (aryl CH), 700 (C-F). ^{1}H NMR (DMSO) = δ 13.90, (s, 1H, NH), 12.76 (s, 1H, HC=NH), 8.22 - 6.81 (aromatic CH), 4.69 (1H, HC=NH) 2.59 (2H, CH$_2$). ^{13}C NMR (DMSO): δ 158.11 (C=O), 147.0 (C-F) 132 (C=N), 131.86 - 128.44 (aromatic carbons), 112.21 (C≡N), 77.96, 77.53 (C$_5$-C$_6$ of 1,2,4-triazine), 40.133 (-CH=NH), 33.63 (CH$_2$).

3-Amino-6(5'-fluoro-2'-triphenylphosphiniminophenyl)-thiazolo[3,2-b][1,2,4]triazine-7-one (11)

Compound **10** (0.5 gm) in DMF (20 ml) warm for 2 h then cooled and poured onto ice. The solid produced filtered off and crystallized from EtOH to give **11** as broom ppt, Yield (66%); m.p. 223°C - 225°C. Analytical data: Found C = 64.51 H = 3.38, F = 3.21 N = 12.55, S = 5.62%. Calculated for C$_{29}$H$_{21}$FN$_5$OPS (537), C = 64.80, H = 3.91, F = 3.53, N = 13.03, S = 5.95%. M/Z, 537 (2%), 370 (2), 226 (2), 168 (100), 140 (60), 114 (30), 62 (18), 70 (18). IR vcm^{-1} = 3348, (b-NH$_2$), 16430 (C=O), 1383 (P=N), 1250 (C-F), 1086 (C-S), 1045 (P-N), 878 (aryl CH). ^{1}H NMR (DMSO) = δ 8.11 (s, 1H, = CH thiaszole), 7.72 - 7.011, 6.98 - 6.80 (aromatic CH), 3.99 - 3.84 (2H-NH$_2$). ^{13}C NMR (DMSO) = 162.54 (C=O), 132.16 (C-F), 132.00 (C=N), 131.99 (C-S), 131.66 - 131.64 (=CH-), 128.61 - 120.55 (aromatic carbons), 77.59, 77.38 (C$_5$-C$_6$ of 1,2,4-triazine), 40.51 (-N-C=N).

3-(4'-Fluoro benzoyl)amino-6-(5'-fluoro-2'-triphenylphosphiniminophenyl)-thiazolo[3,2-b][1,2,4] triazine-7-one (12)

Equimolar mixture of **11** and 4-fluorobenzoyl chloride in DMF (20 ml) warm for 10 min then cold and poured onto ice. The resulted solid filtered off and crystallized from EtOH to give **12** as deep-Yellowish crystals. Yield (75%). m.p. 205°C - 207°C. Analytical data: Found C = 65.19 H = 3.41, F = 5.49 N = 10.51, S = 4.59%. Calculated for C$_{36}$H$_{25}$F$_2$N$_5$O$_2$PS (660), C = 65.55, H = 3.80, F = 5.76, N = 10.60, S = 4.80. IR vcm^{-1} = 3342, (b-NH), 1651 (b, 2C=O), 1381 (P=N), 1326 (NCSN) 1230 (C-F), 1086 (C-S), 1045 (P-N), 879 (aryl CH). ^{1}H NMR (DMSO) = δ 13.70 (s, 1H, NH), 9.89 (s, = CH of thiazole) 8.411, 8.17, 8.07 - 8.05, 8.01, 7.99, 7.95, 7.79, 7.66 -

7.64, 7.44 - 7.42, 7.28, 7.27 (aromatic CH). ^{13}C NMR = (DMSO): δ 167.53 (C=O), 162.54 (C=O) 138.59 (C-F) 132.25 (C-N), 132.19 (C=N), 129.33 - 127.27, 117.78 - 115.17, 112.15, 12.09, 110.52, 108.12, 107.95 (aromatic carbons), 77.64. 77.43 (C$_5$-C$_6$ of 1,2,4-triazine).

Schiff base (13)

Equimolar amounts of **11** and 4-fluorobenzaldehyde in absolute ethanol (20 ml) reflux for 30 min then cooled. The solid thus obtain filtered off and crystallized from EtOH to give **13** as Yellowish ppt. Yield (70%); m.p. 248°C - 250°C. Analytical data: Found C = 66.85, H = 3.61, F = 5.75 N = 10.59, S = 4.71%. Calculated for C$_{36}$H$_{24}$F$_2$N$_5$OPS (643), C = 67.18, H = 3.73, F = 5.90, N = 10.88, S = 4.97%. IR vcm^{-1} = 3100, 2880 (aromatic & aliphatic CH), 1700 (C=O), 1600 (C=N), 1483 (C-P), 1370 (P=N), 1230 (C-F), 1200 (C-S), 1045 (P-N), 880, 840, 810 (aryl CH), 650 (C-F) ^{1}H NMR (DMSO) = δ 9.97 (s, 1H, -CH=N-), 8.62 (s, 1H, -CH = thiazole) 8.23, 8.22, 8.09 - 8.00, 7.94 - 7.92, 7.71 - 7.63, 7.56 - 7.53, 7.49 - 7.45, 7.26 - 7.23, 7.12 - 7.10, 7.0 - 6.96, 6.89 - 6.84, 6.81 - 6.79 (aromatic CH).

6-(5'-Fluoro-2'-triphenyl phosphiniminophenyl)-3-oxo-3phenyl-thiazolo[3,2-b][1,2,4]triazine-7-one (14)

A mixture of **2** (0.01 mol) and phenacylbromide (0.01 mol) in ethanolic KOH, (20 ml, 5%) reflux for 2 h, cold then poured onto ice-HCl. The solid produced filtered off and crystallize from dioxan to give **14** as brown ppt. Yield (60%); m.p. > 300°C. Analytical data: Found C = 69.88, H = 3.59, F = 3.01 N, 9.00, S = 5.13%, Calculated for C$_{35}$H$_{24}$FN$_4$OPS (598); C = 70.23, H = 4.01, F = 3.17, N = 9.36 S = 5.35%. IR vcm^{-1} = 3080, 3030 (aromatic CH), 1680 (C=O), 1380 (P=N), 1240 (C-F), 1180 (C-S), 1045 (P-N), 880, 850 (aryl CH).

Diaylthioether (16)

A mixture of **2** (0.01 mol) and Schiff base **15** (0.01 mol) in dry C$_6$H$_6$ (100/ml) reflux 8 h, cold and used petereither 100°C - 120°C to complete precipitation. The solid obtained filtered off and crystallized dioxan to give **16** as Yellowish crystals. Yield (80%); m.p. 204°C - 205°C. Analytical data: Found C = 66.53, H = 4.31, F = 4.44, N = 11.88, S = 3.66%, Calculated for C$_{45}$H$_{36}$F$_2$N$_7$O$_2$PS (807); C = 66.91, H = 4.46, F = 4.70, N = 12.14, S = 3.96%. M/Z = (807.0.0), 580 (5), 515 (4), 462 (8), 423 (10), 370 (5), 339 (20), 282 (20), 225 (20), 207 (60), 176 (100), 149 (56), 119 (38), 85 (90), 58 (100). UV: (λ_{max} EtOH) 323 nm. IR vcm^{-1} = 3332 (NH), 2973, 2886 (aliphatic CH), 1636 (C = O), 1488 (CH$_3$), 1381 (P=N), 1324 (NCSN), 1250 (C-F), 1086 (C-S), 1045, (P-N), 880, 755 (aryl CH). ^{1}H NMR (DMSO) = δ 12.77, (s, 1H, NH), 10.75 (s, 1H, NH), 9.68 (s, 1H, S-CH-Ar), 8.23, 7.83 - 7.28, 7.27 - 7.00, 6.99 - 6.84 (aromatic CH), 3.69 (s, CH$_3$-N) 2.79 (s, CH$_3$-C).^{13}C NMR (DMSO): δ 164.73 (C= O), 163.08 (C=O), 160.63, 155.18 (C-F), 151.9 (C-S), 134.64, 134.25, 134.23 (C=N), 129.487 - 115.492 (aromatic carbons), 77.72, 77.50 (C$_5$-C$_6$ of 1,2,4-triazine), 67.00 (S-CH=NH), 39.95 - 39.81, 39.67 - 39.55 (2 CH$_3$).

2,3-Diaryl–2,3-dihydro-4-thioxo-7-(5'-fluoro-2'-triphenylphosphiniminophenyl)-1,3,5-thiazolo[3,2-b] [1,2,4] triazine-8-one (17)

A mixture of **16** (0.01 mol) and CS$_2$ (5 ml) in DMF (20 ml) reflux for 4 h, cold then powered onto ice. The resultant solid filtered off and crystallized from dioxan to give **17** as yellowish crystals. Yield (75%), m.p. 254°C - 255°C. Analytical data: Found C = 64.88, H = 3.85, F = 4.38, N = 11.40, S = 7.45%, Calculated for C$_{46}$H$_{34}$F$_2$N$_7$O$_2$PS$_2$ (849); C = 65.01, H = 4.00, F = 4.47, N = 11.54, S = 7.53%; M/Z (849, 0.0%), 370 (2), 329 (40), 290 (100), 159 (100), 128 (100), 102 (100), 96 (100), 65 (100). UV: (λ_{max} EtOH) 34.7 nm. IR vcm^{-1} = 2873 (aliphatic CH), 1684 (C=O), 1614. 1593 (C=N) 1475, 1425 (CH$_3$), 1318 (P=N), 1264 (C-F), 1199 (C=S), 1130 (C-S), 1052 (P-N), 985, 899, 854, 814, 732 (aryl CH). ^{13}C NMR (DMSO) = δ 179.70 (C=S), 163.07 (C=O), 155.08 (C-F), 138.62 (C=N, 1,2,4-triazine), 132.23, 132.13 (C=C pyrazole), 129.43 - 115.44, 112.09 - 107.96 (aromatic carbons), 77.71, 77.28 (C$_5$-C$_6$ of 1,2,4-triazine) 66.94 (S-CH-NH), 40.57, 39.76 (2 CH$_3$).

Di-Heteroaryldisulfide (18)

Compound **2** (0.05 gm) and FeCl$_3$ (0.5 gm) in MeOH (20 ml) reflux for 3 h, then filtered. The solid produced filtered off and crystallized from dioxan to give **18** as deep-yellowish crystals. Yield (80%), m.p. 238°C - 240°C. Analytical data: Found C = 64.85, H = 10.89, F = 3.55, N = 10.89, S = 6.22%. Calculated for C$_{54}$H$_{38}$F$_2$N$_8$O$_2$P$_2$S$_2$ (994); C = 65.19, H = 11.26, F = 3.82, N = 11.26, S = 6.43%. IR vcm^{-1} = 3300, 3200 (NH, NH), 1680 (C=O), 1600 (C=N), 1350 (P=N), 1100 (C-S), 1040 (P-N), 900, 850, 800 (aryl CH), 650 (C-F). ^{1}H NMR (DMSO) = δ 14.55, 12.78 (each s, 2H, NH, NH), 8.20 - 6.85 (aromatic CH). ^{13}C CNMR (DMSO): δ 179.90, 179.74 (2C-S), 159 - 66, 158.66 (2C=O), 138.63 (C-F), 132.18 (C=N), 121.12 - 107.95 (aromatic carbons), 77.65, 77.43 (C$_5$-C$_6$ of 1,2,4-triazine).

6-(5'-triphenylphosphiniminophenyl)-5-oxo-2H-1,2,4-triazine-3-sulfonic acid (19)

Compound **2** (0.05 gm) in ethanol (10 ml) and H$_2$O$_2$, (0.5 ml) add with stirring for 2 h. The solid obtained filtered off and crystallized from EtOH to give **19** as yellowish crystals yield (75%); m.p. 258°C - 260°C. Analytical data: Found C = 59.00, H = 3.44, F = 3.25, N = 9.87, S = 5.45%. Calculated for $C_{27}H_{20}FN_4O_4PS$ (546), C = 59.34, H = 3.66, F = 3.47, N = 10.25, S = 5.86%. IR vcm^{-1} = 3300 (NH), 1696 (C=O), 1390 (NCSN), 1360 (P=N), 879, 820, 780, (aryl CH). 670 (C-F). ^{1}H NMR (DMSO) = δ 12.79 (s, 1H, NH), 10.72 (s, 1H, SO_2-OH), 8.13 - 6.76 (aromatic CH). ^{13}C NMR (DMSO): δ 179.75 (-S=O), 163.09 (C=O), 159.67 (C-F), 138.62 (C=N), 132.23 (C-S), 121.08 - 107.95 (aromatic carbons), 77.60, 77.39 (C_5-C_6 of 1,2,4-triazine).

6-(5'-Fluoro-2'-triphenylphosphiniminophenyl)-1,2,4-triazine-3,5-(2H, 4H)dione (20)

Compound 2 (0.05 gm) in ethanol (10 ml) and $KMnO_4$ solution (ethanolic 1%, 1 ml) add drop wise then stirring for 2 h. The produced solid filtered off and crystallized from Et OH to give 20 as yellowish crystals yield (50%); m.p. 273°C - 275°C. Analytical data: Found C = 66.89, H = 4.01, N = 11.35, F = 3.55. Calculated for $C_{27}H_{20}FN_4O_2P$ (482), C = 67.21, H = 4.14, N = 11.61, F = 3.94. M/Z: (482, 0.0%), 370 (10), 206 (101, 148, (16), 128 (24), 110 (35), 96 (55), 83 (78), 68 (100). IR vcm^{-1} = 3426, 3259, 3170 (OH, NH, NH), 1766, 1681 (2C=O), 1619 (C=N) 1452 (C-P) 1301 (P=N), 1252, (C-F) 1048 (P-N), 905, 861, 820, 802, 784, 761 (CH), 687 (C-F), ^{1}H NMR (DMSO) = δ 12.73, 10.82 (each s, NH, OH), 7.88 - 6.84 (aromatic CH). ^{13}C NMR = δ 153.58 (C=O), 152.78 (C=O), 147.84 (C-F). 132.58 (C=N) 130.51 - 111. (Aromatic carbons), 77.85, 77.60 (C_5-C_6 of 1,2,4-triazine).

3. Results and Discussion

3.1. Chemistry

A recent work on the synthesis and chemistry of bioactive sulfur bearing 1,2,4-triazinone moiety was reported [16] [19]. In continuation of this attitude the present investigation reports the synthesis of fluorine and phosphorus-substituted 6-amino-phenyl-3-thioxo-1,2,4-triazin-5-(2H, 4H) one (1) and study that behavior towards various alkylating agents. Treatment of 5-fluoroisatin with thiosemicarbozide in alkaline medium [14] [15] produced 6-(2'-amino-5'-fluorophenyl-3-thioxo-1,2,4-triazin-5-(2H'4H) one (1). Warm compound 1 with triphenylphosphine in acetonitrile produced the yield 2 (Scheme 1).

In the imino [yield, 2] a negative charge of nitrogen is bonded to positive charge of phosphorus stabilized by partial overlap of the filled N-P orbital. This stabilization increase due to the charge on the α-carbon atom is spread by 1,2,4-triazine resonance. Abdel-Rahman [14] [15] reported that N-alkyl of 3-thioxo-1,2,4-triazinones exhibited a wide biological spectrum anti HIV and anticancer properties. Similarly, hydroxyl methylation of compound 2 by boil with formaldehyde-methanolproduced 2,4-di(hydroxylmethyl)-6-(5'fluoro-2'-triphenylphosphiniminophenyl)-3-thioxo-1,2,4-triazin-5-one (3). Also, reflux of compound 2 with secondary and primary amines such as piper dine, 4-fluoroaniline and 4-amino-antipyrine in the presence of formaldehyde methanol, furnished the Mannich bases 4 and 5 (Scheme 2).

Formation of 3 and 4 was may be as (Figure 1).

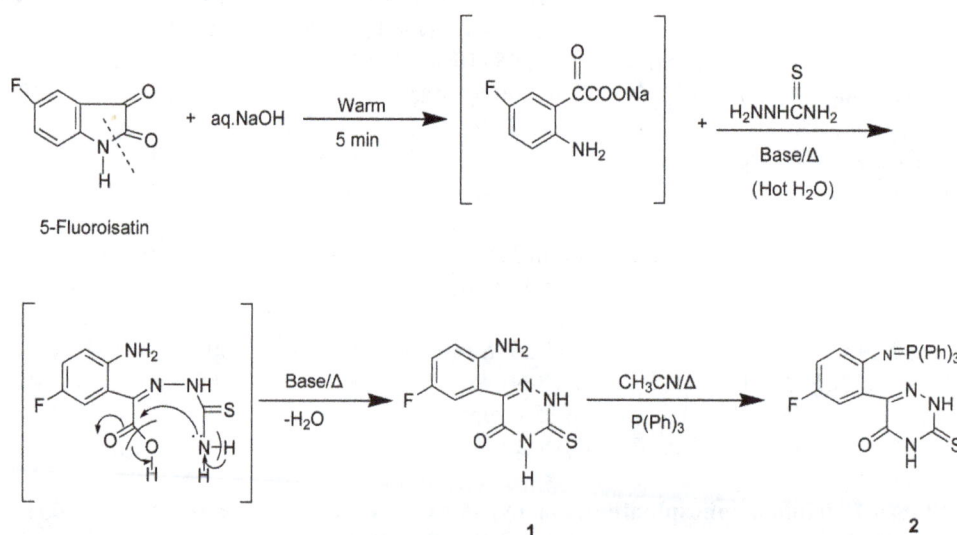

Scheme 1. Formation of compounds 1 & 2.

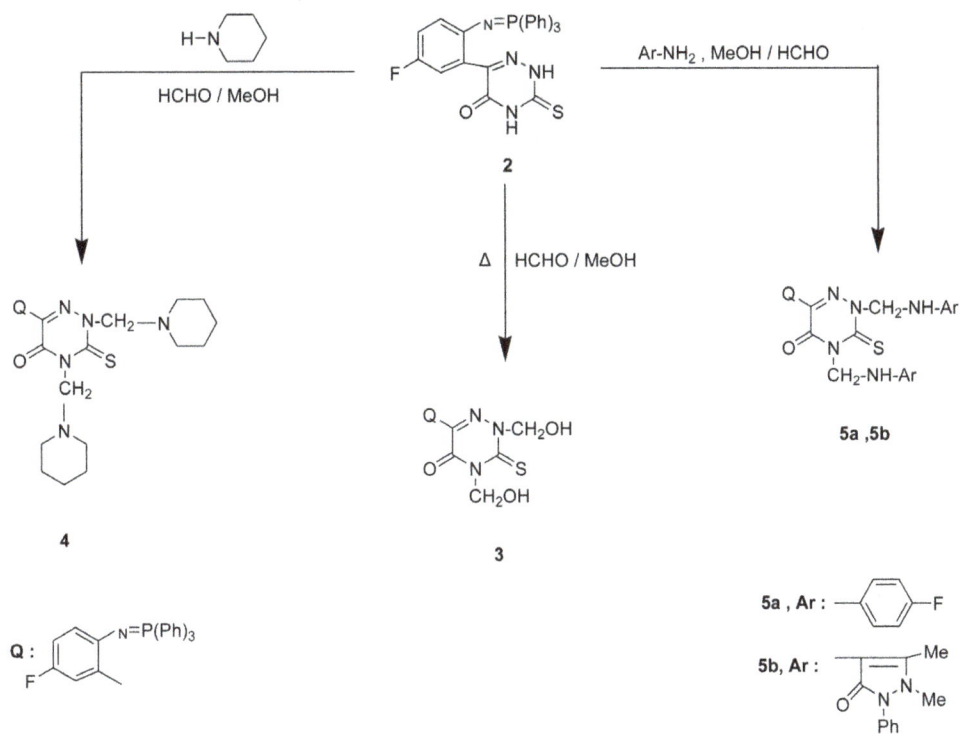

Scheme 2. Formation of compounds 3 - 5.

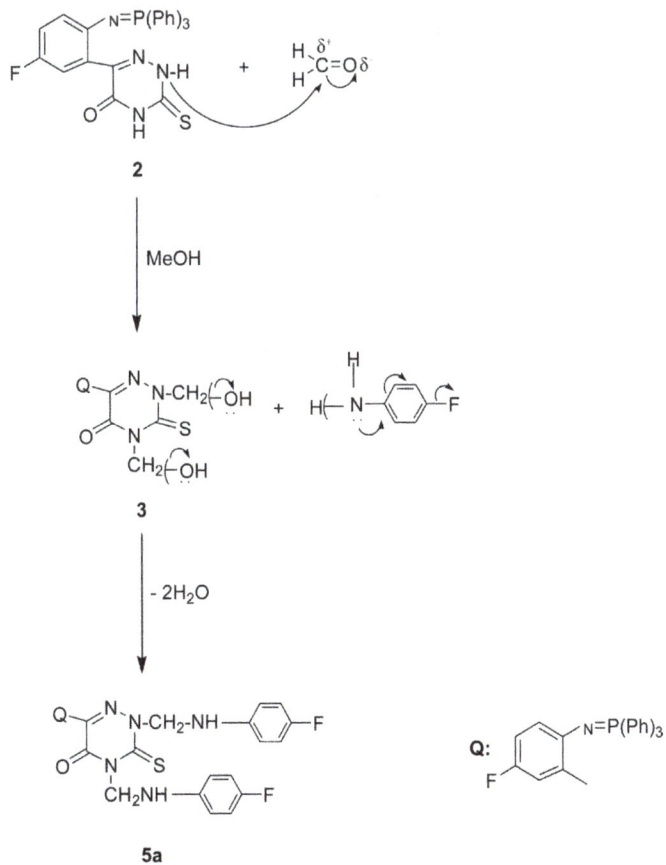

Figure 1. Formation of compounds 3 & 5a.

Due to a higher nucleophilicity of sulfur atoms, the direct displacement of an acidic proton of mercapto group by a simple electrophile can be easily occur via treatment of compound **2** with haloacetic acids. Thus treatment of compound **2** with halo aliphatic acids such as mono/di/trichloroacetic acids in DMF afforded the substituted thiaacetic acids **6-8** (**Scheme 3**).

The multicomponent reaction (MCR) was considered as powerful synthetic tool for preparing target molecules of biological relevance in an efficient manner. Thus, treatment of compound **2** with active methylene reagents as chloroacetonitrile in warm DMF [20] produced 3-cyanomethyl thai-6-iminophosphorane-1,2,4-trinazin-5-(2H)one (**10**). The latter compound **10** use for the synthesis of thiazolo [3,2-b][1,2,4]triazinones (**11-13**) systems (**Scheme 4**). Acidic hydrolysis of **10** by warm with diluted HCl for short time (10 min) yielded the compound **6**. Boil compound 6 with DMF along time afforded 6-iminophosphorane-2,3-dihydoro-thiazolo [3,2-b] [1,2,4] triazine-3,7-dione (**9**) (**Scheme 4**).

Heat compound **10** on heating with DMF a long time (2 hours), produced 3-aminothiazolo-1,2,4-triazine **11**. Presence of an amino group in structure **11** was deduced from treat with 4-fluorobenzoylchloride (DMF) and/or with 4-fluorobenzaldehyde (EtOH) yield the anilido **12** and/or Schiff's base **13** (**Scheme 4**). Treatment of compound **2** with α, β-bifunctional oxygen-halogen reagents as phenacyl bromide in ethanolic KOH, yielded 3-phenyl-6-iminophosphorane-thiazolo [3,2-b][1,2,4]triazin-7-one (**14**) (**Scheme 5**). The nitrogen-sulfur containing fused heterobicyclic structures have demonstrated a high degree of binding affinity when they serve as Ligands for various biological receptors [12] [13]. Thus addition of Mercator group (as nucleophilic) of compound **2** to an Schiff's base **15** in boil dry dioxan yielded the thioether**16**, which upon ring closure reaction by reflux with CS_2 in DMF furnished 2,3-diaryl-2,3-dihydro-7-iminophosphorane-4-thioxo-1,3,5-thiadiazino[3,2-b] [1,2,4]-triazin-8-one (**17**) (**Scheme 5**).

Abdel-Rahman *et al.* [21]-[25] reported that thioethers, sulfide and sulfonic acid bearing a 1,2,4-triazine moieties. Exhibited a very interesting medicinal activity as anti-HIV and anticancer agents. Recently, Slawinski *et al.* [25] synthesized 2-mercaptobenzene sulfonamide bearing a 1,2,4-trinzines exhibited a significant activity against cell lines of colon cancer, renal cancer, and melanoma, as well as good selectivity toward non-small cell lung cancer. Similarly, oxidation of compound **2** via treatment with $FeCl_3$ in boiling methanol and/or with H_2O_2 in ethanol by stirred at room temperature furnished the disulfide **18** and/or 3-sulfonic-1,2,4-triazinone **19**. Finally, treatment of **2** with ethanolic $KMnO_4$ at room temperature [21] led to the direct formation of 6-(5'-fluoro-2-triphenylphosphiniminophenyl)-1,2,4-triazin-3,5(2H, 4H)dione (**20**) (**Scheme 5**).

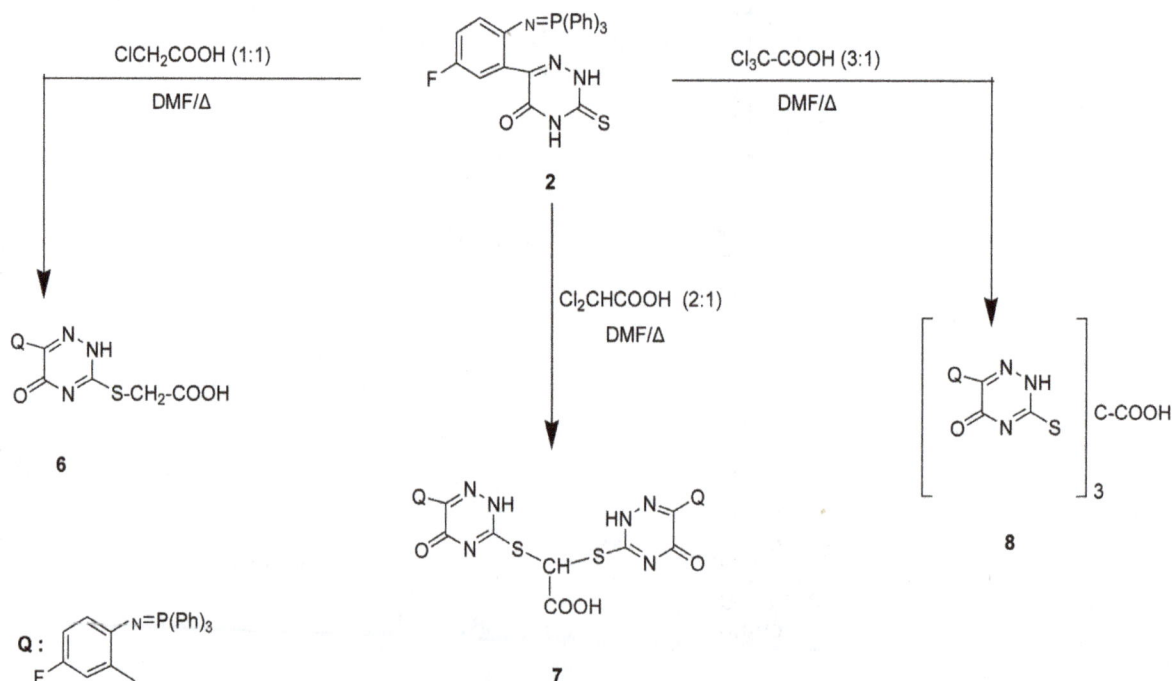

Scheme 3. Formation of compounds 6 - 8.

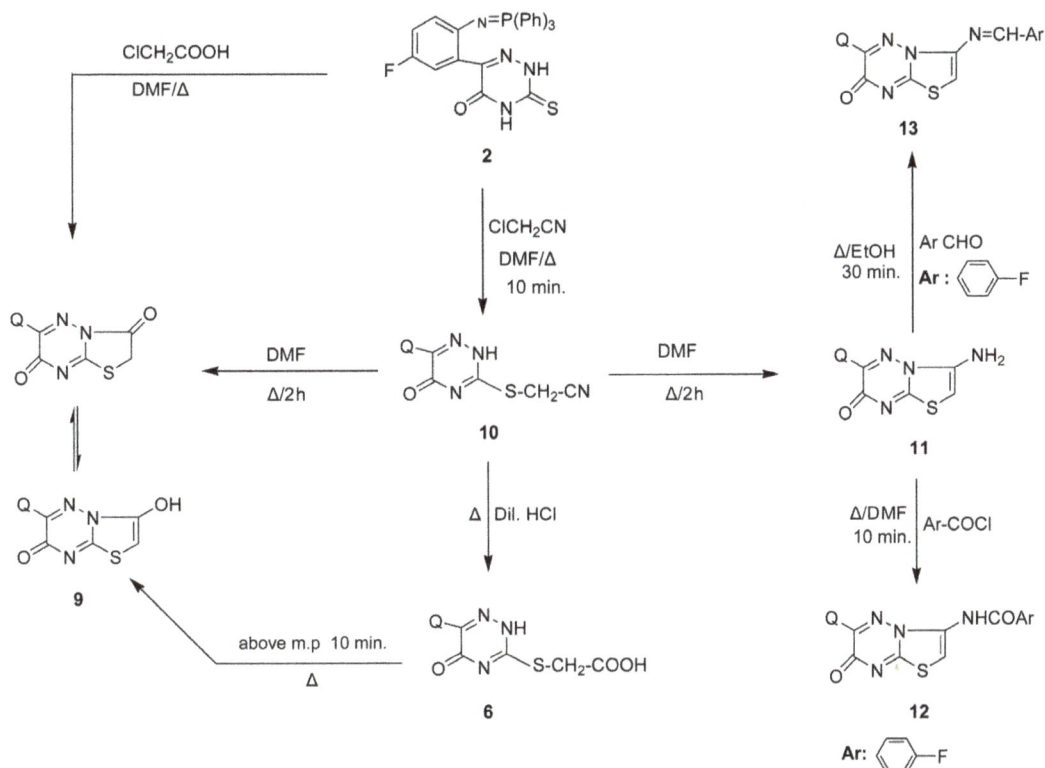

Scheme 4. Formation of compounds 9 - 13.

3.2. Elucidation the Former Structures

3.2.1. UV Spectra

The electronic conjugated molecule of compound **2** exhibited λ_{max} at 310 nm while that of compounds **3** (363), **5a** (364), **8** (359) and **16** (323) nm. A higher absorption bands of new acyclic systems than that of **2** confirm that N- and S-substitution were formed. On the other hand, the absorption bands of fused heterobicycle compounds **9** (352), **17** (347) and **10** (321) nm is higher than the start **2** (310) nm. This is attributing to extension of hetero-conjugation of heterobicylic systems through a type of cylization.

3.2.2. IR Spectra

The new compounds obtained recorded the absorption bands at 1380 - 1390, 1250 - 1230 cm^{-1} due to presence of both P=N and C-F functional groups. Compounds **3-5** showed a lack of band at 3200 - 3100 cm^{-1} for NH=OH of 1,2,4-triazinones, while that of compounds **6-8** and **10** recorded the absorption band at 3343 and 1643 cm^{-1} attributed to presence of ^{4}NH & ^{5}C=O of 1,2,4-triazinone. Only compounds **9-14** showed a lack of the absorption bands at 1200 - 1100 cm^{-1} for C=S, which confirm that heterocyclization. In addition to the compounds **6-9** & **18**, **20** exhibited a two absorption bands at v 1700 and 1665 cm^{-1} due to the presence of two carbonyl groups. Also, IR absorption spectra of compounds **3-8**, **9-10** and **16** recorded the absorption bands at v 2975 and 2885 cm^{-1} attributed to aliphatic functional groups [1] [14] [15] [26].

3.2.3. NMR Spectral Study

1) ^{1}H NMR spectrum of **1** showed a resonated signals at δ 14.6, 12.6 and 10.9 ppm for 3NH with δ 8.6 - 0.80, 7.69 - 7.64, 7.41 - 7.31 ppm for three aromatic protons, while that of **3** exhibited a signals at δ 5.24 and 4.98 ppm attributed to two OH with δ 2.92 - 2.88, 2.62 - 2.58 ppm for two CH$_2$ protons. Compounds **3**, **4** and **5** showed a lack's of ^{4}NH and ^{2}NH of 1,2,4-triazine moiety, while that of **5** recorded additional signals at δ 1.9 and 1.75 ppm of two methyl groups of antipyrine moiety. ^{1}H NMR spectra of **6-8** recorded δ at 12.7, 4.7 ppm for NH and OH protons, while that of **9** showed a signal at δ 10.5 and 8.5 ppm, attributed to OH and CH = of thiazole moiety. In addition to compound **10** recorded a signals at δ 13.90, 2.59 ppm for NH, CH$_2$ protons, while that of

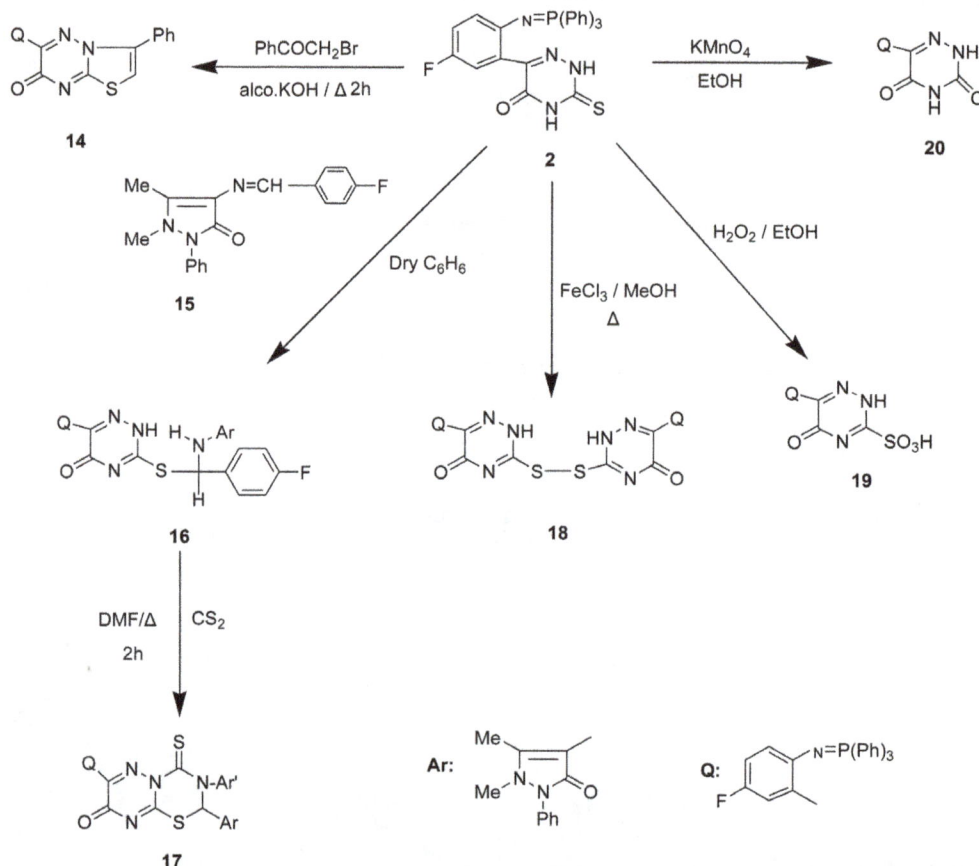

Scheme 5. Formation of compounds 14 - 20.

11 exhibited only signals at δ 8.01 and 3.99 ppm for = CH thiazole and amino-protons. Moreover ^{1}H NMR spectrum of **16** showed a signals at δ 12.76 and 10.75 ppm for two NH of 1,2,4 trinazine while a lacks of these (2NH) protons of **17**, with presence of CH proton of thiadiazine moiety at δ 9.68 ppm. ^{1}H NMR spectra of compounds **18** recorded the presence of δ at 14.55 and 12.79 ppm attributed to 2NH of 1,2,4-triazine protons, while that of **19** exhibited a signals at δ 12.8 and 10.7 p pm for NH and CH. (SO$_2$-OH) protons, with signals of aromatic protons. Finally, compound **20** exhibited δ at 12.73 and 10.82 ppm attributed to NH and OH protons [14] [15] [19] [26].

2) ^{13}C NMR spectra of all the synthesized compounds showed a resonated signals at δ 180, 165 - 163, 140 - 138, 135 - 121 and 112 p pm attributed to C=S, C=O, C=N, aromatic and C-F carbons. Also, ^{13}C NMR spectra of compounds **3-6**, **9** and **10** recorded signals at δ 39 - 33 ppm for CH$_2$ carbons. Only the compound **10** showed an additional signal at δ 112 p pm for C≡N carbon. Finally, ^{13}C NMR spectra of the entire compound exhibited a resonated signals at 77 - 75 ppm for C5-C6 of 1,2,4-triazine [27] (**Figure 2**).

3) ^{19}F NMR spectral study recorded a signal at δ −126 to −125 ppm.

4) ^{31}P NMR spectral study exhibited a signal at δ 30 - 29 p pm attributed to P=N [17].

3.2.4. Mass Fragmentation Study

Mass fragmentation pattern study of some selective synthesized compounds indicated that fused heterobiycyclic systems 11 have a more base peak, while that of acyclic structures 1and 16 have only base peak which indicate that their less stability. A higher stability of fused heterobicyclic systems is due to the delocalization of net charge over all the active centers (**Figure 3** to **Figure 5**).

4. Molluscicidal Activity

Based upon the earlier work by Abdel-Rahman *et al.* [7] [16] on the synthesis of phosphono substituted-1,2,4-

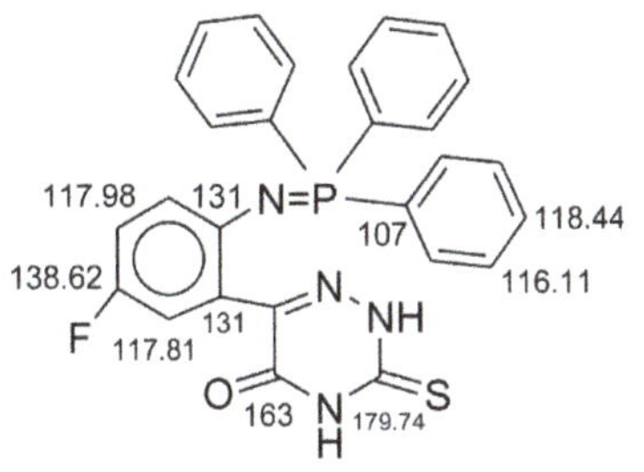

Figure 2. ^{13}C NMR data of compound 2.

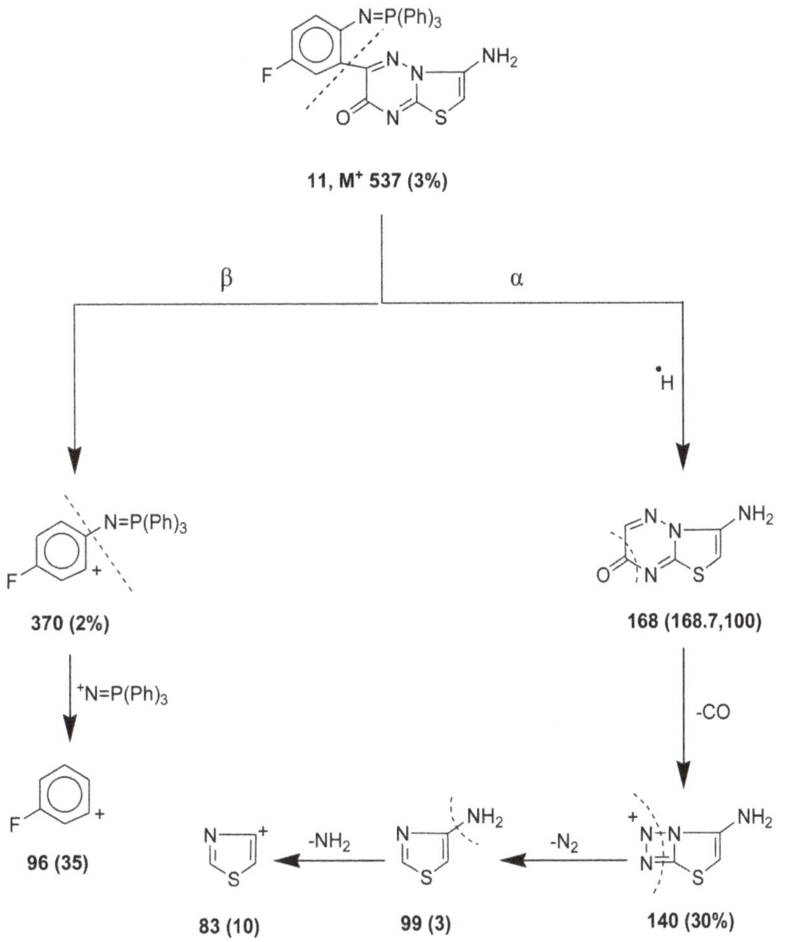

Figure 3. Mass fragmentation pattern of compound 11.

triazine derivative and their molluscicdal activities against Biomophalaria Alexandrina Snails responsible for Bilharziasis diseases, the prepared compounds were tested as killing of that snails (shell in diameter 5 - 8). The intermediate host of sohistosomamausoni in Giza Govern state that was not treated with molluscicides. The snails were adapted to laboratory conditions for two weeks before being used in toxicity tests to be sure that the snails are strong and healthy. Snails were kept in plastic aguaria filled with de chlorinated tap water at room temperature (25°C - 27°C). Stock solution (500 μg·ml^{-1}) of the tested compounds were synthesized in the least volume of ethanol and completed of the least volume of ethanol and completed to the required volume with de chlorinated tap water on the basis of weight volume. A series of more diluted solutions were then prepared following the instructions given by WHO organization [28] [29]. The result given in (**Table 1**) revealed that the high activity towards snails in the following sequences:

18 > **2** > **20** > **3** > **8** and **9** > **10** > **6** > **17** >> **5a** and **5b** > **7** > **14** at 100 ppm in compared with Baylucide as

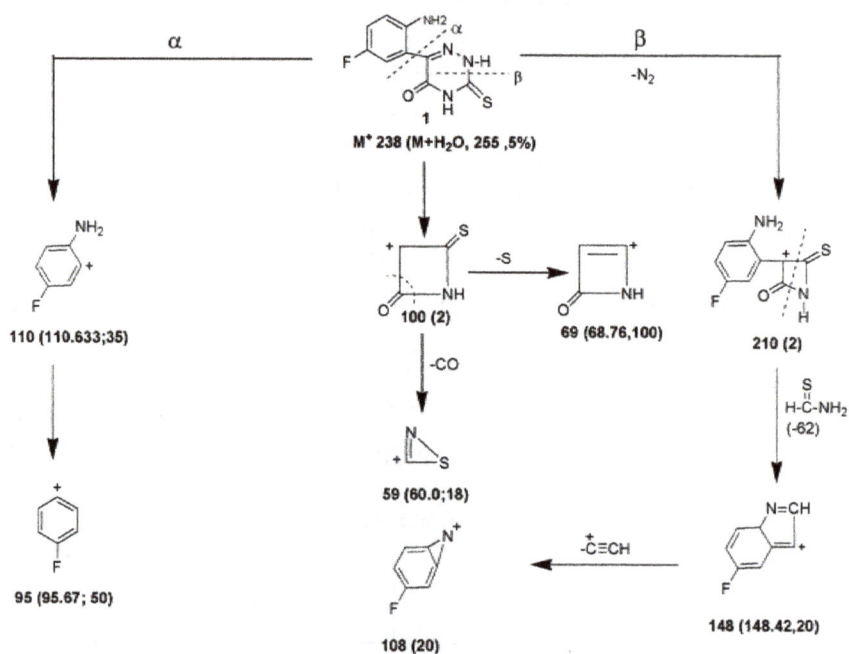

Figure 4. Mass fragmentation pattern of compound 1.

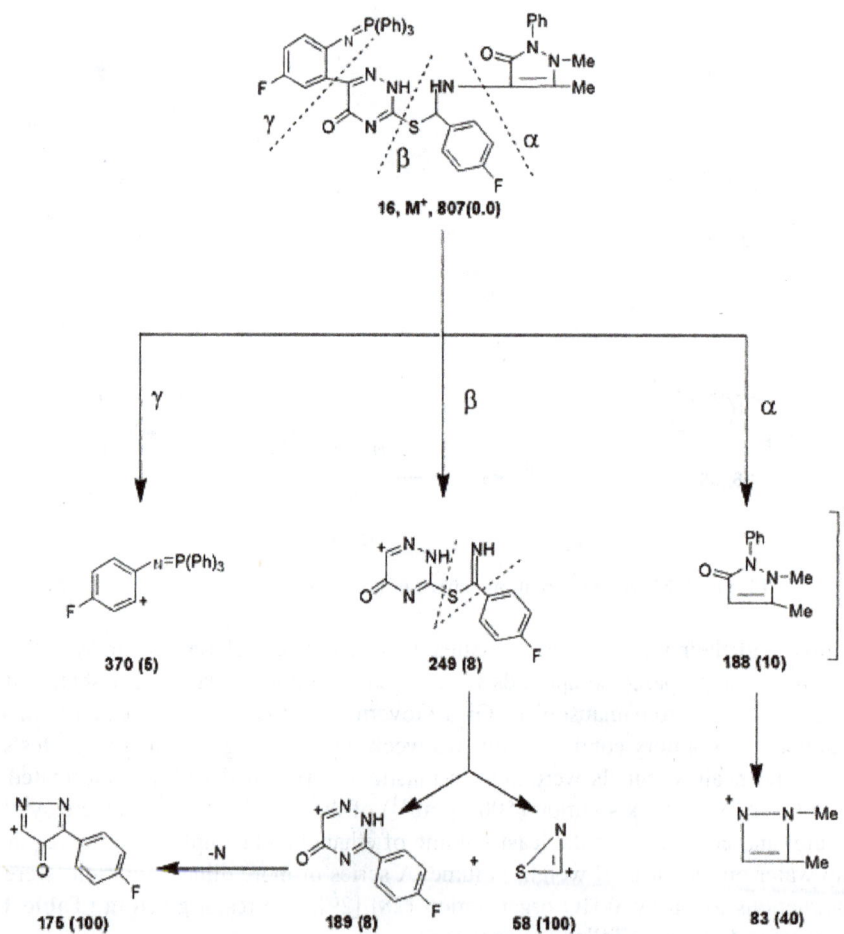

Figure 5. Mass fragmentation pattern of compound 16.

Table 1. The molluscicidal activity of the synthesized systems (2 - 20) mortality of snails various concentration (ppm).

Comp. No.	25 ppm	50 ppm	100 ppm
2	30	60	80
3	30	50	80
4	20	30	60
5a	20	40	70
5b	10	20	50
6	20	40	70
7	20	30	50
8	30	50	80
9	30	50	70
10	30	50	70
12	10	20	30
13a	10	20	30
13b	10	20	30
14	10	30	40
16	10	20	30
17	20	40	70
18	40	60	90
19	30	40	50
20	30	60	80
Reference standard, Baylucide		100	100

standard reference. In general, the strong effect of the compounds **2**, **3**, **8**, **18** and **20** is due to presence both the S-S, S-H and O-H functional groups which agree with bio-oxidation-reduction processes. The moderate effect of the compounds **5a**, **6**, **9**, **10** and **17** is attributed to thioether and cyclic sulfur nitrogen systems. Finally, the lethal effect of the compounds **4**, **5b**, **7**, **11** and **14** may be to absence of SH and/or OH of Mannich base and for thiazolotriazine systems which led to the inhibition of delocalization electron-density over all the center of systems. Also, presence of hetero-elements (F, P, S, O) and N elements in corporated with 1,2,4-trinazines led to increases of electro-negativity, over all the molecular structure and enhance the electrostatic force and hydrophobic properties [17] [18] [31]-[33]. Thus, total electron-barrier of molecular distribution of the evaluated systems synthesized led to highly inhibition of the enzymatic effect on the living processes for the tested snails by causing break of a vital cyclic of that snails, and enhance the possibility killing of these snails. QSAR study of the obtained resulted from (**Table 1**), and based on the introduction of P, S and F in the synthesized 1,2,4-triazines, in compared with the mortality of tested snails, indicated that, increases of P and S percent % led to increase of mortality, while, increase of F percentage % led to decrease of mortality of snails. Also, very high electronegative of fluorine atom can modify the electronic distribution in the molecule affecting its absorption distribution and metabolism. In conclusion, 3-thioxo-1,2,4-triazine-5-ones bearing an P, S and F elements and their related S-alkyl derivatives, enhance the mortality of snails, which cause Bilharziasis Diseases than that their non-fluorinated and non-phosphinated systems. Also, increases of P and S percentage % led to higher mortality of the tested snails, in hope to obtain more clean water from waste water.

5. Conclusion

New fluorine substituted 6-(5'-fluoro-2'-triphenylphosphiniminophenyl) 3-thioxo-1,2,4-triazin-5 (2H, 4H) one (**2**) was obtained via Wittig's reaction of the corresponding 6-(5'-fluoro-2'-aminophenyl)-3-thioxo-1,2,4-triazinone (**1**). 3-thioxo-1,2,4-triazine-5-ones bearing an P, S and F elements and their related S-alkyl derivatives, enhance the mortality of snails, which cause Bilharziasis Diseases than that their non-fluorinated and non-phosphinated systems. Also, increases of P and S percentage % led to higher mortality of the tested snails, in hope to obtain more clean water from waste water.

Acknowledgements

The authors are thankful to Prof. M. M. El-Sayed for helping in testing the molluscicidal activity in Theodor Bilharz Research institute, Giza, Egypt.

References

[1]   Abdel-Rahman, R.M. and Ali, T.E. (2013) Synthesis and Biological Evaluation of Some New Polyfluorinated 4-Thiazolidinone and α-Aminophosphonic Acid Derivatives. *Monatshefte fur Chemie*, **144**, 1243-1252. http://dx.doi.org/10.1007/s00706-013-0934-6

[2]   Makki, M.S.T., Bakhotmah, D.A. and Abdel-Rahman, R.M. (2012) Highly Efficient Synthesis of Novel Fluorine Bearing Quinoline-4-Carboxylic Acid and the Related Compounds as Amylolytic Agents. *International Journal of Organic Chemistry*, **2**, 49-55. http://dx.doi.org/10.4236/ijoc.2012.21009

[3]   Makki, M.S.T., Bakhotmah, D.A., Abdel-Rahman, R.M. and El-Shahawy, M.S. (2012) Designing and Synthesis of New Fluorine Substituted Pyrimidine-Thion-5-Carbonitrles and the Related Derivatives as Photochemical Probe Agent for Inhibition of Vitiligo Disease. *International Journal of Organic Chemistry*, **2**, 311-320. http://dx.doi.org/10.4236/ijoc.2012.223043

[4]   Abdel-Rahman, R.M., Makki, M.S.T. and Bawazir, W.A. (2011) Synthesis of Some New Fluorine Heterocyclic Nitrogen Systems Derived from Sulfa Drugs as Photochemical Probe Agents for Inhibition of Vitiligo Disease Part I. *E-Journal of Chemistry*, **8**, 405-414.

[5]   Abdel-Rahman, R.M., Makki, M.S.T. and Bawazir, W.A. (2010) Syntheisis of Fluorine Heterocyclic Nitrogen Systems Derived from Sulfa Drugs as Photochemical Probe Agents for Inhibition of Vitiligo Disease Part II. *E-Journal of Chemistry*, **7**, 593-5102.

[6]   Abdel-Rahman, R.M. (2003) Synthesis of New Phosphaheterobicyclic Systems Containing 1,2,4-Triazine Moiety—Part IX: Straight forward Synthesis of New Fluorine Bearing 5-Phospha-1,2,4 Triazine/1,2,4-Triazepine-3-Thiones; Part X: Synthesis of New Phosphaheterobicyclic Systems, Containing a 1,2,4-Triazine Moiety. *Trends in Heterocyclic Chemistry*, **8**, 187-195.

[7]   Ali, T.E., Abdel-Rahman, R.M., Hanafy, F.J. and El Edfawy, S.M. (2008) Synthesis and Molluscicidal Activity of Phosphorus—Containing Heterocyclic Compounds Derived from 5,6-Bis (4-brome phenyl)-3-hydrazino-1,2,4-triazine. *Phosphorus, Sulfur, and Silicon*, **183**, 2565-2577.

[8]   Abdel-Rahman, R.M., Ibrahim, M.A. and Ali, T.E. (2010) 1,2,4-Triazine Chemistry Part II, Synthetic Approaches for Phosphorus Containing 1,2,4-Triazine Derivatives. *European Journal of Chemistry*, **1**, 388-396. http://dx.doi.org/10.5155/eurjchem.1.4.388-396.154

[9]   Blakley, B., Brousseau, P., Fournier, M. and Voccia, I. (1999) Immunotoxicity of Pesticides. *Toxicology Industrial Health*, **15**, 119-132. http://dx.doi.org/10.1177/074823379901500110

[10] Sengupta, A.K., Bajaj, O.P. and Agarwal, K.C. (1980) Synthesis and Insecticide Activity of N^4-Aryl-N^1-(0.0-Dialkylthiophosphoryl) Piperazines. *Journal of the Indian Chemical Society*, **57**, 1170-1171.

[11] Du, Y.M., Tian, J., Liao, H., Bai, C.J., Yan, X.L. and Liu, G.D. (2009) Aluminum Tolerance and High Phosphorus Efficiency Helps Stylosanthes Better Adapt to Low-P Acid Soils. *Annals of Botony*, **103**, 1239-1247. http://dx.doi.org/10.1093/aob/mcp074

[12] Abdel-Rahman, R.M. (2001) Role of Uncondensed 1,2,4-Triazine Derivatives as Biological Plant Protection Agents. *Pharmazie*, **56**, 195-212.

[13] Abdel-Rahman, R.M. (2001) Role of Uncondensed 1,2,4-Triazine Compounds and Related Heterobicyclic Systems as Therapeutic Agents. *Pharmazie*, **56**, 18-30.

[14] Abdel-Rahman, R.M. (1991) Synthesis and Anti Human Immune Virus Activity of Some New Fluorine Containing Substituted 3-Thixo-1,2-4-Triazin-5-Ones. *Farmaco*, **46**, 379-389.

[15] Abdel-Rahman, R.M. (1992) Synthesis of Some New Fluorine Bearing Tri-Substituted 3-Thioxo-1,2,4-Triazine-5-

Ones as Potential Anti Cencer Agent. *Farmaco*, **47**, 319-326.

[16] Makki, M.S.T., Abdel-Rahman, R.M. and El-Shahawi, M.S. (2012) Synthesis and Voltammetric Study of Some New Macrocylis of Arsenic (III) in Wastewater and as Molluscicidal Agents against Biomophalaria Alexandrina Snails. *Comptes Rendus Chimie*, **15**, 617-626.

[17] Basavaiah, D., Chandrashekar, V., Das, U. and Reddy, G.J. (2005) A Study toward Understanding the Role of a Phosphorus Stereogenic Center in (5*S*)-1,3-Diaza-2-Phospha-2-oxo-3-Phenylbicyclo(3.3.0)Octane Derivatives as Catalysts in the Borane-Mediated Asymmetric Reduction of Prochiral Ketones. *Tetrahedron: Asymmetry*, **16**, 3955-3962. http://dx.doi.org/10.1016/j.tetasy.2005.10.038

[18] Ali, P., Pamakanth, P. and Meshram, J. (2010) Exploring Microwave Synthesis for Co-Ordination: Synthesis, Spectral Characterization and Comparative Study of Fluorine Substituted Transition Metal Complexes with Binuclear Core Derived from 4-Amino-2,3-Dimethyl-1-Phenyl-3-Pyrazolin-5-One. *Journal of Coordination Chemistry*, **63**, 323-329. http://dx.doi.org/10.1080/00958970903305437

[19] Abdel-Rahman, R.M. (2000) Chemistry of Uncondensed 1,2,4-Triazines: Part II-Sulfur Containing 5-oxo-1,2,4-Triazin 3-yl Moiety. *Phosphorus, Sulfur and Silicon*, **166**, 315-357. http://dx.doi.org/10.1080/10426500008076552

[20] Abdel-Rahman, R.M. and Islam, I.E. (1993) Synthesis and Reactions of Acetonitrile Derivatives Bearing a 5,6-Dipheny-1,2,4-Triazin-3-yl Moiety. *Indian Journal of Chemistry, Section B*, **32**, 526-529.

[21] Abdel-Rahman, R.M., Seada, M., Fawzy, M.M. and El-Baz, I. (1993) Synthesis and Anti-Canceranti Human Immune Virus Activities of Some New Thioether Bearing a 1,2,4-Triazine-3-Hydrazones. *Farmaco*, **48**, 397-406.

[22] EL-Gendy, Z., Morsy, J., Allimony, H.A., Abdel-Monem, W.R. and Abdel-Rahman, R.M. (2003) Synthesis of Heterobicyclic Nitrogen System Bearing the 1,2,4-Triazine Moiety as Anti-HIV and Anti-Cancer Drugs, Part II. *Phosphorus, Sulfur and Silicon*, **178**, 2055-2071. http://dx.doi.org/10.1080/10426500390228738

[23] Abdel-Rahman, R.M. and El-Mahdy, K. (2012) Biological Evaluation of Pyramidpyrimidines as Multi-Targeted Small Molecule Inhibitors and Resistance Modifying Agents. *Heterocycles*, **85**, 2391-2414. http://dx.doi.org/10.3987/REV-12-745

[24] Zaki, M.T., Abdel-Rahman, R.M. and El Sayed, A.Y. (1995) Use of Arylidenrhodanines for the Determination of Cu(II) Hg(II) and CN$^-$. *Analytica Chimica Acta*, **307**, 127-138. http://dx.doi.org/10.1016/0003-2670(95)00048-5

[25] Slawinski, J. and Gdaniec, M. (2005) Synthesis, Molecular Structure, and *in Vitro* Antitumor Activity of New 4-Chlore-2-Mercaptobenzenesulfonamide Derivatives. *European Journal of Medicinal Chemistry*, **40**, 377-389. http://dx.doi.org/10.1016/j.ejmech.2004.11.014

[26] Abdel-Rahman, R.M. and Abdel-Malik, N.S. (1990) Synthesis of Some New 3,6-Diheteroarryl-1,2,4-Triazine-5-Ones and Their Effect on Amylolytic Activity of Some Fungi. *Pakistan Journal of Science and Industrial Research*, **33**, 142-147.

[27] Ebraheem, M.A., Abdel-Rahman, R.M., Abdel-Haleem, A.M., Ibrahim, S.S. and Allimony, H.A. (2008) Synthesis and Antifungal Activity of Novel Polyheterocyclic Compound Containing Fused 1,2,4-Triazine Moiety. *Arkivoc*, **21**, 202-213.

[28] WHO (1953) Expert Committee on Bilharziasis, **65**, 33.

[29] WHO (1965) Snail Control Information of Bilharziasis Monograph Series. **50**, 124.

[30] Smarti, B.E. (2001) Fluorine Substituent Effect (on Bioactivity). *Journal of Fluorine Chemistry*, **109**, 3-11. http://dx.doi.org/10.1016/S0022-1139(01)00375-X

[31] Billard, T., Gille, S., Ferry, A., Bartelemy, A., Christophe, C. and Langlois, B.R. (2005) From Fluoral to Heterocycles: A Survey of Polyfluorinated Iminiums Chemistry. *Journal of Fluorine Chemistry*, **126**, 189-196. http://dx.doi.org/10.1016/j.jfluchem.2004.08.007

[32] Isanbor, C. and O'Hagan, D. (2006) Fluorine in Medicinal Chemistry: A Review of Anti-Cancer Agents. *Journal of Fluorine Chemistry*, **127**, 303-319. http://dx.doi.org/10.1016/j.jfluchem.2006.01.011

[33] Sandford, G. (2007) Elemental Fluorine in Organic Chemistry (1997-2006). *Journal of Fluorine Chemistry*, **128**, 90-104. http://dx.doi.org/10.1016/j.jfluchem.2006.10.019

6

Highly Efficient, One Pot Synthesis and Oxidation of Hantsch 1,4-Dihydropyridines Mediated by Iodobenzene Diacetate (III) Using Conventional Heating, Ultrasonic and Microwave Irradiation

Khalid Hussain[1*], Deepak Wadhwa[2]

[1]Mewat Engineering College (WAKF), Nuh, India
[2]Department of chemistry, Kurukshetra University, Kurukshetra, India
Email: *khalidchem83@yahoo.co.in

Abstract

A mild, general, convenient, and efficient one-pot synthesis of 4-arylpyridines (4) is described using conventional heating, ultrasound and microwave irradiation. Aryl aldehydes (2) were efficiently condensed with ethylacetoacetate (1) and ammonium acetate in acetonitrile to give dihydropyridine intermediates (4). The latter underwent a smooth Iodobenzene Diacetate (III) mediated aromatization reaction in the same pot to afford 4-arylpyridines (4) in good to excellent yields.

Keywords

Dihydropyridines, Iodobenzene Diacetate, Ultrasonic, Microwave, One-Pot, Formylpyrazoles

1. Introduction

Development of highly efficient synthetic methodologies for the construction of biologically important compounds is one of the challenges to medicinal and organic chemist. One of the most relevant approaches to synthetic efficiency is based on multi component reactions (MCRs) which combine three or more substrates, either simultaneously, leading to domino processes [1], or through the sequential reactions without isolating interme-

*Corresponding author.

diate species. MCRs offer several advantages such as atom economy, minimized waste generation, because of the reduction in the number of work-up, extraction and purification stages.

Use of ultrasonic irradiation and microwave irradiation as alternative sources of energy has proved to be one of the stepping stone towards the green syntheses as it offers advantage of enhanced reactivity, shorter reaction times and higher yields of pure products compared to the traditional heating methods [2] [3]. The efficiency and expediency of MCRs leading to heterocyclic scaffolds can be increased to several times using one pot reaction profiles, greener catalysts, ultrasonic, and microwave irradiation.

1,4-Dihydropyridines (1,4-DHPs) belong to a class of nitrogen containing heterocycles having a six-membered ring. Much attention has been devoted to explore their pharmacological activities. A considerable portion of today's efforts in dihydropyridine chemistry is expanded in synthesizing reduced form of nicotinamide adenine dinucleotide (NADH) mimics, exploring the reactions and mechanisms of these compounds, and utilizing them in a variety of synthetic reactions. Newly synthesized substituted 1,4-DHPs possess other pharmacological activities such as antitumor [4], bronchodilating [5], antidiabetic [6], neurotropic [7], antianginal [8] and P-glyco protein Inhibitors [9]. The benign environmental character and easy commercial availability makes hypervalent iodine (III) reagents increasingly important for the oxidation of organic molecules [10]-[18]. These days much work has been done to explore the oxidation ability, their electrophilic properties and to develop novel reaction using hypervalent iodine compounds [19].

In view of numerous biological properties associated with 1,4-DHP and the biological importance of the oxidation step of 1,4-DHP [20], we became interested in the synthesis of some new 1,4-DHPs and their corresponding pyridine derivatives via iodine (III) mediated oxidation. In continuation of our work on the utility of iodine (III) reagents for the synthesis of various types of heterocyclic compounds, we herein carried out the synthesis of 4-arylpyridines using IBD as a greener oxidant at room temperature, ultrasonic and microwave irradiation.

2. Result and Discussion

The present manuscript reports the synthesis of targeted diethyl 2,6-dimethyl-4-aryl-pyridine-3,5-dicarboxylate (4a-4m), by a one pot domino process. The main aim of this manuscript is to develop efficient synthetic methodology which requires lesser reaction time and reduces the number of steps involved in the synthesis of 4a-4m. In order to achieve our aim, synthesis of diethyl 2,6-dimethyl-4-((1,3-diphenyl)-1H-pyrazol-4-yl) pyridine-3,5-dicarboxylate was carried out by a two-step reaction. We first synthesized diethyl 2,6-dimethyl-4-((1,3-diphenyl)-1H-pyrazol-4-yl) pyridine-3,5-dicarboxylate (3a). For this a pseudo four component reaction of ethyl acetoacetate (1) (2.0 mmol), 1,3-diphenyl-1H-pyrazole-4-carbaldehyde (2) (1.0 mmol) and ammonium acetate (2.2 mmol) was carried out in ethanol and the reaction mixture was allowed to reflux for 25 - 35 min. The progress of the reaction was checked by TLC using petroleum ether: ethyl acetate (85: 15, v/v) as eluent. After completion of the reaction as evident from TLC, the reaction mixture was cooled down to room temperature and the solid separated was filtered under suction to afford 2,6-dimethyl-4-((1,3-diphenyl)-1H-pyrazol-4-yl)-1,4-di-hydropyridine-3,5-dicarboxylate (1a) as the desired product in 75% yield. The product so obtained was subjected to oxidation using IBD (1.2 mmol) in DCM at room temperature. The reaction was completed in 5 min as evident from TLC (petroleum ether: ethyl acetate (85: 15, v/v)). After the completion of reaction as evident from TLC, the reaction mixture was washed with aqueous NaHCO$_3$ solution. Organic phase was then separated, dried and concentrated on water bath. Crude product, thus obtained, was purified by silica gel column chromatography (petroleum ether: ethyl acetate, 97:3, v/v) to afford diethyl 2,6-dimethyl-4-((1,3-diphenyl)-1H-pyrazol-4-yl) pyridine-3,5-dicarboxylate (4a) as a pure product in 68% yield. In order to improve the overall yield of 4a and to reduce the time required, this method can be further expanded to a modular one-pot synthesis, without isolation of intermediate 3a, in contrast to our previous protocols. We decided to try a domino process for the synthesis of 4a without the isolation of 3a in one pot. To achieve this, firstly a mixture of ethyl acetoacetate (1) (2.0 mmol), 1, 3-diphenyl-1H-pyrazole-4-carbaldehyde (2) (1.0 mmol) and ammonium acetate (2.2 mmol) was taken in 50 mL round bottomed flask using acetonitrile as solvent. The content of the flask was heated to reflux for 20 - 25 min. The progress of the reaction was monitored by TLC using petroleum ether: ethyl acetate (85: 15, v/v) as eluent. After formation of 3a as confirmed by TLC, IBD (1.2 mmol) was then added to the same reaction mixture and the reaction mixture was refluxed again for 5 - 8 min. After the completion of reaction as evident from TLC, the reaction mixture was cooled at room temperature and washed with aqueous NaHCO$_3$ solution. Organic phase was then separated, dried and concentrated on water bath. Crude product, thus obtained, was purified by silica gel column chromatography to afford pure diethyl 2,6-dimethyl-4-((1,3-diphenyl)-1*H*-pyrazol-4-yl)pyridine-3,5-dicarboxylate **(4a)** in 85% yield. The generality of the optimized protocol was checked by carrying out the reactions using different 3-(aryl)-1-phenyl-1*H*-pyrazole-4-carbaldehyde and also substituted benzaldehydes containing both electron withdrawing and electron releasing substituents yielded the corresponding diethyl 2,6-dimethyl-4-arylpyridine-3,5-dicarboxylate **(4a-4m)** in excellent yields. (**Scheme 1**, **Table 1**).

The efficiencies of this protocol prompted us to explore this protocol further using Ultrasonic and microwave irradiation to reduce the present serious energy crisis in the environment. All the reactions proceeded successfully and yielded the corresponding products in excellent yield. The results obtained using ultrasonic and microwave irradiation is summarized in **Table 1**.

3. Conclusion

We have synthesized a series of 2,6-dimethyl-4-aryl-pyridine-3,5-dicarboxylate derivatives **(4a-4m)** by one-pot domino process in acetonitrile and IBD as a greener oxidant using conventional heating, ultrasonic and microwave irradiation. This method thus provides a one pot, facile, rapid and efficient synthesis of compounds **4a-4m** which are otherwise accessible through a two-step process.

4. Experimental

Structures of all the compounds were identified by their spectral data. Silica gel 60 F$_{254}$ (Precoated aluminium plates) from Merck were used to monitor reaction progress. Melting points were determined on a melting point apparatus and are uncorrected. IR (KBr) spectra were recorded on buck scientific IR M-500 spectrophotometer and the values are expressed as v_{max} cm^{-1}. The ^{1}H NMR spectra were scanned on a Bruker (300 MHz) spectrometer in CDCl$_3$ using tetramethylsilane as an internal standard. Mass spectral data were recorded on a Waters micromass Spectrometer running under Mass Lynex version 4.0 software and equipped with an ESI source. The chemical shift values are recorded on δ scale and the coupling constants (J) are in Hz. Ultrasonic bath (54 KHz, 300 W, 1 Lt, capacity) of Through clean ultrasonic Pvt. Ltd. (India) was used for reactions under ultrasonic irradiation. CEM discover microwave reactor was used for reactions under microwave irradiation. Pyrazole aldehydes **(3)** were synthesized according to the literature method [21].

4.1. Preparation of (4a-4m) under Conventional Heating

A mixture of ethylacetoacetate **(1)** (1.0 mmol), aryl aldehyde **(2)** (1.0 mmol), and ammonium acetate (2.2 mmol) was dissolved in 5 mL of acetonitrile in a 50 mL round-bottomed flask. The reaction contents were refluxed on water bath for 20 - 25 min. The progress of the reaction was monitored by TLC using petroleum ether: ethyl acetate (85: 15, v/v) as eluent. After formation of 3a as evident from TLC after 25 min, IBD (1.2 mmol) was then added to the above reaction mixture and the reaction mixture was refluxed for another 5 - 8 min. After consumption of **3a** as evident from TLC, the reaction mixture was cooled to room temperature and washed with aqueous NaHCO$_3$ solution. Organic phase was then separated, dried and concentrated on water bath. Crude product, thus obtained, was purified by silica gel column chromatography to afford pure diethyl 2,6-dimethyl-4-aryl-pyridine-3, 5-dicarboxylate **(4a-4m)**. All the compounds **4a-4m** are characterized by ^{1}H NMR, ^{13}C NMR, IR and mass data.

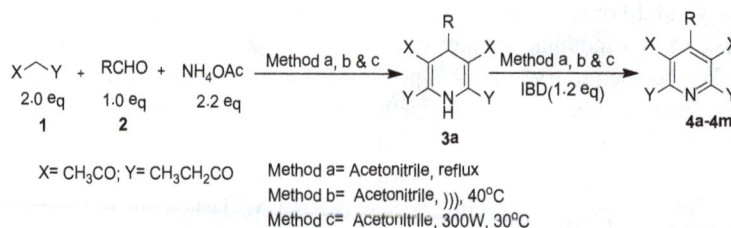

Scheme 1. Synthesis of diethyl 2,6-dimethyl-4-aryl-pyridine-3,5-dicarboxylate (4a-4m) using one pot domino protocol.

Table 1. Synthesis of diethyl 2,6-dimethyl-4-aryl-pyridine-3,5-dicarboxylate (4a-4m) using conventional heating, ultrasonic and microwave irradiation.

S.no	Ar	Method A		Method B		Method C	
		Time (min)	Yield (%)[a]	Time (min)	Yield (%)[a]	Time (Sec)	Yield (%)[a]
4a	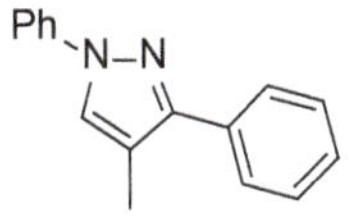 1-Ph-4-methyl-3-phenyl-pyrazole	30	87	18	88	120	84
4b	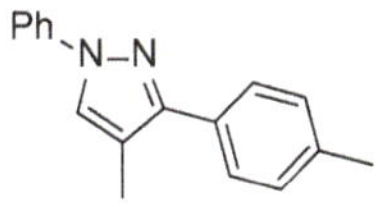 1-Ph-4-methyl-3-(4-methylphenyl)-pyrazole	32	85	16	84	110	86
4c	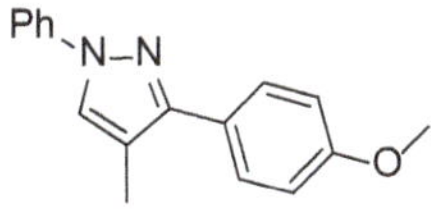 1-Ph-4-methyl-3-(4-methoxyphenyl)-pyrazole	34	84	14	86	100	88
4d	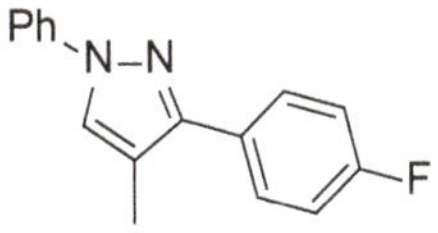 1-Ph-4-methyl-3-(4-fluorophenyl)-pyrazole	30	87	16	90	160	85
4e	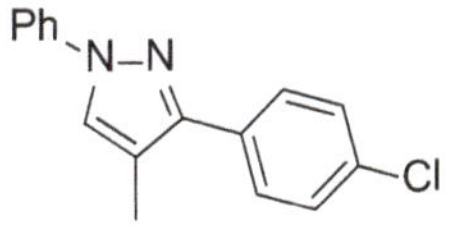 1-Ph-4-methyl-3-(4-chlorophenyl)-pyrazole	28	88	18	90	150	86
4f	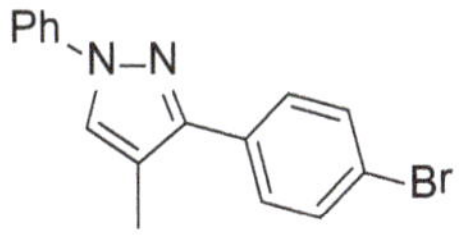 1-Ph-4-methyl-3-(4-bromophenyl)-pyrazole	33	89	19	92	140	90
4g	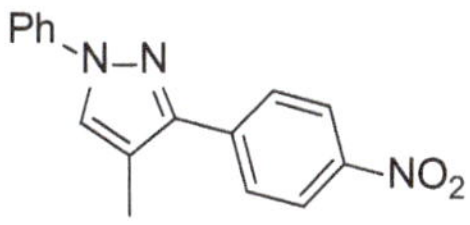 1-Ph-4-methyl-3-(4-nitrophenyl)-pyrazole	35	92	20	94	175	90
4h	C_6H_5	35	92	16	95	90	95
4i	$4\text{-}CH_3C_6H_4$	35	90	15	93	85	93
4j	$3\text{-}BrC_6H_4$	45	88	19	90	100	90
4k	$4\text{-}ClC_6H_4$	50	92	17	95	100	95
4l	$4\text{-}OCH_3C_6H_4$	30	96	14	95	80	95
4m	$4\text{-}NO_2C_6H_4$	55	94	19	91	105	91

[a]Isolated yield.

4.2. Preparation of (4a-4m) under Ultrasonic Irradiation

A mixture of ethylacetoacetate (**1**) (1.0 mmol), aryl aldehyde (**2**) (1.0 mmol), and ammonium acetate (2.2 mmol) was dissolved in 5 mL of acetonitrile in a 50 mL round-bottomed flask. The reaction contents were sonicated at 40°C for appropriate time as mentioned in **Table 1**. The progress of the reaction was monitored by TLC using petroleum ether: ethyl acetate (85: 15, v/v) as eluent. After formation of **3a** as evident from TLC after 8 min, IBD (1.2 mmol) was then added to the above reaction mixture and the reaction mixture sonicated at room temperature for another 3 - 4 min. After consumption of **3a** as evident from TLC the reaction mixture was washed with aqueous NaHCO₃ solution. Organic phase was then separated, dried and concentrated on water bath. Crude product, thus obtained, was purified by silica gel column chromatography to afford pure diethyl 2,6-dimethyl-4-aryl-pyridine-3,5-dicarboxylate (**4a-4m**). All the compounds **4a-4m** are characterized by ¹H NMR, ¹³C NMR, IR and mass data.

4.3. Preparation of (4a-4m) under Microwave Irradiation

A mixture of ethylacetoacetate (1) (1.0 mmol), aldehyde (2) (1.0 mmol), and ammonium acetate (2.2 mmol) was dissolved in 5 mL of acetonitrile in a sealed vial and placed in a CEM Discover microwave reactor. The vial was subjected to microwave irradiation, programmed at 30°C and 300 W. After formation of 3a as evident from TLC after 1 min, IBD (1.2 mmol) was then added to the above reaction mixture and the reaction mixture irradiated for another 30 sec. After completion of the reaction as evident from TLC, the reaction mixture was washed with aqueous $NaHCO_3$ solution. Organic phase was then separated, dried and concentrated on water bath. Crude product, thus obtained, was purified by silica gel column chromatography to afford pure diethyl 2,6-dimethyl-4-aryl-pyridine-3,5-dicarboxylate (4a-4m). All the compounds 4a-4m are characterized by ^1H NMR, ^{13}C NMR, IR and mass data.

4.4. Characterization Data

4.4.1. 4-(1,3-Diphenyl-1H-Pyrazol-4-yl)-2,6-Dimethyl-Pyridine-3,5-Dicarboxylic Acid Diethyl Ester (4a)

Mp: 111°C; IR (v_{max}, cm^{-1}, KBr): 1736, 1233; ^1H NMR (300 MHz, CDCl$_3$, δ, ppm): 0.95 (t, 6H, CH$_3$), 2.613 (s, 6H, CH$_3$), 3.910 - 4.07 (m, 4H, OCH$_2$), 7.110 - 7.313 (m, 4H), 7.817 (s, 1H), 7.581 - 7.690 (m, 6H); ^{13}C NMR (75 MHz, DMSO-d_6) 166.8, 155.7, 147.8, 139.3, 138.2, 133.4, 131.2, 130.2, 129.2, 128.8, 128.0, 127.2, 118.5, 115.7, 61.7, 23.0, 13.5; Elemental analysis: Calcd for C$_{28}$H$_{27}$N$_3$O$_4$: C 71.64, H 5.76, N 8.95; found: C 71.63, H 5.79, N 8.93; MS (m/z): 470.20 (M$^+$ + 1).

4.4.2. 2,6-Dimethyl-4-(1-Phenyl-3-P-Tolyl-1H-Pyrazol-4-Yl)-Pyridine-3,5-Dicarboxylic Acid Diethyl Ester (4b)

Mp: 105°C; IR (v_{max}, cm^{-1}, KBr): 1720, 1234; ^1H NMR (300 MHz, CDCl$_3$, δ, ppm): 0.93 (t, 6H, CH$_3$), 2.611 (s, 6H, CH$_3$), 3.810 (s, 3H), 3.99 (q, 4H, OCH$_2$), 6.84 (d, 2H, J = 8.7 Hz), 7.280 - 7.501 (m, 5H), 7.732 - 7.759 (d, 2H, J = 8.7 Hz), 7.905 (s, 1H); ^{13}C NMR (75 MHz, DMSO-d_6); 167.2, 155.7, 147.9, 139.5, 137.9, 137.2, 129.8, 128.3, 127.9, 127.6, 127.3, 127.1, 119.2, 116.2, 61.2, 34.4, 23.1, 13.4; Elemental analysis: Calcd for C$_{29}$H$_{29}$N$_3$O$_4$: C 72.05, H 6.00, N 8.70; found: C 72.06, H 6.05, N 8.70; MS (m/z): 484.40 (M$^+$ + 1).

4.4.3. 4-[3-(4-Methoxy-Phenyl)-1-Phenyl-1H-Pyrazol-4-yl]-2,6-Dimethyl-Pyridine-3,5-Dicarboxylic Acid Diethyl Ester (4c)

Mp: 136°C; IR (v_{max}, cm^{-1}, KBr): 1740, 1034; ^1H NMR (300 MHz, CDCl$_3$, δ, ppm): 0.95 (t, 6H, CH$_3$), 2.612 (s, 6H, CH$_3$), 3.808 (s, 3H), 4.00 (q, 4H, OCH$_2$), 6.84 (d, 2H, J = 8.7 Hz), 7.311–7.501 (m, 5H), 7.74 (d, 2H, J = 8.7 Hz), 7.905 (s, 1H); ^{13}C NMR (75 MHz, DMSO-d_6); 167.1, 155.9, 148.5, 139.6, 137.8, 137.3, 129.6, 128.4, 127.9, 127.8, 127.4, 127.2, 119.0, 116.3, 61.4, 44.4, 23.0, 13.5; Elemental analysis: Calcd for C$_{29}$H$_{29}$N$_3$O$_5$: C 69.73, H 5.81, N 8.41; found: C 69.71, H 5.83, N 8.40;   MS (m/z): 500.29 (M$^+$ + 1).

4.4.4. 4-[3-(4-Fluoro-Phenyl)-1-Phenyl-1H-Pyrazol-4-Yl]-2,6-Dimethyl-Pyridine-3,5-Dicarboxylic Acid Diethyl Ester (4d)

Mp: 121°C; IR (v_{max}, cm^{-1}, KBr): 1728, 1236, 1037; ^1H NMR (300 MHz, CDCl$_3$, δ, ppm): 0.94 (t, 6H, CH$_3$), 2.615 (s, 6H, CH$_3$), 3.905 - 4.105 (q, 4H, OCH$_2$), 6.987 - 7.044 (m, 2H), 7.280 - 7.365 (m, 1H), 7.469 - 7.622 (m, 4H), 7.74 (d, 2H, J = 7.8 Hz), 7.923 (s, 1H); ^{13}C NMR (75 MHz, DMSO-d_6) 166.9, 155.6, 148.9, 139.4, 138.1, 133.3, 131.3, 130.1, 129.0, 128.9, 128.0, 127.2, 118.6, 115.8, 61.6, 23.1, 13.6; Elemental analysis: Calcd for C$_{28}$H$_{26}$N$_3$O$_4$F: C 68.99, H 5.38, N 8.62; found: C 68.95, H 5.37, N 8.63; MS (m/z): 488.36 (M$^+$ + 1).

4.4.5. 4-[3-(4-Chlorophenyl)-1-Phenyl-1H-Pyrazol-4-Yl]-2,6-Dimethyl-Pyridine-3,5-Dicarboxylic Acid Diethyl Ester (4e)

Mp: 101°C - 102°C (101°C - 102°C, lit [22]); IR (v_{max}, cm^{-1}, KBr): 3055, 2989, 1747, 1620, 1597, 1461, 1322, 1087, 1002, 952, 850, 836, 698; ^1H NMR (300 MHz, CDCl$_3$, δ, ppm): 0.935 (t, J = 7.2 Hz, 6H, CH$_3$), 2.618 (s, 6H, CH$_3$), 3.898 - 4.118 (m, 4H, OCH$_2$), 7.310 - 7.370 (m, 2H); 7.487 - 7.513 (m, 5H), 7.746 (d, J = 7.8 Hz, 2H), 7.923 (s, 1H); Elemental analysis: Calcd. for C$_{28}$H$_{26}$ClN$_3$O$_4$: C, 66.73; H, 5.20; N, 8.34. Found: C, 61.66; H, 5.29;

N, 8.26.

4.4.6. 4-[3-(4-Bromo-Phenyl)-1-Phenyl-1H-Pyrazol-4-Yl]-2,6-Dimethyl--Pyridine-3,5-Dicarboxylic Acid Diethyl Ester (4f)

Mp: 115°C; IR (ν_{max}, cm^{-1}, KBr): 1734, 1030; ^{1}H NMR (300 MHz, CDCl$_3$, δ, ppm): 0.95 (t, 6H, CH$_3$), 2.617 (s, 6H, CH$_3$), 3.99 (q, 4H, OCH$_2$), 7.200 - 7.495 (m, 7H), 7.74 (d, 2H, J = 7.2 Hz), 7.921 (s, 1H); ^{13}C NMR (75 MHz, DMSO-d_6) 166.5, 155.7, 148.8, 139.6, 138.1, 132.3, 131.2, 130.0, 129.1, 128.8, 128.0, 127.1, 117.6, 115.6, 61.7, 23.4, 13.5; Elemental analysis: Calcd for C$_{28}$H$_{26}$N$_3$O$_4$Br: C 61.42, H 4.75, N 7.68; found: C 61.31, H 4.79, N 7.69. MS (m/z): 548.20, 550.20.

4.4.7. 2,6-Dimethyl-4-[3-(4-Nitro-Phenyl)-1-Phenyl-1H-Pyrazol-4-Yl]-Pyridine-3, 5-Dicarboxylicacid Diethyl Ester (4g)

Mp: 172°C; IR (ν_{max}, cm^{-1}, KBr): 1728, 1234, 1034; ^{1}H NMR (300 MHz, CDCl$_3$, δ, ppm): 0.91 (t, 6H, CH$_3$), 2.632 (s, 6H, CH$_3$), 3.923 - 4.039 (m, 4H, OCH$_2$), 7.279 - 7.410 (m, 3H), 7.499 - 7.769 (m, 4H), 7.960 (s, 1H), 8.19 (d, 2H, J = 7.5 Hz); ^{13}C NMR (75 MHz, DMSO-d_6) 167.9, 155.8, 149.0, 139.8, 138.2, 133.6, 131.5, 130.3, 129.2, 128.9, 128.0, 127.4, 119.6, 116.6, 62.5, 23.4, 13.7; Elemental analysis: Calcd for C$_{28}$H$_{26}$N$_4$O$_6$: C 64.37, H 4.98, N 10.73; found: C 65.34, H 5.08, N 10.87; MS (m/z): 515.26 (M$^+$ + 1).

4.4.8. Diethyl-4-Phenyl-2,6–Dimethylpyridine-3,5-Dicarb Oxylate (4h)

Mp: 62°C - 63°C; IR (ν_{max}, cm^{-1}, KBr): 3026, 2978, 1729, 1592, 1477, 1301, 1212, 1171, 792, 761; ^{1}H NMR (300 MHz, CDCl$_3$, δ, ppm): 1.22 (t, J = 7.11 Hz, 6H, CH$_3$), 4.27 (q, J = 7.11 Hz, 4H, OCH$_2$), 2.67 (s, 6H, CH$_3$), 7.18 - 7.23 (m, 2H), 7.30 - 7.32 (m, 3H); Elemental analysis: Calcd. for C$_{19}$H$_{21}$NO$_4$: C, 69.71; H, 6.47; N, 4.28. Found: C, 69.88; H, 6.55; N, 4.19.

4.4.9. Diethyl-4-(4-Methylphenyl)-2,6-Dimethylpyridine-3,5-Dicarboxylate (4i)

Mp: 70°C - 71°C (71°C -72°C, lit [22]); IR (ν_{max}, cm^{-1}, KBr): 3022, 2978, 1725, 1582, 1444, 1228, 1013, 822, 857, 776 ; ^{1}H NMR (300MHz, CDCl$_3$, δ, ppm): 1.234 (t, J = 7.11 Hz, 6H, CH$_3$), 2.35 (s, 3H, CH$_3$), 2.66 (s, 6H, CH$_3$), 4.28 (q, J = 7.11 Hz, 4H, OCH$_2$), 7.12 (d, J = 6.79 Hz, 2H), 7.23 (d, J = 6.79 Hz, 2H); Elemental analysis: Calcd. for C$_{20}$H$_{23}$NO$_4$: C, 70.36; H, 6.79; N, 4.10. Found: C, 70.44; H, 6.84; N, 4.28.

4.4.10. Diethyl-4-(3-Bromophenyl)-2,6-Dimethylpyr-Idine-3,5-Dicarboxylate (4j)

Mp: 71°C - 73°C (70°C - 72°C, lit [22]); IR (ν_{max}, cm^{-1}, KBr): 3055, 2988, 1727, 1562, 1280, 1102, 1035, 866, 777, 697; ^{1}H NMR (300MHz, CDCl$_3$, δ, ppm): 1.24 (t, J = 7.13 Hz, 6H, CH$_3$), 4.30 (q, J = 7.13 Hz, 4H, OCH$_2$), 2.66 (s, 6H, CH$_3$), 7.20 - 7.44 (m, 4H); Elemental analysis: Calcd. for C$_{19}$H$_{20}$BrNO$_4$ C, 56.17; H, 4.96; N, 3.45. Found: C, 56.32; H, 4.88; N, 3.28.

4.4.11. Diethyl-4-(4-Chlorophenyl)-2,6-Dimethylpyr-Idine-3,5-Dicarboxylate (4k)

Mp: 70°C - 72°C (69°C - 71°C, lit [22]); IR (ν_{max}, cm^{-1}, KBr): 3028, 2991, 1728, 1588, 1232, 1106, 1045, 857, 657; ^{1}H NMR (300MHz, CDCl$_3$, δ, ppm): 1.22 (t, J = 7.11 Hz, 6H, CH$_3$), 4.27 (q, J = 7.11 Hz, 4H, OCH$_2$), 2.70 (s, 6H, CH$_3$), 7.12 (d, J = 8.99 Hz, 2H), 7.32 (d, J = 8.99 Hz, 2H). Elemental analysis: Calcd. for C$_{19}$H$_{20}$ClNO$_4$: C, 63.07; H, 5.57; N, 3.87. Found: C, 62.92; H, 5.66; N, 3.66.

4.4.12. Diethyl-4-(4-Methoxyphenyl)-2,6-Dimethyl Pyridine-3,5-Dicarboxylate (4l)

Mp:50°C - 51°C (51°C - 52°C, lit [22]); IR (ν_{max}, cm^{-1}, KBr): 3034, 2987, 1731, 1599, 1523, 1288, 1107, 856, 834, 772; ^{1}H NMR (300MHz, CDCl$_3$, δ, ppm): d = 1.22 (t, J = 7.12 Hz, 6H, CH$_3$), 4.27 (q, J = 7.12 Hz, 4H, OCH$_2$), 2.69 (s, 6H, CH$_3$), 3.86 (s, 3H, OCH$_3$), 6.91 (d, J = 8.57 Hz, 2H), 7.11 (d, J = 8.57 Hz, 2H); Elemental analysis: Calcd. for C$_{20}$H$_{23}$NO$_5$: C, 67.21; H, 6.49; N, 3.92. Found: C, 67.34; H, 6.54; N, 4.02.

4.4.13. Diethyl-4-(4-Nitrophenyl)-2,6-Dimethylpyri-Dine-3,5-Dicarboxylate (4m)

Mp: 110°C - 112°C (112°C - 113°C, lit [22]); IR (ν_{max}, cm^{-1}, KBr): 3023, 2988, 1726, 1555, 1504, 1351, 1106, 866, 843, 745; ^{1}H NMR (300MHz, CDCl$_3$, δ, ppm):1.23 (t, J = 7.12 Hz, 6H, CH$_3$), 2.63(s, 6H, CH$_3$), 4.25 (q, J = 7.12 Hz, 4H, OCH$_2$), 7.41 (d, J = 8.23 Hz, 2H), 8.22 (d, J = 8.23 Hz, 2H). Elemental analysis: Calcd. for C$_{19}$H$_{20}$N$_2$O$_6$: C, 61.29; H, 5.41; N, 7.53. Found: C, 61.44; H, 5.32; N, 7.65.

Acknowledgements

We are thankful to CSIR, New Delhi for the award of Junior Research Fellowship (JRF) to Khalid Hussain and KUK for Teaching Assistantship to Deepak Wadhwa.

References

[1]   Tietze, L.F. (1996) Domino Reactions in Organic Synthesis. *Chemical Reviews*, **96**, 115-136.
      http://dx.doi.org/10.1021/cr950027e

[2]   Mason, T.J. and Meulenaer, E.C.D. (1998) Practical Considerations for Process Optimisation. Synthetic Organic Sono-
      chemistry. Plenum Press, New York and London.

[3]   Kappe, C.O. (2004) Controlled Microwave Heating in Modern Organic Synthesis. *Angewandte Chemie International
      Edition*, **43**, 6250-6284. http://dx.doi.org/10.1002/anie.200400655

[4]   Tsuruo, T., Iida, H., Nojiri, M., Tsukagoshi, S. and Sakurai, Y. (1983) Circumvention of Vincristine and Adriamycin
      Resistance *in Vitro* and *in Vivo* by Calcium Influx Blockers. *Cancer Research*, **43**, 2905-2910.

[5]   Chapman, R.W., Danko, G. and Siegels, M.I. (1984) Effect of Extra- and Intracellular Calcium Blockers on Histamine
      and Antigen-Induced Bronchospasms in Guinea Pigs and Rats. *Pharmacology*, **29**, 282-291.
      http://dx.doi.org/10.1159/000138024

[6]   Malaise, W.J. and Mathias, P.C.F. (1985) Stimulation of Insulin Release by an Organic Calcium Agonist. *Diabetologia*,
      28, 153-156.

[7]   Krauze, A., Germane, S., Eberlins, O., Sturms, I., Klusa, V. and Duburs, G. (1999) Derivatives of 3-Cyano-6-Phenyl-
      4-(3'-Pyridyl)-Pyridine-2(1*H*)-Thione and Their Neurotropic Activity. *European Journal of Medicinal Chemistry*, **34**,
      301-310. http://dx.doi.org/10.1016/S0223-5234(99)80081-6

[8]   Peri, R., Padmanabhan, S., Singh, S., Rutledge, A. and Triggle, D.J. (2000) Permanently Charged Chiral 1,4-Dihy-
      dropyridines: Molecular Probes of L-Type Calcium Channels. Synthesis and Pharmacological Characterization of
      Methyl (ω-Trimethylalkylammonium) 1,4-Dihydro-2,6-Dimethyl-4-(3-Nitrophenyl)-3,5-Pyridinedicarboxylate Iodide.
      Calcium Channel Antagonists. *Journal of Medicinal Chemistry*, **43**, 2906-2914. http://dx.doi.org/10.1021/jm000028l

[9]   Zhou, X., Zhang, L., Tseng, E., Scott-Ramsay, E., Schentag, J.J., Coburn, R.A. and Morris, M.E. (2005) New 4-Aryl-
      1,4-Dihydropyridines and 4-Arylpyridines as p-Glycoprotein Inhibitors. *Drug Metabolism and Disposition*, **33**, 321-
      328. http://dx.doi.org/10.1124/dmd.104.002089

[10] Karade, N.N., Gampawar, S.V., Kondre, J.M. and Shinde, S.V. (2008) An Efficient Combination of Dess-Martin Peri-
      odinane with Molecular Iodine and KBr for the Facile Oxidative Aromatization of Hantzsch 1,4-Dihydropyridines. *Ar-
      kivoc*, **xii**, 9-16. http://dx.doi.org/10.3998/ark.5550190.0009.c02

[11] Cheng, D.P. and Chen, Z.C. (2002) Hypervalent Iodine in Synthesis. 76. An Efficient Oxidation of 1,4-Dihydropyridines
      to Pyridines Using Iodobenzene Diacetate. *Synthetic Communications*, **32**, 793-798.
      http://dx.doi.org/10.1081/SCC-120002521

[12] Kumar, P. (2009) Solid State Oxidative Aromatization of Hantzsch 1,4-Dihydropyridines to Pyridines Using Iodoben-
      zene Diacetate or Hydroxy(tosyloxy)iodobenzene. *Chinese Journal of Chemistry*, **27**, 1487-1491.
      http://dx.doi.org/10.1002/cjoc.200990250

[13] Kumar, P. (2010) A Novel, Facile, Simple and Convenient Oxidative Aromatization of Hantzsch 1,4-Dihydropyridines to
      Pyridines Using Polymeric Iodosobenzene with KBr. *Journal of Heterocyclic Chemistry*, **47**, 1429-1433.

[14] Lee, K.H. and Ko, K.Y. (2002) Aromatization of Hantzsch 1,4-Dihydropyridines with [Hydroxy(tosyloxy)iodo] Ben-
      zene. *Bulletin of the Korean Chemical Society*, **23**, 1505-1506.

[15] Lee, J.W. and Ko, K.Y. (2004) Aromatization of Hantzsch 1,4-Dihydropyridines with a Polymer-Supported Hyperva-
      lent Iodine Reagent. *Bulletin of the Korean Chemical Society*, **25**, 19-20.

[16] Varma, R.S. and Kumar, D. (1999) Solid State Oxidation of 1,4-Dihydropyridines to Pyridines Using Phenyli-
      odine(III)Bis(trifluoroacetate) or Elemental Sulfur. *Journal of the Chemical Society*, *Perkin Transactions 1*, **24**,
      1755-1757.

[17] Varvoglis, A. (1997) Chemical Transformations Induced by Hypervalent Iodine Reagents. *Tetrahedron*, **53**, 1179-1255.
      http://dx.doi.org/10.1016/S0040-4020(96)00970-2

[18] Wirth, T., Chiai, M., Zhdankin, V.V., Koser, G.F., Tohma, H. and Kita, Y. (2003) Topics in Current Chemistry. Vol.
      224, Springer, Berlin.

[19] Zhdankin, V.V. and Stang, P.J. (2008) Chemistry of Polyvalent Iodine. *Chemical Reviews*, **108**, 5299-5358.
      http://dx.doi.org/10.1021/cr800332c

[20] Böcker, R.H. and Guengerich, F.P. (1986) Oxidation of 4-Aryl- and 4-Alkyl-Substituted 2,6-Dimethyl-3,5-bis(alkoxy-

carbonyl)-1,4-dihydropyridines by Human Liver Microsomes and Immunochemical Evidence for the Involvement of a Form of Cytochrome P-450. *Journal of Chemical Education*, **29**, 1596-1603. http://dx.doi.org/10.1021/jm00159a007

[21] Rajput, A.P. and Rajput, S.S. (2011) A Novel Method for the Synthesis of Formyl Pyrazoles Using Vilsmeier-Haack Reaction. *International Journal of Pharmacy and Pharmaceutical Sciences*, **3**, 346-351.

[22] Kumar, P., Kumar, A. and Hussain, K. (2012) Iodobenzene Diacetate (IBD) Catalyzed an Quick Oxidative Aromatization of Hantzsch-1,4-dihydropyridines to Pyridines under Ultrasonic Irradiation. *Ultrasonics Sonochemistry*, **19**, 729-735. http://dx.doi.org/10.1016/j.ultsonch.2011.12.021

7

Synthesis and Characterization of New Chiral Monoanionic [ON] Ancillary Phenolate Ligands

Pascal Binda[1,2]*, Leslie Glover[2]

[1]Department of Chemistry and Forensic Science, Savannah State University, Savannah, GA, USA
[2]Science Division, Southern Wesleyan University, Central, SC, USA
Email: *bindap@savannahstate.edu

Abstract

Three new chiral monoanionic [ON] ancillary phenolate ligands with varying pendant arms have been synthesized in moderate to high yields (50% - 85%) *via* Mannich-type condensation reaction of chiral substituted phenol, formaldehyde and (+)-bis-[(*R*)-1-phenylethyl]amine. These new organic compounds were fully characterized via nuclear magnetic resonance spectroscopy (^{1}H and ^{13}C) and elemental analysis. The newly synthesized ligands are suitable candidates for metal-catalyzed ring-opening of lactones and asymmetric catalysis.

Keywords

Phenolate Ligands, Mannich Condensation Reaction, Ring-Opening Polymerization of Lactones

1. Introduction

Ligand design has been of great importance in asymmetric catalysis and ring opening polymerization of lactones due to the fact that the reactivity and selectivity of metal catalysts are largely determined by the ancillary ligands [1]-[7]. One of the challenges is to derive efficient chiral catalysts for asymmetric induction in different substrates with subtle variations. Since it is not expected that a single catalyst will work for a wide range of substrates, an efficient strategy towards new catalysts would be the design of a search pathway that provides access to a large number of structurally similar ligands with tunable yet diverse substituents [8]. Indeed, many researchers have sought to develop asymmetric catalysts by screening a large pool of chiral ligands [9]-[12].

---

*Corresponding author.

Aminophenolate ligands have received great attention in metal-catalyzed ring-opening polymerization (ROP) of lactones due to the potential to fine tune the steric and electronic properties by varying the substituent groups and pendant side-arms, as well as their inexpensive synthetic strategies [13]-[22]. In fact, well-defined metal complexes of phenolate as ancillary ligands have been studied intensively to investigate the electronic and steric properties of the central metal and their effects in ROP of cyclic esters [13]-[46]. However, given their wide-spread application, it is somewhat surprising that the chiral variants of aminophenolate ligands are relatively lacking in the literature. The introduction of an aromatic ring in the pendant side arm, the resonance in the backbone is attenuated in comparison with the regular ligands and this may offer some unique opportunities for electronic differentiation and stereocontrol upon coordination [47]. Of particular interest are ligands with chiral pendant substituents that are in close proximity with the open coordination site at the periphery where catalysis occurs. Discussed herein is the synthesis of new chiral [ON] aminophenolate ligands as potential ancillary ligands in asymmetric catalysis and ring-opening polymerization of lactones.

2. Experimental

2.1. General

Deuterated solvents were purchased from Cambridge Isotope Laboratory and used as received. 2,4-di-tert-butylphenol, 2,4-dimethylphenol, 2,4-di-tert-pentylphenol, 37 wt% formaldehyde, and N-methylbenzylamine were purchased from Acros Organic and used as received. 2-tert-butyl-4-methylphenol and (+)-bis-[(R)-1-phenylethyl] amine were purchased from Aldrich while 4-tert-butyl-2-methylphenol was purchased from Fluka and used as received. All ^{1}H and ^{13}C NMR spectra were recorded on a JEOL ECX-300 MHz NMR spectrometer and referenced to CDCl$_3$. Elemental analyses were performed by Midwest Microlab, Indianapolis, IN. Melting points were obtained on a Mel-Temp apparatus and are uncorrected.

2.2. Synthesis of Ligands

2.2.1. HL^a

2,4-Di-tert-butylphenol (1.834 g, 8.87 mmol), 37 wt% formaldehyde (0.266 g, 8.87 mmol), and (+)-bis-[(R)-1-phenylethyl] amine (2.000 g, 8.87 mmol) were dissolved in ethanol (13 mL). The resulting solution was heated at reflux for 18 h and then cooled to room temperature. Solvent and water were removed using high vacuum Schlenk line to obtain pale yellow oily compound, which was purified by column chromatography (5% ethyl acetate and 95% hexane). (2.198 g, 55.9%). Elemental analysis: (Found: C 83.22, H 9.03, N 3.38. C$_{31}$H$_{41}$NO requires C 83.922, H 9.315, N 3.157. ^{1}H NMR (300 MHz; CDCl$_3$; 298 K) 1.30 (s, 9H, Ar^{t}Bu), 1.37 (d, 3H, J = 6.87 Hz, ArCH(Me)N), 1.44 (s, 9H, Ar^{t}Bu), 1.52 (d, 3H, J = 6.87 Hz, ArCH(Me)N), 3.71 (d, 1H, J = 14.76 Hz, ArCH$_2$N), 4.12 (q, 1H, J = 6.87 Hz, ArCH(Me)N), 4.29 (d, 1H, J = 14.76 Hz, ArCH$_2$N), 4.84 (br, 1H, ArCH(Me)N), 6.64 (d, 1H, J = 8.22 Hz, ArH), 7.12 (d, 1H, J = 8.22 Hz, ArH), 7.30 - 7.40 (br, 10H, ArH), 11.12 (s, 1H, ArOH). ^{13}C{H} NMR (75 MHz; CDCl$_3$; 298 K) 24.9 (ArCH(Me)N), 29.7 (ArCMe$_3$), 31.8 (ArCMe$_3$), 34.7 (ArCMe$_3$), 34.9 (ArCMe$_3$), 51.3 (ArCH(Me)N), 53.3 (ArCH$_2$N), 116.0, 123.6, 126.9, 127.1, 128.3, 128.5, 128.6, 142.9, 152.1 (all ArC).

2.2.2. HL^b

2-Tert-butyl-4-methylphenol (2.189 g, 13.31 mmol), 37 wt% formaldehyde (0.400 g, 13.31 mmol), and (+)-bis-[(R)-1-phenylethyl] amine (3.000 g, 13.31 mmol) were dissolved in ethanol (13 mL). The resulting solution was heated at reflux for 18 h and then cooled to room temperature. Solvent and water were removed using high vacuum Schlenk line to obtain pale yellow oily compound, which was dried at 70°C.(2.872 g, 80.6%). Elemental analysis: (Found: C 83.63, H 8.77, N 3.42. C$_{28}$H$_{35}$NO requires C 83.74, H 8.785, N 3.49. ^{1}H NMR (300 MHz; CDCl$_3$; 298 K) 1.34 (d, 6H, J = 6.87 Hz, ArCH(Me)N), 1.47 (s, 9H, Ar^{t}Bu) 2.32 (s, 3H, ArMe), 3.58 (q, 2H, J = 6.87 Hz, ArCH(Me)N), 4.09 (br, 2H, ArCH(Me)N), 6.62 (d, 1H, J = 7.92 Hz, ArH), 6.90 (d, 1H, J = 7.92 Hz, ArH), 7.26 - 7.38 (br, 10H, ArH), 11.05 (s, 1H, ArOH). ^{13}C{H} NMR (75 MHz; CDCl$_3$; 298 K) 16.7 (ArCMe$_3$), 24.9 (ArCMe$_3$), 31.9 (ArCH(Me)N), 34.2 (ArCH(Me)N), 55.4 (ArCH$_2$N), 114.9, 123.8, 124.1, 127.1, 127.3, 128.3, 128.8, 143.1, 145.2, 152.2 (all ArC).

2.2.3. HL^c

4-Tert-butyl-2-methylphenol (1.510 g, 8.87 mmol), 37 wt% formaldehyde (0.266 g, 8.87 mmol), and (+)-bis-

[(R)-1-phenylethyl]amine (2.000 g, 8.87 mmol) were dissolved in ethanol (13 mL). The resulting solution was heated at reflux for 18 h and then cooled to room temperature. Solvent and water were removed using high vacuum Schlenk line to obtain pale yellow oily compound, which was dried at 70°C.(2.133 g, 59.9%). Elemental analysis: (Found: C 83.47, H 8.95, N 3.68. $C_{28}H_{35}NO$ requires C 83.74, H 8.785, N 3.49. ^1H NMR (300 MHz; CDCl$_3$; 298 K) 1.41-1.46 (s, 15H, Ar^{t}Bu, ArCH(Me)N), 2.44 (s, 3H, ArMe), 3.67 (q, 2H, J = 6.87 Hz, ArCH(Me)N), 4.36 (br, 2H, ArCH(Me)N), 6.84 (d, 1H, J = 8.25 Hz, ArH), 7.20 (d, 1H, J = 8.25 Hz, ArH), 7.33 - 0.46 (br, 10H, ArH), 10.52 (s, 1H, ArOH). $^{13}C\{H\}$ NMR (75 MHz; CDCl$_3$; 298 K) 16.7 (ArCMe$_3$), 24.9 (ArCMe$_3$), 31.9 (ArCH(Me)N), 34.2 (ArCH(Me)N), 55.4 (ArCH$_2$N), 114.9, 123.8, 124.1, 127.1, 127.3, 128.3, 128.8, 143.1, 145.2, 152.2 (all ArC).

2.2.4. HL^d

2,4-Di-tert-butylphenol (3.522 g, 17.069 mmol), 37 wt% formaldehyde (0.512 g, 17.069 mmol), and N-methylbenzylamine (2.068 g, 17.069 mmol) were dissolved in ethanol (30 mL). The resulting solution was heated at reflux for 18 h and then cooled to room temperature. Crystallization from the saturated ethanol solution at room temperature yielded white solid, which was dried under high vacuum at 70°C (4.510 g, 77.8%). Mp: 132.6°C - 132.9°C. Elemental analysis: (Found: C 80.97, H 9.67, N 4.15. $C_{23}H_{33}NO$ requires C 81.37, H 9.80, N 4.13%). ^1H NMR (500 MHz; CDCl$_3$; 298 K) 1.43 (s, 9H, Ar^{t}Bu), 1.64 (s, 9H, Ar^{t}Bu), 2.29 (s, 3H, ArCH$_2$NMe), 3.61 (br, 2H, ArCH$_2$NMe), 3.78 (br, 2H, ArCH$_2$NMe), 6.92 (s, 1H, ArH), 7.29–7.40 (br, 6H, ArH), 11.17 (br, 1H, ArOH). $^{13}C\{H\}$ NMR (125 MHz; CDCl$_3$; 298 K) 29.7 (ArCMe$_3$), 31.4 (ArCMe$_3$), 34.2 (ArCMe$_3$), 34.4 (ArCMe$_3$), 41.2 (ArCH$_2$NMe), 60.9 (ArCH$_2$NMe), 62.1 (ArCH$_2$NMe), 121.4, 122.9, 123.4, 1275, 128.5, 129.5, 135.7, 137.3, 140.6, 154.3, 176.6 (all ArC).

2.2.5. HL^e

2,4-Di-tert-pentylphenol (2.000 g, 8.535 mmol), 37 wt% formaldehyde (0.256 g, 8.535 mmol), and N-methylbenzylamine (1.034 g, 8.535 mmol) were dissolved in ethanol (30 mL). The resulting solution was heated at reflux for 18 h and then cooled to room temperature. Crystallization from the saturated ethanol solution at −10°C (freezer) yielded white solid, which was dried under high vacuum at 50°C (2.490 g, 79.4%). Mp: 73.9°C - 74.1°C. Elemental analysis: (Found: C 81.61, H 9.99, N 3.90. $C_{25}H_{37}NO$ requires C 81.69, H 10.15, N 3.81%). ^1H NMR (300 MHz; CDCl$_3$; 298 K) ^1H NMR (500 MHz; CDCl$_3$; 298 K) 0.74 (t, 3H x 2, J = 7.50 Hz, ArCMe$_2$CH$_2$Me), 1.31 (s, 6H, ArCMe$_2$CH$_2$Me), 1.47 (s, 6H, ArCMe$_2$CH$_2$Me), 1.65 (q, 2H, J = 7.50 Hz, ArCMe$_2$CH$_2$Me), 2.03 (q, 2H, J = 7.50 Hz, ArCMe$_2$CH$_2$Me), 2.27 (s, 3H, ArCH$_2$NMe), 3.57 (br, 2H, ArCH$_2$NMe), 3.79 (br, 2H, ArCH$_2$NMe), 6.86 (s, 1H, ArH), 7.16 (s, 1H, ArH), 7.31 - 7.41 (br, 5H, ArH), 11.04 (br, 1H, ArOH). $^{13}C\{H\}$ NMR (125 MHz; CDCl$_3$; 298 K) 9.2 (ArCMe$_2$CH$_2$Me), 9.6 (ArCMe$_2$CH$_2$Me), 27.7 (ArCMe$_2$CH$_2$Me), 28.6 (ArCMe$_2$CH$_2$Me), 33.0 (ArCMe$_2$CH$_2$Me), 37.3 (ArCMe$_2$CH$_2$Me), 38.5 (ArCMe$_2$CH$_2$Me), 41.1 (ArCH$_2$NMe), 60.8 (ArCH$_2$NMe), 62.1(ArCH$_2$NMe), 121.1, 124.1, 125.1, 127.5, 128.4, 129.5, 133.9, 137.6, 138.7 154.0 (all ArC).

2.2.6. HL^f

2,4-Di-methylphenol (10.421 g, 85.3 mmol), 37 wt% formaldehyde (2.56 g, 85.3 mmol), and N-methylbenzylamine (10.342 g, 85.3 mmol) were dissolved in methanol (20 mL). The resulting solution was heated at reflux for 18 h and then cooled to room temperature. Solvent and water were removed using high vacuum Schlenk line to obtain pale yellow oily solid. Recrystallization from ethanol at −10°C (freezer) yielded off white solid, which was dried under high vacuum at room temperature. (20.45 g, 96.5%).Mp: 33.5°C - 33.6°C. Elemental analysis: (Found: C 80.01, H 8.13, N 5.50. $C_{17}H_{21}NO$ requires C 79.960, H 8.289, N 5.485. ^1H NMR (300 MHz; CDCl$_3$; 298 K) 2.22 (s, 3H, ArMe), 2.23 (s, 3H, ArMe), 2.24 (s, 3H, ArCH$_2$NMe), 3.61 (s, 2H, ArCH$_2$NMe), 3.71 (s, 2H, ArCH$_2$NMe), 6.67 (s, 1H, ArH), 6.88 (s, 1H, ArH), 7.29 - 7.37 (br, 5H, ArH), 10.97 (br, 1H, ArOH). $^{13}C\{H\}$ NMR (125 MHz; CDCl$_3$; 298 K) 15.8 (ArMe), 20.6 (ArMe), 41.2 (ArCH$_2$NMe), 61.1 (ArCH$_2$NMe), 61.7 (ArCH$_2$NMe), 120.9, 124.8, 126.8, 127.7, 128.9, 129.5, 130.7, 137.0, 153.6 (all ArC).

3. Result and Discussion

The chiral ligands HL^a, HL^b and HL^c (**Figure 1**) were synthesized *via* Mannich condensation reactions using inexpensive substituted phenols, formaldehyde and (+)-bis-[(R)-1-phenylethyl] amine in refluxing ethanol (**Scheme 1**).

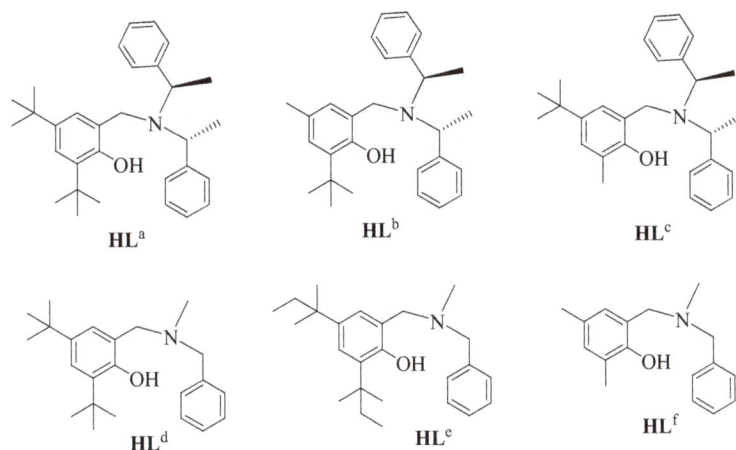

Figure 1. New multidentate ancillary phenolate ligands.

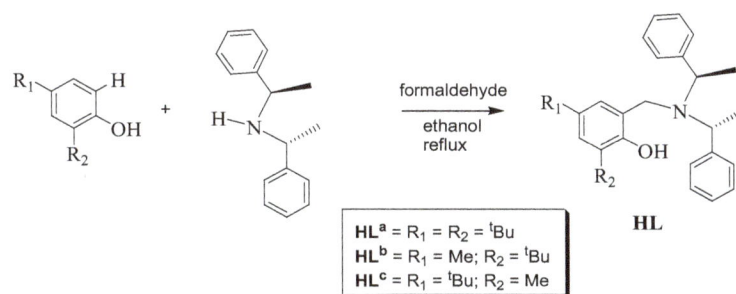

Scheme 1. Synthesis of chiral phenolate ligands *via* Mannich condensation reactions.

These compounds were purified by column chromatography to obtain oily products in moderate to high yields. The corresponding products were then characterized using NMR and elemental analysis to ascertain the structures. The use of different phenolic substituents (methy, butyl and pentyl) will provide a library of compounds suitable for metal catalytic investigations in organic functional group transformations and polymerization reactions.

Meanwhile, the non-chiral ligands HL[d], HL[e] and HL[f] (**Figure 1**) were synthesized in a similar manner using substituted phenols, formaldehyde and N-methylbenzylamine in refluxing ethanol. These compounds were purified by recrystallization to obtain white solids in high yields. The solids were also characterized using NMR and elemental analysis. Yields of the non-chiral ligands were generally higher (77% - 96%), presumably due to the use of a less bulky N-methylbenzylamine compared to the bulky (+)-bis-[(R)-1-phenylethyl] amine and also due to easier purification methods.

Deprotonation of the ligands and attachment to zinc, tin and palladium metals would offer new research opportunities in asymmetric synthesis and metal catalyzed ring-opening polymerization of lactones. There is great interest in investigating the effect of one stereogenic center in conjunction with phenolic bulky substituent on catalytic selectivity. These ligands are expected to be bidentate with the possibility of having a tridentate coordination *via* the phenyl pendant arms.

4. Conclusion

New chiral monoanionic [ON] ancillary phenolate ligands with varying pendant arms have been synthesized and characterized *via* nuclear magnetic resonance spectroscopy (^{1}H and ^{13}C) and elemental analysis. The synthesized ligands are suitable candidates for applications in asymmetric catalysis and ring-opening polymerization of lactones.

Acknowledgements

The authors are grateful to Duke Energy Advance SC for research funding.

References

[1]   Walsh, P.J. and Kozlowski, M.C. (2008) Fundamentals in Asymmetric Catalysis. USB, Sausalito.
      http://dx.doi.org/10.1007/978-3-642-58571-5

[2]   Jacobsen, E.N., Pfaltz, A. and Yamamoto, H. (1999) Comprehensive Asymmetric Catalysis. Springer, Heidelberg.

[3]   Martin, R., Buchwald, S.L. (2008) Palladium-Catalyzed Suzuki-Miyaura Cross-Coupling Reactions Employing Dial-
      kylbiaryl Phosphine Ligands. *Accounts of Chemical Research*, **41**, 1461-1473. http://dx.doi.org/10.1021/ar800036s

[4]   Heitbaum, M., Glorius, F. and Escher, I. (2006) Asymmetric Heterogeneous Catalysis. *Angewandte Chemie Interna-
      tional Edition*, **45**, 4732-4762. http://dx.doi.org/10.1002/anie.200504212

[5]   Miura, M. (2004) Rational Ligand Design in Constructing Efficient Catalyst Systems for Suzuki-Miyaura Coupling.
      Angewandte Chemie International Edition, **43**, 2201-2203. http://dx.doi.org/10.1002/anie.200301753

[6]   Katz, B.A., Cass, R.T., Liu, B., Arze, R. and Collins, N. (1995) Topochemical Catalysis Achieved by Structure-Based
      Ligand Design. *The Journal of Biological Chemistry*, **270**, 31210-31218.

[7]   Falciola, C.A. and Alexakis, A. (2008) Copper-Catalyzed Asymmetric Allylic Alkylation. *European Journal of Or-
      ganic Chemistry*, 3765-3780. http://dx.doi.org/10.1002/ejoc.200800025

[8]   Hoveyda, A.H., Hird, A.W. and Kacprzynski, M.A. (2004) Small Peptides as Ligands for Catalytic Asymmetric Alky-
      lations of Olefins. *Chemical Communications*, 1779-1883. http://dx.doi.org/10.1039/b401123f

[9]   Lu, Y., Johnstone, T.C. and Arndtsen, B.A. (2009) Hydrogen-Bonding Asymmetric Metal Catalysis with α-Amino
      Acids: A Simple and Tunable Approach to High Enantioinduction. *Journal of the American Chemical Society*, **131**,
      11284-11285. http://dx.doi.org/10.1021/ja904185b

[10] Shi, B.-F., Maugel, N., Zhang, Y.-H. and Yu, J.-Q. (2008) Pd^{II}-Catalyzed Enantioselective Activation of C(sp^2)-H and
      C(sp^3)-H Bonds Using Monoprotected Amino Acids as Chiral Ligands. *Angewandte Chemie International Edition*, **47**,
      4882-4886. http://dx.doi.org/10.1002/anie.200801030

[11] Blaser, H.-U., Pugin, B., Spindler, F. and Thommen, M. (2007) From a Chiral Switch to a Ligand Portfolio for Asym-
      metric Catalysis. *Accounts of Chemical Research*, **40**, 1240-1250. http://dx.doi.org/10.1021/ar7001057

[12] Cesar, V., Bellemin-Laponnaz, S., Wadepohl, H. and Gade, L.H. (2005) Designing the "Search Pathway" in the De-
      velopment of a New Class of Highly Efficient Stereoselective Hydrosilylation Catalysts. *Chemistry—A European
      Journal*, **11**, 2862-2873. http://dx.doi.org/10.1002/chem.200500132

[13] Delbridge, E.E., Dugah, D.T., Nelson, C.R., Skelton, B.W. and White, A.H. (2007) Synthesis, Structure and Oxidation
      of New Ytterbium(II) Bis(phenolate) Compounds and Their Catalytic Activity Towards Epsilon-Caprolactone. *Dalton
      Transactions*, No. 1, 143-153.

[14] Binda, P.I. and Delbridge, E.E. (2007) Synthesis and Characterisation of Lanthanide Phenolate Compounds and Their
      Catalytic Activity towards Ring-Opening Polymerisation of Cyclic Esters. *Dalton Transactions*, **41**, 4685-4692.
      http://dx.doi.org/10.1039/b710070a

[15] Kerton, F.M., Whitwood, A.C. and Willans, C.E. (2004) A High-Throughput Approach to Lanthanide Complexes and
      Their Rapid Screening in the Ring Opening Polymerisation of Caprolactone. *Dalton Transactions*, No. 15, 2237-2244.
      http://dx.doi.org/10.1039/b406841f

[16] Binda, P.I., Delbridge, E.E., Abrahamson, H.B. and Skelton, B.W. (2009) Coordination of Substitutionally Inert Phe-
      nolate Ligands to Lanthanide(II) and (III) Compounds—Catalysts for Ring-Opening Polymerization of Cyclic Esters.
      Dalton Transactions, No. 15, 2777-2787. http://dx.doi.org/10.1039/b821770j

[17] Dugah, D.T., Skelton, B.W. and Delbridge, E.E. (2009) Synthesis and Characterization of New Divalent Lanthanide
      Complexes Supported by Amine Bis(Phenolate) Ligands and Their Applications in the Ring Opening Polymerization
      of Cyclic Esters Dalton Transactions, 1436-1445. http://dx.doi.org/10.1039/b816916k

[18] Xu, X., Ma, M., Yao, Y., Zhang, Y. and Shen, Q. (2005) Synthesis, Characterisation of Carbon-Bridged (Diphenolato)
      Lanthanide Complexes and Their Catalytic Activity for Diels-Alder Reactions. *European Journal of Inorganic Chemi-
      stry*, **2005**, 676-684. http://dx.doi.org/10.1002/ejic.200400519

[19] Lendlein, A., Schmidt, A.M., Schroeter, M. and Langer, R. (2005) Shape-Memory Polymer Networks from Oligo
      (ϵ-Caprolactone) Dimethacrylates. *Journal of Polymer Science Part A: Polymer Chemistry*, **43**, 1369-1381.
      http://dx.doi.org/10.1002/pola.20598

[20] Amgoune, A., Thomas, C.M., Roisnel, T. and Carpentier, J.F. (2005) Ring-Opening Polymerization of Lactide with
      Group 3 Metal Complexes Supported by Dianionic Alkoxy-Amino-Bisphenolate Ligands: Combining High Activity,
      Productivity and Selectivity. *Chemistry—A European Journal*, **12**, 169-179.
      http://dx.doi.org/10.1002/chem.200500856

[21] Guo, H., Zhou, H., Yao, Y., Zhang, Y. and Shen, Q. (2007) Synthesis and Structural Characterization of Novel Mixed-

Valent Samarium and Divalent Ytterbium and Europium Complexes Supported by Amine Bis(phenolate) Ligand. *Dalton Transactions*, No. 32, 3555-3561. http://dx.doi.org/10.1039/b705353c

[22] Zats, G.M., Arora, H., Lavi, R., Yufit, D. and Benisvy, L. (2011) Phenolate and Phenoxyl Radical Complexes of Cu(II) and Co(III), Bearing a New Redox Active N,O-Phenol-Pyrazole Ligand. *Dalton Transactions*, **40**, 10889-10896. http://dx.doi.org/10.1039/c1dt10615e

[23] Radano, C.P., Baker, G.L. and Smith, M.R. (2000) Stereoselective Polymerization of a Racemic Monomer with a Racemic Catalyst: Direct Preparation of the Polylactic Acid Stereocomplex from Racemic Lactide. *Journal of the American Chemical Society*, **122**, 1552-1553. http://dx.doi.org/10.1021/ja9930519

[24] Darensbourg, D.J., Choi, W., Karroonnirum, O. and Bhuvanesh, N. (2008) Ring-Opening Polymerization of Cyclic Monomers by Complexes Derived from Biocompatible Metals. Production of Poly(lactide), Poly(trimethylene carbonate) and Their Copolymers. *Macromolecules*, **41**, 3493-3502. http://dx.doi.org/10.1021/ma800078t

[25] Ovitt, T.M. and Coates, G.W. (2002) Stereochemistry of Lactide Polymerization with Chiral Catalysts: New Opportunities for Stereocontrol Using Polymer Exchange Mechanisms. *Journal of the American Chemical Society*, **124**, 1316-1326. http://dx.doi.org/10.1021/ja012052+

[26] Kasperczyk, J.E. (1995) Microstructure Analysis of Poly(lactic acid) Obtained by Lithium Tert-Butoxide as Initiator. *Macromolecules*, **28**, 3937-3939. http://dx.doi.org/10.1021/ma00115a028

[27] Radano, C.P., Baker, G.L. and Smith III, M.R. (2000) Stereoselective Polymerization of a Racemic Monomer with a Racemic Catalyst: Direct Preparation of the Polylactic Acid Stereocomplex from Racemic Lactide. *Journal of the American Chemical Society*, **12**, 1552-1553. http://dx.doi.org/10.1021/ja9930519

[28] Thakur, K.A.M., Kean, R.T., Zell, M.T., Padden, B.E. and Munson, E.J. (1998) An Alternative Interpretation of the HETCOR NMR Spectra of Poly(lactide). *Chemical Communications*, No. 17, 1913-1914. http://dx.doi.org/10.1039/a708911b

[29] Thakur, K.A.M., Kean, R.T., Hall, E.S., Kolstad, J.J. and Lindgren, T.A. (1997) High-Resolution ^{13}C and ^{1}H Solution NMR Study of Poly(lactide). *Macromolecules*, **30**, 2422-2428. http://dx.doi.org/10.1021/ma9615967

[30] Zell, M.T., Padden, B.E., Paterick, A.J., Thakur, K.A.M., Kean, R.T., Hillmyer, M.A. and Munson, E.J. (2002) Unambiguous Determination of the ^{13}C and ^{1}H NMR Stereosequence Assignments of Polylactide Using High-Resolution Solution NMR Spectroscopy. *Macromolecules*, **35**, 7700-7707. http://dx.doi.org/10.1021/ma0204148

[31] Ovitt, T.M. and Coates, G.W. (2000) Stereoselective Ring-Opening Polymerization of Rac-Lactide with a Single-Site, Racemic Aluminum Alkoxide Catalyst: Synthesis of Stereoblock Poly(lactic acid), *Polymer Chemistry*, **38**, 4686-4692. http://dx.doi.org/10.1002/1099-0518(200012)38:1+<4686::AID-POLA80>3.0.CO;2-0

[32] Cheng, M., Attygalle, A.B., Lobkovsky, E.B. and Coates, G.W. (1999) Single-Site Catalysts for Ring-Opening Polymerization: Synthesis of Heterotactic Poly(lactic acid) from *Rac*-Lactide. *Journal of the American Chemical Society*, **121**, 11583-11584. http://dx.doi.org/10.1021/ja992678o

[33] Ovitt, T.M and Coates, G.W. (1999) Stereoselective Ring-Opening Polymerization of *meso*-Lactide: Synthesis of Syndiotactic Poly(lactic acid). *Journal of the American Chemical Society*, **121**, 4072-4073. http://dx.doi.org/10.1021/ja990088k

[34] Chisholm, M.H., Iyer, S.S., McCollum, D.G., Pagel, M. and Werner-Zwanziger, U. (1999) Microstructure of Poly(lactide). Phase-Sensitive HETCOR Spectra of Poly(*meso*-lactide), Poly(*rac*-lactide) and Atactic Poly(lactide). *Macromolecules*, **32**, 963-973. http://dx.doi.org/10.1021/ma9806864

[35] Chisholm, M.H., Gallucci, J.C. and Phomphrai, K. (2005) Comparative Study of the Coordination Chemistry and Lactide Polymerization of Alkoxide and Amide Complexes of Zinc and Magnesium with a β-Diiminato Ligand Bearing Ether Substituents. *Inorganic Chemistry*, **44**, 8004-8010. http://dx.doi.org/10.1021/ic048363d

[36] Zhong, Z., Dijkstra, P.J. and Feijen, J. (2003) Controlled and Stereoselective Polymerization of Lactide: Kinetics, Selectivity, and Microstructures. *Journal of the American Chemical Society*, **125**, 11291-11298. http://dx.doi.org/10.1021/ja0347585

[37] Cai, C.-X., Amgoune, A., Lehmann, C.W. and Carpentier, J.-F. (2004) Stereoselective Ring-Opening Polymerization of Racemic Lactide Using Alkoxy-Amino-Bis(phenolate) Group 3 Metal Complexes. *Chemical Communications*, No. 3, 330-331. http://dx.doi.org/10.1039/b314030j

[38] Coates, G.W. (2000) Precise Control of Polyolefin Stereochemistry Using Single-Site Metal Catalysts. *Chemical Reviews*, **100**, 1223-1252. http://dx.doi.org/10.1021/cr990286u

[39] Nimitsiriwat, N., Marshall, E.L., Gibson, V.C., Elsegood, M.R.J. and Dale, S.H. (2004) Unprecedented Reversible Migration of Amide to Schiff Base Ligands Attached to Tin: Latent Single-Site Initiators for Lactide Polymerization. *Journal of the American Chemical Society*, **126**, 13598-13599. http://dx.doi.org/10.1021/ja0470315

[40] Chisholm, M.H., Eilerts, N.W., Huffman, J.C., Iyer, S.S., Pacold, M. and Phomphrai, K. (2000) Molecular Design of Single-Site Metal Alkoxide Catalyst Precursors for Ring-Opening Polymerization Reactions Leading to Polyoxyge-

nates. 1. Polylactide Formation by Achiral and Chiral Magnesium and Zinc Alkoxides, (η^3-L)MOR, Where L=Trispyra-zolyl- and Trisindazolylborate Ligands. *Journal of the American Chemical Society*, **122**, 11845-11854. http://dx.doi.org/10.1021/ja002160g

[41] Cai, C.X., Toupet, L., Lehmann, C.W. and Carpentier, J.-F. (2003) Synthesis, Structure and Reactivity of New Yttrium Bis(dimethylsilyl) Amido and Bis(trimethylsilyl) Methyl Complexes of a Tetradentate Bis(phenoxide) Ligand. *Journal of Organometallic Chemistry*, **683**, 131-136. http://dx.doi.org/10.1016/S0022-328X(03)00513-8

[42] Chamberlain, B.M., Cheng, M., Moore, D.R., Ovitt, T.M., Lobkovsky, E.B. and Coates, G.W. (2001) Polymerization of Lactide with Zinc and Magnesium β-Diiminate Complexes: Stereocontrol and Mechanism. *Journal of the American Chemical Society*, **123**, 3229-3238. http://dx.doi.org/10.1021/ja003851f

[43] Kowalski, A., Duda, A. and Penczek, S. (1998) Polymerization of L,L-Lactide Initiated by Aluminum Isopropoxide Trimer or Tetramer. *Macromolecules*, **31**, 2114-2122. http://dx.doi.org/10.1021/ma971737k

[44] Dove, A.P., Gibson, V.C., Marshall, E.L., White, A.J.P. and Williams, D. (2001) A Well Defined Tin(II) Initiator for the Living Polymerisation of Lactide. *Chemical Communications*, No. 3, 283-284. http://dx.doi.org/10.1039/b008770j

[45] Grunova, E., Kirillov, E., Roisnel, T. and Carpentier, J.-F. (2008) Group 3 Metal Complexes of Salen-Like Fluorous-Dialkoxy-Diimino Ligands: Synthesis, Structure and Application in Ring-Opening Polymerization of *rac*-Lactide and *rac*-β-Butyrolactone. *Organometallics*, **27**, 5691-5698. http://dx.doi.org/10.1021/om800611c

[46] Binda, P.I., Delbridge, E.E., Dugah, D.T., Skelton, B.W. and White, A.H. (2008) Synthesis and Structural Characterization of Some Potassium Complexes of Some Bis(phenolate) Ligands and Some Novel Heterobimetallic Binuclear Arrays Formed with Trivalent Lanthanoid Ions. *Zeitschrift für anorganische und allgemeine Chemie*, **634**, 325-334. http://dx.doi.org/10.1002/zaac.200700442

[47] Cortright, S.B. and Johnston, J.N. (2002) IAN-Amines: Direct Entry to a Chiral C_2-Symmetric Zirconium(IV) β-Dike-timine Complex. *Angewandte Chemie International Edition*, **41**, 345-348. http://dx.doi.org/10.1002/1521-3773(20020118)41:2<345::AID-ANIE345>3.0.CO;2-U.

8

An Efficient Synthesis of 2,3-Diaminoacid Derivatives Using Phosphine Catalyst

Yohei Oe[1], Hiroaki Kishimoto[2], Nahoko Sugioka[2], Daisuke Harada[2], Yukio Sato[2], Tetsuo Ohta[1], Isao Furukawa[2]

[1]Department of Biomedical Information, Faculty of Life and Medical Sciences, Doshisha University, Kyoto, Japan
[2]Department of Molecular Science and Technology, Faculty of Engineering, Doshisha University, Kyoto, Japan
Email: yoe@mail.doshisha.ac.jp, tota@mail.doshisha.ac.jp

Abstract

Ethyl 2,3-diphthalimidoylpropanoate was effectively synthesized from ethyl propynoate with two equivalents of phthalimide catalyzed by triphenylphosphine in good yield. The choice of reaction media was important for selective synthesis of the desired 2,3-diaminocarboxylic acid derivatives. The reaction is considered to occur through a zwitterionic intermediate derived from the reaction of the α,β-unsaturated ester with triphenylphosphine.

Keywords

Addition, Alkynes, Amino Acids, Catalysis, Multi-Component Reaction

1. Introduction

2,3-Diaminoacids and their derivatives have attracted a great deal of attention due to their application as key structural fragments of biologically active compounds [1]-[7], and/or as ligands for metal complexes [8]-[13]. Various methods to prepare these compounds have been reported [1] [2] [4] [14]-[19].

While studying the PPh$_3$-catalyzed three-component coupling of ethyl propynoate, a nitrogen nucleophile, and an aldehyde [20], we found an efficient synthesis of 2,3-diaminoacid derivatives from ethyl propynoate and phthalimide catalyzed by triphenylphosphine. Further details on this synthesis are described in this report.

2. Results and Discussion

Reaction of ethyl propynoate (**1**) with phthalimide (**2**) in the presence of a stoichiometric amount of triphenylphosphine at room temperature gave ethyl 2,3-diphthalimidoylpropanoate (**3**) together with ethyl 2-phthalimi-doylpropenoate (**4**) (**Scheme 1**). The ratio of **3** and **4** depended on the solvent used (**Table 1**). **3** was produced when highly polar solvents, such as DMSO, DMF, acetonitrile, and ketone were used. In these cases, **3** was obtained at more than 70% with no formation of **4**. When 2-pyrrolidone was used, a mixture of **3** and **4** was obtained in 44 and 19% yields, respectively. However, the dehydroamino acid derivative **4** was mainly produced when a less polar solvent was used [21]. In esters, ethers, halogenated hydrocarbons, and aromatic solvents, **4** was obtained in moderate to good yields with no formation of **3**.

DMSO and 2-butanone were selected to examine the effect of reaction temperature on the reaction (**Table 2**). The reaction in DMSO at room temperature for a short reaction time gave mostly **4**. However, when the reaction time was increased, the yield of **4** began to decrease, while the yield of product **3** began to increase. Compound **3** was obtained in 97% yield at 100°C and 87% yield at room temperature (**Table 2**, entry 8 and 11). In the 2-butanone solvent, it was also found that higher temperatures gave a better yield of **3**.

The concentration of the substrates did not affect the yield of **3**; the reaction of 1 mmol of substrate **1**, 2 mmol of phthalimide (**2**), and 1 mmol of PPh₃ in 1, 2, or 5 mL of DMSO gave **3** in 81%, 87%, and 88% yields, respectively, which were not significantly different.

This reaction also proceeded under other phosphines (**Table 3**). For example, the reaction progressed efficiently, when triphenylphosphine, tri(*p*-fluorophenyl)phosphine, and tri(*p*-methoxyphenyl)phosphine were used. These phosphines were found to act as catalysts [22]. For example, the reaction at room temperature for 24 h using 10 mol% triphenylphosphine resulted in 82% yield of **3**. In contrast, the yield significantly decreased when tributylphosphine was used. Moreover, no reaction occurred when pyridine was used [23].

Other nucleophiles were also investigated. Trost *et al*. reported that **4** was formed in 95% yield from the reaction catalyzed by triphenylphosphine in a mixture of toluene and a buffer solution of acetic acid and sodium acetate at 105°C [24]. They found that using *p*-toluenesulfoamide as a substrate gave a trace amount of **3**. Di-

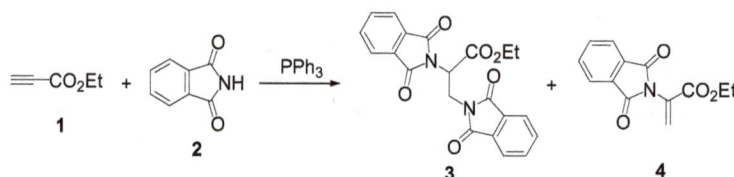

Scheme 1. Reaction of ethyl propynoate with phthalimide catalyzed by PPh₃.

Table 1. Effect of solvent[a].

Entry	Solvent	$\mu(D)^6$	ε_p/°C	Yield[b](%)	
				3	4
1	DMSO	4.30	48.9/20	87	-
2	DMF	3.86	36.71/20	81	-
3	CH₃CN	3.44	37.5/20	73	-
4	2-Butanone	-	-	84	-
5	Acetone	2.69	20.70/25	82	-
6	2-Pyrrolidone	2.30	-	44	19
7	Ethyl Benzoate	1.99	5.98/25	-	72
8	Ethyl Acetate	1.88	6.02/20	-	81
9	THF	1.70	7.58/25	-	82
10	CHCl₃	1.15	4.90/20	-	60
11	CH₂Cl₂	1.14	9.10/20	-	72
12	Diethyl Ether	1.12	4.20/27	-	55
13	1,4-Dioxane	0.45	2.70/25	-	94
14	Toluene	0.40	2.24/20	-	52
15	Benzene	0	2.28/20	-	58

[a]A mixture of ethyl propynoate (1.0 mmol), phthalimide (2.0 mmol), PPh₃ (1.0 mmol) was stirred in a solvent (2.0 mL) at room temperature for 24 h.
[b]Yields were determined by ¹H NMR using internal standard (bibenzyl) method.

Table 2. Effect of reaction time and temperature[a].

Entry	Solvent	Time (h)	Temp. (°C)	Yield[b](%)	
				3	4
1	DMSO	5 (min)	r.t.	4	21
2	DMSO	10 (min)	r.t.	36	32
3	DMSO	0.5	r.t.	66	14
4	DMSO	1	r.t.	76	7
5	DMSO	1.5	r.t.	84	4
6	DMSO	3	r.t.	86	-
7	DMSO	6	r.t.	86	-
8	DMSO	24	r.t.	87	-
9	DMSO	48	r.t.	84	-
10	DMSO	6	100	94	-
11	DMSO	24	100	97	-
12	DMSO	48	100	97	-
13	2-Butanone	24	r.t.	84	-
14	2-Butanone	24	80	89	-
15	2-Butanone	24	reflux	93	-

[a]A mixture of ethyl propynoate (1.0 mmol), phthalimide (2.0 mmol), PPh_3 (1.0 mmol) was stirred in a solvent (2.0 mL) at room temperature. [b]Yields were determined by ^1H NMR using internal standard (bibenzyl) method.

Table 3. Effect of phosphine[a].

Entry	Phosphine/1 (mol%)	Yield[b](%)	
		3	4
1	PPh_3 (10)	82	-
2	PPh_3 (30)	87	-
3	PPh_3 (50)	86	-
4	PPh_3 (100)	87	-
5	$P(p\text{-}C_6H_4F)_3$ (30)	78	-
6	$P(p\text{-}C_6H_4OMe)_3$ (30)	88	-
7	$PMePh_2$ (30)	67	-
8	Bu_3P (30)	14	12

[a]A mixture of ethyl propynoate (1.0 mmol), phthalimide (2.0 mmol), phosphine (0.1 - 1.0 mmol) was stirred in dimethylsulfoxide (2.0 mL) at room temperature for 24 h. [b]Yields were determined by ^1H NMR using internal standard (bibenzyl) method.

acetamide, maleimide, benzylamine, allylamine, N-methyl-p-toluenesulfoamide, dimethyl malonate, and acetylacetone were also investigated, and they were found to give a 1:1 ratio of **3** and **4**. Methyl- and phenyl- substituted propynoates were not converted to the desired 2,3-diaminoacid derivatives.

The reaction of **4** with phthalimide was also examined (**Table 4**). Without the presence of PPh_3, **3** was not produced On the other hand, in the presence of PPh_3, **3** was formed in good yield after 24 h. However, this reaction did not proceed in CH_2Cl_2.

The proposed reaction mechanism is shown in **Scheme 2**. First, triphenylphosphine attacks the propynoate to give zwitterionic intermediate **5** [25]-[39]. The intermediate subsequently remove a proton from phthalimide, giving a vinylphosphonium salt with a phthalimidate anion. The anion adds to the salt via Michael addition, followed by a proton transfer and elimination of phosphine to give **4**. PPh_3 attacks **4** again to give zwitterionic intermediate **7**, which acts as a base to form another phthalimidate anion. This anion then attacks **4** by nucleophilic substitution to afford **3**. Solvent effect is considered to stabilize zwitterionic intermediates **5** and **7**.

3. Conclusion

α,β-Diamino acid derivatives are efficiently prepared by the reaction of propynoate with phthalimide in the

Table 4. Synthesis 3 from the reaction of 4 and phthlimide[a].

Entry	Solvent	PPh₃	Ratio[b](%)	
			3	4
1	DMSO	Yes	75	25
2	DMSO	No	-	100
3	CH₃CN	Yes	79	21
4	CH₂Cl₂	Yes	-	100

[a]A mixture of **4** (1.0 mmol), phthalimide (1.0 mmol), PPh₃ (30 mol% or 0) was stirred in a solvent at room temperature for 24 h. [b]Yields were determined by ¹H NMR using internal standard (bibenzyl) method.

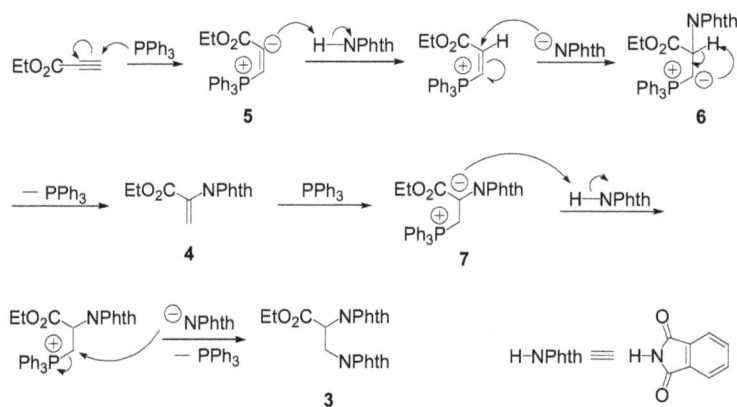

Scheme 2. Proposed mechanism.

presence of catalytic amount of triphenylphosphine. This method is very simple and the yield of the product is almost quantitative. Furthermore, if optically active amino acids could be prepared, this method might become more usable. Our first trial of the reaction using chiral phosphines gave racemic product. But now preliminary results using chiral additives indicate the possibility to yield the optically active product. Now establishment of asymmetric reaction is underway.

4. Experimental

4.1. General

Proton nuclear magnetic resonance (¹H NMR) spectra were measured using a JEOL JNM A-400 (400 MHz) spectrometer using tetramethylsilane as the internal standard. IR spectra were measured on a Shimadzu IR-408 spectrometer. Mass spectral (GC-MS) data were recorded on a Shimadzu QP2000A instrument. Melting points were measured on a Yanako Model MP and were not corrected. All substrates were purchased and used without further purification except triphenylphosphine and phthalimide (recrystalization from methanol). Solvents were purified according to the literature method and stored under Ar atmosphere [40].

4.2. Typical Experimental Procedure

Into a dry 80 mL Schlenk tube were added phthalimide (2.0 mmol), triphenylphosphine (0.3 mmol), and DMSO (2 mL). To this mixture, ethyl propynoate (1.0 mmol) was added dropwise, and the mixture was stirred for 24 h. The yield was determined by internal standard (bibenzyl) method. That is, bibenzyl (0.25 mmol) was added to the reaction mixture, and then concentrated mixture was analyzed by ¹H NMR. The integration area of 5.17 ppm (product **3**), 5.97 ppm (byproduct **4**), and 2.91 ppm (bibenzyl) was used to determining the yields of **3** and/or **4**. Reaction mixture was purified by column chromatography (silica gel 60, 200 - 400 mesh, hexane-ethyl acetate) to give the product.

4.3. Identification of the Products

Ethyl 2,3-diphthalimidoylpropanoate (3): Yellow solid, ¹H NMR (CDCl₃) δ 7.82 - 7.81 (m, 4H), 7.77 - 7.67

(m, 4H), 5.17 (dd, J = 9.2 , 5.6 Hz, 1H), 4.53 - 4.45 (m, 2H), 4.28 (q, J = 7.2 Hz, 2H), 1.27 (t, J = 7.2 Hz, 3H); ^{13}C NMR (CDCl$_3$) δ 167.8, 167.2, 166.8, 134.2, 134.1, 131.5, 62.3, 50.4, 36.8, 14.0; IR (KBr) 2960, 1760, 1724, 1600, 1380, 1220, 1085, 1030, 720, cm^{-1}; GC-MS(m/z) 392; mp. 132°C - 133°C.

Ethyl 2-phthalimidoylpropenoate (4): White solid, ^{1}H NMR (CDCl$_3$) δ 7.91 - 7.87 (m, 2H), 7.79 - 7.75 (m, 2H), 6.66 (s, 1H), 5.97 (s, 1H), 4.26 (q, J = 7.14 Hz, 2H), 1.28 (t, J =7.12 Hz, 3H); ^{13}C NMR (CDCl$_3$) δ 166.5, 162.3, 134.5, 131.8, 129.4, 127.8, 123.9, 61.9, 13.9; IR (KBr) 3150, 2960, 1790, 1724, 1639, 1400, 1380, 1250, 1070, 1030, 720, cm^{-1}; GC-MS(m/z) 245; mp. 71°C - 72°C.

References

[1]   Viso, A., de la Pradilla, R.F., García, A. and Flores, A. (2005) α,β-Diamino Acids: Biological Significance and Synthetic Approaches. *Chemical Reviews*, **105**, 3167-3196. http://dx.doi.org/10.1021/cr0406561

[2]   Viso, A., de la Pradilla, R.F., Tortosa, M., García, A. and Flores, A. (2011) Update 1 of: α,β-Diamino Acids: Biological Significance and Synthetic Approaches. *Chemical Reviews*, **111**, PR1-PR42. http://dx.doi.org/10.1021/cr100127y

[3]   Qian, H., Fu, Z., Huang, W., Zhang, H., Zhou, J., Ge, L., Lin, R., Lin, H. and Hu, X. (2010) Synthesis and Preliminary Biological Evaluation of Capsaicin Derivatives as Potential Analgesic Drugs. *Journal of Medicinal Chemistry*, **6**, 205-210.

[4]   Moura, S. and Pinto, E. (2010) Synthesis of Cyclic Guanidine Intermediates of Anatoxin-a(s) in Both Racemic and Enantiomerically Pure Forms. *SYNLETT*, 967-969. http://dx.doi.org/10.1055/s-0029-1219559

[5]   Ellsworth, B.A., Wang, Y., Zhu, Y., Pendri, A., Gerritz, S.W., Sun, C., Carlson, K.E., Kang, L., Baska, R.A., Yang, Y., Huang, Q., Burford, N.I., Cullen, M.J., Johnghar, S., Behnia, K., Pelleymounter, M.A., Washburn, W.N. and Ewing, W.R. (2007) Discovery of Pyrazine Carboxamide CB1 Antagonists: The Introduction of a Hydroxyl Group Improves the Pharmaceutical Properties and *in Vivo* Efficacy of the Series. *Bioorganic Medicinal Chemistry Letters*, **17**, 3978-3982. http://dx.doi.org/10.1016/j.bmcl.2007.04.087

[6]   Boström, J., Berggren, K., Elebring, T., Greasley, P.J. and Wilstermann, M. (2007) Scaffold Hopping, Synthesis and Structure-Activity Relationships of 5,6-Diaryl-Pyrazine-2-Amide Derivatives: A Novel Series of CB1 Receptor Antagonists. *Bioorganic Medicinal Chemistry Letters*, **15**, 4077-4084. http://dx.doi.org/10.1016/j.bmc.2007.03.075

[7]   Adediran, S.A., Cabaret, D., Flavell, R.R., Sammons, J.A., Wakselman, M. and Pratt, R.F. (2006) Synthesis and β-Lactamase Reactivity of α-Substituted Phenaceturates. *Bioorganic Medicinal Chemistry Letters*, **14**, 7023-7033. http://dx.doi.org/10.1016/j.bmc.2006.06.023

[8]   Huang, Z., Hwang, P. Watson, D.S., Cao, L. and Szoka Jr., F.C. (2009) Tris-Nitrilotriacetic Acids of Subnanomolar Affinity toward Hexahistidine Tagged Molecules. *Bioconjugate Chemistry*, **20**, 1667-1672. http://dx.doi.org/10.1021/bc900309n

[9]   Zangl, A., Kluefers, P., Schaniel, D. and Woike, T. (2009) Photoinduced Linkage Isomerism of {RuNO}^{6} Complexes with Bioligands and Related Chelators. *Dalton Transactions*, 1034-1045. http://dx.doi.org/10.1039/b812246f

[10] Luts, T., Suprun, W., Hofmann, D., Klepel, O. and Papp, H. (2007) Epoxidation of Olefins Catalyzed by Novel Mn(III) and Mo(IV) Salen Complexes Immobilized on Mesoporous Silica Gel. *Journal of Molecular Catalysis A: Chemical*, **261**, 16-23. http://dx.doi.org/10.1016/j.molcata.2006.07.035

[11] Liu, Y., Pak, J.K., Schmutz, P., Bauwens, M., Mertens, J., Knight, H. and Alberto, R. (2006) Amino Acids Labeled with [^{99m}Tc(CO)$_3$]$^+$ and Recognized by the L-Type Amino Acid Transporter LAT1. *Journal of the American Chemical Society*, **128**, 15996-15997. http://dx.doi.org/10.1021/ja066002m

[12] Takashima, H., Hirai, C. and Tsukahara, K. (2005) Selective and Monofunctional Guanosine 5'-Monophosphate Binding by Chloro[3-(2,3-diaminopropionylamino)propionic Acid](Dimethyl Sulfoxide)platinum(II) Complex. *Bulletin of the Chemical Society of Japan*, **78**, 1629-1634. http://dx.doi.org/10.1246/bcsj.78.1629

[13] Rattat, D., Eraets, K., Cleynhens, B., Knight, H., Fonge, H. and Verbruggen, A. (2004) Comparison of Tridentate Ligands in Competition Experiments for Their Ability to Form a [^{99m}Tc(CO)$_3$] Complex. *Tetrahedron Letters*, **45**, 2531-2534. http://dx.doi.org/10.1016/j.tetlet.2004.02.006

[14] Moura, S. and Pinto, E. (2010) Synthesis of Cyclic Guanidine Intermediates of Anatoxin-a(s) in both Racemic and Enantiomerically Pure Forms. *Synlett*, **2010**, 967-969. http://dx.doi.org/10.1055/s-0029-1219559

[15] Becerril, A., León-Romo, J.L., Aviña, J., Castellanos, E. and Juaristi, E. (2002) Diastereoselective Alkylation of a Chiral 1,4-benzodiazepine-2,5-dione Containing the α-Phenethyl Group. Attempted Asymmetric Synthesis of α,β-diaminopropionic Acid. *ARKIVOC*, **2002**, 4-14. http://dx.doi.org/10.3998/ark.5550190.0003.c02

[16] Brown, E.G. and Turan, Y. (1995) Pyrimidine Metabolism and Secondary Product Formation; Biogenesis of Albizziine, 4-Hydroxyhomoarginine and 2,3-Diaminopropanoic Acid. *Phytochemistry*, **40**, 763-771. http://dx.doi.org/10.1016/0031-9422(95)00317-Z

[17] Fouques, D. and Landry, J. (1991) Study of the Conversion of Asparagine and Glutamine of Proteins into Diaminopro-

pionic and Diaminobutyric Acids Using [Bis(trifluoroacetoxy)iodo]benzene Prior to Amino Acid Determination. *Analyst*, **116**, 529-531. http://dx.doi.org/10.1039/an9911600529

[18] Hellmann, H. and Haas, G. (1957) Acylaminomethylation of CH-Acidic Compounds; Syntheses of *β*-Aminocarboxylic Acids. *Chemische Berichte*, **90**, 1357-1363. http://dx.doi.org/10.1002/cber.19570900733

[19] Hellmann, H. (1957) Neuere Methoden der präparativen organischen Chemie II. 8. Amidomethylierungen. *Angewandte Chemie*, **69**, 463-471. http://dx.doi.org/10.1002/ange.19570691305

[20] Oe, Y., Inoue, T., Kishimoto, H., Sasaki, M., Ohta, T. and Furukawa, I. (2012) Three-Component Coupling Catalyzed by Phosphine: Preparation of *α*-amino *γ*-Oxo Acid Derivatives. *International Journal of Organic Chemistry*, **2**, 111-116.

[21] Riddick, J.A. and Bunger, W.B. (1970) Organic Solvent. 3rd ed., Wiley-Interscience, New York.

[22] Ross, J., Chen, W., Xu, L. and Xiao, J. (2001) Ligand Effects in Palladium-Catalyzed Allylic Alkylation in Ionic Liquids. *Organometallics*, **20**, 138-142. http://dx.doi.org/10.1021/om000712y

[23] Nair, V., Sreekanth, A.R. and Vinod, A.U. (2001) Novel Pyridine-Catalyzed Reaction of Dimethyl Acetylenedicarboxylate with Aldehydes: Formal [2+2] Cycloaddition Leading to 2-Oxo-3-benzylidinesuccinates. *Organic Letters*, **3**, 3495-3497. http://dx.doi.org/10.1021/ol016550z

[24] Trost, B.M. and Dake, G.R. (1997) Nitrogen Pronucleophiles in the Phosphine-Catalyzed *γ*-Addition Reaction. *The Journal of Organic Chemistry*, **62**, 5670-5671. http://dx.doi.org/10.1021/jo970848e

[25] Rychnovsky, S.D. and Kim, J. (1994) Triphenylphosphine-Catalyzed Isomerizations of Enynes to (E,E,E)-Trienes: Phenol as a Cocatalyst. *The Journal of Organic Chemistry*, **59**, 2659-2660. http://dx.doi.org/10.1021/jo00088a067

[26] Xu, Z. and Lu, X. (1999) Phosphine-Catalyzed [3+2] Cycloaddition Reactions of Substituted 2-Alkynoates or 2,3-alkenoates with Electron-Deficient Olefins and Imines. *Tetrahedron Letters*, **40**, 549-552. http://dx.doi.org/10.1016/S0040-4039(98)02405-8

[27] Zhang, C. and Lu, X. (1995) Phosphine-Catalyzed Cycloaddition of 2,3-Butadienoates or 2-Butynoates with Electron-Deficient Olefins. A Novel [3+2] Annulation Approach to Cyclopentenes. *The Journal of Organic Chemistry*, **60**, 2906-2908. http://dx.doi.org/10.1021/jo00114a048

[28] Xu, Z. and Lu, X. (1997) Phosphine-Catalyzed [3+2] Cycloaddition Reaction of Methyl 2,3-Butadienoate and *N*-Tosylimines. A Novel Approach to Nitrogen Heterocycles. *Tetrahedron Letters*, **38**, 3461-3464. http://dx.doi.org/10.1016/S0040-4039(97)00656-4

[29] Guo, C. and Lu, X. (1993) Reinvestigation on the Catalytic Isomerization of Carbon-Carbon Triple Bonds. *Journal of the Chemical Society, Perkin Transactions*, **1993**, 1921-1923. http://dx.doi.org/10.1039/p19930001921

[30] Trost, B.M. and Li, C.J. (1994) Phosphine-Catalyzed Isomerization-Addition of Oxygen Nucleophiles to 2-Alkynoates. *Journal of the American Chemical Society*, **116**, 10819-10820. http://dx.doi.org/10.1021/ja00102a071

[31] Trost, B.M. and Kazmaier, U. (1992) Internal Redox Catalyzed by Triphenylphosphine. *Journal of the American Chemical Society*, **114**, 7933-7935. http://dx.doi.org/10.1021/ja00046a062

[32] Trost, B.M. and Dake, G.R. (1997) Nucleophilic *α*-Addition to Alkynoates. A Synthesis of Dehydroamino Acids. *Journal of the American Chemical Society*, **119**, 7595-7596. http://dx.doi.org/10.1021/ja971238z

[33] Trost, B.M. and Li, C.J. (1994) Novel "Umpolung" in C-C Bond Formation Catalyzed by Triphenylphosphine. *Journal of the American Chemical Society*, **116**, 3167-3168. http://dx.doi.org/10.1021/ja00086a074

[34] Kuroda, H., Hanaki, E. and Kawakami, M. (1999) A Convenient Method for the Preparation of Furans by the Phosphine-Initiated Reactions of Enynes Bearing a Carbonyl Group. *Tetrahedron Letters*, **40**, 3753-3756. http://dx.doi.org/10.1016/S0040-4039(99)00601-2

[35] Nozaki, K., Sato, N., Ikeda, K. annd Takaya, H. (1996) Synthesis of Highly Functionalized *γ*-Butyrolactones from Activated Carbonyl Compounds and Dimethyl Acetylenedicarboxylate. *The Journal of Organic Chemistry*, **61**, 4516-4519. http://dx.doi.org/10.1021/jo951828k

[36] Yavari, I. and Mosslemin, M.H. (1998) An Efficient One-Pot Synthesis of Dialkyl 2,5-dihydrofuran-2,3-dicarboxylates Mediated by Vinyltriphenylphosphonium Salt. *Tetrahedron*, **54**, 9169-9174. http://dx.doi.org/10.1016/S0040-4020(98)00554-7

[37] Yavari, I., Hekmat-Shoar, R. and Zonouzi, A. (1998) A New Efficient Rout to 4-Carboxymethylcoumarins Mediated by Vinyltriphenylphosphonium Salt. *Tetrahedron Letters*, **39**, 2391-2392.

[38] Yavari, I. and Baharfar, R. (1997) Vinylphosphonium Salt Mediated One-Pot Synthesis of Functionalized-3-(triphenylphosphoranylidene)butyrolactones. *Tetrahedron Letters*, **38**, 4259-4262.

[39] Caddick, S., Aboutayab, K., Jenkins, K. and West, R.I. (1996) Intramolecular Radical Substitution Reactions: A Novel Approach to Fused [1,2-*a*] Indoles. *Journal of the Chemical Society, Perkin Transactions*, **1996**, 675-682. http://dx.doi.org/10.1039/p19960000675

[40] Perrin, D.D. and Armarego, W.L.F. (1988) Purification of Laboratory Chemicals. 3rd Edition, Pergamon, Oxford.

9

Synthesis of Biotinylated Galiellalactone Analogues

Zilma Escobar[1], Martin Johansson[2], Anders Bjartell[2], Rebecka Hellsten[2], Olov Sterner[1]*

[1]Centre for Analysis and Synthesis, Department of Chemistry, Lund University,
Lund, Sweden
[2]Division of Urological Cancers, Department of Clinical Sciences Malmö, Lund University,
Malmö, Sweden
Email: *olov.sterner@chem.lu.se

Abstract

Two biotinylated derivatives of the fungal metabolite galiellalactone (1) were synthesized in order to facilitate the investigation of the molecular mechanism of action of the galiellalactonoids. Galiellalactone is a STAT3-signaling inhibitor that inhibits growth *in vitro* as well as *in vivo* of prostate cancer cells expressing activated STAT3. To provide a suitable point of attachment for biotin, the 8-hydroxymethyl derivative (3) and its 7-phenyl analogue 4 were synthesized by a modified tandem Pd-catalysed carbonylation and intramolecular vinyl allene Diels-Alder procedure previously developed. The two primary alcohols obtained, 3 and 4, were coupled to biotin as the 6-aminohexanoic acid amide, activated as the acid chloride, yielding the derivatives 5 and 6.

Keywords

STAT3, Galiellalactone, Biotin, Synthesis

1. Introduction

The protein STAT3 (Signal Transducer and Activator of Transcription 3) is a transcription factor that is involved in different cellular processes. It has been shown to be constitutively activated in malignancies and is involved in the proliferation of several types of cancer cells [1] [2]. In normal cells, the activation of STAT3 must consequently be tightly regulated, and STAT3 is today considered to be a relevant target for novel drugs for the treatment of cancer [3]. Galiellalactone (1) is a fungal metabolite that inhibits the STAT3 signaling pathway, presumably by reacting covalently with the thiol group of critical cysteines of STAT3 and thereby blocking the

---

*Corresponding author.

binding of STAT3 to STAT3 specific transcriptional DNA elements [4]. **1** inhibits the growth, both *in vitro* and *in vivo*, of prostate cancer cells expressing activated STAT3 and inhibits the expression of STAT3 regulated genes and proteins, including anti-apoptotic genes [4]. Furthermore, galiellalactone (**1**) inhibits growth and induces apoptosis of prostate cancer stem cell-like cells expressing phosphorylated STAT3 (pSTAT3) [5], and has the potential to be developed into a drug for treating cancers in which STAT3 is activated.

1 contains an α, β-unsaturated lactone moiety, and it has been demonstrated to react with sulfur nucleophiles to produce inactive adducts [6] [7]. It has been proposed that **1** interferes directly with the binding of STAT3 to DNA by reacting with a thiol group of a cysteine in STAT3 that is located in a domain of STAT3 that is involved in the binding to DNA [4], however, no evidence supporting this molecular mechanism of action has yet been presented. Target identification and mechanism of action elucidation are highly challenging yet crucial areas of chemistry and biology, and several methods and strategies have been developed to achieve this. Compounds that covalently react with their target provide an advantage as they can be modified or tagged with moieties that allow for the detection or isolation of the compound-target adducts. Biotin labeling of covalent inhibitors is an effective method for detecting and isolating bound target proteins because of the high affinity biotin displays to streptavidin allowing the capture and detection of biotin labeled inhibitor-target adducts. A main challenge is to attach the biotin label to the inhibitor in such a way that the target affinity is not abolished. We envisioned that a hydroxymethyl group in position 8 of **1** (see **Figure 1**) would allow us to attach a biotin group via a suitable linker so that the STAT3 inhibiting effect was retained. To this end, 8-hydroxymethyl-galiellalactone (**3**) as well as 8-hydroxymethyl-7-phenyl galiellalactone (**4**) was prepared by a modified tandem Pd-catalysed carbonylation and intramolecular vinyl allene Diels-Alder strategy previously developed for the synthesis of 4-*epi*-**1** [8]. **3** and **4** were then coupled with an ester bond to the 6-aminohexanoic acid amide of biotin, yielding the two desired compounds **5** and **6** (**Figure 1**). Both enantiomers of galiellalactone (**1**) have been prepared [9] [10] and found to be an equally potent inhibitors of IL-6 mediated STAT3 signaling [7] [11], consequently **3** and **4** were prepared as racemates while **5** and **6** were obtained as pairs of diastereomers. During our work with analogues of **1** [8], we had indications (unpublished results) that a substituent in position 7 was beneficial for the potency, and to investigate if a phenyl group at C-7 affects the STAT3 inhibiting effect of galiellalactone (**1**) the 7-phenyl analogue **2** was prepared and assayed.

2. Experimental Section

Reagents and solvents were used from commercial sources without purification, except THF and CH_2Cl_2 that were passed through a MBraun SPS-800 solvent system. All reactions were carried out in standard dry glassware and atmospheric surroundings. Analytical thin layer chromatography (TLC) was performed on Kiselgel 60 F254 plates (Merck) and visualized by spraying with vanillin/H_2SO_4 and heating. Silica gel column chromatography was performed on SiO_2 (Matrex LC-gel: 60A, 35-70 MY, Grace). ^{1}H and ^{13}C NMR spectra were rec-

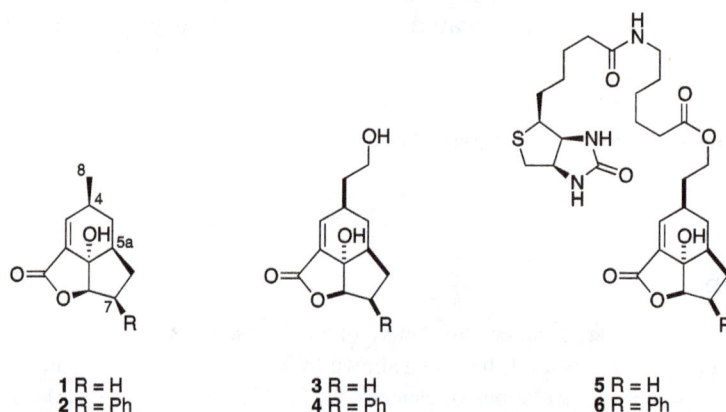

Figure 1. Structures of the fungal metabolite galiellalactone (**1**), (+)-7-phenyl galiellalactone (**2**), the two analogues 8-hydroxymethyl galiellalactone (**3**) and 8-hydroxymethyl-7-phenyl galiellalactone (**4**), and the corresponding *N*-(+)-biotinyl-6-aminohexanoic acid esters **5** and **6**.

orded with a Bruker DRX 400 spectrometer (at 400 MHz for ^{1}H and 100 MHz for ^{13}C) and a Bruker DRX 500 spectrometer (at 500 and 125 MHz, respectively). The spectra were recorded in CDCl$_3$ and CD$_3$OD, and the solvent signals (7.27/77.0 and 3.31/49.0 ppm) were used as reference. The data for the ^{1}H signals are given as chemical shifts in ppm and (number of protons, multiplicity, and coupling constants (J) in Hz), while the ^{13}C data are given as chemical shifts in ppm. HR-ESIMS spectra were recorded with a Waters Q-TOF Micro system.

Hept-6-en-1-yn-3-ol(**8a**): To a solution of ethynyl magnesium bromide (100 ml, 0.5 M in THF, 50 mmol) was slowly added to a solution of 4-pentenal (**7a**, 3.70 g, 44.0 mmol) in 54 ml of dried THF at 0°C. After the addition was complete, the solution was allowed to reach room temperature and stirred overnight. The reaction mixture was quenched by the addition of 50 ml NH$_4$Cl (sat.), the mixture was extracted 3 times with 80 ml EtOAc and the organic phase was dried with MgSO$_4$ and concentrated under reduced pressure. **8a** was obtainedas a yellowish oil (4.75 g, quantitative yield), which was used without further purification in the next step. ^{1}H NMR (400 MHz, CDCl$_3$) 5.69 (1 H, ddt, 17.1, 10.2, 6.7), 4.93 (1 H, ddd, 17.1, 3.4, 1.5), 4.85 (1 H, ddd, 10.2, 3.4, 1.5), 4.25 (1 H, td, 6.7, 2.1), 2.37 (1 H, d, 2.1), 2.10 (2 H, m), 1.67 (2 H, m). ^{13}C NMR (100 MHz, CDCl$_3$) 137.5, 115.4, 84.6, 73.2, 61.7, 36.6, 29.2. HRMS calcd for C$_7$H$_{11}$O [M + H]: 111.0810, found: 111.0834.

4-Phenylhept-6-en-1-yn-3-ol(**8b**): Phenylacetic acid (30.0 g, 220 mmol) was dissolved in 100 ml MeOH and SOCl$_2$(28.8 g, 242 mmol) was added dropwise at 0°C under stirring. The reaction mixture was allowed to reach room temperature overnight and the volatiles were removed under reduced pressure. Phenylacetic acid methyl ester was obtained as a yellowish oil (32.8 g, quantitative yield) and used directly in the next step. ^{1}H NMR (400 MHz, CDCl$_3$) 7.34 (5 H, m), 3.72 (3 H, s), 3.67 (2 H, s). ^{13}C NMR (100 MHz, CDCl$_3$) 171.0, 133.6, 128.6, 127.8, 126.3, 51.0, 40.2. HRMS calcd for C$_9$H$_{10}$O$_2$Na [M + Na]: 173.0578, found: 173.0558. Phenylacetic acid methyl ester (14.0 g, 93 mmol) was dissolved in 150 ml dry THF and added slowly at −78°C over 30 min to a freshly prepared LDA solution [12] 3 g, 120 mmol, diisopropyl amine and 67 ml, 107 mmol, of n-BuLi (1.6 M in hexane)]. The reaction mixture was stirred for 30 min at −78°C before allyl bromide (17.0 g, 141 mmol) was added dropwise at −78°C over 10 min. The reaction mixture was quenched by the addition of 120 ml NH$_4$Cl (sat.), extracted 3 times with 100 ml EtOAc, the combined organic phases were dried (MgSO$_4$) and concentrated by evaporation to afford methyl 2-phenylpent-4-enoate as a yellowish oil (17.7 g, quantitative yield). ^{1}H NMR (400 MHz, CDCl$_3$) 7.27 (3 H, m), 7.22 (2 H, m), 5.68 (1 H, ddt, 17.0, 10.2, 6.8), 5.04 (1 H, ddd, 17.0, 2.9, 1.3), 4.96 (1 H, ddt, 10.2, 2.9, 1.3), 3.61 (1 H, t, 7.8), 3.61 (3 H, s), 2.80 (1 H, m), 2.48 (1 H, m). ^{13}C NMR (100 MHz, CDCl$_3$) 173.8, 138.5, 135.2, 128.6, 127.8, 127.3, 116.9, 51.9, 51.3, 37.5. HRMS calcd for C$_{12}$H$_{14}$O$_2$Na [M + Na]: 213.0891, found: 213.0891. Methyl 2-phenylpent-4-enoate (8.7 g, 46 mmol) in 200 ml of dry CH$_2$Cl$_2$ was reduced to **7b** by slowly adding DIBAL-H (48.0 ml, 1 M in hexane, 48.0 mmol) under N$_2$ atmosphere at -78°C and stirring for 3 h at −78°C. Without work-up and purification, ethynyl magnesium bromide (100 ml, 0.5 M in THF, 50 mmol) was added dropwise to the reaction mixture which was stirred for an additional 30 min at −78°C and left at room temperature overnight before the reaction was quenched by the addition of 120 ml NaHCO$_3$ (sat.). The mixture was extracted 4 times with 80 ml EtOAc and the combined organic phases were dried with MgSO$_4$ and concentrated by evaporation. The crude product was purified by silica gel chromatography, and **8b** (3.5 g, 41%) was obtained as a 1:0.2 mixture of diastereomers. *Diastereomer* a) ^{1}H NMR (400 MHz, CDCl$_3$) 7.24 (5 H, m), 5.64 (1 H, ddd, 17.0; 10.1, 6.2), 5.03 (1 H, ddd, 17.0, 3.1, 1.3), 4.93 (1 H, ddt, 10.1, 3.1, 1.3), 4.48 (1 H, dd, 8.5, 2.2), 2.94 (1 H, dt, 8.5, 6.2), 2.65 (1 H, m), 2.52 (1 H, m), 2.39 (1 H, d, J 2.2). ^{13}C NMR (100 MHz, CDCl$_3$) 139.1, 135.8, 129.1, 128.0, 127.0, 116.6, 82.8, 74.9, 65.3, 50.8, 35.2. *Diastereomer* b) ^{1}H NMR (400 MHz, CDCl$_3$) 7.24 (5 H, m), 5.64 (1 H, ddd, 17.0; 10.1, 6.2), 5.03 (1 H, ddd, 17.0, 3.1, 1.3), 4.93 (1 H, ddt, 10.1, 3.1, 1.3), 4.48 (1 H, ddd, 8.5, 2.2), 2.92 (1 H, dt, 8.5, 6.2), 2.67 (1 H, m), 2.52 (1 H, m), 2.43 (1 H, d, 2.2). ^{13}C NMR (100 MHz, CDCl$_3$) 139.2, 135.9, 128.8, 128.2, 126.9, 116.7, 83.2, 74.6, 65.5, 51.4, 35.1. HRMS calcd for C$_{13}$H$_{14}$ONa [M + Na]: 209.0942, found: 209.0919.

(*Z*)-11-((Triisopropylsilyl)oxy)undeca-1,8-dien-6-yn-5-ol(**10a**): TEA (55 ml, 400 mmol) was added to a mixture of PdCl$_2$(PPh$_3$)$_2$ (0.70 g, 0.99 mmol) and CuI (0.38 g, 1.99 mmol) at 0°C under nitrogen atmosphere and the solution was stirred for 30 minutes. To this solution **9** (7.25 g, 20.5 mmol), prepared according to reference 12 and identical in all aspects with the reported compound, in 40 ml dry THF was added dropwise followed by **8a** (2.19 g, 20.0 mmol) in 40 ml dry THF at 0°C. The reaction mixture was stirred overnight at room temperature, quenched with 50 ml NaHCO$_3$ (sat.), extracted 3 times with 60 ml EtOAc. The combined organic phases were dried with Na$_2$SO$_4$ and concentrated by evaporation. The yield was quantitative and **10a** was used directly in the next step without purification. ^{1}H NMR (400 MHz, CDCl$_3$) 6.04 (1 H, dt, 10.8, 7.0), 5.85 (1 H, ddt, 17.0, 10.2, 7.2), 5.56 (1 H, ddd, 10.8, 3.1, 1.4), 5.08 (1 H, ddd, 17.0, 3.2, 1.5), 5.00 (1 H, dd, 10.2, 3.2), 4.54 (1 H, td, 6.5,

3.1), 3.76 (2 H, t, 7.0), 2.56 (2 H, qd, J 7.0, 1.4), 2.26 (2 H, td, J 7.2, 1.5), 1.83 (2 H, m), 1.13 - 1.03 (21 H, m).^{13}C NMR (100 MHz, CDCl$_3$) 141.1, 137.9, 115.5, 110.0, 94.4, 82.1, 62.7, 62.5, 37.1, 34.4, 29.7, 18.2, 12.2. HRMS calcd for C$_{20}$H$_{37}$O$_2$Si [M + H]: 337.2563, found: 337.2560.

(Z)-Methyl (11-((triisopropylsilyl)oxy)undeca-1,8-dien-6-yn-5-yl) carbonate(**11a**): **10a** (6.70 g, 20.0 mmol) was dissolved in 130 ml of dry CH$_2$Cl$_2$ at 0°C. DMAP (4.86 g, 39.8 mmol) was added and the mixture was stirred for 20 min. Methyl chloroformate (5.6 g, 59.7 mmol) was added dropwise, and the reaction was brought to room temperature overnight before 50 ml NaHCO$_3$ (sat.) was added and the mixture was extracted 2 times with 80 ml CH$_2$Cl$_2$. The combined organic phases were dried over MgSO$_4$ and concentrated under reduced pressure. The crude product was purified by flash chromatography with CH$_2$Cl$_2$/heptane 1:1 as eluent, giving 3.59 g (46%) of **11a** as a yellowish oil. ^{1}H NMR (400 MHz, CDCl$_3$) 6.09 (1 H, dt, 10.8, 7.3), 5.82 (1 H, ddt, 17.0, 10.2, 6.6), 5.56 (1 H, ddd, 10.8, 3.3, 1.7), 5.40 (1 H, td, 6.6, 1.7), 5.07 (1 H, ddd, 17.0, 3.2, 1.3), 5.02 (1 H, ddd, 10.2, 3.2, 1.3), 3.81 (3 H, s), 3.76 (2 H, t, 6.5), 2.55 (2 H, qd, 6.6, 1.3), 2.25 (2 H, m), 1.95 (2 H, m), 1.09 - 1.03 (21 H, m). ^{13}C NMR (100 MHz, CDCl$_3$) 155.2, 142.3, 137.0, 115.9, 109.6, 89.9, 83.7, 68.5, 62.5, 55.1, 34.4, 34.3, 29.4, 18.2, 12.2. HRMS calcd for C$_{22}$H$_{39}$O$_4$NaSi [M + H]: 395.2618, found: 395.2647.

(Z)-4-Phenyl-11-((triisopropylsilyl)oxy)undeca-1,8-dien-6-yn-5-ol(**10b**):**10b** was prepared in the same way as **10a**, from **8b**, and **10b** was obtained in a quantitative yield as a 1:0.2 mixture of diastereomers. *Diastereomer* a) ^{1}H NMR (400 MHz, CDCl$_3$) 7.30 (5 H, m), 6.05 (1 H, dt, 10.8, 7.4), 5.71 (1 H, ddt, 17.1, 10.2, 6.3), 5.55 (1 H, ddd, 10.8, 3.3, 1.5), 5.08 (1 H, ddd, 17.1, 3.2, 1.6), 4.99 (1 H, ddd, 10.2, 3.2, 1.6), 4.74 (1 H, ddd, 7.5, 5.7, 1.5), 3.74 (2 H, t, 6.5), 3.04 (1 H, m), 2.65 (2 H m), 2.51 (2 H, m), 1.77 (1 H, d, 7.5), 1.17 - 1.04 (21 H, m). ^{13}C NMR (100 MHz, CDCl$_3$) 141.2, 139.4, 136.1, 129.2, 128.2, 127.2, 116.7, 109.6, 92.4, 83.8, 66.4, 62.4, 51.4, 35.6, 34.1, 18.0, 12.0. *Diastereomer* b) ^{1}H NMR (400 MHz, CDCl$_3$) 7.30 (5 H, m), 6.03 (1 H, dt, 10.8, 7.4), 5.71 (1 H, ddt, 17.1, 10.2, 6.3), 5.57 (1 H, ddd, 10.8, 3.3, 1.5), 5.06 (1 H, ddd, 17.1, 3.2, 1.6), 4.99 (1 H, ddd, 10.2, 3.2, 1.6), 4.77 (1 H, ddd, 7.5, 5.7, 1.5), 3.78 (2 H, t, 6.5), 3.01 (1 H, m), 2.70 (2 H m), 2.51 (2 H, m), 1.78 (1 H, d, 7.5), 1.17 - 1.04 (21 H, m). ^{13}C NMR (100 MHz, CDCl$_3$) 141.3, 139.6, 136.2, 128.9, 128.4 127.1, 116.8, 109.7, 92.8, 83.5, 66.5, 62.4, 52.1, 35.6, 34.2, 18.0, 12.0.HRMS calcd for C$_{26}$H$_{40}$O$_2$NaSi [M + Na]: 435.2695, found: 435.2663.

(Z)-Methyl (4-phenyl-11-((triisopropylsilyl)oxy)undeca-1,8-dien-6-yn-5-yl) carbonate(**11b**):**11b** was prepared in the same way as **11a**, from **10b**, and **11b** was obtained in a 75% yield as a 1:0.2 mixture of diastereomers. *Diastereomer* a) ^{1}H NMR (400 MHz, CDCl$_3$) 7.28 (5 H, m), 6.04 (1 H, dt, 11.5, 6.9), 5.65 (1 H, ddd, 17.1, 10.9, 6.7), 5.58 (1 H, dd, 4.8, 2.1), 5.48 (1 H, ddd, 11.5, 2.1, 1.6), 5.04 (1 H, ddd, 17.1, 3.2, 1.5), 4.96 (1 H, ddd, 10.9, 3.2, 1.5), 3.78 (3 H, s), 3.68 (2 H, t, 6.4), 3.15 (1 H, m), 2.71 (1 H, m), 2.56 (1 H, m), 2.37 (2 H, dtd, 6.9, 6.4, 1.6), 1.17 - 0.95 (21 H, m). ^{13}C NMR (100 MHz, CDCl$_3$) 154.9, 142.1, 139.2, 135.4, 128.8, 128.2, 127.1, 117.0, 109.2, 88.5, 84.6, 71.7, 62.3, 54.9, 49.4, 35.5, 34.0, 18.0, 11.9. *Diastereomer* b) ^{1}H NMR (400 MHz, CDCl$_3$) 7.28 (5 H, m), 6.06 (1 H, dt, 11.5, 6.9), 5.66 (1 H, ddd, 17.1, 10.9, 6.7), 5.58 (1 H, dd, 4.8, 2.1), 5.50 (1 H, ddd, 11.5, 2.1, 1.6), 5.05 (1 H, ddd, 17.1, 3.2, 1.5), 4.97 (1 H, ddd, 10.9, 3.2, 1.5), 3.76 (3 H, s), 3.70 (2 H, t, 6.4), 3.15 (1 H, m), 2.73 (1 H, m), 2.57 (1 H, m), 2.50 (2 H, dtd, 6.9, 6.4, 1.6), 1.17 - 0.95 (21 H, m). ^{13}C NMR (100 MHz, CDCl$_3$) 154.8, 142.3, 138.9, 135.2, 128.8, 128.2, 127.1, 116.9, 109.3, 88.46, 84.7, 71.8, 62.3, 54.8, 49.7, 34.9, 34.1, 18.0, 11.9. HRMS calcd for C$_{28}$H$_{42}$O$_4$NaSi [M + Na]: 493.2750, found: 493.2773.

rac-(6R,7aR)-Methyl 6-(2-((triisopropylsilyl)oxy)ethyl)-2,6,7,7a-tetrahydro-1H-indene-4-carboxylate(**12a**): To a dried autoclave flask was added Pd(OAc)$_2$ (0.41 g, 1.8 mmol) and DPPP (0.75 g, 1.8 mmol) under a nitrogen atmosphere, followed by 9 ml dry toluene and **11a** (3.59 g, 9.1 mmol) dissolved in 9 ml toluene/MeOH (1:1). CO was bubbled though the solution before the autoclave was pressurized with CO to 5 bar. The reaction mixture was stirred for 48 h at 5 bar at room temperature before being diluted with 15 ml EtOAc and filtered through a celite plug. The filtered solution was concentrated and purified by silica gel chromatography with heptane/Et$_2$O (94:6) to afford 0.55 g (16%) of **12a** as a colourless oil. ^{1}H NMR (400 MHz, CDCl$_3$) 6.82 (1 H, d, 2.1), 6.27 (1 H, dd, 6.1, 2.6), 3.82 (2 H, m), 3.78 (3 H, s), 2.73 (1 H, m), 2.72 (1 H, dddd, 16.4, 8.8, 5.2, 2.1), 2.44 (2 H, ddd, 11.5, 4.5, 2.6), 2.14 (1 H, m), 2.12 (1 H, m), 1.77 (1 H, td, 13.5, 6.6), 1.60 (1 H, td, 13.5, 5.2), 1.37 (1 H, ddd, 12.5, 11.5, 3, 6.1), 1.10 (1 H, m), 1.08 - 1.04 (21 H, m). ^{13}C NMR (100 MHz, CDCl$_3$) 167.2, 145.7, 137.8, 127.0, 126.9, 61.2, 51.7, 44.7, 38.6, 36.4, 35.2, 32.4, 31.2, 18.2, 12.2. HRMS calcd for C$_{22}$H$_{38}$O$_3$NaSi [M + Na]: 401.2518, found: 401.2488.

Methyl 2-phenyl-6-(2-((triisopropylsilyl)oxy)ethyl)-2,6,7,7a-tetrahydro-1H-indene-4-carboxylate(**12b**):**12b** was prepared in the same way as **12a**, starting from **11b** (6.55 g, 13.9 mmol). **12b** was obtained as colourless oil as a 1:0.8 inseparable mixture of two diastereomers (total yield 2.64 g, 42%). *Diastereomer* a) ^{1}H NMR (400 MHz,

CDCl$_3$) 7.24 (5 H, m), 6.94 (1 H, t, 2.5), 6.32 (1 H, m), 4.02 (1 H, m), 3.85 (2 H, m), 3.79 (3 H, s), 2.86 (1 H, m), 2.75 (1 H, m), 2.60 (1 H, dt, 12.3, 7.0), 2.16 (1 H, dd, 10.2, 2.5), 2.00 (1 H, dt, 10.2, 9.2), 1.81 (1 H, tt, 13.2, 6.5), 1.65 (1 H, dt, 13.2, 6.5), 1.37 (1 H, dt, 12.3, 10.2), 1.21 - 0.91 (21 H, m). ^{13}C NMR (100 MHz, CDCl$_3$) 166.9, 146.8, 146.0, 138.3, 130.4, 129.4, 128.4, 127.5, 126.6, 126.1, 60.9, 51.6, 51.3, 50.3, 44.6, 43.2, 42.2, 39.8, 38.4, 36.2, 35.2, 18.0, 12.0. *Diastereomer* b) ^{1}H NMR (400 MHz, CDCl$_3$) 7.24 (5 H, m), 6.40 (1 H, m), 6.32 (1 H, m), 4.02 (1 H, m), 3.85 (2 H, m), 3.78 (3 H, s), 2.96 (1 H, m), 2.75 (1 H, m), 2.60 (1 H, dt, 12.3, 7.0), 2.16 (1 H, dd, 10.2, 2.5), 2.00 (1 H, dt, 10.2, 9.2), 1.81 (1 H, tt, 13.2, 6.5), 1.61 (1 H, m), 1.37 (1 H, dt, 13.2, 10.2), 1.21 - 0.91 (21 H, m). ^{13}C NMR (100 MHz, CDCl$_3$) 166.9, 146.7, 145.6, 138.4, 130.4, 129.3, 128.4, 127.2, 126.6, 126.0, 60.9, 51.6, 51.3, 50.3, 44.6, 43.2, 42.2, 39.8, 38.3, 36.1, 35.0, 18.0, 12.0. HRMS calcd for C$_{28}$H$_{43}$O$_3$Si [M + H]: 455.2981, found: 455.2956.

rac-(1a*S*,3a*R*,5*R*,7a*S*)-Methyl 5-(2-((triisopropylsilyl)oxy)ethyl)-1a,2,3,3a,4,5-hexahydroindeno [1,7a-*b*]oxirene-7-carboxylate(**13a**): **12a** (0.55 g, 1.4 mmol) was dissolved in 30 ml dry CH$_2$Cl$_2$ under nitrogen atmosphere and *m*-CPBA (0.32 g, 1.9 mmol) was added at 0°C. The reaction mixture was stirred at this temperature for 1 h before it was quenched by the addition of 10 ml Na$_2$S$_2$O$_3$ (sat.). The phases were separated and organic phase was washed with NaHCO$_3$ (sat.). The water phases were extracted with 3 times with 20 ml dry CH$_2$Cl$_2$ and the combined organic phases were dried over Na$_2$SO$_4$, filtered and evaporated. Flash chromatography (SiO$_2$, heptane/Et$_2$O 85:15) afforded the required isomer **13a** as a colourless oil in 79% (0.45 g) yield. ^{1}H NMR (400 MHz, CDCl$_3$) 7.29 (1 H, d, 2.5), 4.41 (1 H, s), 3.84 (2 H, m), 3.72 (3 H, s), 2.70 (1 H, dddd, 16.5, 8.3, 6.1, 2.5), 2.05 (1 H, dd, 13.9, 7.2), 1.97 (1 dddd, 16.5, 13.2, 6.8, 3.3), 1.89 (1 H, m), 1.83 (1 H, ddd, 13.2, 7.2, 5.3), 1.66 (1 H, m), 1.62 (1 H, s), 1.58 (1 H, m), 1.30 (1 H, td, 13.2, 8.3), 1.12 (1 H, m), 1.08 - 1.04 (21 H, m). ^{13}C NMR (100 MHz, CDCl$_3$) 165.4, 153.9, 126.3, 63.6, 63.3, 60.7, 51.5, 39.7, 38.0, 35.8, 30.1, 27.5, 23.2, 18.0, 11.9. HRMS calcd for C$_{22}$H$_{38}$O$_4$NaSi [M + Na]: 417.2437, found: 417.2413.

rac-(1a*S*,2*S*,3a*S*,5*R*,7a*S*)-Methyl 2-phenyl-5-(2-((triisopropylsilyl)oxy)ethyl)-1a,2,3,3a,4,5-hexahydroindeno [1,7a-*b*]oxirene-7-carboxylate(**13b**): **13b** was prepared in the same way as **13a**, starting from **12b** (2.64 g, 5.8 mmol). The crude product (2.71 g) was purified by flash chromatography (SiO$_2$, heptane/Et$_2$O 85:15). **13b** was obtained as a colourless oil (0.43 g, 16%). ^{1}H NMR (400 MHz, CDCl$_3$) 7.42 (2 H, dd, 8.2, 1.3), 7.34 (1 H, d, 0.9), 7.31 (2 H, td, 8.2, 7.3), 7.24 (1 H, td, 7.3, 1.3), 4.57 (1 H, s), 3.85 (2 H, m), 3.72 (3 H, s), 3.16 (1 H, dd, 12.1, 7.2), 2.75 (1 H, m), 2.12 (1 H, m), 2.04 (1 H, m), 1.92 (1 H, dt, 12.1, 7.2), 1.86 (1 H, dt, 15.2, 6.8), 1.67 (1 H, ddd, 15.2, 10.8, 6.8), 1.42 (1 H, m), 1.32 (1 H, dd, 12.1, 11.5), 1.17 - 1.03 (21 H, m). ^{13}C NMR (100 MHz, CDCl$_3$) 165.4, 154.1, 141.6, 128.4, 127.7, 126.6, 126.1, 66.9, 62.3, 60.6, 51.6, 45.9, 40.2, 38.0, 35.6, 32.3, 30.0, 18.0, 12.0. HRMS calcd for C$_{28}$H$_{42}$O$_4$SiNa [M + Na]: 493.2750, found: 493.2773.

rac-(2a*S*,4*R*,5a*R*,7a*R*)-2a-Hydroxy-4-(2-hydroxyethyl)-4,5,5a,6,7,7a-hexahydroindeno [1,7-*bc*]furan-2(2a*H*)-one(**3**): **13a** (0.45 g, 1.2 mmol) was dissolved in 5 ml THF and a solution of LiOH·H$_2$O (0.12 g, 2.8 mmol) in 5 ml H$_2$O was added at room temperature. The reaction mixture was stirred for 3 days until a TLC analysis revealed that the starting material was consumed. After the addition of 5 ml THF and 5 ml of H$_2$SO$_4$ (10%), the mixture was stirred at room temperaturefor 3 days. The reaction mixture was quenched by the addition of 3 ml NaHCO$_3$ (sat.) and extracted 3 times with 20 ml EtOAc, the combined organic phases were dried with Na$_2$SO$_4$ and concentrated by evaporation. The crude product was purified by chromatography (SiO$_2$, CHCl$_3$/MeOH 9:1) to afford 0.13 g (50%) of **3** as a colourless oil. ^{1}H NMR (500 MHz, CD$_3$OD) 7.11 (1 H, d, 3.1), 4.69 (1 H, dd, 7.5, 2.7), 3.71 (2 H, t, 6.4), 2.70 (1 H, m), 2.43 (1 H, dtd, 10.5, 7.9, 5.0), 2.27 (1 H, m), 2.10 (1 H, ddt, 14.3, 10.5, 7.5), 1.87 (1 H, m), 1.76 (1 H, td, 13.2, 6.4), 1.69 (1 H, dddd, 14.3, 7.5, 5.0, 2.7), 1.65 (1 H, m), 1.18 (1 H, m), 1.10 (1 H, m). ^{13}C NMR (125 MHz, CD$_3$OD) δ 172.2, 149.7, 132.9, 91.9, 83.0, 60.9, 44.4, 39.1, 32.7, 32.6, 32.5, 32.4. HRMS calcd for C$_{12}$H$_{16}$O$_4$Na [M + Na]: 247.0946, found: 247.0932.

rac-(2a*S*,4*R*,5a*S*,7*S*,7a*R*)-2a-Hydroxy-4-(2-hydroxyethyl)-7-phenyl-4,5,5a,6,7,7a-hexahydroindeno [1,7-*bc*] furan-2(2a*H*)-one(**4**): **4** was prepared in the same way as **3**, starting from **13b** (0.13 g, 0.4 mmol). The crude product was purified by chromatography (SiO$_2$,CHCl$_3$/MeOH 9:1) to afford **3** (60 mg, 73%) as a colourless oil. ^{1}H (500 MHz, CDCl$_3$) 7.24 (2 H, t, 7.4), 7.17 (1 H, td, 7.4, 2.3), 7.12 (1 H, d, 3.4), 7.09 (2 H, dt, 7.4, 2.3), 4.82 (1 H, dd, 6.3, 1.0), 3.72 (2 H, m), 3.47 (1 H, dt, 13.2, 6.3), 2.78 (1 H, m), 2.56 (1 H, dddd, 12.0, 8.1, 6.4, 3.8), 2.32 (1 H, ddd, 13.5, 8.1, 7.8), 2.06 (1 H, dddd, 13.2, 6.4, 6.3, 1.0), 1.72 (2 H, m), 1.36 (1 H, td, 13.2, 12.0), 1.04 (1 H, ddd, 13.5, 12.0, 3.8). ^{13}C NMR (125 MHz, CDCl$_3$) 170.3, 148.4, 136.8, 131.4, 128.3, 128.1, 126.7, 90.9, 81.2, 59.8, 48.2, 42.2, 37.7, 37.4, 31.2, 30.1. HRMS calcd for C$_{18}$H$_{20}$O$_4$Na [M + Na]: 323.1259, found: 323.1283.

2-(-2a-Hydroxy-2-oxo-2,2a,4,5,5a,6,7,7a-octahydroindeno [1,7-*bc*]furan-4-yl)ethyl 6-(5-((3a*S*,4*S*,6a*R*)-2-oxo-hexahydro-1*H*-thieno [3,4-*d*]imidazol-4-yl)pentanamido)hexanoate(**5**): To a dried flask containing the *N*-(+)-

biotinyl-6-aminohexanoic acid (50 mg, 0.13 mmol) was added thionyl chloride (0.50 ml, 6.3 mmol) under a nitrogen atmosphere and the mixture was stirred for 40 min at room temperature. The excess thionyl chloride was subsequently removed under reduced pressure and the crude product was used directly in the next step. **3** (19.2 mg, 0.09 mmol) was dissolved in 1.5 ml of freshly distilled MeCN and the previously prepared acid chloride dissolved in 1.5 ml MeCN was added dropwise under a nitrogen atmosphere. The reaction mixture was stirred for 5 h at room temperature and then concentrated under reduced pressure. The crude product was purified by chromatography (SiO$_2$, CHCl$_3$/MeOH 97:3) to afford **5** (9 mg, 19%). The two diastereomers have identical NMR data. ^{1}H (500 MHz, CD$_3$OD) 7.11 (1 H, d, 3.2), 4.70 (1 H, dd, 7.3, 2.4), 4.52 (1 H, dd, 7.8, 4.5), 4.33 (1 H, dd, 7.8, 4.5), 4.24 (2 H, m), 3.23 (1 H, m), 3.17 (2 H, t, 6.9), 2.94 (1 H, dd, 12.8, 4.5), 2.72 (1 H, d, 12.8), 2.66 (1 H, tdd, 12.2, 7.1, 3.2), 2.44 (1 H, dtd, 10.5, 7.5, 2.4), 2.35 (2 H, t, 7.8), 2.29 (1 H, m), 2.20 (2 H, t, 7.4), 2.11 (1 H, ddt, 14.5, 10.5, 7.3), 1.89 (1 H, td, 14.5, 7.5), 1.81 (1 H, m), 1.73 (1 H, m), 1.77 - 1.55 (7 H, m), 1.52 (2 H, dt, 12.2, 6.9), 1.44 (2 H, m), 1.35 (2 H, m), 1.16 (1 H, m), 1.09 (1 H, m). ^{13}C NMR (125 MHz, CD$_3$OD) δ 176.2, 175.4, 172.0, 166.2, 148.9, 133.4, 91.9, 83.0, 63.7, 63.6, 62.0, 57.1, 44.5, 41.1, 40.4, 36.9, 35.1, 35.0, 34.8, 32.9, 32.9, 32.7, 32.5, 30.2, 29.9, 29.6, 27.6, 27.1. HRMS calcd for C$_{28}$H$_{42}$N$_3$O$_7$S [M + H]: 564.2743, found: 564.2744.

2-(-2a-Hydroxy-2-oxo-7-phenyl-2,2a,4,5,5a,6,7,7a-octahydroindeno [1,7-*bc*]furan-4-yl)ethyl 6-(5-((3a*S*,4*S*, 6a*R*)-2-oxohexahydro-1*H*-thieno [3,4-*d*]imidazol-4-yl)pentanamido)hexanoate(**6**): **6** was prepared and purified in the same way as **5**, starting from **4** (10 mg, 0.033 mmol) to afford **6** (9.7 mg, 45%) as a colourless oil. The two diastereomers have identical NMR data. ^{1}H (500 MHz, CD$_3$OD) 7.27 (2 H, m), 7.20 (1 H, m), 7.17 (2 H, dd, 6.1, 2.7), 7.15 (1 H, s), 4.86 (1 H, d, 6.5), 4.47 (1 H, dd, 7.9, 4.8), 4.28 (1 H, m), 4.28 (2 H, dd, 13.1, 5.3), 3.58 (1 H, dt, 13.5, 6.5), 3.21 (1 H, m), 3.16 (2 H, dd, 8.9, 7.4), 2.92 (1 H, dt, 12.8, 4.8), 2.79 (1 H, m), 2.70 (1 H, dd, 12.8, 3.8), 2.62 (1 H, dddd, 10.0, 8.1, 6.5, 3.6), 2.36 (3 H, m), 2.19 (2 H, td, 7.4, 2.1), 2.09 (1 H, ddd, 13.5, 6.5, 5.5), 1.95 (1 H, td, 13.1, 6.2), 1.86 (1 H, m), 1.73 (1 H, m), 1.69 - 1.55 (6 H, m), 1.51 (2 H, m), 1.44 (2 H, m), 1.36 (2 H, m), 1.20 (1 H, ddd, 13.2, 8.1, 3.9). ^{13}C NMR (125 MHz, CD$_3$OD) 176.1, 175.4, 172.0, 166.3, 148.9, 138.9, 133.2, 129.7, 129.3, 127.8, 92.6, 82.7, 63.7, 63.5, 61.7, 57.2, 49.6, 43.6, 41.2, 40.3, 38.7, 37.0, 35.2, 35.0, 33.1, 30.82, 30.2, 29.9, 29.6, 27.6, 27.1, 25.9. HRMS calcd for C$_{34}$H$_{46}$N$_3$O$_7$S [M + H]: 640.3056, found: 640.3051.

(*S*)-*N*-Methoxy-*N*-methyl-2-phenylpent-4-enamide(**17**):To a solution of (*S*)-2-phenylpent-4-enoic acid (**16**) (6.8 g, 38 mmol), prepared from phenylacetic acid (**14**) according to reference 16 and identical in all aspects with the reported compound, in dry CH$_2$Cl$_2$ (4 ml) was added freshly distilled SOCl$_2$ (11.4 g, 96 mmol) at 0°C. The resulting brown mixture was stirred overnight at room temperature, where after the volatiles were removed under reduced pressure. The acid chloride, obtained as a black solid (7.8 g), was then dissolved in dry CH$_2$Cl$_2$ (40 ml) and cooled to 0°C before the addition of *N*-methyl-*O*-methylhydroxylamine hydrochloride (4.3 g, 44 mmol) under N$_2$. After stirring at 0°C for 2 h, pyridine (6.7 g, 84 mmol) was added and the mixture was stirred at room temperature overnight. The solvent was removed under vacuum, and the remaining pale yellow solid was dissolved in a 1:1mixture of Et$_2$O/CH$_2$Cl$_2$ (50 mL) and washed with brine (50 ml).The organic layer was dried over MgSO$_4$, and the solvents were removed under vacuum. The crude product was purified using flash chromatography (SiO$_2$, heptane/EtOAc 8:2) to afford 3.06 g (37% over two steps from **14** via **16**) of **17** as a colourless oil. [α]$_D^{20}$ +75.0 (*c* 0.2 in CDCl$_3$). ^{1}H (400 MHz, CDCl$_3$) 7.37 - 7.21 (5 H, m), 5.75 (1 H, ddt, J 17.0, 10.2, 6.9), 5.06 (1 H, ddd, J 17.0, 3.3, 1.5), 4.99 (1 H, ddd, J 10.2, 3.3, 1.0), 4.09 (1 H, m), 3.47 (3 H, s), 3.17 (3 H, s), 2.85 (1 H, m), 2.47 (1 H, m). ^{13}C (100 MHz, CDCl$_3$) 177.60, 139.64, 136.12, 128.56, 128.17, 126.99, 116.56, 61.29, 51.20, 38.21, 32.26. HRMS calcd for C$_{13}$H$_{18}$NO$_2$ [M + H]: 220.1338, found: 220.1343

(*S*)-4-Phenylhept-6-en-1-yn-3-one(**18**):To a solution of **17** (3.1 g, 14 mmol) in 20 ml of dry THF at −78°C was added dropwise a solution of ethynylmagnesium bromide (110 ml, 0.5 M in THF, 55 mmol) under N$_2$. The mixture was stirred at −78°C for 30 min and then allowed to warm to room temperature overnight. The reaction was quenched with 70 ml of NH$_4$Cl (sat). The layers were separated and the aqueous phase was extracted 3 times with Et$_2$O. The combined organic extracts were dried over MgSO$_4$ and concentrated under reduced pressure. Flash chromatography (SiO$_2$, heptane/EtOAc 7:3) afford 2.6 g (quantitative yield) of the corresponding ynone **18**. [α]$_D^{20}$ +73.3 (*c* 0.3 in CDCl$_3$). ^{1}H (400 MHz, CDCl$_3$) 7.37 (2 H, td, J 8.0, 1.9), 7.32 (1 H, tt, J 8.0, 1.9), 7.27 (2 H, dd, J 8.0, 1.9), 5.70 (1 H, ddt, J 17.0, 10.2, 7.1), 5.08 (1 H, ddd, J 17.0, 3.1, 1.5), 5.02 (1 H, ddd, J 10.2, 3.1, 1.1), 3.86 (1 H, t, J 7.1), 3.19 (1 H, s), 2.93 (1 H, dtd, J 14.4, 7.1, 1.1), 2.56 (1 H, dtd, J 14.4, 7.1, 1.5). ^{13}C (100 MHz, CDCl$_3$) 184.61, 134.74, 128.90, 128.69, 127.81, 117.33, 80.17, 77.20, 60.39, 35.38.HRMS calcd for C$_{13}$H$_{13}$O [M + H]: 185.0961, found: 185.0960.

(*S*)-4-Phenylhept-6-en-1-yn-3-ol(**19**):**18** (3.4 g, 18 mmol) was dissolved in dry THF (100 ml) and cooled to

−78°C before the addition of DIBAL (46 ml, 1M in THF) under N$_2$. The reaction mixture was stirred for 3 h at −78°C and then quenched by the addition of 3 ml of MeOH and 20 ml of a solution of Na/K tartrate (sat), extracted with EtOAc (40 ml x 3), dried over Na$_2$SO$_4$ and concentrated under reduced pressure. **19** (3.2 g, 92%) was obtained as a 1:0.24 diasteromeric mixture and used in the next step without further purification. *Diastereomer* a). ^{1}H (400 MHz, CDCl$_3$) 7.39 - 7.28 (5 H, m), 5.69 (1 H, ddd, 17.1, 10.1, 5.8), 5.07 (1 H, ddd, J 17.1, 3.3, 1.5), 4.98 (1 H, ddt, J 10.1, 3.3, 1.1), 4.55 (1 H, dd, J 5.8, 2.2), 2.98 (1 H, dt, J 9.2, 5.8), 2.78 (1 H, ddd, J 14.4, 5.8, 1.5), 2.57 (1 H, dd, 14.4, 9.2, 1.1), 2.53 (1 H, d, J 2.2). ^{13}C (100 MHz, CDCl$_3$) 139.17, 135.99, 128.86, 128.41, 127.22, 116.97, 83.25, 74.73, 65.72, 51.66, 35.35. *Diastereomer* b). ^{1}H (400 MHz, CDCl$_3$) 7.39 - 7.28 (5 H, m), 5.69 (1 H, ddd, 17.1, 10.1, 5.8), 5.09 (1 H, ddd, J 17.1, 3.3, 1.5), 4.98 (1 H, ddt, J 10.1, 3.3, 1.1), 4.59 (1 H, dd, J 5.8, 2.2), 3.03 (1 H, dt, J 9.2, 5.8), 2.78 (1 H, ddd, J 14.4, 5.8, 1.5), 2.57 (1 H, dd, 14.4, 9.2, 1.1), 2.49 (1 H, d, J 2.2). ^{13}C (100 MHz, CDCl$_3$) 139.02, 135.84, 129.19, 128.27, 128.00, 116.92, 83.25, 75.05, 65.58, 50.90, 35.29. HRMS calcd for C$_{13}$H$_{14}$ONa [M + Na]: 209.0942, found: 209.0946.

Methyl ((4S,Z)-4-phenyldeca-1,8-dien-6-yn-5-yl) carbonate(**20**): The Sonogashira reaction was performed in the corresponding way as when **8a/8b** was transformed to **10a/10b**, starting from **19** (3.4 g, 18 mmol) and *cis*-1-bromo-prop-1-ene (6.6 g, 55 mmol), and the crude product was used directly for the preparation of the dienyne carbonate **20** as described for **11a/11b**. **20** (0.70 g, 13%) was obtained as a 1:0.15 diastereomeric mixture after purification by flash chromatography (Si$_2$O, heptane/CH$_2$Cl$_2$ 1:1). *Diasteromer a*. ^{1}H (400 MHz, CDCl$_3$) 7.36 - 7.22 (5 H, m), 6.03 (1 H, dq, J 10.7, 6.9), 5.66 (1 H, ddt, J 17.2, 10.2, 7.0), 5.59 (1 H, dd, J 6.0, 1.9), 5.50 (1 H, ddd, 10.7, 1.7, 1.5), 5.03 (1 H, ddd, J 17.2, 3.4, 1.4), 4.96 (1 H, ddt, J 10.2, 3.4, 1.1), 3.76 (3 H, s), 3.14 (1 H, dt, J 6.0, 5.1), 2.77 (1 H, m), 2.65 (1 H, m), 1.82 (3 H, dd, J 6.9, 1.7). δ C (100 MHz, CDCl$_3$) 154.87, 140.19, 138.92, 135.46, 128.89, 128.25, 127.15, 116.99, 109.12, 88.69, 84.65, 71.89, 54.89, 49.70, 35.08, 16.05. *Diasteromer b*. ^{1}H (400 MHz, CDCl$_3$) 7.36 - 7.22 (5 H, m), 6.00 (1 H, dq, J 10.7, 6.9), 5.66 (1 H, ddt, J 17.2, 10.2, 7.0), 5.59 (1 H, dd, J 6.0, 1.9), 5.45 (1 H, ddd, 10.7, 1.7, 1.5), 5.01 (1 H, ddd, J 17.2, 3.4, 1.4), 4.96 (1 H, ddt, J 10.2, 3.4, 1.1), 3.80 (3 H, s), 3.14 (1 H, dt, J 6.0, 5.1), 2.77 (1 H, m), 2.65 (1 H, m), 1.74 (3 H, dd, J 6.9, 1.7). ^{13}C (100 MHz, CDCl$_3$) 154.87, 139.97, 139.17, 135.46, 128.78, 128.19, 127.11, 116.96, 109.09, 88.69, 84.65, 71.89, 54.89, 49.41, 35.47, 15.95. HRMS calcd for C$_{18}$H$_{20}$O$_3$Na [M + Na]: 307.1310, found: 307.1331

Methyl 6-methyl-(2R)-phenyl-2,6,7,7a-tetrahydro-1H-indene-4-carboxylate(**21**): The diene **21** was obtained following the same procedure as when **12a/12b** was prepared from **11a/11b**, starting with **20** (0.70 g, 2.4 mmol). **21** (0.32 g, 48%) was obtained as a 1:0.7 mixture of two diasteromers as a yellow oil, after flash chromatography (SiO$_2$, heptane/Et$_2$O 95:5). *Diasteromer* a).^{1}H (400 MHz, CDCl$_3$) 7.35 - 7.17 (5 H, m), 6.81 (1 H, t, J 2.8), 6.33 (1 H, t, J 2.7), 4.04 (1 H, dd, J 3.0, 1.6), 3.79 (3H, s), 2.98 (1 H, m), 2.65 (1 H, m), 2.12 (1 H, m), 2.06 (1 H, t, J 5.4), 1.99 (1 H, dt, J 12.7, 9.6), 1.14 (3 H, dd, J 7.3, 5.7), 1.07 (1 H, m). ^{13}C (100 MHz, CDCl$_3$) 166.93, 166.91, 147.73, 145.90, 138.18, 130.52, 128.34, 127.19, 125.96, 51.59, 50.32, 43.34, 39.69, 38.45, 32.92, 20.74. *Diasteromer* b). ^{1}H (400 MHz, CDCl$_3$) 7.35 - 7.17 (5 H, m), 6.81 (1 H, t, J 2.8), 6.41 (1 H, t, J 2.8), 4.00 (1 H, ddt, J 10.5, 3.0, 1.6), 3.80 (3 H, s), 2.86 (1 H, ddt, J 13.0, 9.5, 2.8), 2.65 (1 H, m), 2.59 (1 H, dt, J 12.2, 7.1), 2.06 (1 H, t, J 5.4), 1.37 (1 H, dt, J 12.3, 10.3), 1.16 (3 H, dd, J 7.3, 5.7), 1.07 (1 H, m). ^{13}C (100 MHz, CDCl$_3$) 166.93, 166.91, 147.60, 145.50, 138.00, 129.53, 128.36, 127.43, 126.11, 51.59, 51.27, 44.66, 42.09, 38.34, 33.14, 20.83. HRMS calcd for C$_{18}$H$_{21}$O$_2$ [M + H]: 269.1542, found: 269.1550.

(1aS,2S,3aS,5S,7aS)-Methyl 5-methyl-2-phenyl-1a,2,3,3a,4,5-hexahydroindeno [1,7a-b]oxirene-7-carboxylate (**22**):**22** was prepared following the same procedure as when **13a/13b** was prepared from **12a/12b**, starting from **21** (0.31 g, 1.17 mmol). **22** (0.14 g, 42%) was obtained after flash chromatography (SiO$_2$, heptane/Et$_2$O 85:15). [α]$_D$ ^{20} + 8.5 (c 4.1 in CDCl$_3$). ^{1}H (400 MHz, CDCl$_3$) 7.42 (1 H, m), 7.32 (2 H, m), 7.23 (2 H, m), 7.17 (1 H, dd, J 7.4, 2.2), 4.57 (1 H, s), 3.75 (3 H, s), 3.52 (1 H, m), 2.56 (1 H, m), 2.14 (1 H, m), 1.94 (1 H, m), 1.90 (1 H, m), 1.64 (1 H, m), 1.40 (1 H, ddd, J 14.8, 9.7, 7.5), 1.20 (3 H, dd, J 7.3). ^{13}C (100 MHz, CDCl$_3$) 165.40, 155.17, 141.93, 128.61, 127.49, 126.58, 125.84, 66.40, 64.14, 62.15, 51.71, 44.67, 37.81, 33.76, 32.19, 20.40. HRMS calcd for C$_{18}$H$_{21}$O$_3$ [M + H]: 285.1491, found: 285.1472.

(2aS,4S,5aS,7S,7aR)-2a-Hydroxy-4-methyl-7-phenyl-4,5,5a,6,7,7a-hexahydroindeno [1,7-bc]furan-2(2aH)-one(**2**):**2** was prepared in the same way as when **13a/13b** was transformed to **3/4**, starting from **22** (0.14 g, 0.5 mmol). The crude product was purified by chromatography (SiO$_2$, heptane/EtOAc 60:40.) to afford **2** (30 mg, 22%) as a colourless oil. [α]$_D$ ^{20} +59 (c 12.5 in CDCl$_3$). ^{1}H (400 MHz, CDCl$_3$) 7.30 (2 H, m), 7.23 (1 H, m), 7.13 (2 H, m), 7.10 (1 H, d, J 3.2), 4.90 (1 H, dd, J 6.7, 1.4), 3.52 (1 H, dt, J 13.4, 6.7), 2.77 (1 H, m), 2.62 (1 H, m), 2.36 (1 H, dt, J 14.0, 7.6), 2.12 (1 H, dtd, J 12.7, 6.7, 1.4), 1.45 (1 H, dd, J 13.4, 12.7), 1.25 (3 H, d, J 7.3), 1.12 (1 H, m). ^{13}C (100 MHz, CDCl$_3$) 170.0, 150.3, 136.8, 130.7, 128.4, 128.2, 126.8, 90.5, 81.3, 48.2, 42.2, 37.6, 31.6, 29.1,

20.6. HRMS calcd for $C_{17}H_{19}O_3$ [M + H]: 271.1334, found: 271.1351

WST-1 cell proliferation assay

The functional activity of **1**, **2**, **3** and **4** were evaluated using a WST-1 proliferation assay with DU145 cells (ATCC, American Type Culture Collection, LGC Standards AB, Borås, Sweden) which expresses constitutively activated STAT3, as previously described [6]. In short, DU145 cells were cultured in 96-well plates (2000 cells/well in 200 μl of medium) and allowed to set for 24 h. The cells were treated with the test items for 72 h. Samples were made in triplicate. 20 μl WST-1 solution (Roche Applied Science) was added per well and incubated at 37°C for 4 h. The absorbance of each well was measured using a scanning multi-well spectrophotometer, ELISA reader at a wavelength of 450 nm and reference wavelength of 690 nm. The results are presented as per cent of untreated control cells.

3. Results and Discussion

Our aim was to introduce a suitable functional group on the C-4 methyl group, as a handle to attach biotin via a linker. Initial attempts to introduce an alkyne or an azide to introduce the biotin group by "click chemistry" [12] failed. Instead a primary hydroxyl group was chosen as the functional group handle to attach biotin through an ester coupled linker, and the synthetic procedures for obtaining the galiellalactone analogues **3** and **4** as well as their biotinylated derivatives **5** and **6** are summarized in **Scheme 1**.

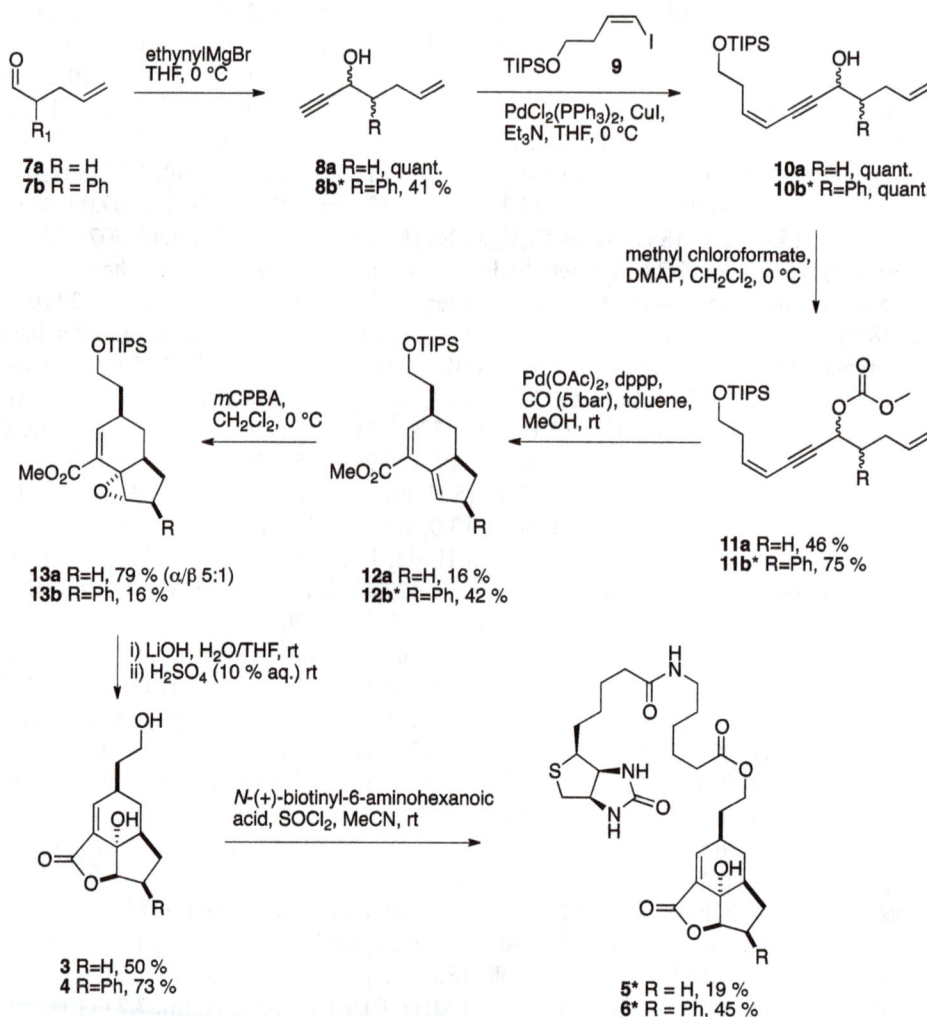

Scheme 1. Synthesis of 8-hydroxymethyl galiellalactone biotin 6-aminohexanoic acid amide ester (**5**) and 8-hydroxymethyl-7-phenyl galiellalactone biotin 6-aminohexanoic acid amide ester (**6**). *Intermediates **8b**, **10b**, **11b** and **12b** as well as the end products **5** and **6** were obtained as mixtures of diastereomers.

Ethynyl magnesium bromide was added to 5-pentenal (**7a**) to give the propargylic alcohol **8a** in a quantative yield. A Sonogashira coupling of the terminal alkyne of **8a** with the known *cis*-iodoalkene **9** [13] containing a TIPS-protected hydroxyl group gave the alkenyne **10a**. This was treated with methyl chloroformate in the presence of DMAP to give the carbonate **11a**, the key substrate for the tandem palladium catalyzed carbonylation/intramolecular Diels-Alder reaction. The ring forming transformation using Pd(OAc)$_2$ and 1,3-bis(diphenylphosphino)propane under a CO atmosphere [8] gave the bicyclic product **12a** with the TIPS protected hydroxyl group intact. Regioselective epoxidation of the electron-rich double bond with *m*-CPBA gave the desired epoxide as a 5:1 α:β diastereomeric mixture, from which the desired isomer **13a** could be isolated. **13a** was subjected to the conditions bringing about the lactone formation [8], in this step the TIPS protecting group was lost and 8-hydroxymethyl galiellalactone **3** was isolated in 50% yield. The synthesis of **3** proceeded with good or acceptable yields, except for the tandem palladium catalyzed carbonylation/intramolecular Diels-Alder reaction (from **11a** to **12a**), but this transformation gives poorer yields when the cyclopentene ring is unsubstituted. [8]

N-(+)-biotinyl-6-aminohexanoic acid had previously been described as a suitable reagent for making biotin conjugates of a biologically active compound [14]. However, attempts to couple *N*-(+)-biotinyl-6-aminohexanoic acid using standard reagents for esterification failed, presumably because a base is used and the tricyclic galiellalactone system is sensitive to basic conditions. Galiellalactone (**1**) is relatively stable in acidic conditions and towards acid chlorides, so by first treating *N*-(+)-biotinyl-6-aminohexanoic acid with thionyl chloride to generate the corresponding acid chloride and then adding the alcohol **3** in acetonitrile the C-8 biotin conjugate **5** was prepared and could be isolated in 19% yield.

The corresponding biotin labeled 7-phenyl substituted derivative **6** was synthesized in an analogous manner. Starting from aldehyde **7b**, prepared by a DIBAL-H reduction of the corresponding 2-phenyl pent-4-ene acid ester, the key intermediate **11b** could be obtained in three steps as a mixture of diastereomers. As we previously have observed, the tandem carbonylation/intramolecular Diels-Alder reaction gives a higher yield when a substituent in position 7 is present [8], and the cyclization of **11b** proceeds in a 42% yield although the product is obtained as a mixture of two diastereomers. The epoxide **13b** was isolated as a pure diastereomer but in a relatively poor yield, due to reduced stereoselectivity in the epoxidation, but the lactone formation to afford **4** runs smoother. The relative configuration of **3** as well as **4** was established by NMR NOESY experiments. Finally, the synthesis of **6** was accomplished by the esterification with *N*-(+)-biotinyl-6-aminohexanoic acid chloride.

To investigate the effect of a phenyl group in position 7, an asymmetric synthesis of 7-phenyl galiellalactone (**2**) was devised. The key objective was to obtain the substituted propargylic alcohol **19** with the absolute stereochemistry of C-4 set as shown in **Scheme 2**, as this would influence the stereochemistry of the subsequent stereocenters generated in the cyclization step. By applying a procedure developed for the enantioselective direct alkylation of arylacetic acids [15] using Koga's base (**15**) [16], (2*S*)-phenylpent-4-enoic acid (**16**) [17] was prepared from phenylacetic acid (**14**). By this procedure, **16** is obtained with 88% ee [17], and the material obtained here was identical to that reported. The acid **16** was converted to the corresponding Weinreb amide **17**, which subsequently was reacted with ethynyl magnesium bromide to give the ynone **18**. Reduction of the ketone with DIBAL gave the desired propargylic alcohol **19**. The final steps from **19** via **20**, **21**, and **22** to **2** followed the procedure outlined in **Scheme 2**. **19** and **20** were obtained as mixtures of two diastereomers, in the ratios 1:0.24 and 1:0.15, but both diastereomers of **20** give the same product in the cyclization step. **21** was obtained as a mixture of two diastereomers in the ratio 1:0.7, the major is shown in **Scheme 2** and this was readily epoxidized to **22**.

To determine how the structural changes in **2**, **3** and **4** affected their ability to block STAT3 signaling compared to galiellalactone (**1**), they were assayed for their ability to inhibit proliferation of DU145 prostate cancer cells which express active STAT3 *in vitro*, using a WST-1 assay (see **Figure 2**) [6]. The hydroxyl methyl analogue **3** had significantly reduced ability to inhibit proliferation (GIC$_{50}$ = 32 μM) compared to **1** (GIC$_{50}$ = 3.7 μM), but the biological activity was restored when a phenyl group was added in position 7 as observed with analogue **4** (GIC$_{50}$ = 4.6 μM). Compared with the natural product, the 7-phenyl analogue **2** has a slightly improved potency (GIC$_{50}$ = 2.9 μM).

4. Conclusion

The biotinylated analogues **5** and **6** will be used in various biological studies to elucidate the mechanism of action of galiellalactone, starting with its effect on STAT3 signaling [5]. The fact that both **5** and **6** retain the

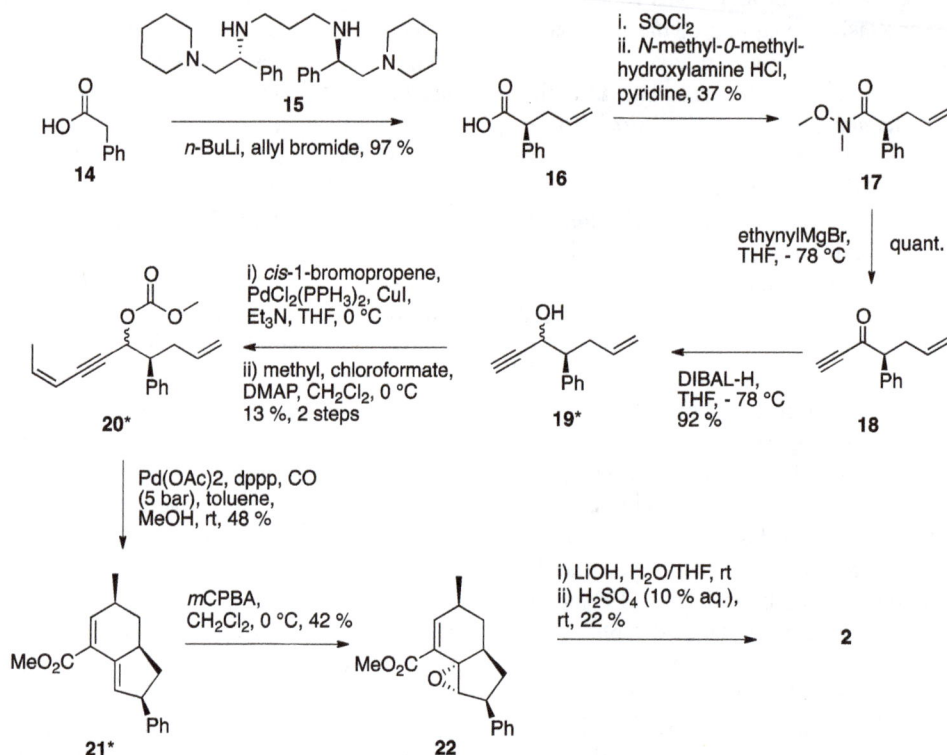

Scheme 2. Asymmetric synthesis of 7-phenyl galiellalactone (**2**). *Intermediates **19**, **20** and **21** were obtained as mixtures of diastereomers.

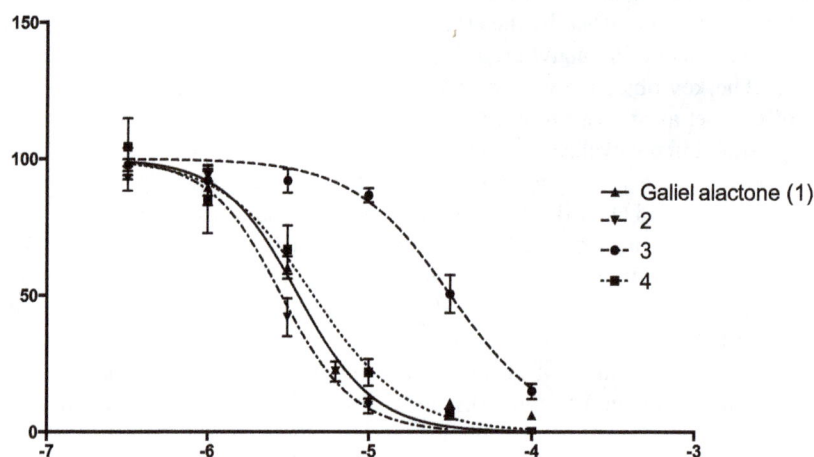

Figure 2. Cell proliferation dose response curve. Galiellalactone (**1**), **2**, **3** and **4** inhibit proliferation of DU145 cells with GIC$_{50}$ values of 3.7 µM, 2.9 µM, 32 µM and 4.6 µM, respectively after 72 h. As discussed below, the biotinylated derivatives **5** and **6** retained the inhibitory effect.

ability to inhibit the proliferation of DU145 cells expressing constitutively active STAT3 (with the GIC$_{50}$ values 6.6 and 14 µM, respectively, in the same WST-1 assay [5]) demonstrates that the attachment of biotin via a linker to position 4 of **1** produces cell permeable covalent STAT3 inhibitors that can be used as important tools for elucidating the mechanism of action of **1**. With the biotinylated analogues synthesized in this work, in combination with biochemical methods, protein isolation and mass spectrometry, it was possible to show that **1** is a direct inhibitor of STAT3 by alkylating critical cysteine residues [6].

Acknowledgements

This work was supported by the Swedish Science Research Council.

References

[1] Peyser, N.D. and Grandis, J.R. (2013) Critical Analysis of the Potential for Targeting STAT3 in Human Malignancy. *Journal of OncoTargets and Therapy*, **6**, 999-1010. http://dx.doi.org/10.1002/jcp.10364

[2] Calò, V., Migliavacca, M., Bazan, V., Macaluso, M., Buscemi, M., Gebbia, N. and Russo, A. (2003) STAT Proteins: From Normal Control of Cellular Events to Tumorigenesis. *Journal of Cellular Physiology*, **197**, 157-168.

[3] Shodeinde, A.L. and Barton, B.E. (2012) Potential Use of STAT3 Inhibitors in Targeted Prostate Cancer Therapy: Future Prospects. *OncoTargets and Therapy*, **5**, 119-125.

[4] Hellsten, R., Johansson, M., Dahlman, A., Sterner, O. and Bjartell, A. (2008) Galiellalactone is a Novel Therapeutic Candidate against Hormone-Refractory Prostate Cancer Expressing Activated Stat3. *Prostate*, **68**, 269-280. http://dx.doi.org/10.1002/pros.20699

[5] Hellsten, R., Johansson, M., Dahlman, A., Sterner, O. and Bjartell, A. (2011) Galiellalactone Inhibits Stem Cell-Like ALDH-Positive Prostate Cancer Cells. *PLoS One*, **6**, e22118. http://dx.doi.org/10.1371/journal.pone.0022118

[6] Don-Doncow, N., Escobar, Z., Johansson, M., Kjellstrom, S., Garcia, V., Munoz, E., Sterner, O., Bjartell, A. and Hellsten, R. (2014) Galiellalactone Is a Direct Inhibitor of STAT3 in Prostate Cancer Cells. *Journal of Biological Chemistry*, **289**, 15969-15978. http://dx.doi.org/10.1074/jbc.M114.564252

[7] Johansson, M. (2002) Biosynthetic and SyntheticStudies of the Fungal Metabolite Gliellalactone. Ph.D. Dissertation, Bioorganic Chemistry, Lund University, Lund.

[8] Gidlof, R., Johansson, M. and Sterner, O. (2010) Tandem Pd-Catalyzed Carbonylation and Intramolecular Vinyl Allene Diels-Alder Reaction toward Galiellalactone Analogues. *Organic Letters*, **12**, 5100-5103. http://dx.doi.org/10.1021/ol101989m

[9] Johansson, M., and Sterner, O. (2001) Synthesis of (+)-Galiellalactone. Absolute Configuration of Galiellalactone. *Organic Letters*, **3**, 2843-2845. http://dx.doi.org/10.1021/ol016286+

[10] Johansson, M. and Sterner, O. (2002) Synthesis of (−)-Galiellalactone. *The Journal of Antibiotics*, **55**, 663-665.

[11] Von Nussbaum, F., Hanke, R., Fahrig, T. and Benet-Buchholz, J. (2004) The High-Intrinsic Diels-Alder Reactivity of (−)-Galiellalactone; Generating Four Quaternary Carbon Centers under Mild Conditions. *European Journal of Organic Chemistry*, **2004**, 2783-2790. http://dx.doi.org/10.1002/ejoc.200400137

[12] Best, M.D. (2009) Click Chemistry and Bioorthogonal Reactions: Unprecedented Selectivity in the Labeling of Biological Molecules. *Biochemistry*, **48**, 6571-6584. http://dx.doi.org/10.1021/bi9007726

[13] La Cruz, T.E. and Rychnovsky, S.D. (2007) A Reductive Cyclization Approach to Attenol A. *Journal of Organic Chemistry*, **72**, 2602-2611. http://dx.doi.org/10.1021/jo0626459

[14] Honda, T., Janosik, T., Honda, Y., Han, J., Liby, K.T., Williams, C.R., Couch, R.D., Anderson, A.C., Sporn, M.B. and Gribble, G.W. (2004) Design, Synthesis, and Biological Evaluation of Biotin Conjugates of 2-Cyano-3,12-dioxooleana-1,9(11)-dien-28-oic Acid for the Isolation of the Protein Targets. *Journal of Medicinal Chemistry*, **47**, 4923-4932. http://dx.doi.org/10.1021/jm049727e

[15] Stivala, C.E. and Zakarian, A. (2011) Highly Enantioselective Direct Alkylation of Arylacetic Acids with Chiral Lithium Amides as Traceless Auxiliaries. *Journal of the American Chemical Society*, **133**, 11936-11939. http://dx.doi.org/10.1021/ja205107x

[16] Frizzle, M.J., Caille, S., Marshall, T.L., McRae, K., Nadeau, K., Guo, G., Wu, S., Martinelli, M.J. and Moniz, G.A. (2007) Dynamic Biphasic Counterion Exchange in a Configurationally Stable Aziridinium Ion: Efficient Synthesis and Isolation of a Koga C_2-Symmetric Tetraamine Base. *Organic Process Research & Development*, **11**, 215-222. http://dx.doi.org/10.1021/op0602371

[17] Alliot, J., Gravel, E., Pillon, F., Buisson, D.A., Nicolas, M. and Doris, E. (2012) Enantioselective Synthesis of Levo-milnacipran. *Chemical Communications*, **48**, 8111-8113. http://dx.doi.org/10.1039/c2cc33743f

10

Dehydrosulfurization of Aromatic Thioamides to Nitriles Using Indium(III) Triflate

Tomoko Mineno*, Yu Takebe, Chiaki Tanaka, Sho Mashimo

Laboratory of Medicinal Chemistry, Faculty of Pharmacy, Takasaki University of Health and Welfare, Takasaki, Gunma, Japan
Email: *mineno@takasaki-u.ac.jp

Abstract

The efficient dehydrosulfurization of thioamides to nitriles was carried out using indium(III) triflate as a catalyst. Based on the results of the initial study, the optimal reaction conditions required 5 mol% of indium(III) triflate with toluene as the practical solvent. Various thioamides were successfully converted to nitriles in high yields.

Keywords

Indium(III) Triflate, Dehydrosulfurization, Thioamide, Nitriles

1. Introduction

Functional group conversions are of significant importance in the field of organic synthesis. Hence, the development of novel functional group conversions is a world-wide pursuit. Organosulfur compounds have been recognized as highlyuseful precursors, or synthons, in order to demonstrate functional grouptransformation [1]. Due to their considerable versatility and unique reactivity, thioamides are receiving increased attention, and many precedent interconverting reactions have been introduced. However, the dehydrosulfurization of thioamides to nitriles is often implemented under harsh reaction conditions that require elongated reaction times. Several studies that have focused on the dehydrosulfurization of thioamides to nitrileshave used reagents such as diphosphorus tetraiodide [2], 2, 4-dichloro-5-nitropyrimidine [3], and aryl chlorothionoformate [4] or a combination of reagents like $S_8/NaNO_2/NH_3$ [5] and benzyl chloride/tetra-n-butylammonium bromide as a phase transfer catalyst [6]. In alternative conversions of thioamides to nitriles, researchers have employed metal reagents by intro-

*Corresponding author.

ducing reactions with silver carboxylates [7] [8], manganese oxide [9], and *n*-butyltin oxide [10]. Ogura *et al.* has reported for the dehydrosulfurization of thioamides to nitriles by using selenium-based and tellurium-based reagents [11]-[15], while Enthaler *et al.* has reported using iron-based and zinc-based reagents [16] [17]. We have studied the catalytic utility of indium reagents, and reported the chemical application of indium reagents, including an efficient conversion of primary amides to nitriles using catalytic indium(III) triflate [18]-[24]. In ongoing research, we have found that the catalytic use of indium(III) triflate facilitates the dehydrosulfurization of thioamides to nitriles. Herein, we describe the details of our study.

2. Results and Discussion

Using 3-methyl thiobenzamide as the starting substrate, a search for the optimal conditions of dehydrosulfuriza-tion was attempted, as presented in **Table 1**. The amount of *N*-methyl-*N*-(trimethylsilyl)trifluoroacetamide (MSTFA) was fixed at 3.5 equivalents, as established in our previous report (**Figure 1**) [24], and thus the cata-lytic amount of indium(III) triflate and appropriate solvents were investigated. The entries using toluene as the solvent afforded the expected products (**Table 1**, entries 1 and 2), whereas the entries using THF ended with no reactions (**Table 1**, entries 3 and 4). The use of 5 mol% of indium(III) triflate was also discovered as one of the key factors for the optimal reaction conditions (**Table 1**, entry 2).

Once the optimal conditions for the dehydrosulfurization of thioamides to nitriles were obtained, we next ex-amined their applicability. Various aromatic thiobenzamides were subjected to dehydrosulfurization reactions. As shown in **Table 2**, the reactions proceeded smoothly to give the corresponding nitriles in good to excellent yields [24]. The alkyl group substituents,such as methyl on the aromatic ring and the *tert*-butyl groups were inert under the reaction conditions of dehydrosulfurization, and gave high yields (**Table 2**, entries 1, 3, and 5) [25] [26]. The dehydrosulfurization reactions of halogenated aromatic thioamides were carried out without dehalo-genation (**Table 2**, entries 2 and 7) [27]. Also, the reaction led to the desired nitriles without damaging methoxy functionality (**Table 2**, entries 4 and 6). Consequently, the scope of the silyl protecting groups was investigated. The conversion reaction was successfully conducted maintaining both *tert*-butyldimethylsilyl (TBDMS) and *tert*-butyldiphenylsilyl (TBDPS) functionalities, and furnished the desired products in excellent order (**Table 2**, entries 8 and 9), which supported the mildness of the reaction conditions by not affecting the acid labile silyl functionalities. Furthermore, these reaction processes were extended to bicyclic aromatic compounds. The de-hydrosulfurization reactions starting with naphthalene-1-thiocarboxamide and naphthalene-2-thiocarboxamide provided the corresponding nitriles in excellent yields (**Table 2**, entries 10 and 11). In addition, the entry using 2, 2-diphenylthioacetamide, which is a non-aromatic, afforded the desired nitrile productin an 84% yield (**Table 2**, entry 12) [28].

Table 1. Reactions in the search for optimal conditions.

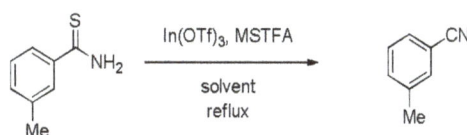

Entry	Solvent	MSTFA	In(OTf)$_3$	Reflux Time (h)	Yield[a] (%)
1	toluene	3.5 eq.	10 mol%	3	86
2	toluene	3.5 eq.	5 mol%	3	quant
3	THF	3.5 eq.	10 mol%	3	N.R.
4	THF	3.5 eq.	5 mol%	3	N.R.

[a]Isolated yields.

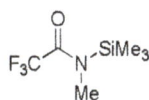

MSTFA

Figure 1. Structure of *N*-methyl-*N*-(trimethylsilyl)trifluoroacetamide (MSTFA).

Table 2. Dehydrosulfurization of aromatic thioamides to nitriles.

Entry	Starting substrate	Product	Reflux Time (h)	Yield[a] (%)
1			3	quant
2			2	84
3			3	52
4			3	88
5			3	85
6			3	quant
7			3	94
8			2	quant
9			3	86
10			3	94
11			2	97
12			2.5	84

3. Conclusion

In conclusion, we have established an efficient method for the dehydrosulfurization of thioamides to nitriles in the presence of a catalytic amount of indium(III) triflate. Many aromatic thioamides were subjected to these reaction conditions, and provided the corresponding nitriles in good to excellent yields. The dehydrosulfurization reaction conditions were sufficiently mild, so as to not cleave the acid labile silyl functionalities. Further mechanistic investigations are ongoing.

4. Experimental

4.1. Materials and Instruments

All reagents were of analytical grade and were purchased commercially and used without further purification. All reactions were carried out under an argon atmosphere using magnetic stirring unless otherwise noted. ^{1}H NMR and ^{13}C NMR spectral data were recorded on a JEOL JMTC-500 spectrometer using TMS as an internal standard.

4.2. General Experimental Procedure

The starting thioamide substrates (1 mmol) and In(OTf)$_3$ (5 mol%) were dissolved in dehydrated toluene (6 mL) contained in a 100 mL flask equipped with a magnetic stirrer and a reflux condenser. MSTFA (3.5 mmol) was added using a syringe at room temperature. The reaction mixture was heated at reflux for 3 h, and was monitored for completion by TLC. After the reaction mixture was cooled to room temperature, the solvent was washed with aqueous solutions and concentrated by rotary evaporation. Flash column chromatography on silica gel furnished the corresponding nitrile product, which was confirmed by spectroscopy [24]-[28].

References

[1]   Satchell, D.P.N. (1977) Metal-Ion-Promoted Reactions of Organo-Sulphur Compounds. *Chemical Society Reviews*, **6**, 345-371. http://dx.doi.org/10.1039/CS9770600345

[2]   Suzuki, H., Tani, H. and Takeuchi, S. (1985) Desulfurization of Thioketones and Thioamides with Diphosphorus Tetraiodide. *Bulletin of the Chemical Society of Japan*, **58**, 2421-2422. http://dx.doi.org/10.1246/bcsj.58.2421

[3]   Kondo, K., Komamura, C., Murakami, M. and Takemoto, K. (1985) 2,4-Dichloro-5-nitropyrimidine as a New Dehydrating or Desulfhydrating Reagent. *Synthetic Communications*, **15**, 171-177. http://dx.doi.org/10.1080/00397918508063784

[4]   Bose, D.S. and Goud, P.R. (1999) Aryl Chlorothionoformate: A New Versatile Reagent for the Preparation of Nitriles and Isonitriles under Mild Conditions. *Tetrahedron Letters*, **40**, 747-748. http://dx.doi.org/10.1016/S0040-4039(98)02361-2

[5]   Sato, R., Itoh, K., Itoh, K., Nishina, H., Goto, T. and Saito, M. (1984) Novel Conversion of Aromatic Thioamides and Aldehydes into Nitriles with Elemental Sulfur and Sodium Nitrite in Liquid Ammonia. *Chemistry Letters*, **11**, 1913-1916. http://dx.doi.org/10.1246/cl.1984.1913

[6]   Funakoshi, Y., Takido, T. and Itabashi, K. (1985) Facile Conversion of Primary Thioamides into Nitriles with Benzyl Chloride under Phase Transfer Conditions. *Synthetic Communications*, **15**, 1299-1303. http://dx.doi.org/10.1080/00397918508077278

[7]   Avalos, M., Babiano, R., Duran, C.J., Jimenez, J.L. and Palacios, J.C. (1994) Reaction of Thioamides with Silver Carboxylates in Aprotic Media. A Nucleophilic Approach to the Synthesis of Imides, Amides, and Nitriles. *Tetrahedron Letters*, **35**, 477-480. http://dx.doi.org/10.1016/0040-4039(94)85085-2

[8]   Avalos, M., Babiano, R., Cintas, P., Duran, C.J., Higes, F.J., Jimenez, J.L., Lopez, I. and Palacios, J.C. (1997) Reaction of Thioamides with Metal Carboxylates in Organic Media. *Tetrahedron*, **53**, 14463-14480. http://dx.doi.org/10.1016/S0040-4020(97)00938-1

[9]   Yamaguchi, K., Yajima, K. and Mizuno, N. (2012) Facile Synthesis of Nitriles via Manganese Oxide Promoted Oxidative Dehydrosulfurization of Primary Thioamides. *Chemical Communications*, **48**, 11247-11249. http://dx.doi.org/10.1039/c2cc36635e

[10] Lim, M.-I., Ren, W.-Y. and Klein, R.S. (1982) Facile Conversion of Primary Thioamides into Nitriles with Butyltin Oxides. *Journal of Organic Chemistry*, **47**, 4594-4595. http://dx.doi.org/10.1021/jo00144a043

[11] Hu, N.X., Aso, Y., Otsubo, T. and Ogura, F. (1985) Mild and Selective Oxidations with Polystyrene-Bound Diaryl Selenoxide. *Chemistry Letters*, **14**, 603-606. http://dx.doi.org/10.1246/cl.1985.603

[12] Hu, N.X., Aso, Y., Otsubo, T. and Ogura, F. (1986) Polymer-Supported Diaryl Selenoxide and Telluroxide as Mild and Selective Oxidizing Agents. *Bulletin of the Chemical Society of Japan*, **59**, 879-884. http://dx.doi.org/10.1246/bcsj.59.879

[13] Hu, N.X., Aso, Y., Otsubo, T. and Ogura, F. (1986) Novel Oxidizing Properties of p-Methoxybenzenetellurinic Acid Anhydride. *Tetrahedron Letters*, **27**, 6099-6102. http://dx.doi.org/10.1016/S0040-4039(00)85408-8

[14] Fukumoto, T., Matsuki, T., Hu, N.X., Aso, Y., Otsubo, T. and Ogura, F. (1990) Benzenetellurinic Mixed Anhydrides as Mild Oxidizing Agents. *Chemistry Letters*, **19**, 2269-2272. http://dx.doi.org/10.1246/cl.1990.2269

[15] Aso, Y., Omote, K., Takagi, S., Otsubo, T. and Ogura, F. (1995) Mild and Efficient Dehydrosulfurization of Thioamides to Nitriles Induced by Tellurium or Selenium Tetrachloride with Triethylamine. *Journal of Chemical Research, Synopses*, No. 4, 152-153.

[16] Enthaler, S. (2011) Straightforward Iron-Catalyzed Synthesis of Nitriles by Dehydration of Primary Amides. *European Journal of Organic Chemistry*, **2011**, 4760-4763. http://dx.doi.org/10.1002/ejoc.201100754

[17] Enthaler, S. and Inoue, S. (2012) An Efficient Zinc-Catalyzed Dehydration of Primary Amides to Nitriles. *Chemistry—An Asian Journal*, **7**, 169-175. http://dx.doi.org/10.1002/asia.201100493

[18] Mineno, T. (2002) A Fast and Practical Approach to Tetrahydropyranylation and Depyranylation of Alcohols Using Indium Triflate. *Tetrahedron Letters*, **43**, 7975-7978. http://dx.doi.org/10.1016/S0040-4039(02)01864-6

[19] Mineno, T. and Kansui, H. (2006) High Yielding Methyl Esterification Catalyzed by Indium(III) Chloride. *Chemical & Pharmaceutical Bulletin*, **54**, 918-919. http://dx.doi.org/10.1248/cpb.54.918

[20] Mineno, T., Nikaido, N. and Kansui, H. (2009) One-Step Transformation of Tetrahydropyranyl Ethers Using Indium(III) Triflate as the Catalyst. *Chemical & Pharmaceutical Bulletin*, **57**, 1167-1170. http://dx.doi.org/10.1248/cpb.57.1167

[21] Mineno, T., Sakai, M., Ubukata, A., Nakahara, K., Yoshimitsu, H. and Kansui, H. (2013) The Effect of Indium(III) Triflate in Oxone-Mediated Oxidative Methyl Esterification of Aldehydes. *Chemical & Pharmaceutical Bulletin*, **61**, 870-872. http://dx.doi.org/10.1248/cpb.c13-00072

[22] Mineno, T., Tsukagoshi, R., Iijima, T., Watanabe, K., Miyashita, H. and Yoshimitsu, H. (2014) Reductive Coupling Reaction of Aldehydes Using Indium(III) Triflate as the Catalyst. *Tetrahedron Letters*, **55**, 3765-3767. http://dx.doi.org/10.1016/j.tetlet.2014.05.079

[23] Mineno, T., Yoshino, S. and Ubukata, A. (2014) Oxone-Mediated Oxidative Esterification of Heterocyclic Aldehydes Using Indium(III) Triflate. *Green and Sustainable Chemistry*, **4**, 20-23. http://dx.doi.org/10.4236/gsc.2014.41004

[24] Mineno, T., Shinada, M., Watanabe, K., Yoshimitsu, H., Miyashita, H. and Kansui, H. (2014) Highly-Efficient Conversion of Primary Amides to Nitriles Using Indium(III) Triflate as the Catalyst. *International Journal of Organic Chemistry*, **4**, 1-6. http://dx.doi.org/10.4236/ijoc.2014.41001

[25] Littke, A., Soumeillant, M., Kaltenbach III, R.F., Cherney, R.J., Tarby, C.M. and Kiau, S. (2007) Mild and General Methods for the Palladium-Catalyzed Cyanation of Aryl and Heteroaryl Chlorides. *Organic Letters*, **9**, 1711-1714. http://dx.doi.org/10.1021/ol070372d

[26] Grossman, O. and Gelman, D. (2006) Novel Trans-Spanned Palladium Complexes as Efficient Catalysts in Mild and Amine-Free Cyanation of Aryl Bromides under Air. *Organic Letters*, **8**, 1189-1191. http://dx.doi.org/10.1021/ol0601038

[27] Yang, C. and Williams, J.M. (2004) Palladium-Catalyzed Cyanation of Aryl Bromides Promoted by Low-Level Organotin Compounds. *Organic Letters*, **6**, 2837-2840. http://dx.doi.org/10.1021/ol049621d

[28] Katritzky, A.R., Akue-Gedu, R. and Vakulenko, A.V. (2007) C-Cyanation with 1-cyanobenzotriazole. *ARKIVOC*, **2007**, 5-12. http://dx.doi.org/10.3998/ark.5550190.0008.302

11

Optimization of Grafted Fibrous Polymer as a Solid Basic Catalyst for Biodiesel Fuel Production

Yuji Ueki*, Seiichi Saiki, Takuya Shibata, Hiroyuki Hoshina, Noboru Kasai, Noriaki Seko

Environment and Industrial Materials Research Division, Quantum Beam Science Center, Sector of Nuclear Science Research, Japan Atomic Energy Agency, Takasaki, Japan
Email: *ueki.yuji@jaea.go.jp

Abstract

Grafted fibrous polymer with quaternary amine groups could function as a highly-efficient catalyst for biodiesel fuel (BDF) production. In this study, the optimization of grafted fibrous polymer (catalyst) and transesterification conditions for the effective BDF production was attempted through a batch-wise transesterification of triglyceride (TG) with ethanol (EtOH) in the presence of a cosolvent. Trimethylamine was the optimal quaternary amine group for the grafted fibrous catalyst. The optimal degree of grafting of the grafted fibrous catalyst was greater than 170%. The optimal transesterification conditions were as follows: The optimal molar quantity of quaternary amine groups, transesterification temperature, molar ratio of TG and EtOH, and primary alkyl alcohol were 0.8 mmol, 80°C, 1:200, and 1-pentanol, respectively. The grafted fibrous catalyst could be applied to BDF production using natural oils. Furthermore, the grafted fibrous catalyst could be used repeatedly after regeneration involving three sequential processes, *i.e.*, organic acid, alkali, and alcohol treatments, without any significant loss of catalytic activity.

Keywords

Biodiesel Fuel, Triglyceride Transesterification, Radiation-Induced Graft Polymerization, Grafted Polymer, Heterogeneous Basic Catalysis

1. Introduction

The global energy consumption has increased every year and has more than doubled between 1970 and 2008

---

*Corresponding author.

(1970: 207 quadrillion British thermal units (Btu); 2008: 505 quadrillion Btu); experts predict that it will increase by another 53% by 2035 to become 770 quadrillion Btu [1] [2]. This growing global demand for energy will likely lead to the depletion of fossil fuels, unprecedented emissions of air pollutants and greenhouse gases, high energy-resource prices, and regional conflicts due to uneven distribution of fossil fuels. Therefore, to avoid and overcome these problems, many researchers worldwide are working on the development of methods to harvest energy from renewable sources such as solar, wind, hydraulic, geothermal, and biomass, together with innovative practical technologies. Among these, biomass energy is garnering attention because it has a high energy density and requires relatively facile handling and storage. Currently, studies on the production of biomethane from cellulose and organic waste [3] [4], production of bioethanol from starch and nonfood crops [5] [6], conversion of vegetable oils, animal fats, and waste oils into biodiesel fuel (BDF) [7]-[15], and production of biofuels from algae [16] [17] are in progress.

BDF is defined as the monoalkyl esters of long-chain fatty acids (carbon chain lengths of 12 - 24), and is generally produced by transesterification of triglyceride (TG), which is the main constituent of vegetable oils and animal fats, with short-chain primary alcohols such as methanol or ethanol (EtOH). BDF has a number of important advantages: It is renewable, environmentally friendly, easily biodegradable, and compatible with current fuel infrastructure, and its production process, transport, and storage are very straightforward. Additionally, BDF can be used in compression-ignition engines instead of petroleum diesel with little or no modifications to the engine components. For these reasons, BDF has attracted significant attention as an alternative fossil fuel. BDF production can be classified into several methods according to the type of catalyst: homogeneous catalyst (alkaline and acid), heterogeneous catalyst (metal oxides, carbonates, zeolites, heteropoly acids, functionalized zirconia or silica, ion-exchange resins, hydrotalcites and alkaline salts) and enzymatic (lipases) catalysts and a noncatalytic supercritical method [7]-[15]. Some methods for the industrial production of BDF from oil and fat have already been developed; currently, an alkali catalyst method using a homogeneous catalyst such as NaOH or KOH is predominantly used because of its relatively fast reaction rates. However, this method has some disadvantages: The alkali catalyst corrodes equipment, the reaction requires a large amount of water during the neutralization and washing processes, undesirable byproducts are produced by a saponification side reaction, and it is difficult to separate the homogenized catalyst from the reaction mixture. These problems increase the BDF production costs.

To solve or minimize these problems, authors proposed the use of grafted fibrous polymer as a solid basic catalyst [18]. Grafted fibrous polymer (grafted fibrous catalyst) was synthesized by radiation-induced graft polymerization which could impart a desired functional group into pre-existing polymeric materials (*i.e.*, trunk polymer) without altering their inherent properties, and the grafted fibrous polymer was used as adsorbents for environmental reclamation and obtaining resources [19]-[25]. In particular, when nonwoven fabric having both large specific surface area and high contact efficiency is selected as a trunk polymer for grafting, the BDF production speed (*i.e.*, transesterification speed) of the grafted fibrous catalyst was more than twice that of commercial granular anion exchange resin, and the grafted fibrous catalyst could efficiently produce BDF within a shorter period of time [18]. This difference in the transesterification speed of the grafted fibrous catalyst and anion exchange resin was mostly attributed to their shape differences: The porous anion exchange resin had numerous micropores that increase the surface area. Thus, most of the reaction sites (*i.e.*, functional groups) of the porous resin were located within its micropores, and hence the reactants were transported to the functional groups by diffusion (*i.e.*, concentration gradient). Therefore, diffusional mass-transfer was the rate-determining factor for BDF production. In contrast, in the grafted fibrous catalyst, the functional groups were immobilized onto the graft chains, and the reactants were easily and immediately transported to the functional groups by convective flow of the reaction mixture. Therefore, the diffusional mass-transfer resistance of the reactants to the functional groups could be neglected. Additionally, the catalyst had further advantages: it was less corrosive than homogeneous alkali catalysts, did not produce soap, and was easily recovered from the BDF, which negated the need for neutralization and washing. For these advantages, authors think that the grafted fibrous catalyst will contribute to simplify and streamline the BDF production process, reduce the BDF production costs, reduce the waste product, and establish the large-scale BDF production process.

In our previous paper, we found that the grafted fibrous catalyst could function as a highly-efficient catalyst for BDF production [18]. The objective of this study is to characterize the catalytic properties of the grafted fibrous catalyst and determine the optimal BDF production conditions using the grafted fibrous catalyst. First, we demonstrated the effectiveness of type of immobilized quaternary amine groups onto the grafted fibrous catalyst, from both perspectives of the ease of amination and the catalytic performance. Second, the effects of the catalyst and transesterification conditions on the catalytic activity, transesterification speed, and production yield of BDF were elucidated by batch-wise transesterification of TG with EtOH in the presence of a cosolvent. Additionally, the effect of the type of natural oil on the transesterification yield and speed was investigated. Finally, the recyclability of the grafted fibrous catalyst and regeneration of the deactivated grafted fibrous catalyst were evaluated.

2. Experimental

2.1. Materials

Polyethylene-coated polypropylene (PE/PP) nonwoven fabric was supplied by Kurashiki Textile Manufacturing Co., Ltd. (Osaka, Japan) and used as a trunk polymer for BDF catalysts. The average fiber diameter of the PE/PP nonwoven fabric was 13 μm. 4-Chloromethylstyrene (CMS; purity > 95%) was purchased from AGC Seimi Chemical Co., Ltd. (Kanagawa, Japan). Polyoxyethylene (20) sorbitan monolaurate (Tween 20), which was used as a nonionic surfactant for preparing the monomer emulsion, was obtained from Kanto Chemical Co., Inc. (Tokyo, Japan), as were trimethylamine (TMA), triethylamine (TEA), tri-n-propylamine (TPA), tri-n-butylamine (TBA), triolein (purity > 60%), and oleic acid (purity > 80%). Sodium hydroxide, methanol, EtOH, 1-propanol, 2-propanol, 1-butanol, 1-pentanol, 1-hexanol, 1-octanol, 1-decanol, 1-dodecanol, acetonitrile, hexane, decane, linseed oil, beef tallow, and citric acid were supplied by Wako Pure Chemical Industries, Ltd. (Osaka, Japan). Rapeseed oil (51% erucic acid) and safflower oil (food-grade, 70% - 80% linoleic acid) were purchased from MP Biomedicals, LLC. (CA, USA). Palm oil (32% - 47% palmitic acid, 34% - 44% oleic acid) was obtained from Spectrum Chemical Mfg. Corp. (NJ, USA). EtOH, 2-propanol, acetonitrile, and hexane were of HPLC grade and all other chemicals were of reagent grade unless otherwise stated. Deionized water obtained from a Milli-Q deionization system (Nihon Millipore K.K., Tokyo, Japan) was used directly for preparing the monomer emulsions and aqueous solutions.

2.2. Synthesis of Grafted Fibrous Catalyst for Biodiesel Fuel Production

Grafted fibrous catalyst for BDF production was synthesized by radiation-induced emulsion grafting technique, as published previously [18]. PE/PP nonwoven fabric was irradiated with electron beam (100 kGy). Then, the irradiated PE/PP nonwoven fabric reacted with a deaerated monomer emulsion, which was composed of 3 wt% CMS, 0.3 wt% tween 20 and 96.7 wt% deionized water, in a glass ampoule and was kept in water bath at 40°C. Afterward, the CMS-grafted nonwoven fabric was recovered from the emulsion and washed repeatedly with water and then methanol to remove residual monomers and homopolymers. The amount of CMS grafted onto the PE/PP nonwoven fabric was evaluated by the degree of grafting (Dg [%]), which was calculated using the following equation:

$$Dg[\%] = 100 \times (W_1 - W_0)/W_0 ,$$

where W_0 and W_1 are the dry weights of the PE/PP nonwoven fabrics before and after grafting, respectively.

To introduce quaternary amine groups into the CMS-graft chains of CMS-grafted nonwoven fabric (catalyst precursor), the catalyst precursor was treated with 0.25 and 2.5 M amine-ethanol solutions at 60°C for 24 h. To eliminate residual amine reagents, the aminated catalyst precursor was then washed with deionized water until it was neutral. The resultant aminated catalyst precursor is referred to as the grafted fibrous catalyst. The quaternary amine-group densities of the grafted fibrous catalysts were estimated by an analysis of their nitrogen content using an elemental analyzer (model: 2400 Series II CHNS/O elemental analyzer, PerkinElmer, Inc., MA, USA). The quaternary amine-group density of the grafted fibrous catalyst and degree of amination (Da [%]) of the CMS-graft chain are defined as follows:

Quaternary amine-group density [*mmol-amine/g-catalyst*] = (N/100)/14 × 1000, and

$$Da[\%] = 100 \times \left[152.6 \times (W_2 - W_0)/(W_1 - W_0) - 152.6 \right]/M_{Amine} ,$$

where N, W_2, and M_{Amine} are the measured nitrogen content (%) of the grafted fibrous catalyst, weight of the grafted fibrous catalyst, and molecular weight of the introduced amine compound, respectively. The molecular weights of TMA, TEA, TPA, and TBA are 59.11, 101.19, 143.27 and 185.35 g/mol, respectively, while the values 14 and 152.6 refer to the molecular masses of N and CMS, respectively.

2.3. Transesterification Procedure with Grafted Fibrous Catalyst

The catalytic performance of the grafted fibrous catalyst was evaluated by the transesterification of TG and alcohols under the following standard conditions, as published previously [18].

The transesterification tests were performed in a batch reactor equipped with a magnetic stirrer. The reactor was initially filled with 1.4 g (1.6 mmol) of triolein as a TG and 3.6 g (78 mmol) of EtOH as an alcohol, followed by the addition of 5.0 g of decane as a cosolvent in order to obtain a homogeneous phase. After stirring for 10 min, approximately 0.2 g of the pretreated grafted fibrous catalyst was immersed in the homogenous reaction solution, and the resultant mixture was heated to the reaction temperature (standard: 50°C) while stirring at 600 rpm. The reaction time is set at zero when the grafted fibrous catalyst was added to the reaction solution. In a typical test, the molar ratio of TG to EtOH was 1:50, and the Dg, type of quaternary amine group, and amine group density of the grafted fibrous catalyst were ~300%, TMA, and 3.6 mmol-amine/g-catalyst, respectively. The molar ratio of TG to quaternary amine groups in the reaction mixture was fixed at 2:1, *i.e.*, 0.8 mmol of quaternary amine group of the grafted fibrous catalyst was in the reaction mixture per test. At pertinent intervals, 0.1 mL aliquots were withdrawn from the reaction solution, quenched to room temperature, and diluted to 10 times the volume with 2-propanol/hexane (5:4, v/v) for compositional analysis. The effect of the molar ratio of TG to quaternary amine group (16:1 - 2:1), type of quaternary amine group, reaction temperature (20°C - 80°C), molar ratio of TG to EtOH (1:3 - 1:500), type of alcohol, and reaction time on the conversion ratio of TG to biodiesel fuel were investigated in detail.

2.4. Compositional Analysis

The compositions of the collected samples were analyzed using high-performance liquid chromatography (HPLC), according to the separation conditions described by Holčapek *et al.* [26]. The HPLC system (Shimadzu, Kyoto, Japan) consisted of a pump (model: LC-20AD) with a quaternary gradient system, degasser (model: DGU-20A5), system controller (model: CBM-20Alite), sample injector (model: 7725i; Rheodyne, CA, USA) with a 1.0 μL sample reservoir, column oven (model: CTO-20AC), and UV-Vis detector (model: SPD-20A). System control and data processing were carried out using Shimadzu LC solution software. A reversed-phase C18 column (model: L-column ODS, column size: 2.1 mm i.d. × 150 mm long, particle diameter: 5 μm, Chemicals Evaluation and Research Institute, Tokyo, Japan) was used for separation. The compositional analysis was conducted with a quaternary mobile phase consisting of deionized water (solvent A), acetonitrile (solvent B), 2-propanol (solvent C), and hexane (solvent D). The quaternary gradient elution program was as follows: 30% A + 70% B at 0 min, 100% B at 5 min, 50% B + 27.8% C + 22.2% D at 10 min, and isocratic elution with 50% B + 27.8% C + 22.2% D for the last 5 min. The flow rate was set at 1.0 mL/min, the effluent was monitored at 205 nm, and the column temperature was maintained at 40°C. The conversion ratio of TG to BDF was calculated from the rate of change of the total HPLC peak areas for TG before and after transesterification.

3. Results and Discussion

3.1. Effect of Type of Quaternary Amine Groups on Catalytic Performance

To produce BDF using grafted fibrous catalyst, OH⁻ ions, which were chemically immobilized on the grafted fibrous catalyst, are the actual catalytic elements, while the polymer matrix acts as a scaffold to immobilize these ions. Although OH⁻ ions can be easily immobilized onto the grafted fibrous catalyst by interionic interaction with the cationic functional groups such as a quaternary amine group, it is considered that the type of quaternary amine group has possibility to have a significant influence on the catalytic performance. In order to select the best quaternary amine group of grafted fibrous catalyst, we discussed the effects of four types of amines with different alkyl chain lengths, such as TMA, TEA, TPA, and TBA, from both perspectives of the reactivity with CMS-grafted chains and the catalytic performance.

A CMS-grafted nonwoven fabric with 170% Dg was used as the catalyst precursor, and each amine concentration was fixed at 0.25 M. As expected, the quaternary amine-group density increased with increasing reaction time, and TMA, which has the shortest alkyl chain length, exhibited the highest reactivity toward the CMS-graft chains. As shown in **Figure 1**, amination with 0.25 M TMA began immediately and completed within 1 h; the quaternary amine-group density and Da reached 3.0 mmol-TMA/g-catalyst and 93%, respectively. However, other amine reagents with longer alkyl chain lengths, such as TEA, TPA, and TBA, were less reactive because of the increased steric hindrance of the amine compounds. Amination of the CMS-graft chains did not completely finish even after 24 h of amination with 0.25 M amine solution; the Da for the products of the reactions with TEA, TPA, and TBA were 81%, 63%, and 65%, respectively. To enhance these values, the CMS-grafted nonwoven fabric was treated with 2.5 M TEA-, TPA-, and TBA-EtOH solutions. After 24 h of amination with 2.5 M TEA, TPA, and TBA, Da values of 91%, 85%, and 88%, respectively, were achieved. These values are comparable to that for TMA. However, longer alkyl chain lengths of the amine compounds led to reduced quaternary amine-group densities per gram of grafted fibrous catalysts, even when the Da values for each amine were comparable to that of TMA. As denoted in **Figure 1**, the quaternary amine-group density for each amine reached 2.6 mmol-TEA/g-catalyst, 2.2 mmol-TPA/g-catalyst, and 2.0 mmol-TBA/g-catalyst after 24 h of amination. The effect of the Dg of the CMS-grafted nonwoven fabric on amination was also investigated, and it was found that the Dg value did not significantly influence the Da: The Da values after 1 h of amination with 0.25 M TMA were almost constant at 93%, regardless of the Dg. The quaternary amine-group densities were 2.6, 3.3, 3.6, and 3.8 mmol-TMA/g-catalyst for Dg values of 100%, 200%, 300%, and 400%, respectively.

Next, the effect of the type of immobilized quaternary amine groups on the catalytic activity was investigated; these results are given in **Figure 2**. As described in Section 2.3, transesterification was conducted in batch mode,

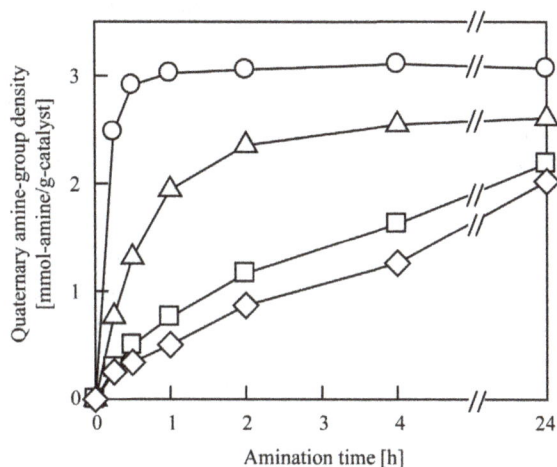

Figure 1. Effect of the type of amine reagents on the quaternary amine-group density of CMS-grafted nonwoven fabric.

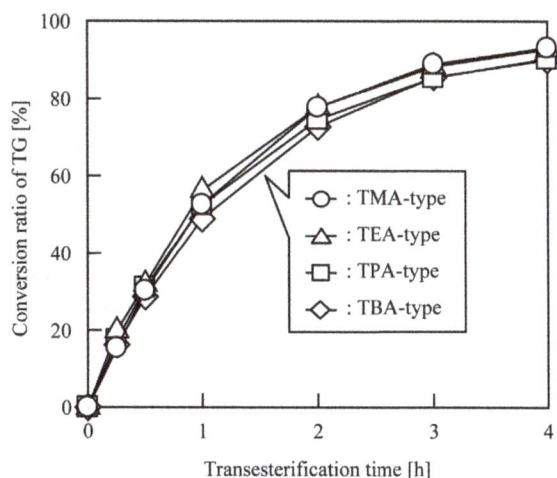

Figure 2. Effect of the type of immobilized quaternary amine group on the catalytic activity.

and the molar quantity of quaternary amine groups of each catalyst in the reaction mixture was fixed at 0.8 mmol. As is evident from **Figure 2**, each catalyst exhibited similar catalytic activity toward BDF production regardless of the type of quaternary amine groups, and hence steric bulk, and transesterification proceeded with all the grafted fibrous catalysts. The conversion ratios of TG using TMA-, TEA-, TPA-, and TBA-type fibrous catalysts were 93%, 93%, 90%, and 90%, respectively, after 4 h of transesterification. The similarity of the transesterification speeds is attributed to the function of the quaternary amine groups, *i.e.*, as a scaffold to immobilize the OH⁻ ions that are the actual catalytic elements; therefore, the quaternary amine groups were not directly involved in transesterification. From the results of the amination of CMS-grafted nonwoven fabric and the effect of the type of quaternary amine groups on the catalytic activity, we concluded that TMA is the optimal quaternary amine group for the grafted fibrous catalyst because it is the smallest trialkylamine and thus could easily be immobilized onto the CMS-graft chains. The order of reactivity of the trialkylamines for the amination of CMS-graft chains was TMA >> TEA > TPA > TBA.

3.2. Effect of Catalyst Conditions on Catalytic Performance

The yield and rate of BDF production were greatly dependent on the catalyst conditions such as the Dg value, the shape and the fiber diameter of the grafted fibrous catalyst, and the transesterification conditions including the molar quantity of quaternary amine groups in the reaction mixture, transesterification temperature, molar ratio of TG to EtOH in the reaction mixture, type of alcohol, and so on. In this section, the effects of the catalyst conditions on catalytic performance were investigated.

First, the effects of the Dg value of the grafted fibrous catalyst on the catalytic activity and yield were investigated. In this experiment, the Dg of the grafted fibrous catalyst was controlled within a range up to 400%, and the molar quantity of TMA in the reaction mixture was 0.8 mmol per test. The other conditions were the same as the typical conditions described in Section 2.3. Regardless of the Dg value, each catalyst exhibited almost the same catalytic activity toward BDF production, except in the case of the grafted fibrous catalysts with 80% Dg or less. The TG conversion ratios using catalysts with 80%, 170%, 200%, 300%, and 400% Dg reached 70%, 87%, 91%, 93%, and 89%, respectively, after 4 h of transesterification. The decline in the transesterification efficiency of grafted fibrous catalyst with less than 80% Dg is attributed to the large volumes of these catalysts, which are sterically bulky, required to introduce 0.8 mmol of TMA into the reaction system. As a result, part of the grafted fibrous catalyst was not immersed in the reaction mixture during transesterification. Based on these results, we concluded that the optimal Dg value of the grafted fibrous catalyst to achieve effective transesterification was greater than 170%. The TMA group density of 170% Dg is 3.1 mmol-TMA/g-catalyst.

The fiber diameter of the grafted fibrous catalyst is also an important factor that affects the catalytic activity; accordingly, the relationship between the fiber diameter of the grafted fibrous catalyst and catalytic activity was examined. In this test, four types of grafted fibrous catalysts with different fiber diameters (*i.e.*, 26, 39, 47, and 67 μm) were used. The grafted fibrous catalysts with different fiber diameters were prepared by controlling the fiber diameter of the nonwoven fabric to be 13, 19, 24, and 34 μm. The Dg and quaternary amine-group density of all grafted fibrous catalysts were fixed at approximately 300% and 3.6 mmol-TMA/g-catalyst, and the other conditions were the same as the typical conditions described in Section 2.3. Each catalyst exhibited almost the same catalytic activity toward BDF production regardless of the fiber diameter of the grafted fibrous catalyst, although the fiber diameter of the trunk polymer had a significant influence on the grafting efficiency. During emulsion grafting, trunk polymers with finer fibers had a larger specific surface area, which increased the speed and efficiency of the graft polymerization. A similar result was reported by Basuki *et al.* [27]. In contrast, during transesterification, the molar quantity of TMA in the reaction mixture was more important than the specific surface area because the reactants could be easily and immediately transported to the functional groups by the convective flow of the reaction mixture. Based on these results, the fiber diameter of the grafted fibrous catalyst did not significantly influence the catalytic activity. The conversion ratios of TG using the catalysts with fiber diameters of 26, 39, 47, and 67 μm were 93%, 92%, 88%, and 90%, respectively, after 4 h of transesterification.

3.3. Effect of Transesterification Conditions on Catalytic Performance

The effect of the molar quantity of quaternary amine group on the transesterification rate was investigated; these results are shown in **Figure 3**. In this experiment, the molar quantities of quaternary amine groups varied from 0.1 to 0.8 mmol TMA, and the other conditions were the same as the typical conditions described in Section 2.3.

As expected, the transesterification rate increased with increasing molar quantity of TMA in the reaction mixture, and the conversion ratios of TG using 0.1, 0.2, 0.4, and 0.8 mmol TMA reached 27%, 48%, 76%, and 93%, respectively, after 4 h of transesterification. In this experiment, we concluded that the optimal molar quantity of quaternary amine groups (*i.e.*, TMA) to achieve effective transesterification was 0.8 mmol per test.

Furthermore, the effect of the transesterification temperature on the transesterification rate of the grafted fibrous catalyst was studied in detail, and the results are shown in **Figure 4(a)**. For comparison, the results for commercial granular anion exchange resin are provided in **Figure 4(b)**. DIAION PA306S (Mitsubishi Chemical Co., Ltd., Tokyo, Japan) was used as the commercial granular anion exchange resin with TMA quaternary amine groups at a density of 4.2 mmol-TMA/g-dry resin with a particle diameter of 150 to 425 μm. In this experiment, the transesterification temperature was controlled within the range of 20°C to 80°C and TMA-type grafted fibrous catalyst was used. The other conditions were the same as the typical conditions described in Section 2.3. As indicated in **Figure 4**, although both the grafted fibrous catalyst and commercial granular resin exhibited good catalytic activities at all investigated transesterification temperatures, the transesterification rate of the grafted fibrous catalyst was faster than that of DIAION PA306S at all temperatures. As expected, higher transesterification temperatures led to faster transesterification rates for both catalysts. In particular, the transesterification was remarkably accelerated at high temperatures for the grafted fibrous catalyst. As seen in **Figure 4(a)**, in the case of the grafted fibrous catalyst, about 3 h of transesterification was required to convert 90% of the initial TG into

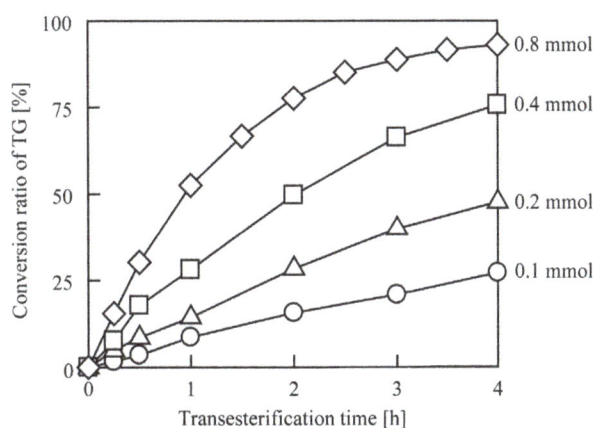

Figure 3. Effect of molar quantity of quaternary amine groups on transesterification speed.

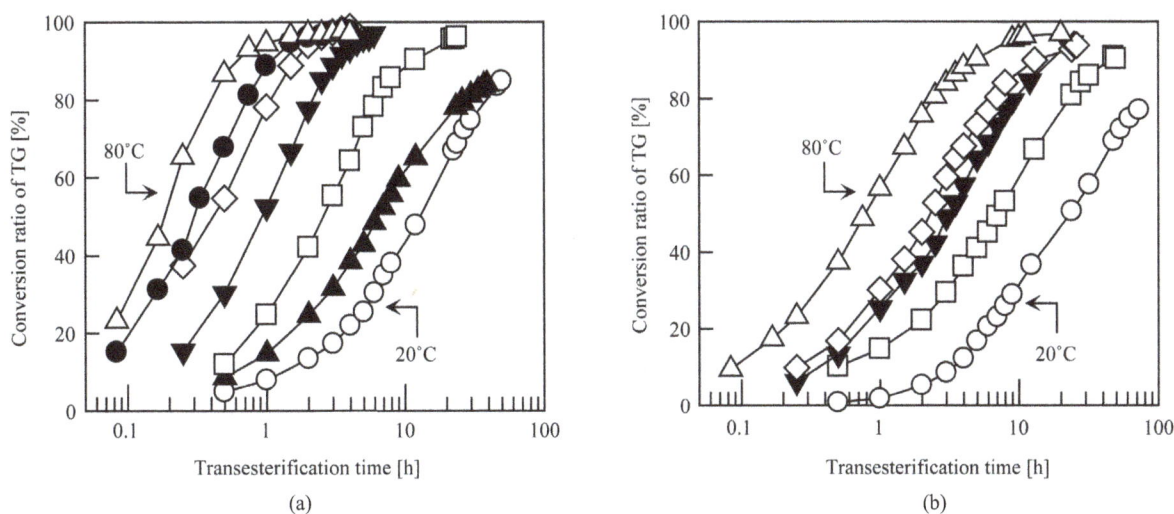

Figure 4. Effect of transesterification temperature on transesterification speed. Catalyst: (a) grafted fibrous catalyst and (b) commercial granular resin. Transesterification temperature: ○, 20°C; ▲, 30°C; □, 40°C; ▼, 50°C; ◇, 60°C; ●, 70°C; △, 80°C.

BDF at 50°C. In contrast, at 80°C, the required time to achieve a conversion ratio of 90% was only 40 min, which is less than one fourth the time required at 50°C. Based on these results, higher reaction temperatures resulted in more effective and efficient transesterification and, therefore, the grafted fibrous catalyst could dramatically reduce the time required for BDF production than the granular resin.

The ratio of TG to alcohol in the reaction mixture is also known to be a controlling factor for BDF production. Therefore, the effects of the molar ratio of TG to EtOH in the reaction mixture on the transesterification yield and rate were investigated; the results are given in **Figure 5**. In this experiment, the molar ratios of TG to EtOH were adjusted to 1:3, 1:10, 1:25, 1:50, 1:100, 1:200, and 1:500, which corresponded to TG/EtOH volume ratios of 1:0.2, 1:0.6, 1:1.5, 1:3, 1:6, 1:12, and 1:30, respectively. The other conditions were the same as the typical conditions described in Section 2.3. As denoted in **Figure 5**, the molar ratio of TG to EtOH had a significant influence on the transesterification yield and rate: Transesterification was accelerated in the presence of excess EtOH. The transesterification times required to convert 90% of the initial TG into BDF were 4.47, 3.18, 2.09, 1.51 and 1.56 h for TG/EtOH molar ratios of 1:25, 1:50, 1:100, 1:200 and 1:500, respectively. TG/EtOH molar ratios of 1:3 and 1:10, the conversion ratio did not reach 90% even after 50 h. There are two potential reasons for these phenomena: One is that the EtOH in the reaction mixture was vaporized by heating during transesterification, and the other is that the relatively low volume proportion of EtOH to TG in the reaction mixture resulted in heterogeneous dispersion of the EtOH. The TG/EtOH molar ratios of 1:3 and 1:10 were corresponded to the TG/EtOH volume ratios of 1:0.2 and 1:0.6, respectively. In contrast, when the percentage of EtOH in the reaction mixture was significantly high, as at a TG/EtOH molar ratio of 1:500, EtOH diluted the reaction mixture and thereby hindered the rate of transesterification. Based on these results, the optimal molar ratio of TG and EtOH in the reaction mixture for the fastest BDF production was 1:200.

Finally, the effects of the type of alcohol on the transesterification yield and speed were investigated. In this test, nine primary alcohols with different linear alkyl chain lengths, *i.e.*, 1 to 12 carbon atoms, were used; these were methanol (C1), EtOH (C2), 1-propanol (C3), 1-butanol (C4), 1-pentanol (C5), 1-hexanol (C6), 1-octanol (C8), 1-decanol (C10), and 1-dodecanol (C12). The molar ratio of TG to primary alkyl alcohol was fixed at 1:50, and the other conditions were the same as the typical conditions described in Section 2.3. **Figure 6** shows HPLC chromatograms of the BDF samples produced from 30 min of transesterification of triolein with each alcohol, and the conversion ratio of TG with each alcohol is plotted versus the transesterification time in **Figure 7**. As shown in **Figure 6**, BDF was produced regardless of the primary alkyl alcohol used; therefore, the grafted fibrous catalyst is applicable to transesterification of TG with a variety of alcohols. Furthermore, the peaks corresponding to BDF gradually shifted to longer retention times as the alkyl chain length of the primary alkyl alcohol increased; this is attributed to the differences in the structures (and hydrophobicity) of the produced BDF, and

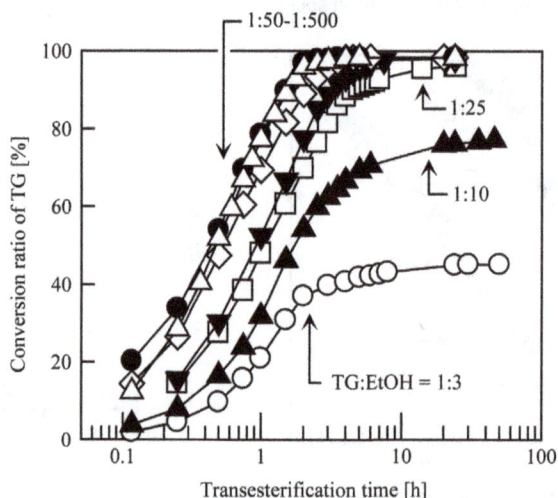

Figure 5. Effects of molar ratio of TG to EtOH in reaction mixture on transesterification yield and speed. Molar ratio of TG to EtOH: ○, 1:3; ▲, 1:10; □, 1:25; ▼, 1:50; ◇, 1:100; ●, 1:200; △, 1:500.

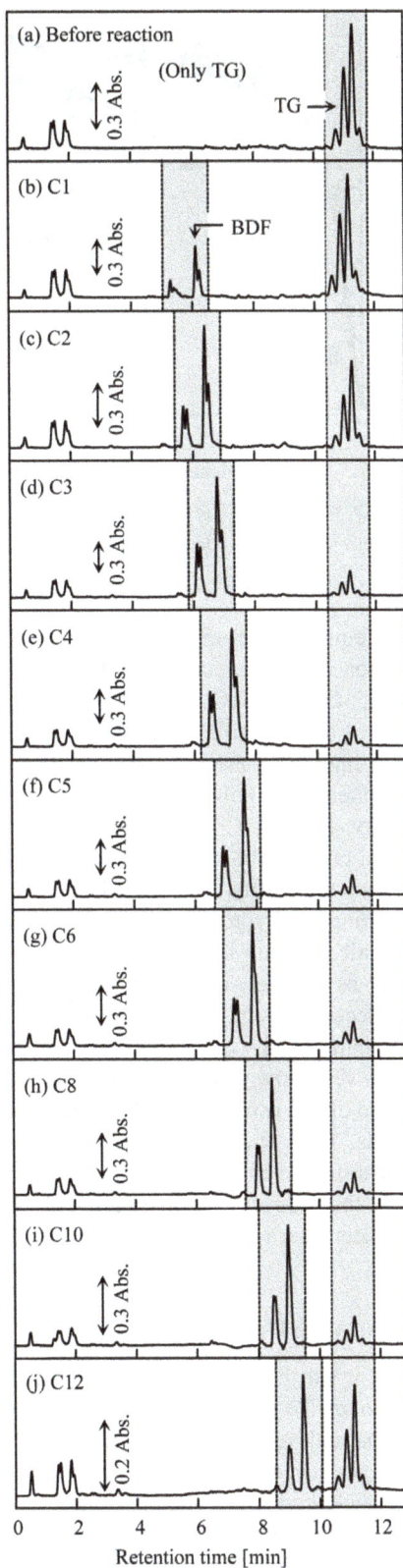

Figure 6. HPLC chromatograms of BDF samples produced by transesterification of triolein with different types of alcohol. Transesterification time: 30 min.

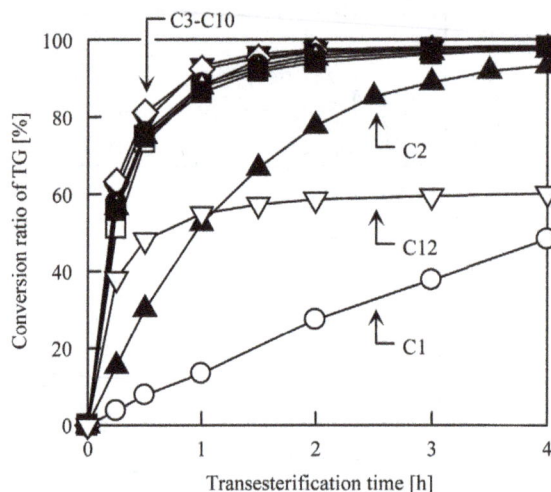

Figure 7. Effects of the type of alcohol on transesterification yield and speed. Alcohol: ○, C1; ▲, C2; □, C3; ▼, C4; ◇, C5; ●, C6; △, C8; ■, C10; ▽, C12.

demonstrated that differed BDFs were produced from different types of alcohol. As can be seen from **Figure 7**, the yield and speed of BDF production varied depending on the type of primary alkyl alcohol: The conversion ratios of TG using C1, C2, C3, C4, C5, C6, C8, C10, and C12 were 8%, 30%, 73%, 79%, 81%, 76%, 75%, 74%, and 48%, respectively, after 30 min of transesterification. These results indicate that transesterification is accelerated when more hydrophobic (*i.e.*, lipophilic) primary alkyl alcohols were used. However, as is evident from the results of using C12, very hydrophobic alcohols resulted in slower transesterification than primary alkyl alcohols with moderate hydrophobicity, such as C4, C5, and C6. Furthermore, transesterification using C12 slowed significantly at the conversion ratio of about 55% during the first 1 h; afterward, transesterification progressed gradually to reach a conversion ratio of 64% after 24 h. This decline in transesterification yield is attributed to the primary alkyl alcohol being too hydrophobic to react with the OH⁻ ions, which are highly hydrophilic, although highly hydrophobic alcohols are highly miscible with TG. In contrast, when primary alkyl alcohols with shorter alkyl chains, such as C1, were used, phase separation occurred before and after transesterification, even if a cosolvent such as decane was added to improve the uniformity of the reaction solution. This phase separation is attributed to the immiscibility of TG and C1. As a result, as shown in **Figure 7**, the trancesterification speed using methanol was very slow, because the reaction occurred only at the interface between the TG and C1 layer. The conversion ratio of TG using C1 reached 96% after 24 h of transesterification. To increase the miscibility of the two compounds, Tang *et al.* suggested that higher pressure and temperature are needed [28]. Based on the above results, the structure of BDF and transesterification speed could be controlled to some extent by the type of alcohol used. In this study, the primary alkyl alcohol that enabled the fastest production of BDF was C5, and the order of transesterification speed was C1 << C2 << C3 < C4 < C5 > C6 > C8 > C10 >> C12.

3.4. BDF Production from Vegetable Oils and Animal Fat

From the above experimental results, it was elucidated that the grafted fibrous catalyst shows good activity for BDF production from TG with an alcohol. To enable practical use of the grafted fibrous catalyst, it must adapt to a wide variety of feed oils such as natural vegetable oils, animal fats, and waste oils, which contain different types of TGs [29] [30], in addition to the reagent-grade triolein used in this study. Thus, the effect of the type of natural oil on the transesterification yield and speed was investigated. In this test, four types of vegetable oils, *i.e.*, linseed oil, safflower oil, rapeseed oil, and palm oil, and one type of animal fat, *i.e.*, beef tallow, were used. EtOH was used as the primary alkyl alcohol, and the reaction mixture was composed of 1.4 g of natural oils, 3.6 g of EtOH, and 5.0 g of decane. The other conditions were the same as the typical conditions described in Section 2.3. **Figure 8** shows the HPLC chromatograms of the BDF samples produced using the different types of natural oils before and after 4 h of transesterification. In the chromatograms of the natural oils, many peaks

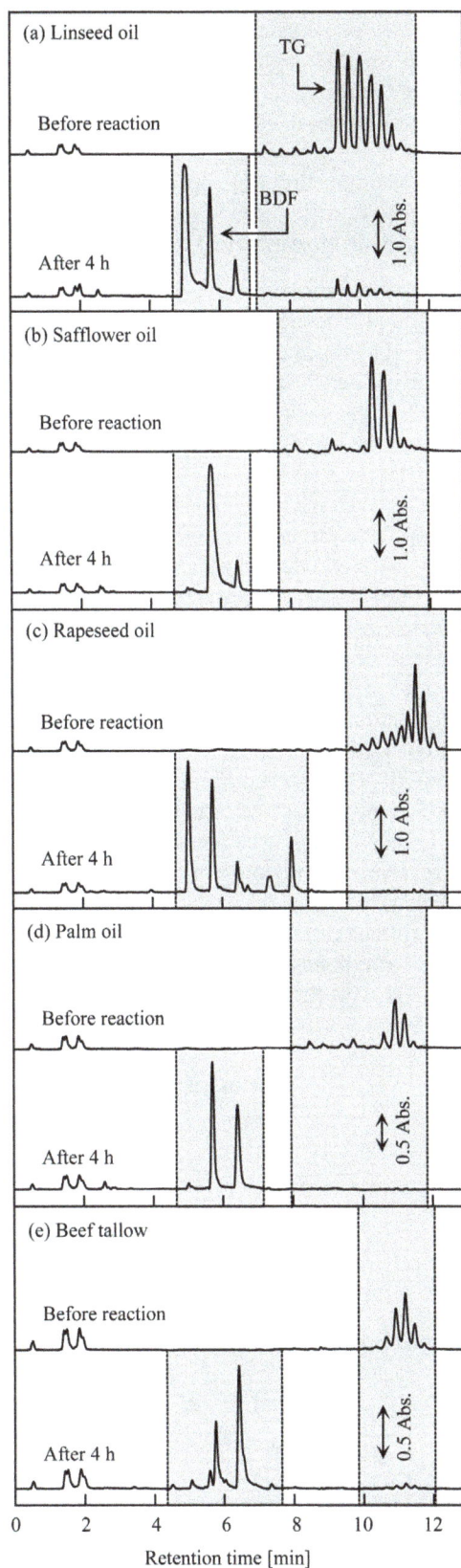

Figure 8. BDF production from vegetable oils and animal fat using grafted fibrous catalyst.

derived from TG, which was contained in the natural oils, were observed before transesterification, unlike in the chromatogram of reagent-grade triolein. Although the transesterification speed differed according to the type of natural oil, the conversion ratios for all the natural oils were greater than 90% after 4 h of transesterification. The conversion ratios for linseed oil, safflower oil, rapeseed oil, palm oil, and beef tallow were 92%, 98%, 97%, 97%, and 93%, respectively, after 4 h of transesterification. Based on these results, it is evident that the grafted fibrous catalyst is applicable for BDF production from many types of natural oils.

3.5. Repeated-Use Stability and Regeneration of Grafted Fibrous Catalyst

Repeated-use stability and ease of regeneration are important aspects of the grafted fibrous catalyst for the production of large amounts of BDF and extended use of the catalyst. Firstly, the repeated-use stability of the grafted fibrous catalyst was evaluated by repeated transesterification of triolein with EtOH, and the immutability of the catalytic activity for each transesterification experiment was monitored. In this test, the grafted fibrous catalyst was only washed with EtOH after each transesterification. The molar ratio of TG to EtOH was fixed at 1:200, and the other conditions were the same as the typical conditions described in Section 2.3. As shown in **Figure 9**, the catalytic activity of the grafted fibrous catalyst gradually decreased with increasing number of transesterification reactions. There are two main reasons for this decay of catalytic activity: One is the removal of OH⁻ ions from the grafted fibrous catalyst, and the other is contamination of the quaternary amine groups by oleate ions, which are catalytically inactive. Oleate ions are generated during the transesterification process by a direct ion-exchange reaction of the OH⁻ ions, which were immobilized onto the graft chains, with the oleic acid group of triolein, diolein, or monoolein. In the presence of organic acid ions, the counterion of the quaternary amine group that was introduced into the graft chain is easily replaced with an organic acid ion instead of a OH⁻ ion. Additionally, rate of decrease of the catalytic activity was not steady: A rapid decrease was observed after the 8th and subsequent transesterifications. The transesterification times required to convert 90% of the initial TG into BDF were 0.99, 1.00, 1.04, 1.10, 1.18, 1.32, 1.44, 1.97, 2.90 and 12.00 h for the first ten reactions. At the 11th reaction, the catalytic activity of the grafted fibrous catalyst was almost negligible; thus, the conversion ratio only reached 9%, even after 24 h of transesterification.

Next, the deactivated grafted fibrous catalyst was regenerated, according to the procedure described by Shibasaki-kitakawa *et al.* [12]. The regeneration process consists of the following three steps: 1) Washing with 0.25 M citric acid solution (solvent: EtOH) to desorb the organic acid ions that cover the active sites, *i.e.*, the quaternary amine groups, of the grafted fibrous catalyst, 2) regenerating with 1 M NaOH aqueous solution to replace the citric acid ions, which formed ionic bonds with the quaternary amine groups, with OH⁻ ions and washing with deionized water, and 3) washing with EtOH to restore the initial swelled condition. When the deactivated grafted fibrous catalyst was treated with the first and third steps and only the third step, the catalytic activities of

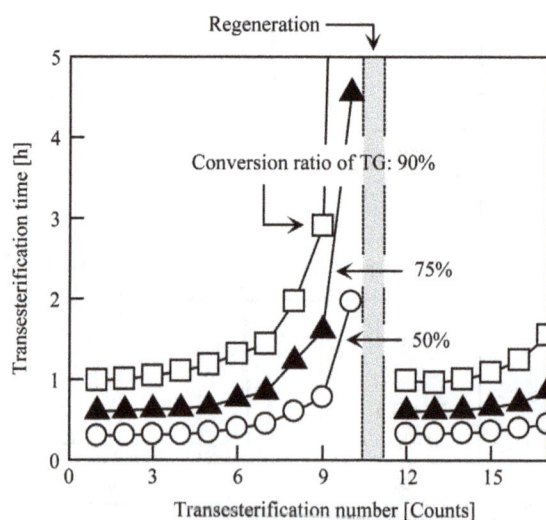

Figure 9. Repeated-use stability and regeneration of grafted fibrous catalyst.

each grafted fibrous catalyst did not recover their original state. The conversion ratios of each grafted fibrous catalyst were almost zero, even after 24 h of transesterification. When the deactivated grafted fibrous catalyst was treated with the second and third steps, the catalytic activity of the grafted fibrous catalyst partially recovered: The conversion ratio after 1 h of transesterification was 25%, which is equivalent to about one-fourth of the original ratio (90% after 1 h of transesterification). Also, the transesterification reaction almost stopped after 4 h, and the conversion ratio after 4 h was about 43%. However, when the deactivated grafted fibrous catalyst was treated with all three steps, the catalytic activity of the grafted fibrous catalyst almost completely recovered to its original state:

The transesterification times required to convert 90% of the initial TG into BDF were 0.99 and 0.99 h for the 1st and 12th reactions, respectively. The conversion ratio and transesterification speed of the regenerated grafted fibrous catalyst almost overlapped with those of the unused grafted fibrous catalyst. As shown in **Figure 9**, the grafted fibrous catalyst could be used repeatedly, although the catalytic activity of the regenerated grafted fibrous catalyst gradually decreased with increasing number of transesterification reactions, as was the case with the unused grafted fibrous catalyst. Additionally, the regenerated grafted fibrous catalyst did not suffer significant physical or chemical damage, even after several transesterification-regeneration cycles. Based on these results, it is evident that the grafted fibrous catalyst could be used up to ten times without any proper regeneration and the BDF production capacity per gram of the grafted fibrous catalyst with 3.6 mmol-TMA/g-catalyst was 70 g because the weight ratio of triolein to grafted fibrous catalyst in the reaction mixture was fixed at ~7:1 (1.4 g/0.2 g) per test. Furthermore, it was confirmed that the three sequential regeneration processes, which involved organic acid, alkali and alcohol treatments, were required to completely recover the catalytic activity of the deactivated grafted fibrous catalyst. Also, the grafted fibrous catalysts are sufficiently stable for repeated use and sufficiently durable for long-term use.

4. Conclusion

TMA is the optimal quaternary amine group for the grafted fibrous catalyst from both perspectives of the ease of amination and the catalytic performance, although the various types of quaternary amine groups with different alkyl chain lengths could be introduced into the CMS-graft chains. The optimal Dg value of the grafted fibrous catalyst to achieve effective transesterification was greater than 170%, andthe quaternary amine-group density of 170% Dg is 3.1 mmol-TMA/g-catalyst. Additionally, the transesterification temperature, molar ratio of TG to EtOH, and type of alcohol significantly influenced the transesterification efficiency, as did the molar quantity of quaternary amine groups in the reaction mixture. The optimal molar quantity of quaternary amine groups, transesterification temperature, molar ratio of TG and EtOH, and primary alkyl alcohol were 0.8 mmol, 80°C, 1:200, and 1-pentanol, respectively. The grafted fibrous catalyst could be applied to BDF production using any type of natural oil as the starting oil. The grafted fibrous catalyst could be used ten times without any proper regeneration, and the BDF production capacity per gram of the grafted fibrous catalyst was ~70 g. The deactivated grafted fibrous catalyst could be recovered to its initial state by three sequential regeneration processes, including organic acid, alkali, and alcohol treatments, without any significant loss of the catalytic activity. Based on these results, the grafted fibrous catalyst, which has sufficient catalytic activity, efficient BDF production capacity, and high repeated-use stability for practical application in BDF production, will contribute to the feasibility of environmentally friendly energy resources, reduce the dependence on petroleum, and improve global environment preservation in the near future. Furthermore, we are firmly convinced that this new type of catalyst will encourage the dissemination of radiation processing and create new possibilities for radiation research.

References

[1]   US Department of Energy, Energy Information Administration (US DOE/EIA) (2001) International Energy Outlook 2001. US Department of Energy, Energy Information Administration, Washington DC.

[2]   US Department of Energy, Energy Information Administration (US DOE/EIA) (2011) International Energy Outlook 2011. US Department of Energy, Energy Information Administration, Washington DC.

[3]   Monlau, F., Sambusiti, C., Barakat, A., Guo, X.M., Latrille, E., Trably, E., Steyer, J.-P. and Carrere, H. (2012) Predictive Models of Biohydrogen and Biomethane Production Based on the Compositional and Structural Features of Lignocellulosic Materials. *Environmental Science & Technology*, **46**, 12217-12225. http://dx.doi.org/10.1021/es303132t

[4]   Li, Y., Zhang, R., Liu, X., Chen, C., Xiao, X., Feng, L., He, Y. and Liu, G. (2013) Evaluating Methane Production

from Anaerobic Mono- and Co-Digestion of Kitchen Waste, Corn Stover, and Chicken Manure. *Energy & Fuels*, **27**, 2085-2091. http://dx.doi.org/10.1021/ef400117f

[5]   Yangcheng, H., Jiang, H., Blanco, M. and Jane, J.-L. (2013) Characterization of Normal and Waxy Corn Starch for Bioethanol Production. *Journal of Agricultural and Food Chemistry*, **61**, 379-386. http://dx.doi.org/10.1021/jf305100n

[6]   Limayem, A. and Ricke, S.C. (2012) Lignocellulosic Biomass for Bioethanol Production: Current Perspectives, Potential Issues and Future Prospects. *Progress in Energy and Combustion Science*, **38**, 449-467. http://dx.doi.org/10.1016/j.pecs.2012.03.002

[7]   Tran, H.-L., Ryu, Y.-J., Seong, D.H., Lim, S.-M. and Lee, C.-G. (2013) An Effective Acid Catalyst for Biodiesel Production from Impure Raw Feedstocks. *Biotechnology and Bioprocess Engineering*, **18**, 242-247. http://dx.doi.org/10.1007/s12257-012-0674-1

[8]   Sagiroglu, A., Ozcan, H.M., Isbilir, S.S., Paluzar, H. and Toprakkiran, N.M. (2013) Alkali Catalysis of Different Vegetable Oils for Comparisons of Their Biodiesel Productivity. *Journal of Sustainable Bioenergy Systems*, **3**, 79-85. http://dx.doi.org/10.4236/jsbs.2013.31011

[9]   Mata, T.M., Sousa, I.R.B.G., Vieira, S.S. and Caetano, N.S. (2012) Biodiesel Production from Corn Oil via Enzymatic Catalysis with Ethanol. *Energy & Fuels*, **26**, 3034-3041. http://dx.doi.org/10.1021/ef300319f

[10] Babajide, O., Musyoka, N., Petrik, L. and Ameer, F. (2012) Novel Zeolite Na-X Synthesized from Fly Ash as a Heterogeneous Catalyst in Biodiesel Production. *Catalysis Today*, **190**, 54-60. http://dx.doi.org/10.1016/j.cattod.2012.04.044

[11] Ilham, Z. and Saka, S. (2012) Optimization of Supercritical Dimethyl Carbonate Method for Biodiesel Production. *Fuel*, **97**, 670-677. http://dx.doi.org/10.1016/j.fuel.2012.02.066

[12] Shibasaki-Kitakawa, N., Honda, H., Kuribayashi, H., Toda, T., Fukumura, T. and Yonemoto, T. (2007) Biodiesel Production Using Anionic Ion-Exchange Resin as Heterogeneous Catalyst. *Bioresource Technology*, **98**, 416-421. http://dx.doi.org/10.1016/j.biortech.2005.12.010

[13] Tsuji, T., Kubo, M., Shibasaki-Kitakawa, N. and Yonemoto, T. (2009) Is Excess Methanol Addition Required to Drive Transesterification of Triglyceride toward Complete Conversion? *Energy & Fuels*, **23**, 6163-6167. http://dx.doi.org/10.1021/ef900622d

[14] Demirbas, A. (2008) Comparison of Transesterification Methods for Production of Biodiesel from Vegetable Oils and Fats. *Energy Conversion and Management*, **49**, 125-130. http://dx.doi.org/10.1016/j.enconman.2007.05.002

[15] Helwani, Z., Othman, M.R., Aziz, N., Kim, J. and Fernando, W.J.N. (2009) Solid Heterogeneous Catalysts for Transesterification of Triglycerides with Methanol: A Review. *Applied Catalysis A: General*, **363**, 1-10. http://dx.doi.org/10.1016/j.apcata.2009.05.021

[16] Zhou, D., Zhang, S., Fu, H. and Chen, J. (2012) Liquefaction of Macroalgae Enteromorpha Prolifera in Sub-/Supercritical Alcohols: Direct Production of Ester Compounds. *Energy & Fuels*, **26**, 2342-2351. http://dx.doi.org/10.1021/ef201966w

[17] Menetrez, M.Y. (2012) An Overview of Algae Biofuel Production and Potential Environmental Impact. *Environmental Science & Technology*, **46**, 7073-7085. http://dx.doi.org/10.1021/es300917r

[18] Ueki, Y., Mohamed, N.H., Seko, N. and Tamada, M. (2011) Rapid Biodiesel Fuel Production Using Novel Fibrous Catalyst Synthesized by Radiation-Induced Graft Polymerization. *International Journal of Organic Chemistry*, **1**, 20-25. http://dx.doi.org/10.4236/ijoc.2011.12004

[19] Madrid, J.F., Ueki, Y. and Seko, N. (2013) Abaca/Polyester Nonwoven Fabric Functionalization for Metal Ion Adsorbent Synthesis via Electron Beam-Induced Emulsion Grafting. *Radiation Physics and Chemistry*, **90**, 104-110. http://dx.doi.org/10.1016/j.radphyschem.2013.05.004

[20] Ueki, Y., Dafader, N.C., Hoshina, H., Seko, N. and Tamada, M. (2012) Study and Optimization on Graft Polymerization under Normal Pressure and Air Atmospheric Conditions, and Its Application to Metal Adsorbent. *Radiation Physics and Chemistry*, **81**, 889-898. http://dx.doi.org/10.1016/j.radphyschem.2012.02.031

[21] Iwanade, A., Kasai, N., Hoshina, H., Ueki, Y., Saiki, S. and Seko, N. (2012) Hybrid Grafted Ion Exchanger for Decontamination of Radioactive Cesium in Fukushima Prefecture and Other Contaminated Areas. *Journal of Radioanalytical and Nuclear Chemistry*, **293**, 703-709. http://dx.doi.org/10.1007/s10967-012-1721-2

[22] Hoshina, H., Kasai, N., Shibata, T., Aketagawa, Y., Takahashi, M., Yoshii, A., Tsunoda, Y. and Seko, N. (2012) Synthesis of Arsenic Graft Adsorbents in Pilot Scale. *Radiation Physics and Chemistry*, **81**, 1033-1035. http://dx.doi.org/10.1016/j.radphyschem.2012.02.018

[23] Hoshina, H., Seko, N., Ueki, Y., Iyatomi, Y. and Tamada, M. (2010) Evaluation of Graft Adsorbent with N-Methyl-D-Glucamine for Boron Removal from Groundwater. *Journal of Ion Exchange*, **21**, 153-156.

[24] Seko, N., Hoshina, H., Kasai, N., Ueki, Y., Tamada, M., Kiryu, T., Tanaka, K. and Takahashi, M. (2010) Novel Sys-

tem for Recovering Scandium from Hot Spring Water with Fibrous Graft Adsorbent. *Journal of Ion Exchange*, **21**, 117-122.

[25] Seko, N., Katakai, A., Hasegawa, S., Tamada, M., Kasai, N., Takeda, H., Sugo, T. and Saito, K. (2003) Aquaculture of Uranium in Seawater by a Fabric-Adsorbent Submerged System. *Nuclear Technology*, **144**, 274-278.

[26] Holčapek, M., Jandera, P., Fischer, J. and Prokeš, B. (1999) Analytical Monitoring of the Production of Biodiesel by High-Performance Liquid Chromatography with Various Detection Methods. *Journal of Chromatography A*, **858**, 13-31. http://dx.doi.org/10.1016/S0021-9673(99)00790-6

[27] Basuki, F., Seko, N. and Tamada, M. (2010) Recovery of Scandium with Phosphoric Chelating Adsorbent Prepared by Direct Radiation Graft Polymerization. *Journal of Ion Exchange*, **21**, 127-130.

[28] Tang, Z., Du, Z., Min, E., Gao, L., Jiang, T. and Han, B. (2006) Phase Equilibria of Methanol-Triolein System at Elevated Temperature and Pressure. *Fluid Phase Equilibria*, **239**, 8-11. http://dx.doi.org/10.1016/j.fluid.2005.10.010

[29] Gunstone, F.D., Hamilton, R.J., Padley, F.B. and Qureshi, M.I. (1965) Glyceride Studies. V. The Distribution of Unsaturated Acyl Groups in Vegetable Triglycerides. *Journal of the American Oil Chemists Society*, **42**, 965-970. http://dx.doi.org/10.1007/BF02632456

[30] Ramos, M.J., Fernández, C.M., Casas, A., Rodríguez, L. and Pérez, Á. (2009) Influence of Fatty Acid Composition of Raw Materials on Biodiesel Properties. *Bioresource Technology*, **100**, 261-268. http://dx.doi.org/10.1016/j.biortech.2008.06.039

12

Synthesis of Highly-Selective Fibrous Adsorbent by Introducing 2-Ethylhexyl Hydrogen-2-Ethylhexylphosphonate for Scandium Adsorption

Hiroyuki Hoshina*, Yuji Ueki, Seiichi Saiki, Noriaki Seko

Environment and Industrial Materials Research Division, Quantum Beam Science Center, Sector of Nuclear Science Research, Japan Atomic Energy Agency, Takasaki, Japan
Email: *hoshina.hiroyuki@jaea.go.jp

Abstract

2-ethylhexyl hydrogen-2-ethylhexylphosphonate (EHEP) is commonly used as a metal extractant because it has a particular affinity for rare-earth metals like Scandium (Sc). To develop a highly-selective adsorbent of Sc(III), EHEP was introduced as a functional group onto a polyethylene fabric with radiation-induced graft polymerization(RIGP). The adsorption performances for Sc(III) were evaluated with aqueous solutions containing Sc(III) and Fe(III) in bath and column tests. As a result of column test, the adsorption capacities of Sc(III) and Fe(III) until the bed volume reached 5000 were 5.22 and 0.12 mg/g, respectively. It means that the amount of collected Sc(III) by the EHEP adsorbent was approximately 44 times higher than that of Fe(III). These results indicate that the grafted adsorbent containing EHEP has an extremely high selectivity for Sc(III) adsorption.

Keywords

Fibrous Adsorbent, Scandium, Radiation-Induced Graft Polymerization, Selective Adsorption, 2-Ethylhexyl Hydrogen-2-Ethylhexylphosphonate

1. Introduction

A fibrous adsorbent containing phosphoric acids which has a predilection for Scandium (Sc)(III) [1] [2], has

*Corresponding author.

been investigated to recover Sc(III) from hot spring water [3]. The Sc(III) recovery was conducted with the adsorbent, which was prepared by introducing 2-hydroxyethyl methacrylate phosphoric acid (HMPA) into a nonwoven polyethylene fabric with radiation-induced graft polymerization (RIGP) [4], in Kusatsu Hot Springs in Gunma Prefecture [5]. The adsorbent containing HMPA could rapidly collect Sc(III) from hot spring water contained a low concentration of Sc(III). However, Fe(III) was also caught by the adsorbent unwillingly because the Fe(III) had dissolved at a 400 times higher than concentration of Sc(III) in hot spring water with acidic condition. To recover Sc(III) efficiently from hot spring water, a highly-selective adsorbent for Sc(III) was required.

2-ethylhexyl hydrogen-2-ethylhexylphosphonate (EHEP) is a superior metal extractant and is widely used in solvent extraction [6] [7]. EHEP has a particular affinity for rare-earth metals like Sc(III), especially in acid solutions [8]-[10]. If EHEP is able to be introduced into a polymeric material while maintaining its efficiency, it is possible to develop a highly-selective adsorbent for Sc(III).

RIGP is a useful technique to functionalize a polymeric material such as cellulose, polypropylene and polyethylene [11]-[13]. By using the RIGP technique, the functional groups can be introduced into the polymeric material without changing its characteristic dramatically. In the case of introducing the EHEP as a functional group, first dodecyl methacrylate (DMA) which has long-chain alkyl groups, was introduced with RIGP onto the polymeric material, and subsequent introduction of EHEP by hydrophobic interaction of long-chain alkyl groups in the DMA structure and that of EHEP (**Figure 1**).

In this article, the highly-selective adsorbent for Sc(III) was developed by introducing EHEP with RIGP technique, and evaluation of its adsorption performance for Sc(III) was carried out with an aqueous solution containing Sc(III) and Fe(III).

2. Experimental

2.1. Materials

A non-woven fabric composed of polyethylene coated polypropylene fiber was purchased from Kurashiki Textile Manufacturing Co., Ltd. to be used as a base material for the adsorbent. EHEP was provided by Daihachi Chemical Industry Co., Ltd.DMA, Tween 20, phosphoric acid and methanol were supplied by Kanto Chemical Co., INC. A monomer of GMA and HMPA were purchased from Tokyo Chemical Industry Co., Ltd. and Kyoeisha Chemical Co., Ltd., respectively. All chemicals were used without further purification.

2.2. Preparation of the Adsorbent for Sc(III) Adsorption

The non-woven fabrics were packed into polyethylene bags. Then, the inside of the bags were made into a nitrogen atmosphere with nitrogen gas, which were irradiated at the maximum dose of 500 kGy with an electron beam accelerator under a dry ice temperature. The irradiated fabrics were placed into glass ampoules and were evacuated. A deoxidized monomer solution, which was a mix of DMA, Tween 20, and deionized water, was transferred into the ampoules. The concentrations of DMA and Tween 20 in the solution were 20% and 2% (w/w), respectively. The grafting of DMA was carried out at 60°C for 6 hours. After grafting, the grafted fabrics were washed with methanol to remove the residual DMA and homopolymer. The degree of grafting (Dg) was calculated by the following equation;

$$\text{Degree of grafting}(Dg)[\%] = (W_1 - W)_0 / W_0 \times 100$$

where W_0 and W_1 were the weight of fabrics before and after graft polymerization.

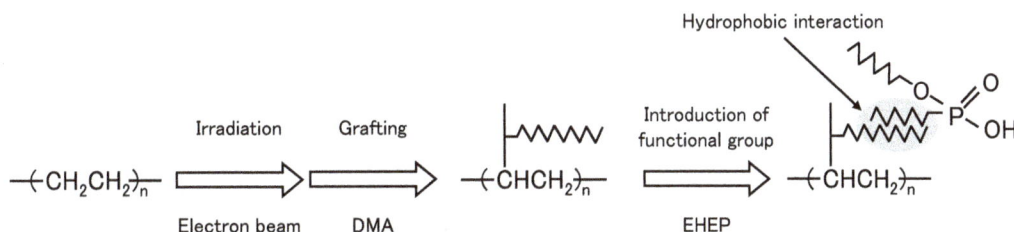

Figure 1. Preparation scheme of graft adsorbent containing EHEP.

To introduce EHEP as a functional group by hydrophobic interaction of long-chain alkyl group in the DMA structure and that of EHEP, the grafted fabrics were soaked in EHEP, and were shaken for 12 hours at room temperature. The adsorbents introduced EHEP (EHEP adsorbent) were obtained after the residual EHEP was removed.

2.3. Evaluation of the EHEP Adsorbent

The adsorption performance of Sc(III) was evaluated with an aqueous solution containing Sc(III) and Fe(III) at concentrations 1 mg/L in batch and column adsorption tests. To compare the adsorption performance for Sc(III), two other types of adsorbents containing phosphoric acid as a functional group were prepared for batch adsorption tests. One of the adsorbents was prepared with RIGP of glycidyl methacrylate (GMA), and followed by chemical modification with phosphoric acid (GMA-PA adsorbent). Another adsorbent was synthesized by introducing HMPA with RIGP technique (HMPA adsorbent). The adsorbents were soaked in 50 mL of the aqueous solutions containing Sc(III) and Fe(III), and were stirred for 12 hours at room temperature. The solutions were adjusted to pH 0, 1, 2 and 3 with nitric acid. After the adsorption tests, the concentrations of Sc(III) and Fe(III) in the solutions were measured with induced coupled plasma atomic emission spectroscopy [ICP-AES, Perkin Elmer Inc., Optima 4300DV]. In the column test, 0.08 mL of the EHEP adsorbent was packed into a 7 mm internal diameter column. The aqueous solution containing Sc(III) and Fe(III) was adjusted to pH 1, then it was delivered to the column at the flow rate of space velocity (SV) 250 h^{-1}.

3. Results and Discussion

3.1. Preparation of the Adsorbent for Sc(III) Adsorption

The RIGP of DMA was conducted with 20% (w/w) of the DMA solution for 6 hours at absorbed dose of 20, 30, 50 100, 200 and 500 kGy. The effect of the absorbed dose on the Dg was shown in **Figure 2**. The Dg increased simultaneously with the absorbed dose. The Dg reached 120% at the absorbed dose of 500 kGy. To introduce EHEP as a functional group, the grafted fabric of Dg 120% was soaked into EHEP for 12 hours and consequently the EHEP adsorbent with the density of 1.0 mmol/g was obtained. The adsorbent was analyzed by using a spectrum one FT-IR spectrometer [Perkin Elmer Co., Ltd.]. As shown in **Figure 3**, the FTIR spectrum of the EHEP adsorbent showed a new characteristic adsorption band of EHEP at around 1240 cm^{-1}, which was due to the double bond between the phosphorus and oxygen, in addition to the same adsorption band of the base material. This result confirmed the presence of EHEP in the adsorbent.

3.2. Evaluation of the EHEP Adsorbent

The effect of pH on the Sc(III) adsorption with Fe(III) coexisting was investigated at pH 0, 1, 2 and 3 with the three types of adsorbent containing phosphoric acid groups. As shown in **Table 1**, the EHEP adsorbent had the high adsorption capacity more than 6.5 mg/g for Sc at any pH. In the case of the GMA-PA adsorbent, the adsorption capacity of Sc(III) increased with decreasing the pH, and the maximum capacity was 3.9 mg/g at pH 0. The HMPA adsorbent had an adsorption capacity of approximately 1 mg/g at the pH of 0 to 4. As a result, the EHEP adsorbent showed a high adsorption capacity of more than double in comparison with the two other types of adsorbent at any pH. The adsorption capacity of Fe(III) increased as the pH increased with every type of adsorbent. The selectivity of Sc(III) adsorption was evaluated as follows,

$$\text{Adsorption ratio}\,(R)\,[\%] = (C_0 - C_1)/C_0 \times 100$$

where C_0 and C_1 were the concentration of the Sc(III) and Fe(III) in solution before and after adsorption.

$$\text{Selectivity of Sc}\,(\text{III})\,(S)\,[-] = R_{Sc} - R_{Fe}$$

where R_{Sc} and R_{Fe} were the adsorption ratio of Sc(III) and Fe(III), respectively.

The selectivity for Sc(III) versus Fe(III) of the EHEP, GMA-PA and HMPA adsorbents were presented in **Figure 4**. In this figure, the Y-axis is the selectivity for Sc(III). For example, if the selectivity of Sc(III) reached 100, all of the Sc(III) in the solution was adsorbed and Fe(III) was not adsorbed at all. If the selectivity of Sc(III) was less than 0, the adsorption ratio of Sc(III) was lower than that of Fe(III). As shown in **Figure 4**, the selectivity of Sc(III) increased by decreasing the pH in all types of adsorbent. Compared with two other types of adsor--

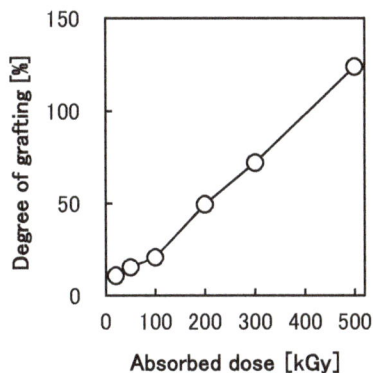

Figure 2. Effect of absorbed dose on the degree of grafting.

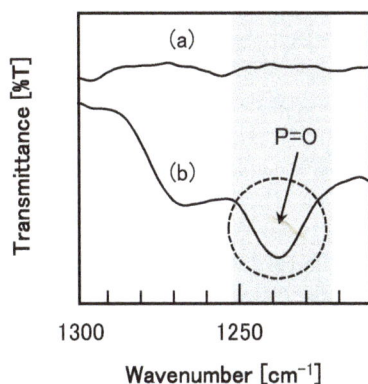

Figure 3. FTIR spectra of base material (a) and the EHEP adsorbent (b).

Table 1. Effect of pH on the adsorption capacity of Sc(III) and Fe(III).

adsorbent	Density of functional groups [mmol/g]	pH							
		0		1		2		3	
		Adsorption capacity		Adsorption capacity		Adsorption capacity		Adsorption capacity	
		Sc(III) [mg/g]	Fe(III) [mg/g]	Sc(III) [mg/g]	Fe(III) [mg/g]	Sc(III) [mg/g]	Fe(III) [mg/g]	Sc(III) [mg/g]	Fe(III) [mg/g]
EHEP	1.0	7.45	0.21	6.52	0.44	7.99	1.79	7.32	3.79
GMA-PA	1.3	3.93	1.67	2018	1.68	1.74	2.21	1.04	2.71
HMPA	1.4	1.42	0.16	0.82	0.15	1.22	0.95	1.03	1.90

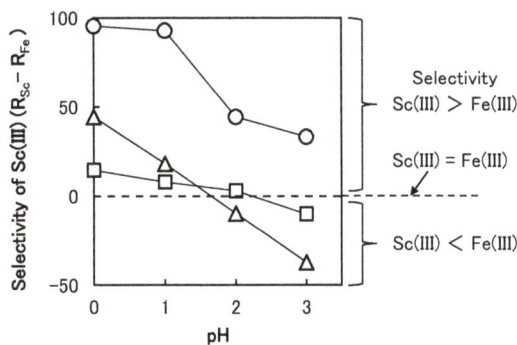

Figure 4. Effect of pH on the adsorption selectivity for Sc(III) versus Fe(III) of EHEP (○), GMA-PA (△) and HMPA (□).

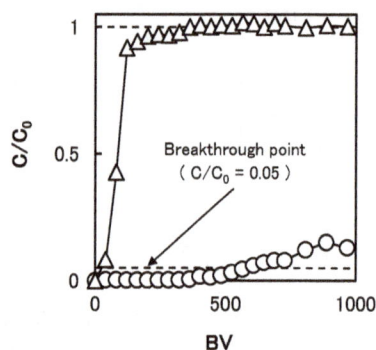

Figure 5. The breakthrough curves of Sc(III) (○) and Fe(III) (△) with the mixture solution of Sc(III) and Fe(III) at pH1.

bents, the EHEP adsorbent showed an extremely high selectivity for Sc(III) at any pH. The selectivity of Sc(III) reached higher than 90 in strongly acidic conditions (pH 0 and 1). Therefore, the column mode adsorption test was carried out with the aqueous solutions containing Sc(III) and Fe(III) adjusted to pH 1.

The breakthrough curves of Sc(III) and Fe(III) adsorption were shown in **Figure 5**. The breakthrough curve was obtained by plotting C/C_0 which was the concentration of Sc(III) and Fe(III) in the feeding solution (C_0) and the effluent (C) versus bed volume (BV). The breakthrough point was defined when C/C_0 reached 0.05. The breakthrough point of Sc(III) adsorption was BV 564. The C/C_0 of Sc(III) gradually increased incrementally in BV, and reached 0.78 at BV 5000. In the case of Fe(III) adsorption, the C/C_0 exceeded the breakthrough point immediately following the start of the adsorption test, and reached 1.00 at BV 362. The adsorption capacities of Sc(III) and Fe(III) until the BV reached 5000 were 5.22 and 0.12 mg/g, respectively. It means that the amount of collected Sc(III) by the EHEP adsorbent was approximately 44 times higher than that of Fe(III). These results established that the EHEP adsorbent could collect Sc(III) selectively from a strongly acidic water media.

4. Conclusion

The highly-selective fibrous adsorbent for Sc(III) was developed by RIGP of DMA and subsequent introduction of EHEP as a functional group by hydrophobic interaction of long-chain alkyl group in the DMA structure and that of EHEP. The EHEP adsorbent exhibited a high adsorption capacity and selectivity for Sc(III) in acidic conditions compared with other types of the adsorbents containing phosphoric acid groups. In the case of column mode adsorption test using an aqueous solution containing Sc(III) and Fe(III) at pH 1, the adsorption capacity of Sc(III) with the EHEP was 44 times higher than that of Fe(III). It meant that the EHEP adsorbent could collect only Sc(III) mostly without Fe(III) adsorption. If the EHEP adsorbent is able to be applied for the Sc(III) recovery from hot spring water, it will be achieved the practical use of the Sc(III) recovery and hot spring will be a new resource of Sc.

References

[1]   Marhol, M., Beranova, H. and Cheng, K.L. (1974) Selective Ion-Exchangers Containing Phosphorus in Their Functional Groups. *Journal of Radioanalytical Chemistry*, **21**, 177-186. http://dx.doi.org/10.1007/BF02520860

[2]   Hubicki, Z. (1990) Studies on Selective Separation of Sc(III) from Rare Earth Elements on Selective Ion-Exchangers. *Hydrometallurgy*, **23**, 319-331. http://dx.doi.org/10.1016/0304-386X(90)90013-R

[3]   Kasai, N., Seko, N., Tamada, M. and Ichikawa, E. (2006) The Recovery Method of Scandium from Hot Spring Water. *JAEA-Review* 2006-042, 46.

[4]   Kabanov, V.Y., Aliev, R.E. and Kudryavtsev, V.N. (1991) Present Status and Development Trends of Radiation-Induced Graft Polymerization. *Radiation Physics and Chemistry*, **37**, 175-192.

[5]   Seko, N., Hoshina, H., Kasai, N., Ueki, Y. and Tamada, M. (2010) Novel System for Recovering Scandium from Hot Spring Water with Fibrous Graft Adsorbent. *Journal of Ion Exchange*, **21**, 117-122.

[6]   Jayachandran, J. and Dhadke, P.M. (1997) Liquid-Liquid Extraction Separation of Ion(III) with 2-Ethyl Hexyl Phosphonic Acid Mono 2-Ethylhexyl Ester. *Talanta*, **44**, 1285-1290. http://dx.doi.org/10.1016/S0039-9140(97)02190-5

[7]   Bandekar, S.V. and Dhadke, P.M. (1998) Solvent Extraction Separation of Tin (IV) with 2-Ethylhexyl Phosphonic

Acid Mono-2-Ethylhexyl Ester. (PC-88A). *Talanta*, **46**, 1181-1186. http://dx.doi.org/10.1016/S0039-9140(97)00364-0

[8]   Surampally, R., Batchu, N.K, Mannepalli, L.K. and Bontha, R.R. (2012) Studies on Solvent Extraction of Dy(III) and Separation Possibilities of Rare Earths Using PC-88A from Phosphoric Acid Solutions. *Journal of the Taiwan Institute of Chemical Engineers*, **43**, 839-844. http://dx.doi.org/10.1016/j.jtice.2012.04.009

[9]   Thakur, N.V., Jayawant, D.V., Iyer, N.S. and Koppiker, K.S. (1993) Separation of Neodymium from Lighter Rare Earths Using Alkyl Phosphonic Acid, PC 88A. *Hydrometallurgy*, **34**, 99-108. http://dx.doi.org/10.1016/0304-386X(93)90084-Q

[10]   Wakui, Y., Matsunaga, H. and Suzuki, T.M. (1989) Selective Recovery of Trace Scandium from Acid Aqueous Solution with (2-Ethylhexyl Hydrogen 2-Ethylhexylphosphonate)-Impregnated Resin. *Analytical Sciences*, **5**, 189-193. http://dx.doi.org/10.2116/analsci.5.189

[11]   Hoshina, H., Takahashi, M., Kasai, N. and Seko, N. (2012) Adsorbent for Arsenic (V) Removal Synthesized by Radiation-Induced Graft Polymerization onto Nonwoven Cotton Fabric. *International Journal of Organic Chemistry*, **2**, 173-177. http://dx.doi.org/10.4236/ijoc.2012.23026

[12]   Bondar, Y., Kim, H.J. Yoon, S.H. and Lim, Y.J. (2004) Synthesis of Cation-Exchange Adsorbent for Anchoring Metal Ions by Modification of Poly (Glycidyl Methacrylate) Chains Grafted into Polypropylene Fabric. *Reactive & Functional Polymers*, **58**, 43-51. http://dx.doi.org/10.1016/j.reactfunctpolym.2003.11.006

[13]   Ueki, Y., Dafader, N.C, Hoshina, H. Seko, N. and Tamada, M. (2012) Study and Optimization on Graft Polymerization under Normal Pressure and Air Atmospheric Conditions, and Its Application to Metal Adsorbent. *Radiation Physics and Chemistry*, **81**, 889-898. http://dx.doi.org/10.1016/j.radphyschem.2012.02.031

13

Synthesis, Characterization and Utilization of Starch Hydroxypropyl Sulphate for Cationic Dye Removal

Aly A. Hebeish, Amal A. Aly

Textile Research Division, National Research Center, Cairo, Egypt
Email: amalahmedali@yahoo.com

Abstract

The best conditions for synthesis of starch hydroxypropyl sulphate as per the dry method were firstly established. This was done through a thorough investigation into factors affecting the synthesis including concentrations of both the NaOH catalyst and the 2-hydroxy-3-chloropropyl sulphate along with duration and temperature of the reaction. The resultant newly synthesized 2-hydroxy-3-chloropropyl sulphate was then reacted with starch to obtain starch hydroxypropyl sulphate ultimately. After being characterized by making use of IR spectroscopy analysis and scanning electronic microscope, the starch hydroxypropyl sulphate samples were submitted to evaluation for cationic dye removal using Methylene Blue (MB). Cationic dye removal was studied under a variety of conditions. Factors involved encompass initial dye concentration, duration, sulphur %, pH and adsorbent dose. 100% dye removal could be achieved under certain conditions which were described in the text.

Keywords

Starch, Cationic Dye, Starch Sulphate, Dye Removal, Methylene Blue

1. Introduction

Wastewater treatment is a major problem around the world. Growing along with the population growth, industries create environmental problems and health hazards for the population. Hence, environmental concerns behoove scientists and engineers to develop materials and methods to lower the extent of pollution of the environment [1].

In modern industrial society, dyes are widely used to color products for textiles, food, printing and dyeing. So they are an integral part of many industrial effluents, they are highly visible and can have undesired effects not only on the environment, but also on living creatures. Because the degradation products of some dyes may be carcinogenic and toxic, they are considered as an important source of water pollution and their treatment becomes a major problem for environmental managers.

The release of colored wastewater from these industries effluent may present an eco-toxic hazard and introduce the potential danger of bioaccumulation, which may eventually affect man through the food chain. There are various conventional methods of removing dyes from waters.

Adsorption is one of the most efficient methods to remove pollutants from wastewater because of its low cost and ease of operation. Many studies have been made on the possibility of using adsorbents based on clay minerals [2]-[4], activated carbon [5], fly ash [6] [7], weeds [8], crosslinked amphoteric starch [9], Indian rosewood sawdust [10], and crosslinked chitosan beads [11].

However, the adsorption capacity of these adsorbents is not very large, so new absorbents are still under development to improve adsorption performance.

Starch and its derivatives represent a cheap and environmentally safe source of raw material for the reparation of low-cost adsorbents that may be useful for the removal of pollutants from wastewater. Because of its particular characteristics (renewable, abundant and biodegradable raw resource) and properties such as its high reactivity and chemical stability, resulting from the presence of chemically reactive hydroxyl groups in its polymer chains [12], making it possible to chemical modify starch according to different requirements, this biopolymer represents an interesting alternative as an adsorbent. After modification, starches have suitable expansion, huge pore volume and high specific area, which can enhance its adsorption ability. It means that modified starches are more suitable as adsorbent for dyes and heavy metals or as catalyst carrier materials. Therefore, the preparation of, renewable, nontoxic, low cost ionized starch is not only of importance in the field of modern pharmacy but also has great prospects in the water treatment field.

In recent years, many research works have covered the preparation of neutral starch and their physicochemical properties [13]. Common neutral starch mainly adsorb physically, so its adsorption and selectively adsorption ability are weak. Ionization of starch can improve its adsorption ability by enhancing the active groups. Anionic starches have high affinity to positively charged drugs, dyes, metals ion [14], thus enhancing the selective adsorption performance by introducing the carboxylate [15]. The carboxylation has been done via saponification of poly (acrlyamide)-starch graft copolymer, or poly(acrylonitrile)-starch graft copolymer, or poly (methylacrylate)-starch graft copolymer, graft polymerization of starch with methacrylic, or acrylic acid [16].

Among polymers of renewable resources; starch and its derivatives such as sulfonic starch [17]-[19] was currently enjoying increased attention. Starch sulfonation could be carried by using starch, dichloroethane and chlorosulphonic acid [20]. Starch sulfate was one of the modified starches which have biological activities. It was reported that starch sulfate had also the activity of anti-HIV, anti-tumour and antivirus. The starch sulfate was a good inhibitor to pepsin. The starch sulfate prepared had a potential application for the medical uses. Potato starch sulfate was obtained by the reaction between potato starch and chlorosulfonic acid in pyridine [21].

Using the mixture of concentrated sulfuric acid and n-propanol as sulphonating agent, the starch sulfate was prepared with degree of substitution 0.1 - 0.6 [22]. Potato starch sulphate was synthesized by the reaction between potato starch and sulphuric acid in ethanol [23].

Methylene Blue (MB) **Scheme 1**, is selected as a model compound for evaluating the waste to remove cationic dye from wastewaters. MB is a thiazine (cationic) dye, which is most commonly used for coloring paper, temporary hair colorant, dyeing cottons, wools and so on. Although MB is not considered to be a toxic dye, it can reveal harmful effects on living things.

Current work is undertaken with a view to synthesize, characterize and use green starch-based materials, namely starch hydroxypropyl sulfate, which are capable of cationic dye removal from aqueous solutions. To achieve the goal, thorough investigations into factors affecting the synthesis of these green starch derivatives, as well as, affecting removal of the cationic dye upon utilization of these newly synthesized green materials. Variables studied in the synthesis comprise concentrations of both the catalyst and the 2-hydroxy-3-chloropropyl sulphate as well as duration and temperature. Meanwhile variables of the utilization in dye removal include concentration of the initial dye, duration, sulphur %, pH, molar mass and adsorbent dose. Methylene blue (MB) is the dye used. The morphology of starch hydroxypropyl sulfate is examined using Scanning Electron Microscopy (SEM) whereas Fourier Transform InfraRed spectroscopy (FTIR) is used for determination of the newly introduced groups in the microstructural organization of starch.

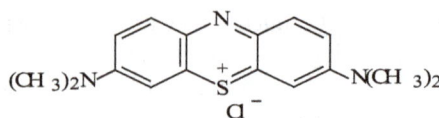

Scheme 1. Structural of methylene blue (MB) dye.

2. Experimental

2.1. Materials

Maize starch was supplied by Cairo Company for Starch and Glucose. Methylen Blue (MB), in commercial purity, were purchased from Aldrich Chemical (Germany), and used without further purification. Sodium bisulphite, hydrochloric acid, sodium hydroxide, epichlorohydrin and ethanol were of analytical grade chemicals.

2.2. Preparation of 2-Hydroxy-3-Chloropropyl Sulphate

This compound was prepared via reacting 0.9 mole of sodium bisulphate with 1 mole epichlorohydrin, until one phase under reflux was formed. After cooling to 60°C - 70°C reaction product was poured into an excess of 96% ethanol. Thus prepared compound was filtered and dried.

Complete transformation of sulphite to sulphate group was effected by oxidizing the product under heating using a ventilating oven.

2.3. Preparation of Starch Hydroxypropyl Sulphate

The starches hydroxypropyl sulphate was prepared by reacting starch with 2-hydroxy-3-chloropropyl sulphate, in presence of sodium hydroxide using the dry state technique according to a previous method [16].

2.4. Preparation of Cationic Dye Solution

The dye stock solutions were prepared by dissolving dyes in distilled water to 500 mg/L. The experimental solutions were obtained by diluting the dye stock solutions in proportions to different initial concentrations

2.5. Dye Sorption

Adsorption experiments were carried out in a rotary shaker at 150 rpm using 250 mL-shaking flasks containing 50 mL of dye solutions at different concentrations and initial pH values of dye solutions. The initial pH values of the solutions were previously adjusted with 0.1 M HCl or 0.1 M NaOH using a DEEP VERSION model (EI) pH meter. The adsorbent (g) was added to each flask, and then the flasks were sealed up to prevent any change of volume of the solution during the experiments. After shaking the flasks for a predetermined time intervals, the samples were withdrawn from the flasks and the dye solutions were separated from the adsorbent by filtration.

The pH values of the separated dye solutions were again measured and changes in pH values were recorded. Dye concentrations in the supernatant solutions were estimated by measuring absorbance at maximum wavelengths of dye with a Shimadzu ultraviolet visspestrophotometer (Japan).

The dye sorption value (m mol/100g sample) of the treated sample was calculated. The sorption value was calculated using the following equation:

Dye sorption value (m mol dye /100 g sample) = Dye sorption value of starch hydroxypropyl sulphate − Dye sorption value of native starch.

% of dye removal = (Dye sorption value (m mol dye/100g sample)/Initial dye concentratin (m mol dye/100g sample) × 100.

2.6. Testing and Analysis

- The sulphite and sulphate contents were determined according to reported methods [24].
- The chlorine content was determined according to reported method [25].
- FTIR spectroscopy.
  The FTIR spectra were obtained from KBr pellets of native and modified starch samples using FTIR spectro-

photometer (JASCOFT/IR-6300, Japan) in the range of 400 - 4000 cm^{-1}.

- Scanning electron microscopy.

By using a scanning electron probe microanalyzer (JXA-840A) JEOL, Tokyo Japan, mages of scanning electron microscopy were studied. The specimens in the form of films were mounted on the specimen stabs and coated with thin film of gold by the sputtering method. The micrographs were taken at magnification of 1000 using 10 kV accelerating voltage.

3. Results and Discussion

3.1. Synthesis of Starch-2-Hydroxypropyl Sulphate

3.1.1. Sodium Hydroxide Concentration
The dependence of the extent of the reaction of starch with 2-hydroxy-3-chloropropyl sulphate (expressed as sulpher %) on the NaOH concentration is shown by **Figure 1(a)**. The reaction was carried out for 180 min at 80°C, using starch (25 m mole), 2-hydroxy-3-chloropropyl sulphate: starch molar ratio 0.5:1, and different sodium hydroxide: 2-hydroxy-3-chloropropyl sulphate molar ratios (3.13 - 25).

It is seen (**Figure 1(a)**) that, increasing sodium hydroxide concentration based on 25 m mole of 2-hydroxy-3-chloropropyl sulphate upto 18.75 m mole/L enhances significantly the extent of the reaction. Further increase in sodium hydroxide concentration has, indeed, a negative impact on the extent of the reaction, expressed as sulphur percent. The enhancement in the extent of the reaction by increasing sodium hydroxide concentration is rather logical and implies that lower sodium hydroxide concentrations are not sufficient enough to drive the reaction to its maximum. Once the latter is attained higher sodium hydroxide concentrations act in favor of alkaline hydrolysis of the functional group (Equation (3)) of 2-hydroxy-3-chloropropyl sulphate and/or splitting off of the 2-hydroxy-3-chloropropyl sulphate moieties from the starch via alkaline hydrolysis of the chemical bonds linking the starch molecules with these moieties (Equation (4)). Thus, it can be concluded that sodium hydroxide concentrations determine the magnitudes of the desirable reaction (reaction 2) and the side undesirable reactions (reactions 3 and 4) as well (**Scheme 2**).

3.1.2. Concentration of 2-Hydroxy-3-Chloropropyl Sulphate
Figure 1(b) shows the sulphur % of the starch when it was reacted with 2-hydroxy-3-chloropropyl sulphate at different concentrations. The 2-hydroxy-3-chloropropyl sulphate: starch molar ratios used were (0.125 - 1.0) with starch (25 m mole) in presence of NaOH: 2-hydroxy-3-chloropropyl sulphate molar ratio (1.5: 1) at 80°C for 240 minutes.

Figure 1(b) discloses that the extent of the reaction increases sharply as the 2-hydroxy-3-chloropropyl sulphate: starch molar ratio increases upto 0.5 molar ratio. Further increase in concentration, specifically beyond 1.0 is accompanied by a slight increase in sulphur percent. This trend could be interpreted in terms of 1) structural changes in starch which diminish   susceptibility of the latter towards further reactions; 2) shortage of accessible starch hydroxyls through their involvement in the reaction with 2-hydroxy-3-chloropropyl sulphate and 3) the possibility that 2-hydroxy-3-chloropropyl sulphate at higher concentrations are more susceptible to alkaline hydrolysis. It follows from this that using 2-hydroxy-3-chloropropyl sulphate at increasing concentration upto 0.5 molar ratio creates conditions where all these possibilities are far from achieving their negative effect after attaining the observed balance.

3.1.3. Durations and Temperature of the Reaction
Table 1 demonstrates the effect of duration and temperature on the extent of the reaction (expressed as sulphur %) occurring between 2-hydroxy-3-chloropropyl sulphate and starch in presence of sodium hydroxide. The starch (25 m mole) was reacted with 2-hydroxy-3-chloropropyl sulphate (12.5 m mole) for varying lengths of durations (30 - 300 min.) at different temperatures (60°C - 90°C). The reaction was effected under the influence of NaOH (18.75 m mole).

Results of **Table 1** shed insight on the effect of duration and temperature of the reaction of starch with 2-hydroxy-3-chloropropyl sulphate. Obviously the magnitude of the extent of the reaction is determined by the duration and temperature of the reaction. For a given temperature, the extent of the reaction increases by prolonging duration of the reaction upto 180 min. Thereafter, the extent of the reaction tends to level off provided that the temperature does not exceed 70°C. With higher temperatures (80°C, 90°C) the extent of reaction de-

(a)

(b)

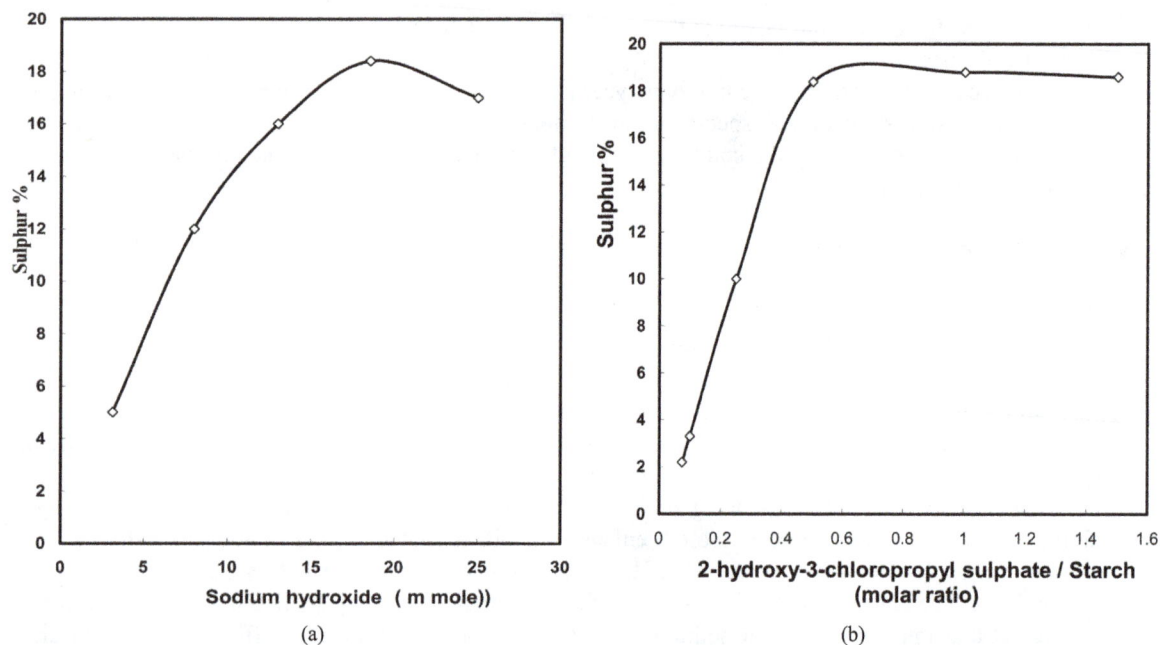

Figure 1. (a) Effect of sodium hydroxide concentration on the extent of reaction of starch with 2-hydroxy-3-chloropropyl sulphate, expressed as sulphur %. Maize starch (25 m mole), 2-hydroxy-3-chloropropylsulphate (12.5 m mole) at 180 min for 80°C. (b) Effect of 2-hydroxy-3-chloropropylsulphate concentration on the extent of the reaction of starch with 2-hydroxy-3-chloropropylsulphate, expressed as sulphur %. Maize starch (25 m mole), sodium hydroxide (18.75 m mole) at 80°C for 240 min.

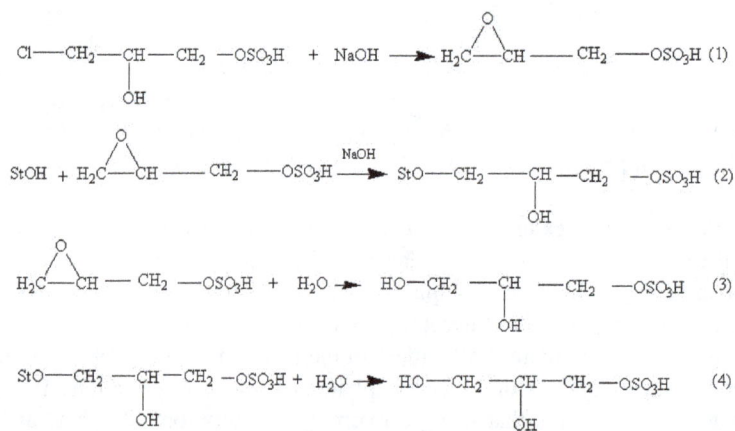

Scheme 2. Sodium hydroxide concentration.

Table 1. Effect of duration and temperature on the sulphur percent.

Duration (min) Temp. °C	Sulphur %			
	60°C	70°C	80°C	90°C
30	15	16.6	17.2	17.6
60	16.8	17.5	18.1	18.4
90	17.2	17.6	18.2	18.1
120	17.6	17.8	18.1	17.9
180	18.0	17.5	17.9	15.4
240	18.2	17.4	17.6	-
300	18.0	-	-	-

Starch, 25 m mole; 2-hydroxy-3-chloropropyl sulphate, 12.5 m mole; sodium hydroxide, 18.75 m mole.

creases by prolonging duration beyond 180 minutes. At any event, however, the highest sulphur percent is observed with modified starch sample prepared at 90°C for 60 min. Sulphur percent values that are very comparable to this highest value could be achieved at 80°C for 90 min. and at 60°C for 240 min. This state of affairs reflects the combined effect of both time and temperature on allowing more reactions between starch and 2-hydroxy-3-chloropropyl sulphate. In concomitant with this is the splitting off of the bond linking starch with the 2-hydroxy-3-chloropropyl sulphate moiety and starch as well as deactivation of the etherifying agents in question through hydrolysis under the action of alkali and heat.

3.2. Characterization of the Starch-2-Hydroxypropyl Sulphate

3.2.1. Infra Red Spectroscopy

Native starch (NS) and starch hydroxypropyl sulphate (SS) samples having sulphur percent 18.4, were characterized by infrared (IR) spectroscopy analysis. From the IR data shown in the **Figure 2**, it can be seen that the band at 829 cm^{-1} with a shoulder at 940 cm^{-1} for a symmetrical C-O-S. These peaks cannot be found in the spectrum of the native starch sample. This can obviously powers the vibration assigned to a C-O-SO$_3$ groups on starch. The shoulder peak at 940 cm^{-1} indicated the presence of axial sulphate ester [26]-[28].

3.2.2. Scanning Electron Microscopy

Scanning electron microscopy is the right techniques used for direct observation of microstructure of spherulites of size in the range between 0.1 - 10 μm like native maize starch sample shown in **Figure 3(a)**. It is evident that the sample clearly exhibits its granular structure. Modification of starch via reaction with 2-hydroxy-3-chloropropyl sulphate, changes in contour of the granules of starch as presented in **Figure 3(b)**. It can be seen that individual granules of starch have joined through the modification process.

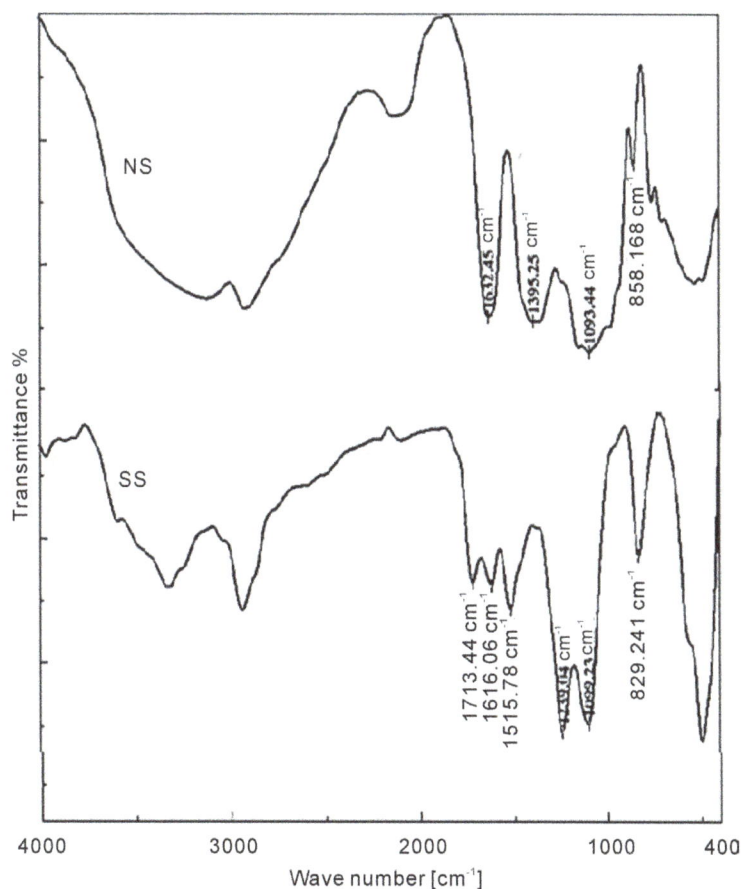

Figure 2. FTIR of native starch (NS) and starch hydroxypropyl sulphate (SS).

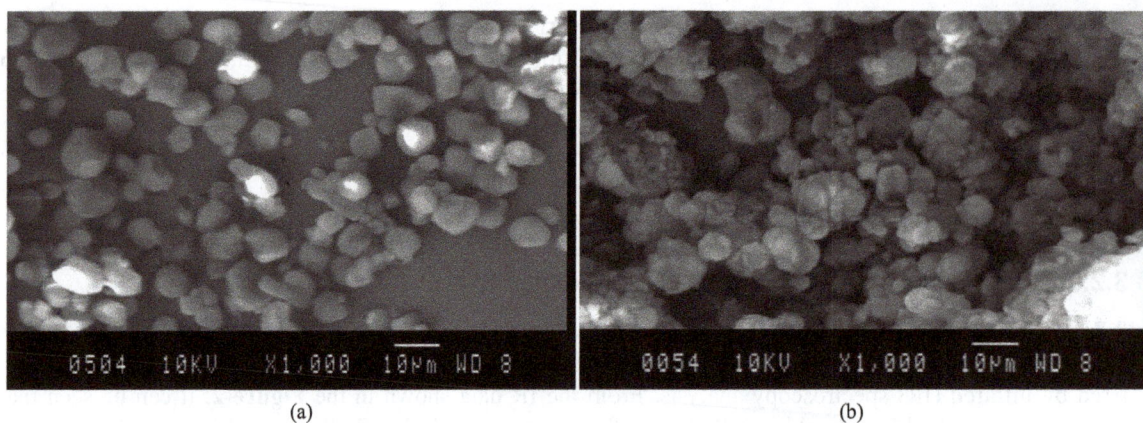

Figure 3. Scanning electron microscopy. (a) Native starch; (b) Starch hydroxypropyl sulphate.

3.3. Utilization of the Starch Hydroxypropyl Sulphate in Cationic Dye Removal

3.3.1. Initial pH

Figure 4 depicts the effect of initial pH on dye sorption value using starch hydroxypropyl sulphate. Dye adsorption was carried out using starch (0.4 g/50ml) having sulphur content of 18.4% during 70 min. duration. Dye concentration (100 ppm) was examined over of pH value ranging from 2 to 10. Results of **Figure 4** signify that the percent value of dye removal increases sharply as the pH increases from 2 to 7 then decreases also sharply thereafter. This is rather logical since at low pH, higher adsorption is associated with increased protonation by the neutralization of the negative charges at the surface of the adsorbent, which, in turn, facilitates the diffusion process and provides more active sites for the adsorbent [7]. At high pH, the dye becomes protonated and the electrostatic repulsion between the protonated dyes and positively charged adsorbent sites is established and results in decreased adsorption.

3.3.2. Contact Time of Duration and Initial Dye Concentration

The uptake of MB onto starch hydroxypropyl sulphate as a function of dye concentration was studied using starch (0.4 g/50ml) acquiring sulpher % of 18.4 and pH 7. The results obtained are summarized in **Table 2**. By and large the effect of increasing the dye concentration is to decrease the % dye removal. On the contrary prolonging the duration of contact between the dye and the starch derivative in question increases the % dye removal. For instance, 100% dye removal could be attained with an initial dye concentration of 75 ppm after 70 min. On the other hand, 100% dye removal could be achieved after 90 min. and 120 min. with initial dye concentrations of 100 ppm and 125 ppm respectively. This full (100%) removal of the dye could be observed with initial dye concentrations higher than 125 ppm even after 120 min. contact time.

The general tendency of the % dye removal to decrease particularly when the contact time between MB and starch hydroxypropyl sulphate is longer than 40 min., is a manifestation of the increase in the deriving force of the concentration gradient with the increase in the initial dye concentration [29]. The results (**Table 2**) reveal also that the uptake of MB onto starch hydroxypropyl sulphate increases by increasing the contact time (10 - 120 min.). It is understandable that initial adsorption is rapid because the adsorption of the dye occurs initially onto the exterior surface, after that dye molecules enter into pores (interior surface), a relatively slow process [30]. 100% dye removal could be arrived at when initial dye concentrations of 75, 100 and 125 ppm were allowed to be in contact with starch hydroxypropyl sulphate for 70, 90 and 120 min. respectively.

3.3.3. Adsorbent Dose

Table 3 shows the % dye removal as function of adsorbent concentration. The adsorption of the dye on starch hydroxypropyl sulphate at different concentrations of the latter (0.1 - 0.5 g in 50 mL) was investigated. Starch hydroxypropyl sulphate at 70 min. was used. Dye concentration (75 ppm) with sulpur percent 18.4 and at pH 7 were employed for performing dye adsorption. Results of **Table 3** show that % of dye removal increases as the adsorbent concentration increases due to the availability of larger surface area with more active functional groups at higher adsorbent dosages.

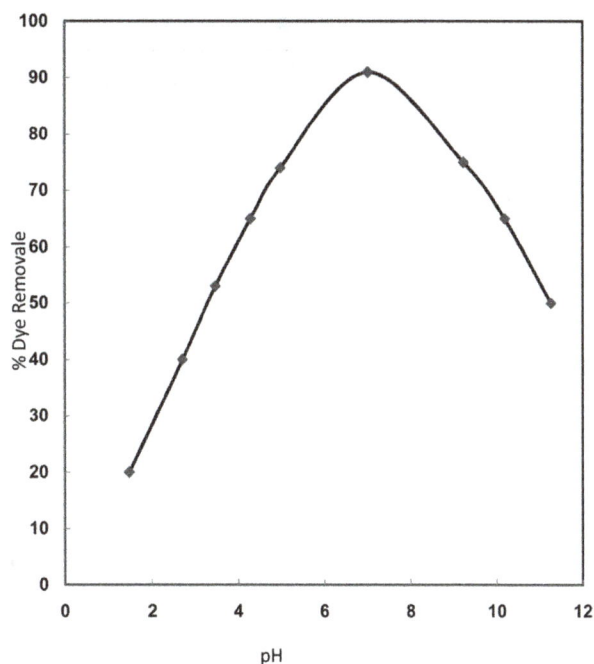

Figure 4. Effect of pH on the dye removal %. Starch hydro-xypropyl sulphate (0.4 g/50ml) acquiring sulphur % of 18.4 after 70 min, dye concentration (100 ppm).

Table 2. Effect of contact time of duration and initial dye concentration on % of dye removal.

concentration Duration (min)	% of dye removal at 75 (ppm)	% of dye removal at 100 (ppm)	% of dye removal at 125 (ppm)	% of dye removal at 150 (ppm)	% of dye removal at 175 (ppm)	% of dye removal at 200 (ppm)
10	45	48	52	55	60	63
20	56	58	60	60	61	60
30	66	66	66	64	63	62
40	77	79	73	68	68	68
50	86	83	79	72	70	69
60	94	87	84	75	72	70
70	100	91	89	78	74	71
80	-	95	93	80	75	72
90	-	100	96	82	76	73
100	-	-	98	84	77	74
110	-	-	99	86	78	75
120	-	-	100	87	79	75

Starch hydroxypropyl sulphate at pH 7 and sulphur % 18.4.

Table 3. Effect of adsorbent dose on % of dye removal.

adsorbent concentration (g in 50 ml)	% of dye removal
0.1	60
0.2	70
0.3	80
0.4	90
0.5	100

Starch hydroxypropyl sulphate at 70 min, dye concentration (75 ppm) with sulphur percent 18.4 and at pH 7.

Table 4. The effect of sulphur percent of on % of dye removal.

Sulphur percent	% of dye removal
6.5	45
10.4	60
14.1	75
16.2	89
18.4	100

Starch hydroxypropyl sulphate at pH 7, 70 min, dye concentration (75 ppm).

3.3.4. Sulphur Percent

Table 4 displays the % dye removal using starch hydroxypropyl sulphate acquiring different sulphur percents at 70 min, dye concentration (75 ppm) and at pH 7.

Results of **Table 4** feature that the dye removal % increases by increasing the sulphur percent of the adsorbent. This is unequivocally due to the presence of more active functional groups *i.e.* hydroxypropyl sulphate, in the molecular structure of starch. Needless to say that the sulphur percent is a measure of these functional groups; the higher the sulphur percent is, the higher the functionality of the starch derivatives in question is.

4. Conclusions

Starch hydroxypropyl sulphate was synthesized using the dry process under a variety of conditions. Factors studied included catalyst concentration and 2-hydroxy-3-chloropropyl sulphate concentration as well as duration and temperature. Based on the results obtained, the best conditions for synthesis of starch hydroxypropyl sulphate derivatives were established; they were 2-hydroxy-3-chloropropyl sulphate: starch molar ratio 0.5:1, sodium hydroxide: 2-hydroxy-3-chloropropyl sulphate molar ratio 1.5:1, at 90°C for 60 min. Under such conditions, starch hydroxypropyl sulphate with sulphur percent 18.4 could be achieved. Characterization of thus modified starch via IR spectroscopy analysis and scanning electronic microscope was investigated and, difference in contour of the starch granules was verified.

The as-prepared starch hydroxypropyl sulphate derivatives acted as cationic dye removal. The goal was realized starch hydroxypropyl sulphate which had sulphur percent of 18.4 at pH 7. 100% dye removal could be achieved at dye concentrations of 75, 100 and 125 ppm after a contact time of 70, 90 and 120 min respectively in presence of the newly synthesized adsorbent at a concentration of 0.5g in 50 mL.

References

[1]   Brostow, W., Pal, S. and Singh, R.P. (2007) A Model of Flocculation, *Materials Letters*, **61**, 4381-4384. http://dx.doi.org/10.1016/j.matlet.2007.02.007

[2]   Wang, C.C., Juang, L.C., Hsu, T.C., Lee, C.K., Lee, J.F. and Huang, F.C.(2004) Adsorption of Basic Dyes onto Montmorillonite. *Journal of Colloid and Interface Science*, **273**, 80-86. http://dx.doi.org/10.1016/j.jcis.2003.12.028

[3]   Tsai, W.T., Chang, Y.M., Lai, C.W. and Lo, C.C. (2005) Adsorption of Basic Dyes in Aqueous Solution by Clay Adsorbent from Regenerated Bleaching Earth. *Applied Clay Science*, **29**, 149-154. http://dx.doi.org/10.1016/j.clay.2004.10.004

[4]   Bilgic, C. (2005) Investigation of the Factors Affecting Organic Cation Adsorption on Some Silicate Minerals. *Journal of Colloid and Interface Science*, **281**, 33-38. http://dx.doi.org/10.1016/j.jcis.2004.08.038

[5]   Wang, S., Zhua, Z. and Coomes, A. (2005) The Physical and Surface Chemical Characteristics of Activated Carbons and the Adsorption of Methylene Blue from Wastewater. *Journal of Colloid and Interface Science*, **284**, 440-446. http://dx.doi.org/10.1016/j.jcis.2004.10.050

[6]   Kumar, K.V., Ramamurthi, V. and Sivanesan, S. (2005) Modeling the Mechanism Involved during the Sorption of Methylene Blue Onto Fly Ash. *Journal of Colloid and Interface Science*, **284**, 14-21. http://dx.doi.org/10.1016/j.jcis.2004.09.063

[7]   Janos, P., Buchtova, H. and Ryznarova, M. (2003) Sorption of Dyes from Aqueous Solutions onto Fly Ash. *Water Research*, **37**, 4938-4944. http://dx.doi.org/10.1016/j.watres.2003.08.011

[8]   Inthorn, D., Singhtho, S., Thiravetyan, P. and Khan, E. (2004) Decolorization of Basic Direct and Reactive Dyes by Pre-Treated Narrow-Leaved Cattail (*Typha angustifolia* Linn.). *Bioresource Technology*, **94**, 299-306. http://dx.doi.org/10.1016/j.biortech.2004.01.002

[9]   Xu S.M., Wang, J.L., Wu, R.L., Wang, J.D and Li, H. (2006) Adsorption Behaviors of Acid and Basic Dyes on Cros-
      slinked Amphoteric Starch. *Chemical Engineering Journal*, **117**, 161-167. http://dx.doi.org/10.1016/j.cej.2005.12.012

[10] Garg, V.K., Amita, M. and Kumar, R. (2004) Basic Dye (Methylene Blue) Removal from Simulated Wastewater by
      Adsorption Using Indian Rosewood Sawdust: A Timber Industry Waste. *Dyes Pigments*, **63**, 243-250.
      http://dx.doi.org/10.1016/j.dyepig.2004.03.005

[11] Chiou, M.S. and Li, H.Y. (2002) Equilibrium and Kinetic Modeling of Adsorption of Reactive Dye on Cross-Linked
      Chitosan Beads. *Journal of Hazardous Materials*, **93**, 233-248. http://dx.doi.org/10.1016/S0304-3894(02)00030-4

[12] Delval, F., Crini, G., Bertini, S., Filiatre, C. and Torri, G. (2005) Preparation, Characterization and Sorption Properties
      of Crosslinked Starch-Based Exchangers. *Carbohydrate Polymers*, **60**, 67-75.
      http://dx.doi.org/10.1016/j.carbpol.2004.11.025

[13] Adebowale, K.O. and Lawal, O.S. (2003) Functional Properties and Retrogradation Behaviour of Native and Chemi-
      cally Modified Starch of Mucuna Bean (*Mucuna pruriens*). *Journal of the Science of Food and Agriculture*, **83**, 1541-
      1546. http://dx.doi.org/10.1002/jsfa.1569

[14] Zhan, G.P., Huang, K.L. and Zhang, F.W. (2005) The Study on Synthesis of Anion Starch Microspheres. *New Chemi-
      cal Materials*, **33**, 44-46.

[15] Zhao, Y. and Wen, J. (2006) Synthesis of Carboxymethyl Starch for Water Treatment. *Technology of Water Treatment*,
      32, 76-77.

[16] Khalil, M.I. and Aly, A.A. (2002) Preparation and Evaluation of Some Anionic Starch Derivatives as Flocculants.
      Starch-Stärke, **54**, 132-139. http://dx.doi.org/10.1002/1521-379X(200204)54:3/4<132::AID-STAR132>3.0.CO;2-E

[17] Fa, C. and Gui, H. (2006) Synthesis and Properties of Sulfonated Starch as Superplasticizer. *Fine Chemicals*, **23**, 711-
      716.

[18] Vieira, M.C., Klemm, D., Einfeldt, L. and Albrecht, G. (2005) Dispersing Agents for Cement Based on Modified Poly-
      saccharides. *Cement and Concrete Research*, **35**, 883-890. http://dx.doi.org/10.1016/j.cemconres.2004.09.022

[19] Zhang, D.F., Ju, B.Z., Zhang, S.F. and Yang, J.Z. (2008) The Study on the Synthesis and Action Mechanism of Starch
      Succinate Half Ester as Water-Reducing Agent with Super Retarding Performance. *Carbohydrate Polymers*, **71**, 80-84.
      http://dx.doi.org/10.1016/j.carbpol.2007.05.020

[20] El-Rehim, H.A. and Diaa, D.A. (2012) Radiation-Induced Eco-Compatible Sulfonated Starch/Acrylic Acid Graft Co-
      polymers for Sucrose Hydrolysis. *Carbohydrate Polymers*, **87**, 1905-1912.
      http://dx.doi.org/10.1016/j.carbpol.2011.09.072

[21] Cui, D., Liu, M., Wu, L. and Bi, Y. (2009) Synthesis of Potato Starch Sulfate and Optimization of the Reaction Condi-
      tions. *International Journal of Biological Macromolecules*, **44**, 294-299.
      http://dx.doi.org/10.1016/j.ijbiomac.2009.01.003

[22] Lin, J., Wu, X., Fang, L., Lin, D. and Yao, S. (2009) Preparation of Starch Sulphate Using Concentrated Sulfuric Acid
      and Propanol. *Journal of Chemical Engineering of Chinese Universities*, **23**, 455-459.

[23] Cui, D., Liu, M., Zhang, B., Gong, H. and Bi, Y. (2011) Optimization of Reaction Conditions for Potato Starch Sul-
      phate and Its Chemical and Structural Characterization. *Starch-Stärke*, **63**, 354-363.
      http://dx.doi.org/10.1002/star.201000033

[24] Klimov, V.A. (1975) Basic Methods of Organic Microanalysis. Chapter 5, Mir. Publishers, Moscow.

[25] Vogel, A.I. (1972) A Text-Book of Quantitative Inorganic Analysis Including Elementary Instrumental Analysis. 3rd
      Edition, Longman Group Ltd., London, 402-404.

[26] Chiovitti, A., Bacic, A., Craik, D.J., Kraft, G.T., Liao, M.L., Falshaw, R., *et al.* (1998) A Pyruvated Carrageenan from
      Australian Specimens of the Red Alga *Sarconema filiforme*. *Carbohydrate Research*, **310**, 77-83.
      http://dx.doi.org/10.1016/S0008-6215(98)00170-0

[27] Xing, G.X., Zhang, S.F., Ju, B.Z. and Yang, J.Z. (2006) Study on Adsorption Behavior of Crosslinked Cationic Starch
      Maleate for Chromium(VI). *Carbohydrate Polymers*, **66**, 246-251. http://dx.doi.org/10.1016/j.carbpol.2006.03.005

[28] Zhong, K., Lin, Z.T., Zheng, X.L., Jiang, G.B., Fang, Y.S., Mao, X.Y. and Liao, Z.W. (2013) Starch Derivative-Based
      Superabsorbent with Integration of Water-Retaining and Controlled-Release Fertilizers. *Carbohydrate Polymers*, **92**,
      1367-1376. http://dx.doi.org/10.1016/j.carbpol.2012.10.030

[29] Hameed, B.H., Mahmoud D.K. and Ahmad A.L. (2008) Sorption of Basic Dye from Aqueous Solution by Pomelo(*Ci-
      trus grandis*) Peel in a Batch System. *Colloids and Surfaces A: Physicochemical and Engineering Aspects*, **316**, 78-84.
      http://dx.doi.org/10.1016/j.colsurfa.2007.08.033

[30] Mall, I.D., Srivastava, V.C. and Agarwal, N.K. (2006) Removal of Orange-G and Methyl Violet Dyes by Adsorption
      onto Bagasse Fly Ash—Kinetic Study and Equilibrium Isotherm Analyses. *Dyes & Pigments*, **69**, 210-223.
      http://dx.doi.org/10.1016/j.dyepig.2005.03.013

14

Effects of Yttrium Doping on the Performance of Ru-Based Catalysts for Hydrogenation of Fatty Acid Methyl Ester

Qi Lin*, Huirong Zheng, Guocai Zheng, Xinzhong Li, Benyong Lou

Department of Chemistry and Chemical Engineering, Minjiang University, Fuzhou, China
Email: *qlin1990@163.com

Abstract

The highly dispersed supported ruthenium-yttrium (Ru-Y) bimetallic catalysts were prepared by impregnation method and their catalytic performance for hydrogenation of ester was fully investigated. The catalyst was characterized by X-ray diffraction and field emission scanning electron microscopy. The results show that the average particle diameter of the bimetallic crystallites was less than 10 nm. The effects of the reaction temperature, the hydrogen pressure, the amount of catalyst and the proportion of yttrium in catalyst on the hydrogenation of ester were studied. The experimental results show that the introduction of yttrium not only changed the chemical and textural properties of ruthenium-based catalyst but also controlled the formation of Ru-Y alloy. The Ru-Y catalyst (Ru-2%Y/TiO$_2$) exhibited high catalytic activity and good selectivity towards the higher alcohols. Under optimal reaction conditions of 240°C and 5 MPa hydrogen pressure, the conversion of palm oil esters was above 93.4% while the selectivity towards alcohol was above 99.0%.

Keywords

Yttrium-Doped, Ruthenium-Based Catalyst, Fatty Acid Methyl Ester, Hydrogenation

1. Introduction

Fatty alcohols (FAlcs) and their derivatives are widely used as surfactants, lubricants, solvents, synthetic detergents, antifoaming agents, perfumes, cosmetics and pharmaceuticals, and as additives in many industrial products [1]. Naturally, derived FAlcs approximately accounts for 65% of the world's steadily growing fatty alcohol

---

*Corresponding author.

demand. They are special products obtained from natural fats and oils by the hydrogenation of fatty acids or fatty acid methyl esters (FAME) [2]. The best method of converting an acid to the corresponding alcohol is through the esters because esters are less corrosive and can be reduced to alcohols with considerably higher yields, and esters can be obtained from acids in nearly quantitative yields.

Since the last century, the natural-fatty-alcohol-based surfactants have gained growing significance in the detergent market due to their excellent washing properties and superior biodegradability. Therefore, the hydrogenation of FAME to the corresponding FAlcs is of great industrial importance. And the concern of catalyst research is raised because catalyst is the key technology for the effective production of FAlcs and the successful utilization of natural FAME. Many efforts have been made to develop high active catalysts [3] [4]. Most of these studies concentrated on the characterization of the catalyst or mechanism of the reaction but as lower catalytic activity or selectivity towards FAlcs.

During the last decade, the development of biodiesel has been emphasized. Biodiesel is defined as a fatty acid methyl ester obtained by the trans-esterification route of renewable biological sources with methanol. It can offer abundant feedstock for the hydrogenation to obtain FAlcs. Therefore, the research subject is to focus on improving the selectivity and yield of FALcs to the highest possible. It is reported that depositing the amorphous alloys on support [3] [5] [6] and adding promoters such as Au and Ag to the amorphous [7] are useful ways to improve their thermal stability and catalytic activity. In this study, the ruthenium-yttrium (Ru-Y) bimetallic catalysts were prepared through impregnation method and their performances of the hydrogenation of methyl palmitate to fatty alcohol were investigated. The effects of reaction conditions such as the reaction temperature, the hydrogen pressure, the proportion of yttrium, the amount of catalyst and the solvent on the hydrogenation of methyl palmitate were studied. The conversion of methyl palmitate could reach high at mild reaction conditions.

2. Experiment

2.1. Chemicals

$RuCl_3$ (Sinopharm Chemical Reagent Co., Ltd., >99.6%), $Y(NO_3)_3$ (Sinopharm Chemical Reagent Co., Ltd., >99.9%), Methyl palmitate (Aladdin Reagent Co. Ltd., >99.9%), TiO_2 (100 - 160 mero), hydrogen (99.99%) and the other reagents were used as received.

2.2. Preparation of Catalyst

All the catalysts were prepared by the impregnation method in aqueous suspensions. The metallic precursors (such as ruthenium chloride and yttrium nitrate) were dissolved in deionic water. Then, the titania was impregnated with above solution for overnight. And water was employed by evaporation at 333 K for 5 h. Next, the samples were dried at 393 K for 12 h. The solids were subjected to calcination treatment for 4 h at 673 K under air flow and reduced by hydrogen for 3 h at 453 K.

In order to eliminate the surface acidity of the catalyst, the catalyst was treated with barium nitrate and washed with deionic water until the absence of barium cations. The solids were dried for 6 h under vacuum. For the bimetallic catalysts, with moninal concentration of 5 wt% in Ru and 2.5 wt% in Y was employed, designated as 5% Ru - 2.5% Y/TiO_2.

2.3. Characterization of the Supported Catalyst

The powder XRD patterns were recorded at room temperature on a Philips X'Pert Pro MPD X-ray diffraction with Cu-K_α radiation at 50 kV and 35 mA. The 2θ angles were scanned from 15° to 75° at a rate of 0.1(°)/s. High-resolution TEM (HRTEM) images were obtained on a Philips TECNAIF-30 FEG instrument at an accelerating voltage of 300 V. The sample was dispersed with dry ethanol.

2.4. Investigation on the Catalytic Activity

The reaction was performed in a 60 mL stainless autoclave with a glass linear and magnetic stirrer. A typical procedure for hydrogenation of methyl palmitate (**Scheme 1**) is as follows: the appropriate amounts of palm oil esters, catalyst and cyclehexane were directly introduced into the autoclave, followed by a purge with hydrogen several times. Hydrogen was introduced until the desired hydrogen pressure was reached. The reaction was carried out under the designed conditions for a desired time. After the reaction, the autoclave was quickly cooled in a water bath and vented. The liquid products were separated from the reaction mixture by decantation and analyzed by GC 4002A (SUPELCOWAX™10 0.25 mm × 0.25 μm × 30 m), incorporating an FID detector. The solid residue obtained was reused in the next reaction.

3. Results and Discussion

3.1. Characterization of the Supported Catalyst

In order to further identify the crystalline phase of Ru in the catalyst, power X-ray diffraction patterns of the six solids with different metal loadings were obtained, as shown in **Figure 1**. The XRD pattern of titania exhibit the typical reflections of titania at 18°. The XRD patterns of Ru/TiO_2 exhibit the typical reflections at 29°, 55° and 64°. As the yttrium dopes on the Ru-based catalyst, the typical reflection of yttrium at 38° and 48° were found. Meanwhile, the characteristic peak of ruthenium (29°) shifts slightly from 29° to 25° while the other peaks are still observed. As the yttrium loadings increase from 1wt% to 2.5 wt%, the ruthenium peaks are remarkably weakened and the other low-angle reflection peaks almost disappear. Further increasing the yttrium loadings, the characteristic peaks of ruthenium and yttrium are still observed again. This indicates that Ru and Y species are highly dispersed in the support, which is possible due to the not calcinated catalyst. This is probably caused by yttrium entering into the interspaces of titania and forming the Ru-Y alloy at the surface of the catalyst, which is in agreement with other reports [2]. According to the above analysis, it is assumed that the size distribution of the highly dispersed Ru and Y crystallites is small when $w(Y)$ was 2.5 wt%. The deduction is also confirmed by the TEM images shown in **Figure 2**.

3.2. Catalytic activity of Ru-Y/TiO₂

3.2.1. Effect of Y Doping Amounts on the Catalytic Activity

To evaluate the catalytic activity of $Ru-Y/TiO_2$, the hydrogenation of methyl palmitate was chosen as a probing reaction. In all the runs, the selectivity for 1-hexadecanol in the products was more than 99%, only trace amounts of by-products palmitic acid were detected by GC-MS.

In order to investigate the effect of the amount of Y in catalyst on the hydrogenation, several kinds of catalysts with different amount of Y were used in the hydrogenation of methyl palmitate. It is clear from **Table 1** that the amount of Y shows a significant influence on the catalytic activity for hydrogenation of methyl palmitate. It is evident that conversion rates increase with Y doping amounts and selectivity is all above 99.9%. When

$$CH_3(CH_2)_{14}COOCH_3 + H_2 \longrightarrow CH_3(CH_2)_{14}CH_2OH + CH_3OH$$

Scheme 1. Hydrogenation of methyl palmitate.

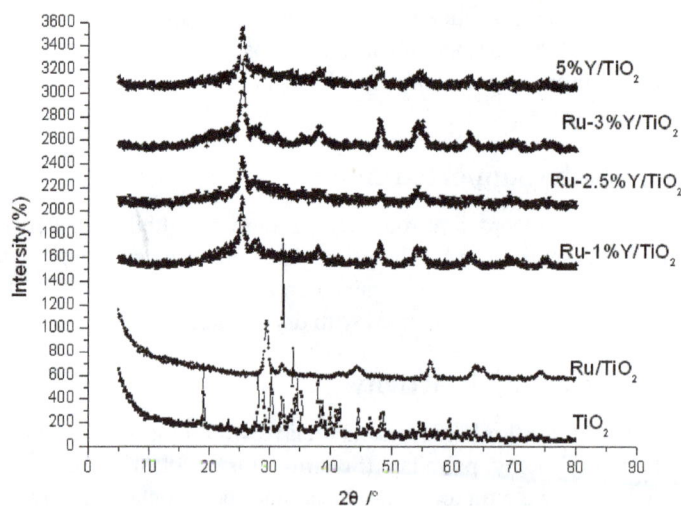

Figure 1. The XRD patterns of the catalysts.

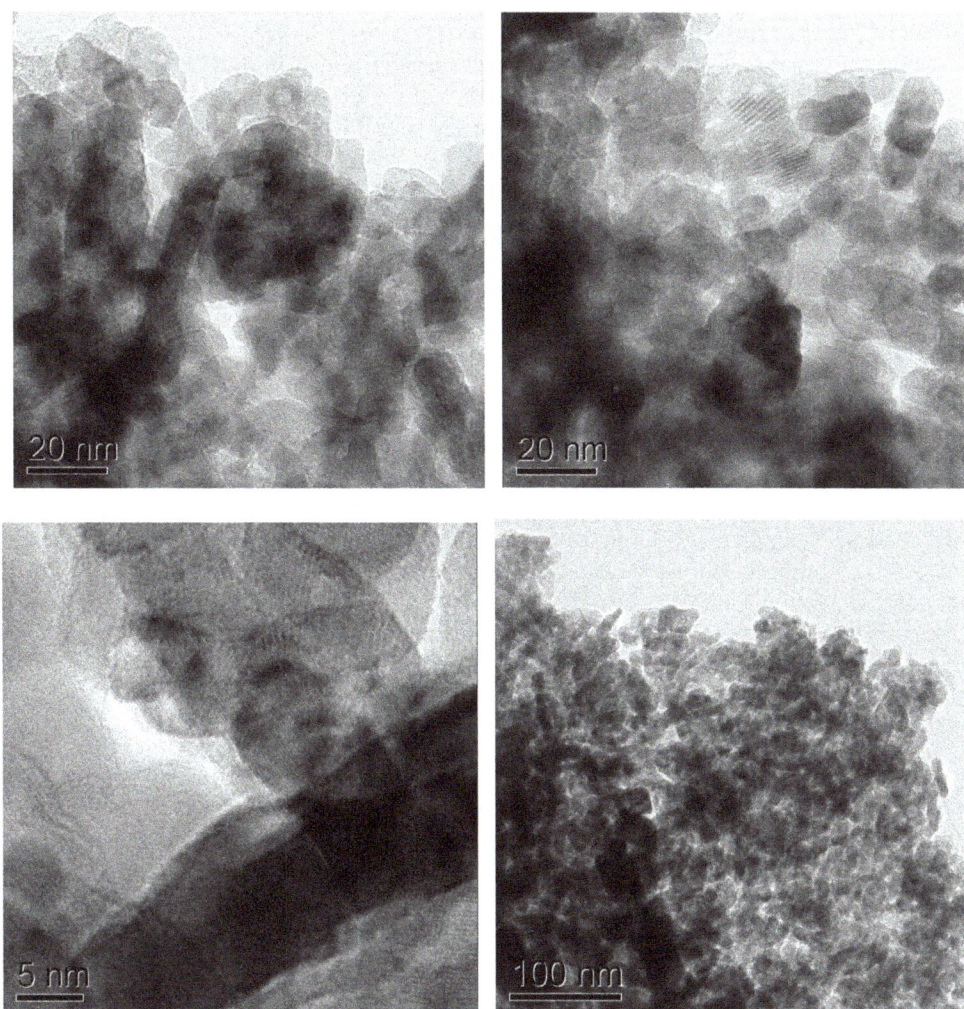

Figure 2. The TEM pictures of Ru-Y/TiO$_2$ catalysts.

Table 1. Effect of the amount of yttrium on the catalytic activity.

The amount of yttrium %	0	0.5	1	1.5	2	5[*]
Conversion (%)	10.4	29.3	72.5	75.8	93.4	0

Reaction conditions: catalyst 30 mg, methyl palmitate 1.85 mmol, cyclohexane 1 mL, 220°C, 5 MPa, 10 h. [*]The supported catalyst was Y/TiO$_2$.

Y doping amount is 0.5%, the conversion rate is just 29.3%. However, when Y doping amount is 2%, the conversion rate reached up to 93.4%. The doping of rare earth metals is one of the efficient methods to improve the activity of supported metal catalysts. This is probably due to the 4f electrons in Y electronic structure. Its outermost electronic structure is $4f^{1+n}5d^{0-1}6s^2$. Most of the earth metals have no 5d electron, so they could easily lose two 6s electrons, one d electron or f electron to form trivalent cations whose outermost electron configuration is $4f^n5s^26p^2$. From the electronic structure, the 5d orbital is empty, which provides a good electron transferring orbital and could be used as an electron transferring station with catalytic action. With the increasing of Y doping amounts, Y scrambles for electrons to reduce the composite of electron hole pair on TiO$_2$ surface so that more O^{2-} or O^{2-}OH are generated on the catalyst surface which has higher catalytic activity. Meanwhile, the specific electronic structure and larger radius of earth metals lead to lattice expansion on Ru/TiO$_2$ surface, which is beneficial to the generation of catalytic active center, increasing of catalytic active center numbers and improving of adsorption capacity of catalysts to hydrogen so as to increase conversion rates.

3.2.2. Effect of Temperature on the Catalytic Activity

The effect of temperature on the catalytic activity is shown in **Table 2**. It is clear from **Table 2** that the temperature also shows a significant influence on the catalytic activity for the hydrogenation of ester. The results indicate that the activity increases when the temperature rises from 180°C to 240°C. The tendency is similar to that of the hydrogenation of ester catalysed by Ru-based catalyst.

3.2.3. Effect of Hydrogen on the Catalytic Activity

Under the same conditions, the effect of the hydrogen pressure on the hydrogenation reaction of methyl palmitate was studied by altering the hydrogen pressure. When using 1.85 mmol methyl palmitate, 30 mg catalyst and 1 mL cyclohexane at 220°C for 10 h, the results are listed in **Table 3**. It indicates that the activity increases when the pressure rises from 3 MPa to 7 MPa. Higher pressure could increase the concentration of hydrogen in the solution and thereby accelerates the hydrogenation.

3.2.4. Recycling of the Catalyst

To investigate the stability of catalyst, the reaction mixture was separated from the catalyst by decantation after the reaction and analyzed by GC. The catalyst was directly reused for the next run. The results of hydrogenation are shown in **Table 4**. No dramatic fall in the catalytic activity is observed at any point of time during the seven experiments, which further confirms the fact that the active metals are immobilized at the surface of titania. It is noteworthy that the catalytic activity of Ru-Y/TiO$_2$ increases to some extent in the recycling experiments, which further indicates that there is an induction time to form the catalytical active species. In the successive recycling runs, no induction time is needed for the high catalytic activity.

4. Conclusion

The supported ruthenium-yttrium bimetallic catalysts were prepared by impregnation and their catalytic performances for hydrogenation of ester were evaluated. The XRD and TEM results reveal that the active metals are highly dispersed and the alloy is formed by Ru with Y. The catalytic results indicate that the catalytic activity of Ru-Y/TiO$_2$ depends on the loading of yttrium. The ruthenium-yttrium catalyst (Ru-2%Y/TiO$_2$) exhibits high catalytic activity and good selectivity towards the higher alcohols compared with Ru/TiO$_2$ and Y/TiO$_2$. Furthermore, the catalyst can be easily separated from the organic products and reused for seven times without significant decrease of activity and selectivity.

Acknowledgements

This work was supported financially by the Great Science & Technology Research Program from Fujian Province

Table 2. Effect of temperature on the catalytic activity.

Temp. (°C)	180	200	220	230	240	250
Conversion (%)	3.9	7.5	23.2	29.3	50.3	65.7

Reaction conditions: catalyst 25 mg, methyl palmitate 1.85 mmol, cyclohexane 1 mL, 5 MPa, 10 h.

Table 3. Effect of hydrogen on the catalytic activity.

P_{H2} (MPa)	3	4	5	6	7
Conversion (%)	38.0	56.2	72.5	74.8	78.2

Reaction conditions: catalyst 25 mg, methyl palmitate 1.85 mmol, cyclohexane 1 mL, 220°C, 10 h

Table 4. Recycling experiment of the catalyst.

Cycle	1	2	3	4	5	6	7
Conversion (%)	43.1	44.5	47.4	49.3	56.9	52.1	50.6

Reaction conditions: catalyst 25 mg, methyl palmitate 1.85 mmol, cyclohexane 1 mL, 5 MPa, 6 h, 240°C.

of China (Project No. 2011H6021), the Nature Science Foundation of Fujian Province of China (Project No. 2013J01053), the Science & Technology Research Program from Education Office of Fujian Province (Project No. JA12268) and the Science & Technology Research Program of Fuzhou Municipal, China (Project No. 2012-G-128).

References

[1]   Weissermel, K. and Arpe, H.-J. (1994) Industrielle Organische Chemie. 4th Edition, VCH Verlagsgesellschaft mbH, Deerfield Beach.

[2]   Echeverri, D.A., Marín, J., Restrepo, G. and Riso, L. (2009) Characterization and Carbonylic Hydrogenation of Methyl Oleate over Ru-Sn/Al$_2$O$_3$: Effects of Metal Precursor and Chlorine Removal. *Applied Catalysis A: General*, 366, 342-347. http://dx.doi.org/10.1016/j.apcata.2009.07.029

[3]   Yuan, P., Liu, Z.Y., Sun, H.J. and Liu, S.C. (2010) Influence of Calcination Temperature on the Performance of Cu-Al-Ba Catalyst for Hydrogenation of Esters to Alcohols. *Acta Physico-Chimica Sinica*, 26, 2235-2241. http://dx.doi.org/10.3866/PKU.WHXB20100642

[4]   Figueiredo, F.C.A., Jordão, E. and Carvalho, W. (2008) Adipic Ester Hydrogenation Catalyzed by Platinum Supported in Alumina, Titania and Pillared Clays. *Applied Catalysis A: General*, 351, 259-266. http://dx.doi.org/10.1016/j.apcata.2008.09.027

[5]   He, L., Li, X., Lin, W., Cheng, H., Yu, Y., Fujita, S., Arai, M. and Zhao, F. (2014) The Selective Hydrogenation of Ethyl Stearate to Stearyl Alcohol over Cu/Fe Bimetallic Catalysts. *Journal of Molecular Catalysis A: Chemical*, 392, 143-149. http://dx.doi.org/10.1016/j.molcata.2014.05.009

[6]   Vigier, K., Pouilloux, Y. and Barrault, J. (2012) High Efficiency CoSn/ZnO Catalysts for the Hydrogenation of Methyl Oleate. *Catalysis Today*, 195, 71-75. http://dx.doi.org/10.1016/j.cattod.2012.04.027

[7]   Zheng, J., Lin, H., Wang, Y., Zheng, X., Duan, X. and Yuan, Y. (2013) Efficient Low-Temperature Selective Hydrogenation of Esters on Bimetallic Au-Ag/SBA-15 Catalyst. *Journal of Catalysis*, 297, 110-118. http://dx.doi.org/10.1016/j.jcat.2012.09.023

15

Reaction of Nitrilimines with 2-Aminopicoline, 3-Amino-1,2,4-Triazole, 5-Aminotetrazole and 2-Aminopyrimidine

Rami Y. Morjan[1], Basam S. Qeshta[1], Hussein T. Al-Shayyah[1], John M. Gardiner[2], Basam A. Abu-Thaher[1], Adel M. Awadallah[1*]

[1]Department of Chemistry, Islamic University of Gaza, Gaza, Palestine
[2]Manchester Institute of Biotechnology, School of Chemistry and EPS, The University of Manchester, Manchester, UK
Email: [*]awada@iugaza.edu.ps

Abstract

The reaction of picoline derivatives 3-6 with hydrazonoylhalide 1a produced imidazo[1,2-a]pyridines 7-10, while the reaction of the same picoline derivatives with hydrazonoylhalide 1b afforded imidazo[1,2-a]pyridine-2-ones 11-13. The reaction of 1b, c with 3-amino-1,2,4-triazole 14 produced the acyclic adducts 18 and 19, respectively. Reaction of 1b, 1c with 5-aminotetrazole 20 produced the acyclic products 23 and 24, respectively. Finally, the reaction of 1b with 4, 6-dimethyl-2-aminopyrimidine 27 afforded compound 29 rather than its isomeric structure 28. The structure of the products was confirmed by the different spectroscopic analytical methods including IR, MS, ^{1}HNMR and ^{13}CNMR.

Keywords

Nitrilimines, Picolines, Imidazo[1,2-a]Pyridine, Triazoles, Tetrazoles, Pyrimidines

1. Introduction

Nitrilimines—generally generated *in situ* from hydrazonoyl halides are 1,3-dipolar species that are widely used for the synthesis of different heterocyclic systems *via* 1,3-dipolar cycloaddition reactions [1] [2], or cyclocondensation reactions [2]. Nitrilimines with C-acetyl or ester moiety represent a dielectrophilic system that may react with dinucleophilic heterocyclic systems at the carbonyl carbon of the acetyl or ester group, and at the terminal carbon of the dipole leading to fused heterocyclic systems. Recent examples include; the synthesis of im-

---

[*]Corresponding author.

idazo[1,2-a]pyridines [3] [4] and their 2-one derivatives [4] from the reaction of nitrilimines with 2-amino-pyridine and imidazo[1,2-a]pyrazine from the reaction with 2-aminopyrazine [5]. Their reaction with 2-cyanomethylben-zimidazole gave pyrrolo[1,2-a]benzimidazole [6] [7]. The reaction of ethyl pyridine-2-acetate with nitrilimines having a C-acetyl moiety afforded the corresponding pyrrolo[1,2-a]pyridine, while their reaction with nitrilimines having a C-ester moiety afforded a cyclic adducts [8]. Similarly, the reaction of nitrilimines with a C-ester moiety with 2-cyanomethylbenzimidazole, and 2-amniobenzimidazole afforded a cyclic adducts [8]. In this work, we will investigate the reaction of nitrilimines with different amino heterocycles hoping to prepare new fused heterocyclic compounds that may be further tested for their biological activity. The following amino heterocycles will be used; aminopicolines, aminotriazole, aminotetrazole, aminoisoxazole and aminopyrimidine.

2. Materials and Methods

2.1. Experimental

^{1}H-NMR spectra were recorded at 300 or 400 MHz and ^{13}C-NMR spectra at 75 or 100 MHz on a Bruker AC300 or AC400 spectrometer. Chemical shifts are denoted in (ppm) relative to the internal solvent standard, TMS. ES-MS and HRMS were recorded on a Micromass LCT orthogonal acceleration time-of-flight mass spectrometer (positive and negative ion mode) with flow injection via a Waters 2790 separation module autosampler. FTIR spectra were recorded using Shimadzu 8201 spectrophotometer with KBr technique in region 4000 - 400 cm^{-1} that was calibrated by polystyrene. Melting point determinations were made using a Stuart Scientific SMP1 apparatus and are uncorrected. All chemicals and solvents were used without further purification, as supplied by Sigma-Aldrich unless otherwise stated. Hydrazonoylhalides **1a-c** were prepared using Japp-Klingman method according to reported literature procedure [9].

2.1.1. General Procedure for Syntheses of Compounds 7-10 and 11-13

Triethylamine (0.6 g, 0.006 mol) in dry THF (10 ml) was dropwise added to a stirred solution of hydrazonoyl chlorides **1a** or **1b** (0.005 mol) and substituted picolines **3 - 6** (0.65 g, 0.006 mol) in 25 ml THF at room temperature producing imidazopyridines **(7 - 10)** and **(11 - 13)**, respectively. Stirring was continued overnight, and the solvent was then evaporated in *vacuo*. The residual solid was washed with water to remove the triethylamonium salt, and the crude products were then recrystallized from the appropriate solvents to give the title compounds.

1) 3-[(4-Chlorophenyl)diazenyl]-2,8-dimethylimidazo[1,2-a]pyridine 7

Yield 93%, Yellow solid, mp 152°C - 155°C; IR (KBr): v_{max} cm^{-1} = 1490, 1416, 1298, 1282, 1221, 1200, 1096, 811 and 774 cm^{-1}. MS: m/z C$_{15}$H$_{13}$ClN$_4$ (284/286 M$^+$, chlorine isotope effect), HRMS (Calculated 284.0902, Found 284.0898). ^{1}H NMR (300 MHz, DMSO-d$_6$): δ 9.62 (d, J = 7.0 Hz, 1H), 7.89 (d, J = 8.7 Hz, 2H), 7.60 (d, J = 8.7 Hz, 2H), 7.51 (d, J = 7.0 Hz, 1H), 7.20 (t, J = 7.0 Hz, 1H), 2.74 (s, 3H, CH$_3$), 2.58 (s, 3H, CH$_3$). ^{13}C NMR (100 MHz, DMSO-d$_6$): δ 166.2, 151.4, 138.2, 136.1, 135.4, 131.2, 129.0, 127.3, 124.0, 123.0, 119.2, 15.2, 10.3

2) 3-[(4-Chlorophenyl)diazenyl]-2,7-dimethylimidazo[1,2-a]pyridine 8

Yield 78%, Yellow solid, mp 174°C - 177°C; IR (KBr): v_{max} cm^{-1} = 2993, 1520, 1473, 1462 and 796 cm^{-1}; MS: m/z C$_{15}$H$_{13}$ClN$_4$ (284/286 M$^+$., chlorine isotopes effect); HRMS (284.0902, Found 284.0901); ^{1}H NMR (300 MHz, DMSO-d6): δ 9.66 (d, J = 6.9 Hz, 1H), 7.88 (d, J = 8.7 Hz, 2H), 7.59 (d, J = 8.8 Hz, 2H + 1H pyr.), 7.16 (d, J = 7.0 Hz, 1H), 2.71 (s, 3H, CH$_3$), 2.49 (s, 3H, CH$_3$); ^{13}C NMR (100 MHz, DMSO-d6): δ 151, 149, 139, 136, 131, 129, 129, 124, 121, 119, 111, 21, 10.

3) 3-[(4-Chlorophenyl)diazenyl]-2,6-dimethylimidazo[1,2-a]pyridine 9

Yield 94%, green solid, mp. 145°C - 147°C; MS: m/z C$_{15}$H$_{13}$ClN$_4$ (284/286 M$^+$·, chlorine isotopes effect); HRMS (Calculated 284.0902, Found 284.0897); ^{1}H NMR (300 MHz, DMSO-d$_6$): δ 9.55 (s, 1H), 7.91 (d, J = 8.7 Hz, 2H), 7.67 (d, J = 9.0 Hz, 1H), 7.60 (d, J = 8.7 Hz, 2H), 7.54 (d, J = 9.0 Hz, 1H), 2.70 (s, 3H, CH$_3$), 2.43 (s, 3H, CH$_3$); ^{13}C NMR (100 MHz, DMSO-d$_6$): (δ 161.0, 151.0, 138.0, 136.0, 131.0, 130.0, 129.0, 126.0, 124.0, 119.0, 111.0, 17.0, 10.0).

4) 3-[(4-Chlorophenyl)diazenyl]-2,5-dimethylimidazo[1,2-a]pyridine 10

Yield 93%, orange solid, mp 157°C - 159°C; MS: m/z C$_{15}$H$_{13}$ClN$_4$ (284/286 M$^+$·, chlorine isotopes effect); HRMS (Calculated 284.0902, Found 284.0908); ^{1}H NMR (300 MHz, DMSO-d$_6$): δ 7.74 (d, J = 9.1 Hz, 2H), 7.60 (d, J = 9.1 Hz, 2H), 7.45 - 7.60 (m, 2H, overlapped), 7.12 (d, J = 6.8 Hz, 1H), 3.00 (s, 3H, CH$_3$), 2.71 (s, 3H, CH$_3$).

5) 8-Methylimidzao[1,2-a]pyridine-2,3-dione-3-[4-chlorophenyl)hydrazone] 11

Yield, 30%, Yellow solid, mp > 300˚C; MS m/z $C_{14}H_{11}ClN_4O$ (286/288 M$^+$, chlorine isotopes effect). HRMS (Calculated 286.0695, Found 286.0685). ^{1}H NMR (400 MHz, DMSO-d$_6$): δ 12.46 (s, 1H, NH), 8.39 (d, J = 6.9 Hz, 1H), 7.64 (d, J = 6.9 Hz, 1H), 7.60 (d, J = 6.9 Hz, 2H), 7.42 (d, J = 6.9 Hz, 2H), 6.88 (t, J = 6.9 Hz, 1H), 2.25 (s, 3H, CH$_3$).

6) 6-Methylimidzao[1,2-a]pyridine-2,3-dione-3-[4-chlorophenyl)hydrazone] 12

Yield 78%, red solid, mp 96˚C - 98˚C; MS *m/z* $C_{14}H_{11}ClN_4O$ (286/288 M$^+$, chlorine isotopes effect); HRMS (Calculated. 286.0695, Found 286.0700) ^{1}H NMR (400 MHz, DMSO-d$_6$): δ 15.55 (s, 1H, NH), 8.23 (s, 1H), 7.66 (d, J = 8. Hz, 1H), 7.60 (d, J = 8.4 Hz, 2H), 7.54 (d, J = 8.4 Hz, 2H), 6.87 (d, J = 8.4 Hz, 1H), 2.27 (s, 3H, CH$_3$). ^{13}C NMR (100 MHz, DMSO-d$_6$): δ 202.0, 150.0, 149.0, 140.0, 136.0, 133.0, 131.0, 130.0, 127.0, 112.0, 111.0, 18.0.

7) 5-Methylimidzao[1,2-a]pyridine-2,3-dione-3-[4-chlorophenyl)hydrazone] 13

Yield 30%, Yellow solid, mp 232˚C - 235˚C; MS *m/z* $C_{14}H_{11}ClN_4O$ (286/288 M$^+$, chlorine isotopes effect); HRMS (Calculated 286.0695, Found 286.0698). ^{1}H NMR (300 MHz, DMSO-d$_6$): δ 13.17 (s, 1H, NH), 7.76 (t, J = 8.0 Hz, 1H), 7.46 (br.s, 4H, not splitted), 7.08 (d, J = 8.0 Hz, 1H, Py), 6.88 (d, J = 8.0 Hz, 1H, Py), 2.82 (s, 3H, CH$_3$).

2.1.2. General Procedures for Syntheses of Compounds 18, 19, 23, 24 and 29

Triethylamine (0.6 g, 0.006 mol) in THF (30 ml) and methanol (10 ml) was drop-wise added to a stirred solution of hydrazonoyl chloride **1b** or **1c** (0.006 mol) and the appropriate azole (0.005 mol) at room temperature. Stirring was continued for 24 hours. The solvent was then removed and the solid precipitate was washed with water 50 ml and then washed with cold ethanol. The crude solid product was collected, dried and crystallized from hot ethanol.

1) Methyl 2-(3-amino-4H-1,2,4-triazol-4-yl)-2-(2-(4-chlorophenyl)hydrazono)acetate 18

Yield 50%, yellow solid, mp 255˚C - 257˚C; MS: m/z $C_{11}H_{11}ClN_6O_2$ (294/295 M$^+$, chlorine isotopes effect). ^{1}H NMR (300 MHz, DMSO-d$_6$): δ 10.88 (s, 1H, NH), 7.95 (s, 1H, N=CHN, triazole ring), 7.35 (s, 4H, aromatic ring), 5.95 (s, 2H, NH$_2$), 3.74 (s, 3H, OCH$_3$). ^{13}C NMR (100 MHz, DMSO) δ 161.5 (C=O), 154.0 (N=CN, triazol ring), 142.28 (C, aromatic ring), 138.76 (CH, NN=CHN, triazol ring), 129.32 (C, aromatic ring), 126.42 (s), 117.45 (CH, aromatic ring), 116.57 (CH, aromatic ring), 52.7 (OCH$_3$).

2) 1-(3-amino-4H-1,2,4-triazol-4-yl)-1-(2-(2,5-dimethylphenyl)hydrazono)propan-2-one19

Yield 40%, brown solid, mp194˚C - 196˚C; MS: m/z $C_{13}H_{16}N_6O$ (272 M$^+$) HRMS (Calculated 272.1386, found 272.1275). ^{1}H NMR (300 MHz, DMSO): δ 9.76 (s, 1H, NH), 7.89 (d, J = 1.3 Hz, 1H, CCHC, aromatic ring), 7.27 (s, 1H, N=CHN, trizole ring), 7.06 (d, J = 7.5 Hz, 1H, CH, aromatic ring), 6.84 (d, J = 7.5 Hz, 1H, CH, aromatic ring), 5.92 (s, 2H, NH$_2$), 2.46 (s, 3H, CH$_3$), 2.30 (s, 3H, CH$_3$), 2.25 (s, 3H, CH$_3$). ^{13}C NMR (100 MHz, DMSO) δ 190.5 (C=O), 153.3 (N=CN, triazol ring), 140.3 (C, aromatic ring), 138.8 (CH, NN=CHN, triazol ring), 135.8 (C, aromatic ring), 130.9 (CH, aromatic ring), 125.9 (aromatic ring), 124.5 (CH, aromatic ring), 123.1 (C, aromatic ring), 117.8 (CH, aromatic ring), 24.7 (CH$_3$), 20.8 (CH$_3$), 17.1 (CH$_3$).

3) Methyl 2-(5-amino-1H-tetrazol-1-yl)-2-(2-(4-chlorophenyl)hydrazono)acetate 23

Yield 30%, Yellow solid, mp 180˚C - 182˚C; IR (KBr) v_{max} cm^{-1}: 3214, 3168, 3111 and 1748. MS: m/z $C_{10}H_{10}ClN_7O_2$ (296) [M+H]$^+$. ^{1}H NMR (300 MHz, DMSO), δ ppm: 11.12 (s, 1H, NH), 7.43 (d, J = 9.0 Hz, 2H, aromatic ring), 7.37 (d, J = 9.0 Hz, 2H, aromatic ring), 7.12 (s, 2H, NH$_2$), 3.79 (s, 3H, OCH$_3$). ^{13}C NMR (100 MHz, DMSO) δ 160.88 (C=O), 156.5 (C=N), 141.9, 129.7, 127.3, 116.9, 116.5 (aromatic ring), 53.1 (OCH$_3$).

4) 1-(5-amino-1H-tetrazol-1-yl)-1-(2-(2,5-dimethylphenyl)hydrazono)propan-2-one 24

Yield 48%, pale Yellow solid, mp 161˚C - 162˚C; IR (KBr):v_{max} cm^{-1} = 3327.56, 2980.65, 2360, 1660 cm^{-1}. MS: m/z $C_{12}H_{17}N_7O$ (273). ^{1}H NMR (400 MHz, DMSO) δ 10.24 (s, 1H, NH), 7.26 (s, 1H, CH, aromatic ring), 7.08 (d, J = 7.6 Hz, 1H, CH, aromatic ring), 6.92 (s, 2H, NH$_2$), 6.88 (d, J = 7.7 Hz, 1H, CH, aromatic ring), 2.47 (s, 3H, CH$_3$), 2.30 (s, 3H, CH$_3$), 2.25 (s, 3H, CH$_3$). ^{13}C NMR (100 MHz, DMSO) δ 189.3 (C=O), 155.96 (NC=N, tetrazol ring), 140.04 (C, aromatic ring), 135.91(N=CN), 131.09 (CH, aromatic ring), 125.12 (CH, aromatic ring), 124.75 (C, aromatic ring), 123.82 (C, aromatic ring), 118.66 (CH, aromatic ring), 24.65 (CH$_3$), 20.70 (CH$_3$), 17.25 (CH$_3$).

5) Methyl-2-[(4-chlorophenyl) hydrazono]-2-(2'-imino-4',6'-dimethylpyrimidin-1'-(2'H)-yl)acetate 29

Yield 52%, brown solid, mp 130˚C - 132˚C MS: m/z $C_{15}H_{16}ClN_5O_2$(333/335 M$^+$, chlorine isotopes effect).^{1}H NMR (400 MHz, DMSO) δ 9.80 (s, 1H), 8.04 (s, 1H), 7.73 (s, 1H), 7.70 (d, 2H), 7.59 (d, 2H), 3.86 (s, 3H), 2.27

(s, 3H), 1.91 (s, 3H).

3. Results and Discussion

The reaction of picoline derivatives **3-6** with hydrazonoyl chloride **1a** in THF in the presence of Et₃N as a base at room temperature produced imidazo[1,2-a]pyridine **7-10** in 85% - 90% yield, while the reaction of hydrazonyl-chloride **1b** with substituted picolines **3-6** under the same reaction conditions afforded the expected imida-zo[1,2-a]pyridine-2-ones **11-13** in 30% - 70% yield (**Scheme 1**). The structure of the products was confirmed by the different spectroscopic analytical methods including IR, MS, ¹HNMR and ¹³CNMR.

The reaction of **1** with 3-amino-1,2,4-triazole **14** in refluxing ethanol was reported by Shawali *et al.* to give imidazo[1,2-b] triazole **15** in very low yield [10]. On the other hand, and under the same reaction conditions, Graf reported that the reaction of **14** with hydrazonylchloride **1** (X = Ar) afforded the acyclic adducts **16** [11]. Treatment of **16** with refluxing AcOH/AcONa led to formation of the cyclic fused product **18** *via* loss of a mo-lecule of NH₃. The structure of **16** and **17** were determined by X-ray crystallography.

In this work, the reaction of hydrozonylchlorides **1b, c** with 3-amino-1,2,4-triazole **14** in the presence of Et₃N and THF at room temperature (**Scheme 2**) afforded the acyclic adducts **18** and **19**, respectively. This as-

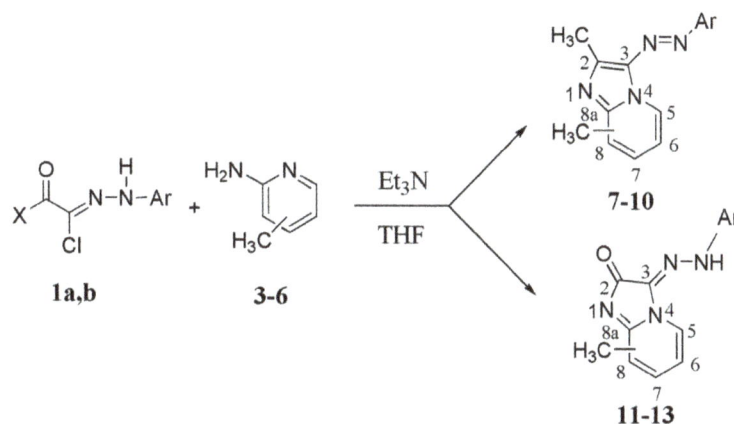

Scheme 1. Reaction of hydrazonylchloride 1a, b with substituted picoline 3-6.

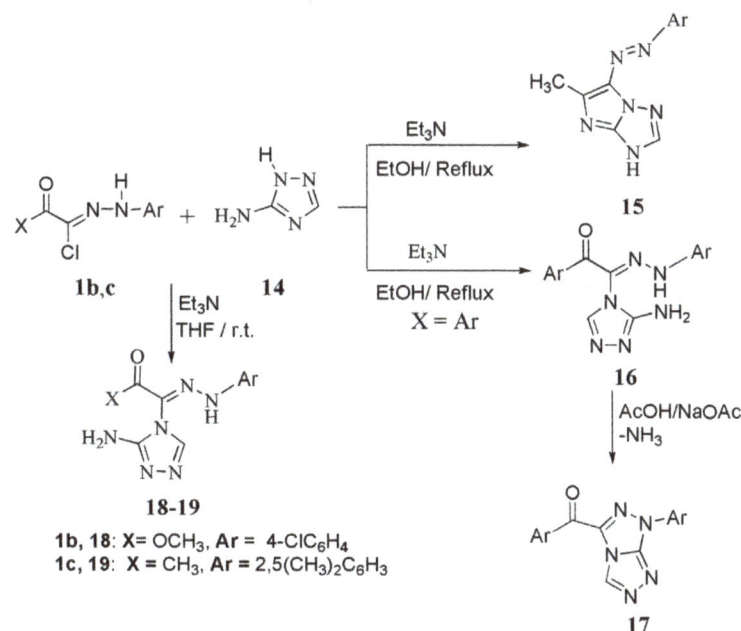

Scheme 2. Reaction of hydrazonylhalides 1b, c with 3-aminotriazole 14.

signment is based on the appearance of the NH₂ group and the C=O group in both IR and NMR spectra. Similar to Graf product, the reaction is proposed to occur at the N4 of the triazole **14**.The fused heterocyclic system reported by Shawali [10] was not obtained. Attempt to cyclize compounds **18** and **19** using AcOH/NaOAc (Graf method) was unsuccessful.

Graf reported [12] that the reaction of hydrazonylhalides **1** with 5-aminotetrazole **20** in refluxing ethanol produced compounds **22** *via* the rearrangement accompanied by loss of hydrazoic acid (HN₃) from the intermediate cycloaddition product **21**. In this work; the reaction of hydrazonyl halides **1b, 1c** with 5-aminotetrazole **20** in THF in the presence of Et₃N at room temperature afforded the acyclic adducts **23** and **24**, respectively (**Scheme 3**).

The reaction of hydrazonylhalide **1a** with 2-aminopyrimidine **25** was reported by Awadallah *et al.* [8] to produce the pyrimido [2,1-d] 1,2,3,5-tetrazine **26**. The reaction of **1b** with 4, 6-dimethyl-2-aminopyrimidine **27** in THE in presence of Et₃N as a base at room temperature produced however, the acyclic adduct, **29**. Two different possible isomeric structures are possible **28** and **29** (**Scheme 4**). The NMR data obtained provided an unambiguous confirmation that the actual obtained product is in fact compound **29** rather than compound **28**. The assignment is based on the appearance of two CH₃ groups at the pyrimidine ring. In compound **28** these two CH₃ are identical as they are in the same chemical and electronic environment, so they should appear as one peak

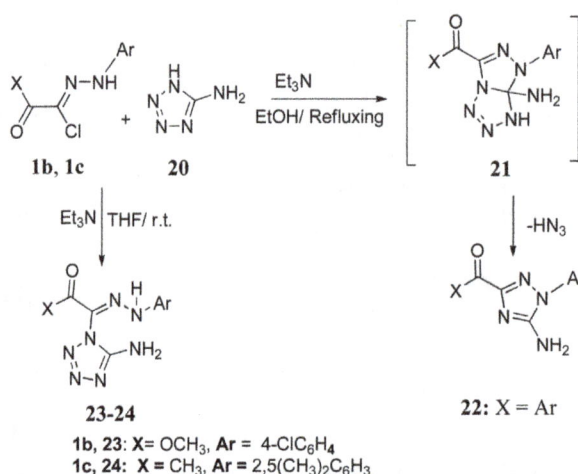

Scheme 3. Reaction of hydrazonylhalides 1b,1c with 5-amino-tetrazole 20.

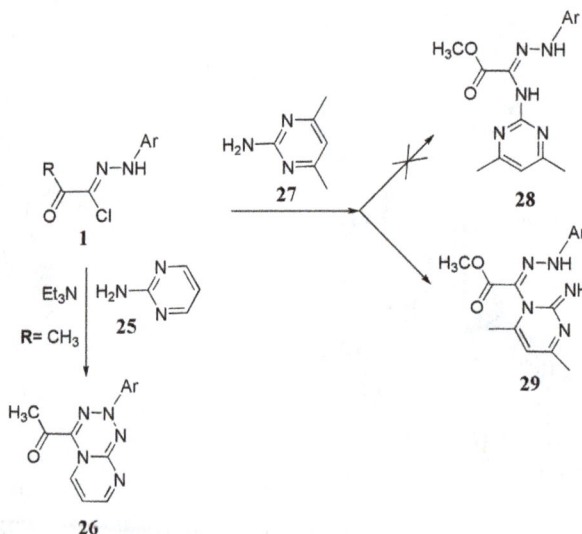

Scheme 4. Reaction of hydrazonylhalides 1b with 4, 6-dimethyl 2-aminopyrimidin 27.

with an integration value corresponding to 6H. On the other hand, the two CH_3 groups in compound **29** are different and two peaks with equal integration are expected. ^{1}HNMR spectra obtained supported structure **29** rather than **28**. The mass spectra form compound **29** in both positive mode 234 $[M+1]^+$ and negative mode 232 $[M-1]^+$ is shown in **Figure 1.**

4. Conclusion

A new series of heterocyclic compounds was synthesized *via* the reaction of Nitrilimines with 2-aminopicoline, 3-amino-1,2,4-triazole, 5-aminotetrazole and 2-aminopyrimidine. The structures of the products were characterized by IR, ^{1}H-NMR, ^{13}C-NMR and MS. The melting points of the synthesized compounds are listed in **Table 1.**

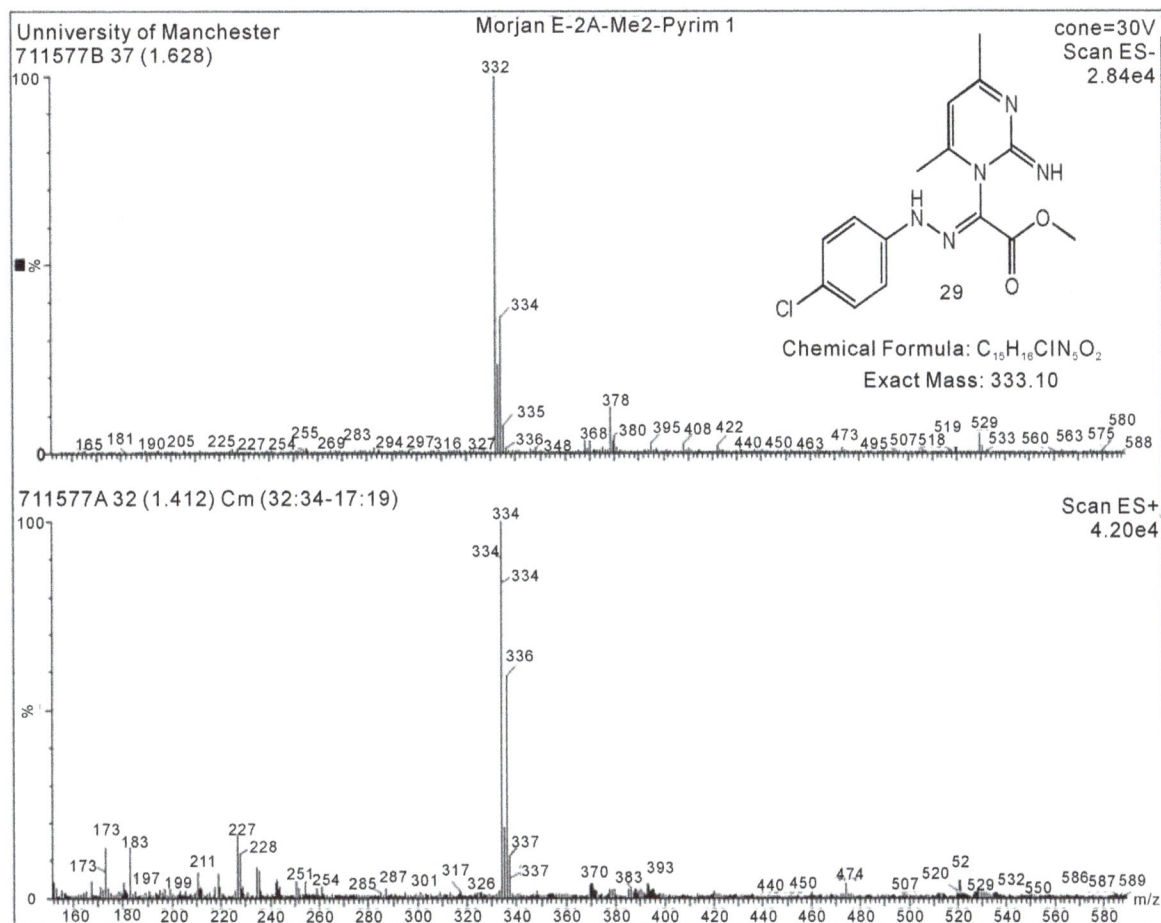

Figure 1. Mass spectra for compound 29 in both positive and negative mode.

Table 1. The melting points of the synthesized compounds.

Comp.	M.F.	M.W	M.P°C	Comp.	M.F.	M.W	M.P°C
7	$C_{15}H_{13}ClN_4$	284	152 - 155	13	$C_{14}H_{11}ClN_4O$	286	232 - 235
8	$C_{15}H_{13}ClN_4$	284	174 - 177	18	$C_{11}H_{11}ClN_6O_2$	294	255 - 257
9	$C_{15}H_{13}ClN_4$	284	145 - 147	19	$C_{13}H_{16}N_6O$	272	194 - 196
10	$C_{15}H_{13}ClN_4$	284	157 - 159	23	$C_{10}H_{10}ClN_7O_2$	295	180 - 182
11	$C_{14}H_{11}ClN_4O$	286	>300	24	$C_{12}H_{17}N_7O$	273	161 - 162
12	$C_{14}H_{11}ClN_4O$	286	96 - 98	29	$C_{15}H_{16}ClN_5O_2$	333	130 - 132

Acknowledgements

The authors would like to thank the Deanship of Scientific Research at the Islamic University of Gaza for their financial support. Thanks are due to Dr. Adeeb-El-Dhashan for his scientific contribution and help in spectroscopic analyses.

References

[1]   Padwa, A. (1984) 1,3-Diploar Cycloaddition Chemistry. Wiley, New York.

[2]   Ferwanah, A.R.S. and Awadallah, A.M. (2005) Reaction of Nitrilimines and Nitrile Oxides with Hydrazines, Hydrazones and Oximes. *Molecules*, **10**, 492-507. http://dx.doi.org/10.3390/10020492

[3]   Thaher, B. (2003) Synthesis and Characterization of Some New Substituted Imidazo[1,2-a]Pyridines. *Abhath Al-Yarmouk Journal (Series of Natural Studies)*, **12**, 555-561.

[4]   Thaher, B. (2005) Synthesis of Some New Substituted Imidazo(1,2-a)Pyrazines. *The Islamic University Journal (Series of Natural Studies and Engineering)*, **13**, 109-115.

[5]   Thaher, B. (2006) Synthesis of Some New Substituted Imidazo(1,2-a)Pyridines and Their 2-One Derivatives. *The Islamic University Journal (Series of Natural Studies and Engineering)*, **14**, 31-38.

[6]   Awadallah, A.M., Seppelt, K. and Shorafa, H. (2006) Synthesis and X-Ray Crystal Structure of Pyrrolo[1,2-a]Benzimidazoles. *Tetrahedron*, **62**, 7744-7746. http://dx.doi.org/10.1016/j.tet.2006.05.071

[7]   Elwan N. (2004) A Facile Synthesis of Pyrrolo[1,2-a]Benzimidazoles and Pyrazolo[3,4:4,3']Pyrrolo[1,2-a]Benzimidazole Derivatives. *Tetrahedron*, **60**, 1161-1166. http://dx.doi.org/10.1016/j.tet.2003.11.068

[8]   Awadallah, A.M. and Zahra J. (2008). Reaction of Nitrilimines with 2-Substituted Aza-Heterocycles. Synthesis of Pyrrolo[1,2-a]Pyridine and Pyrimido[2,1-d]1,2,3,5-Tetrazine. *Molecules*, **13**, 170-176. http://dx.doi.org/10.3390/molecules13010170

[9]   El-Abadelah, M.M., Hussein A.Q. and Thaher, B. (1991) Heterocycles from Nitrile Imines. Part IV. Chiral 4,5-Dihydro1,2,4-Triazin-6-Ones. *Heterocycles*, **32**, 1879-1895. http://dx.doi.org/10.3987/COM-90-5637

[10] Shawali, A.S. (1993) Reactions of Heterocyclic Compounds with Nitrilimines and Their Precursors. *Chemical Reviews*, **93**, 2731-2777. http://dx.doi.org/10.1021/cr00024a007

[11] Abdelhamid, A.O., Hassaneen, H.M., Shawaki, A.S. and Pārkānyāj̦, C. (1983) Reactions of α-Ketohydrazidoyl Halides with Some Heterocyclic Amines. Facile Synthesis of Arylazo Derivatives of Fused Heterocycles with a Bridgehead Nitrogen Atom. *Journal of Heterocyclic Chemistry*, **20**, 639-643. http://dx.doi.org/10.1002/jhet.5570200326

[12] Graf, H. and Klebe, G. (2006) 3-Aroyl-1-Aryl-1H-[1,2,4]Triazolo[3,4-c]-1,2,4-Triazole, Untersuchung des Bildungswegs durch Röntgenstrukturanalyse und Molecular Modelling. *Chemische Berichte*, **120**, 965-977. http://dx.doi.org/10.1002/cber.19871200614

16

Process for the Preparation of Chromones, Isoflavones and Homoisoflavones Using Vilsmeier Reagent Generated from Phthaloyl Dichloride and DMF

Santosh Kumar Yadav

Department of Organic Chemistry & FDW, Andhra University, Visakhapatnam, India
Email: skgoityadav@gmail.com

Abstract

Vilsmeier reagent formed from phthaloyl dichloride and DMF was found to be very effective for converting 2-hydroxyacetophenones, deoxybenzoins and dihydrochalcones into corresponding chromones, isoflavones and homoisoflavones with excellent yield. This method offers significant advantages such as efficiency and mild reaction conditions with shorter reaction time.

Keywords

Phthaloyl Dichloride, Dimethylformamide, Chromones, Isoflavones, Homoisoflavones, $BF_3 \cdot Et_2O$, Vilsmeier Reagent

1. Introduction

In recent years, scientific interest towards chromones (2), isoflavones (9) and homoisoflavones (10) has increased. It is due to the limited distribution of these compounds in the plant kingdom and the possible health effect these compounds exhibit. The development of new methodologies for the synthesis of these compounds is important. It is known that certain natural and synthetic chromone derivatives possess important biological activities such as antitumor [1], antihepatotonic, antioxidant [2], anti-inflammatory [3], antispasmolytic, estrogenic [4] and antibacterial activities [5]. Isoflavones are a privileged class of natural products which are produced by plants mainly in the species of Leguminosae family to protect themselves from environmental stress and are present in dietary components such as fruits, cabbage, soybeans, grains, hops and redwines. Isoflavones possess many biological activities such as estrogenic [6], anticancer [7], antibacterial [8], antimicrobial [9], antiulcer [10] and protein tyrosine kinase inhibitor [11]. Search for new methodologies for the synthesis of isoflavones continues to be of great interest for organic chemists. The two most popular pathways for the synthesis of isoflavones are the deoxybenzoin and the chalcone routes. In the first route, isoflavones are synthesized by the ring closure of deoxybenzoin with C_1 unit by using different reagents such as ethoxalylchoride in pyridine [12], triethylorthoformate with pyridine and piperidine [13], N, N-dimethylformamide and $BF_3 \cdot Et_2O$ with $MeSO_2Cl$ [14] [15] or $POCl_3$ [16], anhydrous ethyl formate and powdered sodium [17], acetic anhydride and sodium acetate [18] [19], acetic-formic anhydride [20], N-formylimidazole in anhydrous THF [21] and phenyliodine (III)bis (trifluoroacetate) [22] [23]. In the second route, isoflavones are synthesized by oxidative rearrangement of a chalcone using reagents like thallium(III)nitrate [24]-[29] and thallium(III)acetate [30]-[32]. Similarly, rearrangement of chalcone epoxide with $BF_3 \cdot Et_2O$ is followed by catalytic hydrogenation [33] [34]. Other routes include the conversion of flavanones into isoflavones by thallium(III)nitrate in a mixture of CH_3OH and $CHCl_3$ [35], arylation of 4-chromanones with 4, 5-dimethoxy-o-benzoquinone in anhydrous DMSO followed by acidification [36] [37], tetrakis (triphenylphosphine) palladium (0) catalyzed cross-coupling reactions of 3-iodo-chromone with arylboronic acid [38], etc. However, the reported syntheses of many isoflavones including daidzein and formononetin are time-consuming (Bass, 1976 [44], Baker et al., 1953 [12], Farkas et al., 1971 [43], Pelter and Foot, 1976 [50], Yoder et al., 1954 [51]).

Homoisoflavonoids are a class of naturally occurring oxygen containing heterocyclic compounds. Both natural and synthetic homoisoflavonoids exhibit numerous biological activities [39]-[41] like antifungal, hypocholesterolemic, antimutagenic, antirhinovirus, antiallergic, angio productive activity, antihistaminic activity, anti-inflammatory, antioxidant, antiviral, cough relief, inhibition of platelet aggregation etc. Homoisoflavonoids can be synthesized either by the condensation of 4-chromanones with arylaldehdes in methanol by passing HCl gas or by using piperidine as a base followed by isomerisation of the double bond using Pd/C at 250°C [42] or by the extension of one carbon in dihydrochalcone using ethylformate/sodium [43] or $BF_3 \cdot Et_2O$ and DMF with $MeSO_2Cl$ [44] or PCl_5 [45] etc.Both the methods have disadvantages; while the first method has multiple steps, in the second method, the phenolic hydroxyls have to be protected to get chalcones in good yield.

However, most of the methods reported for the synthesis of chromones, isoflavones and homoisoflavones suffer from harsh reaction conditions, poor substituent tolerance, long reaction times, and low to moderate yields. Therefore, developing a milder and more general procedure for chromones, isoflavones, and homoisoflavones is still highly desirable. It was reported that when DMF was treated with phthaloyl dichloride in 1,4-dioxane at 40°C for 3 h precipitated only vilsmeier reagent as a solid form while the co-product phthalic anhydride was dissolved in a solvent [46]. The precipitates were collected by filtration through a glass-filter funnel under nitrogen atmosphere. The residue was dried in vacuo to give white crystals, which was identified as the vilsmeier reagent by comparison with authentic supplied by Aldrich Chemical Co. Vilsmeier reagent is well known as a versatile synthetic tool for the formylation of electron-rich aromatics, chlorination of alcohols, conversion of carboxylic acid into the corresponding acid chloride and so on [47] [48]. A series of 2-hydroxyacetophenone, deoxybenzoin and dihydrochalcone was cyclized with a one carbon unit by using this reagent. The reaction requires a short reaction time, mild reaction conditions and easy work-up. Products obtained by this methodology do not have contaminants such as sulphur or phosphine obtained through DMF with $MeSO_2Cl$ [44] or PCl_5 [45]. Naturally, occurring isoflavones such as formononetin (9c), daidzein (9d) and retusin (9h) was synthesized by applying this methodology. To the best of researcher knowledge, the synthesis of chromones, isoflavones and homoisoflavones using vilsmeier reagent formed from phthaloyl dichloride and DMF has not been reported.

2. Results and Discussion

The method first involved the preparation of vilsmeier reagent, for this, to a mixture of DMF in 1,4-dioxane was added phthaloyl dichloride at room temperature, and then the whole mixture was stirred at 40°C for 3 h (**Scheme 1**). The white precipitates of (chloromethylene) dimethyliminiumchloride (VR) were isolated by filtration under a nitrogen atmosphere.

First application came with its usage in preparation of chromones in which 1 equiv. of substituted 2-hydroxyacetophenone was dissolved in3 equiv. of $BF_3 \cdot Et_2O$ and DMF was added drop wise with stirring at 10°C. Then whole reaction mixture was transferred slowly with continuous stirring into 1 equiv. of vilsmeier reagent. The reaction mixture was stirred at 50°C for 30 minutes (**Scheme 2**). The completion of reaction was monitored by TLC. The reaction mixture was poured into 3N HCl, extracted with EtOAc, dried over Na_2SO_4 and concen-

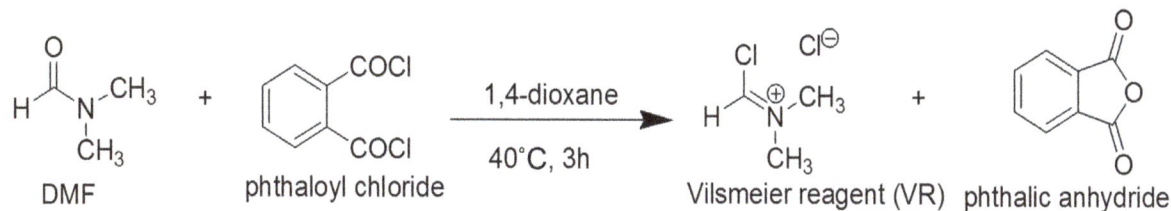

Scheme 1. Preparation of vilsmeier reagent from phthaloyl dichloride and DMF.

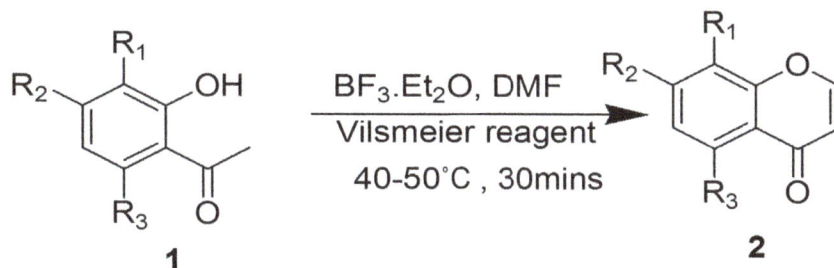

Scheme 2. Synthesis of chromones.

trated. The pure compound was then harvested with column chromatography.

The next success came with its usage in the preparation of isoflavones and homoisoflavones. For this, deoxybenzoins (7) and dihydrochalcones (8) were prepared by the published procedure [49] from phenyl acetic acid and 3-phenylpropanoic acid respectively with substituted phenols by Friedel-Crafts acylation using $BF_3 \cdot Et_2O$ which served as the Lewis acid for the acylation as well as the solvent for the reaction. The acylation was carried out at 85°C - 90°C. The completion of reaction was monitored by TLC. In most cases, the reaction was completed within 90 minutes. However, the substitution pattern as well as the presence of unprotected hydroxyl groups on the aromatic rings influenced the reaction time and the product yield. Conversion of these intermediate deoxybenzoin and dihydrochalcone into their respective isoflavones and homoisoflavones can be carried out either directly by treating with vilsmeier reagent, a minimum of 5 equivalents of $BF_3 \cdot Et_2O$ was required (Method A) or these intermediates were isolated, purified and then cyclised with vilsmeier reagent, for this a minimum of 3 equivalents of $BF_3 \cdot Et_2O$ was required (Method B) (**Scheme 3**). In all cases, the reaction was completed in 30-40 mins and the products were characterized by their spectral data (IR, NMR, and mass spectrometry).

To explain the formation of chromones (2), isoflavones (9), and homoisoflavones (10), a suggested mechanism is shown in **Figure 1**. The mechanism involves the addition of a vilsmeier reagent (I) to the acetophenone. BF_3 complex (II) to form (III), with subsequent nucleophilic attack of the hydroxyl group of 2-hydroxyacetophenone to form (IV), which is deaminated to form the final product (V).

3. Conclusion

A variety of chromones, isoflavones and homoisoflavones were synthesized in excellent yields using vilsmeier reagent generated from phthaloyl dichloride and DMF as the key reagent. The ready availability, low cost of phthaloyl dichloride and DMF, high activity of isolated vilsmeier reagent, the short reaction time, the mild reaction conditions, and the easy purification of the products make this an attractive new method for the synthesis of chromones, isoflavones and homoisoflavones, etc.

4. Experimental

4.1. General Remarks

All synthesized compound melting points were recorded on a Mel-Temp melting point apparatus in open capillaries and are uncorrected. Reactions requiring anhydrous conditions were performed in flame-dried glassware, and cooled under an argon or nitrogen atmosphere. Acme silica gel G and silica gel (100 - 200 mesh) were used for analytical thin-layer chromatography and column chromatography. Visualization of the resulting chroma-

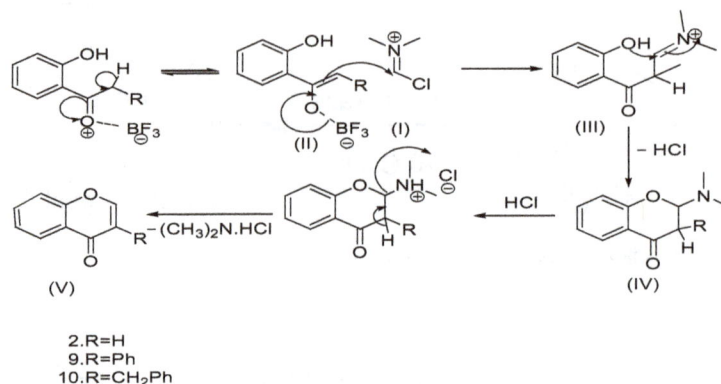

2.R=H
9.R=Ph
10.R=CH₂Ph

Figure 1. Plausible mechanism for the formation of 2, 9, and 10.

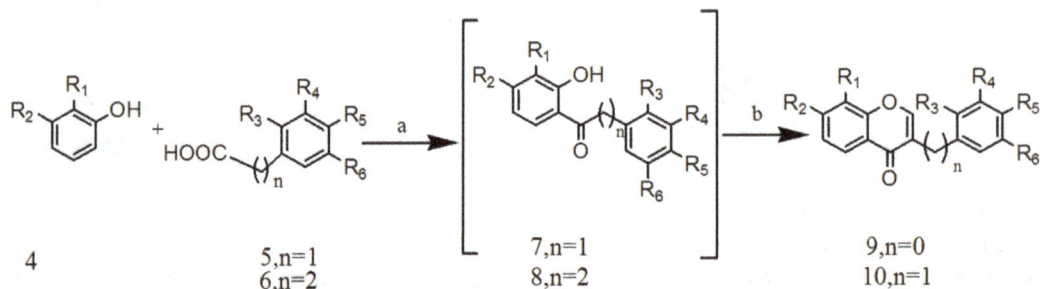

4

5,n=1
6,n=2

7,n=1
8,n=2

9,n=0
10,n=1

Reagents and conditions:(a)BF₃·Et₂O,90˚C,90 min,(b)DMF,10˚C,Vilsmeier reagent,60˚C,30-40 min

Scheme 3. Synthesis of isoflavones (9a-9h) and homoisoflavones (10a-10f).

tograms was done by looking under an ultraviolet lamp ($\lambda = 254 \, nm$). IR spectra were recorded on a Perkin-Elmer BX1 FTIR spectrophotometer and ^{1}H NMR (400 MHz) and ^{13}C NMR (100 MHz) spectra were recorded on a Bruker AMX 400 MHz. NMR spectrometer using TMS as the internal standard and the values for chemical shifts (_) being given in parts per million and coupling constants (J) in hertz. Mass spectra were recorded on an Agilent 1100 LC/MSD.

4.2. Preparation of Vilsmeier Reagent from Phthaloyl Dichloride and DMF in 1,4-Dioxane

A mixture of DMF 30 g (0.41 mol) and phthaloyl dichloride 90 g (0.44 mol) in 1,4-dioxane (330 mL) was stirred at 40˚C for 3 h. The white precipitates of (chloromethylene)dimethyliminium chloride (VR) that formed were collected by filtration under a nitrogen atmosphere, washed with 1,4-dioxane (100 mL × 2) and hexane (100 mL), and dried under reduced pressure, 41 gm (78% yield).

4.3. Preparation of Chromones (2a-2d)

DMF (4.6 mL) was added to a stirred solution of 2-hydroxyacetophenone (3 mmol) in BF₃·Et₂O (7.5 mmol) at 10˚C for 5 min. The reaction mixture was then added to the vilsmeier reagent (4.5 mmol) drop wise with stirring at room temperature. After completion of addition, the reaction mixture was stirred at 50˚C for 30 - 40 mins and poured into boiling dilute HCl slowly and cooled. The solution was extracted with ethyl acetate (30 mL × 2) and the combined organic layer was dried over anhydrous Na₂SO₄. The crude obtained after evaporation of the solvent was chromatographed over silica gel column using chloroform-methanol mixtures as eluent to give 2a-2d.

4.4. Chromen-4-One (2a) (Table 1, Entry 1)

Colorless solid; yield 376 mg (80%); mp 55˚C - 58˚C. ^{1}H NMR (400 MHz, DMSO-d₆)_: 6.32 (d, J = 5.6 Hz, 1H), 7.35 - 7.43 (m, 2H), 7.64 (t, J = 7.6 Hz, 1H), 7.86 (d, J = 5.6 Hz, 1H), 8.18 (d, J = 7.6 Hz, 1H). ^{13}C NMR (100 MHz, DMSO-d₆)_= 111.9, 117.2, 123.9, 124.3, 124.7, 132.8, 154.5, 155.5, 176.6. LC-MS: m/z: 147 [M +

Table 1. Synthesis of chromones (2a-2d).

S. No.	Entry	R₁	R₂	R₃	2 Yield (%)
1	a	H	HH	H	80
2	b	H	OH	H	85
3	c	OH	OH	H	87
4	d	H	OH	OH	89

1]$^+$. Anal.calcd. for $C_9H_6O_2$: C 73.97, H 4.14; Found: C 73.94, H 4.19.

4.5. 7-Hydroxy-4H-Chromen-4-One (2b) (Table 1, Entry 2)

Pale brown solid; yield 412 mg (85%); mp 206°C - 208°C. ^{1}H NMR (400 MHz, DMSO-d$_6$)_: 6.21 (d, J = 6.0 Hz, 1H), 6.84 (d, J = 2.4 Hz, 1H), 6.91 (dd, J = 2.4, 8.4 Hz, 1H), 7.87 (d, J = 8.4 Hz, 1H), 8.14 (d, J = 6.0 Hz, 1H), 10.76 (s, 1H). ^{13}C NMR (100 MHz, DMSO-d$_6$)_= 102.3, 111.9, 115.0, 117.0, 126.6, 156.0, 157.7, 162.5, 175.5. LC-MS: m/z: 161 [M-1]-. Anal.calcd. for $C_9H_6O_3$: C 66.67, H 3.73; found: C 66.65, H 3.75.

4.6. 7,8-Dihydroxy-4H-Chromen-4-One (2c) (Table 1, Entry 3)

Brown solid; yield 466 mg (87%); mp 205°C - 208°C. ^{1}H NMR (400 MHz, DMSO-d$_6$)_: 6.17 (d, J = 6.0 Hz, 1H), 6.93 (d, J = 6.8 Hz, 1H), 7.37 (d, J = 6.8 Hz, 1H), 8.19 (d, J = 6.0 Hz, 1H), 9.40 (s, 1H), 10.29 (s, 1H). ^{13}C NMR (100 MHz, DMSO-d$_6$)_= 111.3, 114.0, 115.0, 117.8, 132.9, 146.9, 150.0, 155.7, 176.0. LC-MS: m/z: 177 [M-1]$^-$. Anal.calcd. for $C_9H_6O_4$: C 60.68, H 3.39; found: C 60.63, H 3.43.

4.7. 5,7-Dihydroxy-4H-Chromen-4-One (2d) (Table 1, Entry 4)

Brown solid; yield 474 mg (89%); mp 268°C - 270°C. ^{1}H NMR (400 MHz, DMSO-d$_6$)_: 6.20 (d, J = 2.0 Hz, 1H), 6.27 (d, J = 6.0 Hz, 1H), 6.36 (d, J = 2.0 Hz, 1H), 8.17 (d, J = 6.0 Hz, 1H), 10.85 (s, 1H), 12.69 (s, 1H). ^{13}C NMR (100 MHz, DMSO-d$_6$)_= 93.9, 98.9, 104.8, 110.4, 149.8, 157.3, 157.7, 164.2, 181.2. LC-MS: m/z: 177 [M-1]$^-$. Anal.calcd. for $C_9H_6O_4$: C 60.68, H 3.39; found: C 60.65, H 3.41.

5. General Experimental Procedure for Isoflavones (9a-9h)

5.1. Method A

A mixture of substituted phenol (3 mmol), phenylacetic acid (3 mmol), and BF₃·Et₂O (15 mmol) was refluxed at 90°C for 90 min under Nitrogen atmosphere. The reaction mixture was then cooled to 10°C and DMF (4.6 mL) was added drop wise. The above reaction mixture was then added drop wise with stirring into vilsmeier reagent (4.5 mmol) at room temperature. After completion of addition, the reaction mixture was stirred at 60°C for 30 - 40 min and poured into boiling dilute HCl slowly and cooled. The solution was extracted with ethylacetate (30 mL × 2) and the organic layer was dried over anhydrous Na₂SO₄. The crude obtained after evaporation of the solvent was chromatographed over a silica gel column using chloroform-methanol mixtures as eluent to give isoflavones (9a-9h).

5.2. Method B

A mixture of substituted phenol (3 mmol), phenylacetic acid (3 mmol), and BF₃·Et₂O (9 mmol) was refluxed at 90°C for 90 min under Nitrogen atmosphere. The mixture was then poured into NaOAc solution (100 mL, 10%) and allowed to stand for 4 hr and the solution was extracted with EtOAc (3 × 100 mL). The combined organic layer was washed with water (20 mL) and brine (20 mL) and dried over anhydrous Na₂SO₄. The crude obtained after evaporation of the solvent was chromatographed over a silica gel column using hexane-EtOAc mixtures as eluent to give deoxybenzoins (7f-7h). The purified materials were then used for the synthesis of isoflavones. A mixture of deoxybenzoin (3 mmol) and BF₃·Et₂O (7.5 mmol) was cooled to 10°C and DMF (4.6 mL) was added drop wise. The cyclization procedure and work up are similar to method A.

5.3. 7-Hydroxy-3-Phenyl-4H-Chromen-4-One (9a) (Table 2, Entry 1)

White solid; yield (method A) 627 mg (88%); mp 210˚C - 213˚C. 1H NMR (400 MHz, DMSO-d6)_: 6.88 (d, J = 2.4 Hz, 1H), 6.96 (dd, J = 2.4, 8.4 Hz, 1H), 7.34 - 7.44 (m, 3H), 7.57 (d, J = 7.2 Hz, 2H), 7.99 (d, J = 8.8 Hz, 1H), 8.36 (s, 1H), 10.80 (s, 1H). ^{13}C NMR (100 MHz, DMSO-d$_6$)_ = 102.1, 115.2, 116.6, 123.5, 127.2, 127.6, 128.0, 128.8, 132.1, 153.6, 157.4, 162.6, 174.3. LC-MS: *m/z*: 237 [M-1]-. Anal.calcd. for $C_{15}H_{10}O_3$: C 75.62, H 4.23; found: C 75.60, H 4.27.

5.4. 7-Hydroxy-3-(3-Methoxyphenyl)-4H-Chromen-4-One (9b) (Table 2, Entry 2)

Pale pink solid; yield (method A) 715 mg (89%); mp 215˚C - 217˚C. ^{1}H NMR (400 MHz, DMSO-d$_6$)_: 3.78 (s, 3H), 6.88 (s, 1H), 6.93 - 6.96 (m, 2H), 7.13 - 7.15 (m, 2 H), 7.33 (t, J = 8.0 Hz, 1H), 7.98 (d, J = 8.8 Hz, 1H), 8.38 (s, 1H), 10.79 (s, 1H) ^{13}C NMR (100 MHz, DMSO-d$_6$)_ = 55.06, 102.1, 113.2, 114.6, 115.2, 116.6, 121.1, 123.3, 127.2, 129.0, 133.4, 153.8, 157.3, 159.0, 162.6, 174.3. LC-MS: *m/z*: 267 [M-1]-. Anal.calcd. for $C_{16}H_{12}O_4$: C 71.64, H 4.51; found: C 71.60, H 4.56.

5.5. 7-Hydroxy-3-(4-Methoxyphenyl)-4H-Chromen-4-One (9c) (Table 2, Entry 3)

Off-white solid; yield (method A) 723 mg (90%); mp 257˚C - 258˚C. ^{1}H NMR (400 MHz, DMSO-d$_6$)_: 3.78 (s, 3H), 6.88 (d, J = 2.0 Hz, 1H), 6.94 (dd, J = 2.4, 8.8 Hz, 1H), 6.98 (d, J = 8.8 Hz, 2H), 7.49 (d, J = 8.8 Hz, 2H), 6.97 (d, J = 8.8 Hz, 1H), 8.31 (s, 1H). ^{13}C NMR (100 MHz, DMSO-d$_6$)_ = 55.1, 102.0, 113.6, 114.1, 115.0, 116.6, 123.1, 124.1, 127.2, 130.0, 153.0, 157.4, 158.9, 162.3, 174.6. LC-MS: *m/z*: 267 [M-1]-. Anal.calcd. for $C_{16}H_{12}O_4$: C 71.64, H 4.51; found: C 71.60, H 4.55.

5.6. 7-Hydroxy-3-(4-Hydroxyphenyl)-4H-Chromen-4-One (9d) (Table 2, Entry 4)

Pale brown powder; yield (method A) 670 mg (88%); mp 310˚C - 312˚C. ^{1}H NMR (400 MHz, DMSO-d$_6$)_: 6.79 (d, J = 8.4 Hz, 2H), 6.83 (d, J = 2.0 Hz, 1H), 6.91 (dd, J = 2.0, 8.0 Hz, 1H), 7.36 (d, J = 8.4 Hz, 2H), 7.96 (d, J = 8.0 Hz, 1H), 8.28 (s, 1H), 9.55 (s, 1H), 10.83 (s, 1H). ^{13}C NMR (100 MHz, DMSO-d$_6$)_ = 102.0, 114.9, 115.1, 122.5, 123.5, 127.2, 130.0, 157.2, 157.3, 162.4, 174.7. LC-MS: *m/z*: 253 [M-1]-. Anal.calcd. for $C_{15}H_{10}O_4$: C 70.86, H 3.96; found: C 70.85, H 3.98.

5.7. 7-Hydroxy-3-(2, 4-Dimethoxyphenyl)-4H-Chromen-4-One (9e) (Table 2, Entry 5)

Off white solid; yield (method A) 795 mg (89%); mp 265˚C - 270˚C. ^{1}H NMR (400 MHz, DMSO-d$_6$)_: 3.70 (s, 3H), 3.80 (s, 3H), 6.56 (d, J = 8.0 Hz, 1H), 6.63 (s, 1H), 6.86 (s, 1H), 6.93 (d, J = 8.8 Hz, 1H), 7.13 (d, J = 8.0 Hz, 1H), 7.92 (d, J = 8.8 Hz, 1H), 8.11 (s, 1H), 10.73 (s, 1H) ^{13}C NMR (100 MHz, DMSO-d$_6$)_ = 55.2, 55.5, 98.6, 102.1, 104.6, 113.5, 114.9, 116.5, 121.5, 127.1, 132.0, 153.8, 157.4, 158.4, 160.6, 162.4, 174.3. LC-MS: *m/z*: 297 [M-1]$^-$. Anal. calcd. for $C_{17}H_{14}O_5$: C 68.45, H 4.73; found: C 68.46, H 4.75.

Table 2. Synthesis of isoflavones (9a-9h).

S. No.	Entry	R_1	R_2	R_3	R_4	R_5	R_6	9 Yield (%)[a]
1	a	H	OH	H	H	H	H	88
2	b	H	OH	H	OCH$_3$	H	H	89
3	c	H	OH	H	H	OCH$_3$	H	90
4	d	H	OH	H	H	OH	H	88
5	e	H	OH	OCH$_3$	H	OCH$_3$	H	89
6	f	OH	OH	H	H	H	H	75, 89[b]
7	g	OH	OH	H	OCH$_3$	H	H	70, 85[b]
8	h	OH	OH	H	H	OCH$_3$	H	75, 90[b]

[a]Unoptimized condition; [b]Deoxybenzoins were isolated and converted into isoflavones.

5.8. 7,8-Dihydroxy-3-Phenyl-4H-Chromen-4-One (9f) (Table 2, Entry 6)

Pale brown solid; yield (method A) 571 mg (75%), yield (method B) 677 mg (89%); mp 200°C - 205°C. ^{1}H NMR (400 MHz, DMSO-d_6)_: 6.98 (d, J = 8.8 Hz, 1H), 7.37 - 7.44 (m, 3H), 7.49 (d, J = 8.8 Hz, 1H), 7.59-7.57 (m, 2H), 8.43 (s, 1H), 9.46 (s, 1H), 10.33 (s, 1H). ^{13}C NMR (100 MHz, DMSO-d_6)_ = 114.2, 115.6, 117.4, 123.0, 127.5, 128.0, 128.9, 132.2, 132.9, 146.7, 150.1, 153.5, 174.8. LC-MS: m/z: 253 [M-1]$^-$. Anal.calcd. for $C_{15}H_{10}O_4$: C 70.86, H 3.96; found: C 70.84, H 4.00.

5.9. 7,8-Dihydroxy-3-(3-Methoxyphenyl)-4H-Chromen-4-One (9g) (Table 2, Entry 7)

Brown solid; yield (method A) 597 mg (70%), yield (method B) 725 mg (85%); mp 216°C - 218°C. ^{1}H NMR (400 MHz, DMSO-d_6)_: 3.78 (s, 3H), 6.94 (d, J = 7.6 Hz, 1H), 6.97 (d, J = 8.8 Hz, 1H), 7.14 - 7.17 (m, 2H), 7.33 (t, J = 7.6 Hz, 1H), 7.49 (d, J = 8.8 Hz, 1H), 8.44 (s, 1H) 9.41 (s, 1H), 10.32 (s, 1H). ^{13}C NMR (100 MHz, DMSO-d_6)_ = 55.0, 113.1, 114.2, 114.7 115.7, 117.4, 121.2, 122.7, 129.0, 132.9, 133.5, 146.6, 150.1, 153.6, 158.9, 174.7. LC-MS: m/z: 283 [M-1]$^-$. Anal. calcd. for $C_{16}H_{12}O_5$: C 67.60, H 4.25; found: C 67.58, H 4.28.

5.10. 7,8-Dihydroxy-3-(4-Methoxyphenyl)-4H-Chromen-4-One (9h) (Table 2, Entry 8)

Pale Brown solid; yield (method A) 640 mg (75%), yield (method B) 768 mg (90%); mp 252°C - 254°C. ^{1}H NMR (400 MHz, DMSO-d_6)_: 3.79 (s, 3H), 6.96 (d, J = 8.8 Hz, 1H), 6.99 (d, J = 8.8 Hz, 2H), 7.48 (d, J = 8.8 Hz, 1H), 7.52 (d, J = 8.8 Hz, 2H), 8.38 (s, 1H), 9.42 (s, 1H), 10.29 (s, 1H). ^{13}C NMR (100 MHz, DMSO-d_6)_ = 55.1, 113.5, 114.1, 115.6, 117.4, 122.6, 124.4, 130.0, 132.8, 146.7, 149.8, 150.0, 152.8, 158.9, 175.0. LC-MS: m/z: 283 [M-1]$^-$. Anal. calcd. for $C_{16}H_{12}O_5$: C 67.60, H 4.25; found: C 67.59, H 4.28.

6. General Experimental Procedure for Homo-Isoflavones (10a-10f)

6.1. Method A

A mixture of substituted phenol (3 mmol), 3-phenylpropanoic acid (3 mmol), and BF$_3$·Et$_2$O (15 mmol) was refluxed at 90°C for 90 min under Nitrogen atmosphere. The reaction mixture was then cooled to 10°C and DMF (4.6 mL) was added drop wise. The above reaction mixture was then added drop wise with stirring into vilsmeier reagent (4.5 mmol) at room temperature. After completion of addition, the reaction mixture was stirred at 60°C for 30 - 40 min and poured into boiling dilute HCl slowly and cooled. The solution was extracted with ethylacetate (30 mL × 2) and the organic layer was dried over anhydrous Na$_2$SO$_4$. The crude obtained after evaporation of the solvent was chromatographed over a silica gel column using chloroform-methanol mixtures as eluent to give homo-isoflavones (10a-10h).

6.2. Method B

A mixture of substituted phenol (3 mmol), 3-phenylpropanoic acid (3 mmol), and BF$_3$·Et$_2$O (9 mmol) was refluxed at 90°C for 90 min under Nitrogen atmosphere. The mixture was then poured into NaOAc solution (100 mL, 10%) and allowed to stand for 4 hr and the solution was extracted with EtOAc (3 × 100 mL). The combined organic layer was washed with water (20 mL) and brine (20 mL) and dried over anhydrous Na$_2$SO$_4$. The crude obtained after evaporation of the solvent was chromatographed over a silica gel column using hexane-EtOAc mixtures as eluent to give dihydrchalcones (8d-8f). The purified materials were then used for the synthesis of homoisoflavones. A mixture of dihydrochalcone (3 mmol) and BF$_3$·Et$_2$O (7.5 mmol) was cooled to 10°C and DMF (4.6 mL) was added drop wise. The cyclization procedure and work up are similar to method A.

6.3. 3-Benzyl-7-Hydroxy-4H-Chromen-4-One (10a) (Table 3, Entry 1)

Pale pink solid; yield (method A) 635 mg (84%); mp 210°C - 214°C. ^{1}H NMR (400 MHz, DMSO-d_6)_: 3.67 (s, 2H), 6.82 (d, J = 2.4 Hz, 1H), 6.89 (dd, J = 2.4, 8.8 Hz, 1H), 7.14 - 7.18 (1H), 7.23 - 7.29 (m, 4H), 7.87 (d, J = 8.8 Hz, 1H), 8.17 (s, 1H), 10.72 (s, 1H). ^{13}C NMR (100 MHz, DMSO-d_6)_ = 30.6, 102.1, 114.9, 116.2, 122.8, 125.9, 126.7, 128.1, 128.4, 139.7, 153.2, 157.7, 162.4, 175.4. LC-MS: m/z: 251 [M-1]$^-$. Anal. calcd. for $C_{16}H_{12}O_3$: C 76.18, H 4.79; found: C 76.14, H 4.82.

Table 3. Synthesis of homoisoflavones (10a-10f).

S. No.	Entry	R_1	R_2	R_3	R_4	R_5	R_6	10 Yield (%)^a
1	a	H	OH	H	H	H	H	84
2	b	H	OH	H	H	OCH$_3$	H	87
3	c	H	OH	H	H	OH	H	89
4	d	H	OH	OCH$_3$	H	H	OCH$_3$	88
5	e	OH	OH	H	H	OCH$_3$	H	78, 88^b
6	f	OH	OH	OCH$_3$	H	H	OCH$_3$	80, 85^b

^{a}Unoptimized condition. ^{b}Dihydrochalcones were isolated and converted into homoisoflavones.

6.4. 7-Hydroxy-3-(4-Methoxybenzyl)-4H-Chromen-4-One (10b) (Table 3, Entry2)

Pale brown solid; yield (method A) 736 mg (87%); mp 161˚C - 165˚C. ^{1}H NMR (400 MHz, DMSO-d$_6$)_: 3.59 (s, 2H), 3.69 (s, 3H), 6.81 - 6.83 (m, 3H), 6.89 (dd, J = 2.0, 8.8 Hz, 1H), 7.20 (d, J = 8.4 Hz, 2H), 7.87 (d, J = 8.8 Hz, 1H), 8.12 (s, 1H), 10.71 (s, 1H). ^{13}C NMR (100 MHz, DMSO-d$_6$)_ = 29.8, 54.9, 102.1, 113.6, 114.9, 116.2, 123.3, 126.7, 129.0, 129.5, 131.5, 153.0, 157.6, 157.7, 162.4, 175.4. LC-MS: m/z: 281 [M-1]-. Anal.calcd. for C$_{17}$H$_{14}$O$_4$: C 72.33, H 5.00; found: C 72.30, H 5.05.

6.5. 7-Hydroxy-3-(4-Hydroxybenzyl)-4H-Chromen-4-One (10c) (Table 3, Entry3)

Colorless solid; yield (method A) 715 mg (89%); mp 210˚C - 212˚C. ^{1}H NMR (400 MHz, DMSO-d$_6$)_: 3.52 (s, 2H), 6.64 (d, J = 8.4 Hz, 2H), 6.79 (d, J = 2.0 Hz, 1H), 6.89 (dd, J = 8.8, 2.0 Hz, 1H), 7.07 (d, J = 8.4 Hz, 2H), 7.86 (d, J = 8.8 Hz, 1H), 8.05 (s, 1H), 9.85 (s, 1H), 10.75 (s, 1H) ^{13}C NMR (100 MHz, DMSO-d$_6$)_ = 30.2, 102.1, 114.5, 115.1, 116.3, 123.9, 126.7, 128.9, 129.4, 152.3, 155.5, 157.8, 162.3, 176.3. LC-MS: m/z: 267 [M-1]. Anal.calcd. for C$_{16}$H$_{12}$O$_4$: C 71.64, H 4.51; found: C 71.62, H 4.54.

6.6. 7-Hydroxy-3-(2, 5-Dimethoxybenzyl)-4H-Chromen-4-One (10d) (Table 3, Entry4)

Light brown solid; yield (method A) 823 mg (88%); mp 184˚C - 188˚C. ^{1}H NMR (400 MHz, DMSO-d$_6$)_: 3.59 (s, 2H), 3.69 (s, 3H), 3.75 (s, 3H), 6.70 - 6.75 (m, 2H), 6.80 (d, J = 2.0 Hz, 1H),), 6.85 (d, J = 8.8 Hz, 1H), 6.90 (dd, J = 2.0, 8.8 Hz, 1H), 7.88 (s, 1H), 7.89 (d, J = 8.8 Hz, 1H), 10.68 (s, 1H). ^{13}C NMR (100 MHz, DMSO-d$_6$)_ = 25.2, 55.2, 55.7, 102.1, 111.3, 111.6, 114.9, 116.1, 116.4, 121.7, 126.7, 128.0, 151.1, 151.2, 152.9, 153.2, 157.7, 162.4, 175.4. LC-MS: m/z: 311 [M-1]$^-$. Anal.calcd. for C$_{18}$H$_{16}$O$_5$: C 69.22, H 5.16; found: C 69.19, H 5.19.

6.7. 7,8-Dihydroxy-3-(4-Methoxybenzyl)-4H-Chromen-4-One (10e) (Table 3, Entry 5)

Colorless solid; yield (method A) 698 mg (78%), yield (method B) 785 mg (88%); mp 250˚C - 253˚C. ^{1}H NMR (400 MHz, DMSO-d$_6$)_: 3.60 (s, 2H), 3.68 (s, 3H), 6.81 (d, J = 8.4 Hz, 2H), 6.93 (d, J = 8.8 Hz, 1H), 7.20 (d, J = 8.4 Hz, 2H), 7.39 (d, J = 8.8 Hz, 1H), 8.19 (s, 1H), 9.34 (s, 1H), 10.20 (s, 1H). ^{13}C NMR (100 MHz, DMSO-d$_6$)_ = 29.8, 54.9, 113.6, 114.0, 115.1, 117.1, 122.7, 129.4, 131.6, 132.8, 147.0, 149.8, 152.8, 157.6, 175.9. LC-MS: m/z: 297 [M-1]$^-$. Anal.calcd. for C$_{17}$H$_{14}$O$_5$: C 68.45, H 4.73; found: C 68.42, H 4.79.

6.8. 7,8-Dihydroxy-3-(2, 5-Dimethoxybenzyl)-4H-Chromen-4-One (10f) (Table 3, Entry 6)

Pale brown solid; yield (method A) 786 mg (80%), yield (method B) 836 mg (85%); mp 226˚C - 230˚C. ^{1}H NMR (400 MHz, DMSO-d$_6$)_: 3.60 (s, 2H), 3.64 (s, 3H), 3.75 (s, 3H), 6.72 - 6.75 (m, 2H), 6.88 (d, J = 8.4 Hz, 1H), 6.93 (d, J = 8.8 Hz, 1H), 7.39 (d, J = 8.8 Hz, 1H), 7.98 (s, 1H), 9.32 (s, 1H), 10.20 (s, 1H). ^{13}C NMR (100 MHz, DMSO-d$_6$)_ = 25.2, 55.2, 55.8, 111.3, 111.5, 114.0, 115.1, 116.4, 116.9, 121.1, 128.1, 132.8, 147.0, 149.8, 151.2, 152.9, 153.0, 175.9. LC-MS: m/z: 327 [M-1]$^-$. Anal. calcd. For C$_{18}$H$_{16}$O$_6$: C 65.85, H 4.91; found: C 65.81, H 4.96.

Acknowledgements

The author is grateful to the Department of Organic Chemistry & FDW, Andhra University, Visakhapatnam, India for giving him the opportunity to pursue his PhD.

References

[1]   McClure, J.W., Harborne, J.B., Mabry, T.J. and Mabry, H. (1975) The Flavonoids. Chapman and Hall, London, 970.

[2]   Attassi, G., Briet, P., Berthelon, J.P. and Collonges, F. (1985) Synthesis and Antitumor Activity of Some (8-Substituted 4-oxo-4H-1-benzopyrans). *Journal of Medicinal Chemistry*, **20**, 393-402.

[3]   Middleton Jr., E., Kandaswami, C. and Arborne, J.B. (1994) The Flavonoids Advances in Research Since 1986. Chapman and Hall, London, 619.

[4]   Bruneton, J. (1995) Pharmacognosy, Phytochemistry and Medicinal Plants, English Translation by Hatton, C. K., Lavoisier Publishing, Paris, 265.

[5]   Harborne, J.B. and Williams, C.A. (2000) Advances in Flavonoid Research Since 1992. *Phytochemistry*, **55**, 481-504. http://dx.doi.org/10.1016/S0031-9422(00)00235-1

[6]   Miksicek, R.J. (1995) Estrogenic Flavonoids: Structural Requirements for Biological Activity. *Proceedings of the Society for Experimental Biology and Medicine*, **208**, 44-50. http://dx.doi.org/10.3181/00379727-208-43830

[7]   Peterson, G. and Branes, S. (1991) Genistein Inhibition of the Growth of Human Breast Cancer Cells: Independence from Estrogen Receptors and the Multi-Drug Resistance Gene. *Biochemical and Biophysical Research Communications*, **179**, 661-667. http://dx.doi.org/10.1016/0006-291X(91)91423-A

[8]   Bandyukova, V.A., Cherevatye, V.S., Ozimina, I.I., Andreeva, O.A., Lebedava, A.L., Davydov, V.S., Vashchenko, T. N. Postnikova, N.V., (1987) Antibacterial Activity of Flavonoids of Some Flowering Plant Species. *Rastitel'nye Resursy*, **23**, 607; Chemical Abstract, 108, 1990, 71937v.

[9]   El-Gammal, A.A. and Mansour, R.M., (1986) Antimicrobial Activities of Some Flavonoid Compounds. *Zentralblatt für Mikrobiologie*, **141**, 561-565; Chemical Abstract, 106, 1987, 135070a.

[10]   Takai, M., Yamaguchi, H., Saitoh, T. and Shibata, S. (1972) *Chemical and Pharmaceutical Bulletin*, **20**, 2488-2490. http://dx.doi.org/10.1248/cpb.20.2488

[11]   Akiyama, T., Ishida, J., Nakagawa, S., Ogawara, H., Watanabe, S., Itoh, N., Shihuya, M. and Fukami, Y. (1987) Genistein, a Specific Inhibitor of Tyrosine-Specific Protein Kinases. *Journal of Biological Chemistry*, **262**, 5592-5595.

[12]   Baker, W., Chadderton, J., Harborne, J.B. and Ollis, W.D. (1953) A New Synthesis of Isoflavones. Part II. 5: 7: 2'-Trihydroxyisoflavone. *Journal of the Chemical Society*, 1860-1864. http://dx.doi.org/10.1039/jr9530001860

[13]   Sathe, V.R. and Venkataraman, K. (1949) A New Reaction for the Synthesis of Chromones and Isoflavones. *Current Science*, **18**, 373; Chemical Abstract, 44, 1950, 8916.

[14]   Wahala, K. and Hase, T.A. (1991) Expedient Synthesis of Polyhydroxyisoflavones. *Journal of the Chemical Society, Perkin Transactions 1*, 3005-3008. http://dx.doi.org/10.1039/p19910003005

[15]   Balasubramanian, S., Ward, D.L. and Nair, M.G. (2000) The First Isolation and Crystal Structure of a Boron Difluoro Complex (Isoflavone Yellow). Biologically Active Intermediates Produced during Isoflavone Synthesis. *Journal of the Chemical Society, Perkin Transactions 1*, 567-569. http://dx.doi.org/10.1039/a908915b

[16]   Kagal, S.A., Nair, P.M. and Venkataraman, K. (1962) A Synthesis of Isoflavones by a Modified Vilsmeier-Haack Reaction. *Tetrahedron Letters*, **3**, 593.

[17]   Wessely, F., Kornfeld, L. and Lechner, F. (1953) Über die Synthese von Daidzein und von 7-Oxy-4'-Methoxy-Isoflavon. *Berichte der deutschen chemischen Gesellschaft (A and B Series)*, **66**, 685-687. http://dx.doi.org/10.1002/cber.19330660515

[18]   Bognar, R. and Levai, A. (1973) Synthesis of 4-β-Glucosyloxydeoxybenzoins and Their Conversion into 7-β-Glucosyloxyisoflavones. *Acta Chimica Academiae Scientiarum Hungaricae*, **77**, 435.

[19]   Levai, A. and Bognar, R. (1974) Preparation of Deoxybenzoin Glycosides and Their Conversion into Isoflavone Glycosides. *Kemiai Kozlemenyek*, **41**, 17-25.

[20]   Liu, D.F. and Cheng, C.C. (1991) A Facile and Practical Preparation of 5,7-Dihydroxy-3-(4-nitrophenyl)-4H-1-benzopyran-4-one. *Journal of Heterocyclic Chemistry*, **28**, 1641-1642. http://dx.doi.org/10.1002/jhet.5570280632

[21]   Krishnamurty, H.G. and Prasad, J.S. (1977) A New Synthesis of Isoflavones Using "Active Formate". *Tetrahedron Letters*, **18**, 3071-3072. http://dx.doi.org/10.1016/S0040-4039(01)83160-9

[22]   Miki, Y., Fujita, R. and Matsushita, K.I. (1998) Oxidative Rearrangement of Pentaalkoxychalcones with Phenyliodine(III) Bis(trifluoroacetate) (PIFA): Synthesis of (±)-10-Bromopterocarpin and (±)-Pterocarpin. *Journal of the Che-

mical Society, Perkin Transactions 1, 2533-2536. http://dx.doi.org/10.1039/a803561j

[23] Kawamura, Y., Maruyama, M., Tokuoka, T. and Tsukayama, M. (2000) Synthesis of Isoflavones from 2'-Hydroxy-chalcones Using Poly[4-(diacetoxy)iodo]styrene or Related Hypervalent Iodine Reagent. *Synthesis*, 2490-2496.

[24] Faskas, L., Gottsegen, A., Nogradi, M. and Antus, S. (1972) Direct Conversion of 2'-Hydroxychalcones into Isofla-vones Using Thallium(III) Nitrate: Synthesis of (±)-Sophorol and (±)-Mucronulatol. *Journal of the Chemical Society, Chemical Communications*, 825-826. http://dx.doi.org/10.1039/c39720000825

[25] Farkas, L., Gottsegen, A., Nogradi, M. and Antus, S. (1974) Synthesis of Sophorol, Violanone, Lonchocarpan, Claus-sequinone, Philenopteran, Leiocalycin, and Some Other Natural Isoflavonoids by the Oxidative Rearrangement of Chalcones with Thallium(III) Nitrate. *Journal of the Chemical Society, Perkin Transactions 1*, 305-312. http://dx.doi.org/10.1039/p19740000305

[26] Farkas, L., Antus, S. and Nogradi, M. (1974) The Oxidative Rearrangement of Chalcones by Thallium(III) Nitrate, II. New Synthesis of Flemichapparin-B and Flemichapparin-C. *Acta Chimica (Academiae Scientiarum) Hungaricae*, **82**, 225-230.

[27] Antus, S., Farkas, L., Kardos-Balogh, Z. and Nógrádi, M. (1975) Oxidative Umlagerung von Chalkonen mit Thallium(III)-nitrat, IV. Synthese des Dalpatins, Fujikinins, Glyciteins und anderer natürlicher Isoflavone. *Chemische Berichte*, **108**, 3883-3893. http://dx.doi.org/10.1002/cber.19751081221

[28] Antus, S., Farkas, L., Gottsegen, A., Kardos-Balogh, Z. and Nogradi, M. (1976) Oxidative Umlagerung von Chalconen mit Thallium(III)-nitrat, VI Synthese der Isoflavonoide Jamaicin und Leiocarpin. *Chemische Berichte*, **109**, 3811-3816. http://dx.doi.org/10.1002/cber.19761091208

[29] Antus, S., Boross, F. and Nógrádi, M. (1977) Alkali-Catalysed Alkoxy Exchange, Alcohol Elimination, and Hydrolysis of Acetals Having a Dissociable α-Proton. *Journal of the Chemical Society, Chemical Communications*, 333-334. http://dx.doi.org/10.1039/c39770000333

[30] Ollis, W.D., Ormand, K.L. and Sutherland, I.O. (1968) The Oxidative Rearrangement of Chalcones by Thallic Acetate: A Chemical Analogy for Isoflavone Biosynthesis. *Journal of the Chemical Society, Chemical Communications*, 1237-1238.

[31] Ollis, W.D., Ormand, K.L. and Sutherland, I.O. (1970) The Oxidative Rearrangement of Olefins by Thallium(III) Ace-tate. Part I. Oxidative Rearrangement of Chalcones. *Journal of the Chemical Society C*, 119-124.

[32] Ollis, W.D., Ormand, K.L., Roberts, R.J. and Sutherland, I.O. (1979) The Oxidative Rearrangement of Olefins by Thallium(III) Acetate. Part II. Synthesis of Isoflavones. *Journal of the Chemical Society C*, 125-128.

[33] Grover, S.K., Jain, A.C. and Seshadri, T.R. (1963) A Convenient Synthesis of Isoflavones and Di-O-Methylangolensin Using Aryl Migration. *Indian Journal of Chemistry*, **1**, 517.

[34] Jain, A.C., Lal, P. and Seshadri, T.R. (1969) Synthesis of 3'-Hydroxyformononetin. *Indian Journal of Chemistry*, **7**, 305.

[35] Kinoshita, T., Ichinose, K. and Sankawa, U. (1990) One-Step Conversion of Flavanones into Isoflavones: A New Facile Biomimetic Synthesis of Isoflavones. *Tetrahedron Letters*, **31**, 7355-7356. http://dx.doi.org/10.1016/S0040-4039(00)88565-2

[36] Weber-Schilling, C.A. and Wanzlick, H.W. (1971) Synthese von 2'.4'.5'-trihydroxylierten Isoflavonen. *Chemische Be-richte*, **104**, 1518-1523. http://dx.doi.org/10.1002/cber.19711040520

[37] Crombie, L., Freeman, P.W. and Whiting, D.A. (1973) A New Synthesis of Rotenoids. Application to 9-Demethyl-munduserone, Mundeserone, Rotenonic Acid, Dalpanol, and Rotenone. *Journal of the Chemical Society, Perkin Trans-actions 1*, 1277-1285. http://dx.doi.org/10.1039/p19730001277

[38] Yokoe, I., Sugita, Y. and Shirataki, Y. (1989) Facile Synthesis of Isoflavones by the Cross-Coupling Reaction of 3-Io-dochromone with Arylboronic Acids. *Chemical and Pharmaceutical Bulletin*, **37**, 529-530.

[39] Qualig, M.G., Desideri, N., Bossu, E., Sgro, R. and Conti, C. (1999) Enantioseparation and Anti-Rhinovirus Activity of 3-Benzylchroman-4-Ones. *Chirality*, **11**, 495-500. http://dx.doi.org/10.1002/(SICI)1520-636X(1999)11:5/6<495::AID-CHIR23>3.0.CO;2-5

[40] Desideri, N., Olivieri, S., Stein, M.L., Sgro, R., Orsi, N. and Conti, C. (1997) Synthesis and Anti-Picornavirus Activity of Homo-Isoflavonoids. *Antiviral Chemistry and Chemotherapy*, **8**, 545-555.

[41] Siddaiah, V., Rao, V.C., Venkateswarlu, S., Krishnaraju, A.V. and Subbaraju, G.V. (2006) Synthesis, Stereochemical Assignments, and Biological Activities of Homoisoflavonoids. *Bioorganic & Medicinal Chemistry*, **14**, 2545-2551. http://dx.doi.org/10.1016/j.bmc.2005.11.031

[42] Malhotra, S., Sharma, V.K. and Parmar, V.S. (1988) Synthesis of Three New Dihydropyranochalcones: Structural Re-vision of Crotmadine, an Antifungal Constituent of Crotalarla Madurensis. *Journal of Natural Products*, **51**, 578-581.

[43] Farkas, L., Gottsegen, Á., Nógrádi, M. and Strelisky, J. (1971) Synthesis of Homoisoflavanones-II: Constituents of *Eu-*

comis autumn alis and *E. Punctata. Tetrahedron*, **27**, 5049-5054.

[44] Bass, R.J. (1976) Synthesis of Chromones by Cyclization of 2-Hydroxyphenyl Ketones with Boron Trifluoride-Diethyl Ether and Methanesulphonyl Chloride. *Journal of the Chemical Society, Chemical Communications*, 78-79. http://dx.doi.org/10.1039/c39760000078

[45] Siddaiah, V., Rao, C.V., Venkateswarlu, S. and Subbaraju, G.V. (2006) A Concise Synthesis of Polyhydroxydihydrochalcones and Homoisoflavonoids. *Tetrahedron*, **62**, 841-846.

[46] Kimura, Y., Matsuura, D., Hanawa, T. and Kobayashi, Y. (2012) New Preparation Method for Vilsmeier Reagent and Related Imidoyl Chlorides. *Tetrahedron Letters*, **53**, 1116-1118. http://dx.doi.org/10.1016/j.tetlet.2011.12.087

[47] Vilsmeier, A. and Haack, A. (1927) Über die Einwirkung von Halogenphosphor auf Alkyl-formanilide. Eine neue Methode zur Darstellung sekundärer und tertiärer p-Alkylamino-benzaldehyde. *Berichte der deutschen chemischen Gesellschaft, Serie B*, **60**, 119-122. http://dx.doi.org/10.1002/cber.19270600118

[48] Marson, C.M. (1992) Reactions of Carbonyl Compounds with (Monohalo) Methyleniminium Salts (Vilsmeier Reagents). *Tetrahedron*, **48**, 3659-3726. http://dx.doi.org/10.1016/S0040-4020(01)92263-X

[49] Panasenko, A., Polyanska, N.L. and Starkov, S.P. (1994) *Zhurnal Obshchei Khimii*, **64**, 673-676.

[50] Pelter, A. and Foot, S. (1976) A New Convenient Synthesis of Isoflavones. Synthesis, 326.

[51] Yoder, L., Cheng, E. and Burroughs, W. (1954) Synthesis of Estrogenic Isoflavone Derivatives. *Journal of the Iowa Academy of Science*, **61**, 271-276.

17

Theoretical and Experimental Study of the Mechanism of Reaction Formation of Cyanine Dyes

Boshkayeva Assyl, Omarova Roza, Pichkhadze Guram, Shalpykova Nasiba

Asfendiyarov Kazakh National Medical University, Almaty, Kazakhstan
Email: kenes65@mail.ru

Abstract

To study the theoretical bases of the mechanism of reaction formation of cyanine dyes, special importance is gained by methods of quantum chemistry. The use of these methods is provided with the known molecular and dynamic HyperChem program. The purpose of studying was quantum and chemical studying of features of a geometrical and electronic structure of model molecules of penicillin acid, of derivative of glutaconic dialdehyde (DGD), and the cyanine dyes, and also an assessment of power of process of their formation. For studying was carried out the experimental mark of chemism of course of reaction with theoretical justification of the mechanism of reaction formation of cyanine dyes. The results show that all studied model molecules are thermodynamic steady systems to what values of enthalpies of their formation, rather high on an absolute value, testify negative on a sign. Reaction goes by the mechanism of nucleophilic addition.

Keywords

Cyanine Dye, β-Lactam, Quantum-Chemical Method, Penicillin Acid, Derivative of Glutaconic Dialdehyde, Electronic Density, The Binding Energy, HyperChem

1. Introduction

As is known, the proximity of chemical structure, similarity methods of obtaining and the presence of certain structural groupings of cyanine dyes form the basic principles of their classification. According to the standard classification of dyes, based on the principle of generality of chromorphic systems, cyanines are organic compounds containing two hetero-cyclic residues connected by chain of methine groups. In organic chemistry, chromorphic system of cyanine dyes consists of a chain of free or substituted methine groups with electron-donors and electron-acceptor substituents at the ends; where in the substituents and a part of methine groups may be a part of the aromatic or heterocyclic radicals, [1]-[9]. Cyanine dyes are formed by reaction of either the opening of the pyridine cycle, pyrimidine and furan rings, or when some condensation reactions.

One of the proposed varieties of using condensation reaction types in the analysis of drugs is the reaction of forming the cyanine dyes. The reaction is based on a combination of derivative of glutaconic dialdehyde (DGD), as one of the cleavage products of the pyridine ring, with the products of hydrolytic cleavage of β-lactam. Theoretical grounding of the reaction mechanism is considered from the perspective of existing classical electronic effects.

One of the formed acid hydrolysis products of penicillin salts reacts with DGD. Unlike the thiazolidine ring, β-lactam ring of penicillin is readily cleaved under the action of alkalis, acids and other substances. This explains, for example, the formation of cleavage products of β-lactam, [10]. Features of the molecular structure of penicillin may be the cause of its behavior in the presence of various reagents. Under the action of 25% hydrochloric acid, penicillins are inactivated with forming corresponding penillic and penicillin acids [11] [12].

Based on the structural features of β-lactam, the appearance of coloring reaction product with DGD facilitate: a chain of alternating single and double bonds (with the chain involved in a double bond between carbon and nitrogen); the presence of groups or atoms strongly attracting electrons to the overall electronic system of the molecule; planar arrangement of atoms. Some fragments in molecule of benzylpenicillin sodium leave the plane, but they do not affect the structural changes which cause the appearance coloring cyanine dyes. Carbon atoms chain bounded to each other by alternating single and double bonds has a direct value for the structure of the colored compound. Expected colored compound in this reaction is subjected to selective absorption of light energy with wavelengths in the range of the visible spectrum.

Electronic density redistribution in penicillin acid molecule as one of the products of hydrolytic cleavage of penicillin has a significance for revealing reaction capacity of β-lactam, son a qualitative level.

The study of reaction formation of cyanine dyes has given confirmation of condensation reaction of DGD with penicillin acid based on the selected conditions of this reaction. Effect of substituents on the electronic density distribution in the molecule of penicillin acid was evaluated by electronic effects-inductive (I) and mesomeric (M). In the last molecule, the electron density distribution is considered (**Figure 1**). The direction of shift of electronic density is designated by a curved arrow.

In penicillin acid molecule, a reactive center is nitrogen of an iminogroup having electronegative character (electronegativity of nitrogen-3.0) as compared with sp2-hybridized carbon atom (carbon electronegativity-2.8). Due to the lone pairs of electrons, nitrogen may bind electrophilic reagent. Due to nitrogen of iminogroup (-I), inductive effect is determined. As a result of its action, penicillin acid molecule may have a deficiency and of electronic density on the nitrogen of an iminogroup-its excess. NH has electron donor impact somewhat increasing the electron density in the molecule due to p, p-conjugation (+M effect). It is also noted that when looking at the colors, depending on the state of the electrons in the molecule of penicillin acid, p-electrons displacement occurs along the entire system of conjugated double bonds in the molecule. As described above, the mobility of p-electrons relates to the connection between two carbon atoms and those structures where there are p-bond between carbon and nitrogen.

Thus, the transition of the molecule in the excited state, the polarization of the molecule, the mobility of p-electrons, produce of continuous distribution of positive and negative charges in the molecule are responsible for the appearance of coloring for condensation product DGD with penicillin acid.

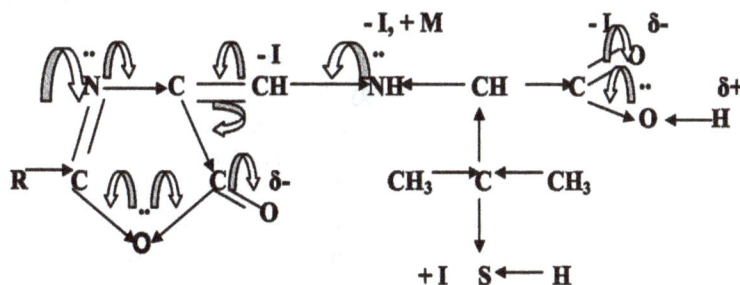

Figure 1. The electronic density distribution in the molecule of penicillin acid.

2. Materials and Methods

The used methods are: Portable HyperChem 8.0.7, MOPAC-packages of quantum chemical programs. In this paper we used the PM3-parametric method 3 (a semiempirical approach), which is a version of the AM1 method and differs from this method by parameter values only. Parameters for PM3 were obtained by comparing the number and type of experiments with the results of calculations.

3. Results and Discussion

According to theoretical calculations, we have proven that the interaction of penicillin acid with pyridinium salt (the rodanopiridiny chloride formed at action on pyridine of the splitting reagents (chloroamine B and thiocyanate of ammonium)) in the first stage reaction there occurs nucleophilic addition of mesomeric anion (nitrogen of secondary amino group) to the resulting rodanopiridinium chloride. In the stage of formation of cyanine dyes there occurs splitting of cycle pyridine diethylamide of nicotinic acid (DENA) to form DGD, which in third stage is condensed with a molecule of penicillin acid (product absorbing in the visible region at a wavelength λ_{max} = 565 nm is formed).

Conjugation of secondary amino group nitrogen of penicillin acid to rodanopiridiniumchloride contributes to dynamic factor determining the stability of intermediate species formed during the reaction products (there are particles with extended conjugated systems, and it is known that coupling stabilizes not only the molecules, but also the intermediate particles, and this is what determines the ease of formation of the intermediate particles and determines the entire course of the reaction). The reaction takes place only in the direction that provides the energy gain. Product formed during the reaction, staining in purple, is a molecule with a very large conjugated system.

In practical experiments, the study of the absorption spectra of DGD in the UV region shows that in the acidic environment a colorless form glutaconic dialdehyde is formed. In weak acidic medium with the cleavage of 25%, 6% of DENA solution yellow enol form of glutaconic dialdehyde is formed. In an unstable condition yellow enol form becomes red monoanilid of glutaconic dialdehyde, on which is drawn attention on reaching the optimal reaction conditions of forming cyanine dyes. The reaction proceeds in a stoichiometric ratio between the penicillin acid and derivative of glutaconic dialdehyde (DGD). Penicillin acid following acid hydrolysis of benzylpenicillin sodium salt is combined with the DGD, which is formed by splitting the pyridine cycle of diethyl-nicotinic acid under the action of thiocyanate chloride at pH 2.0 - 3.0. The last one previously prepared by interaction of chloramine B and thiocyanate ammonium. Hydrolysis of penicillin and cephalosporin antibiotics is performed by using aqueous solutions of hydrochloric acid.

Penicillin and cephalosporin antibiotics undergo hydrolytic degradation for 5 - 10 minutes at a water bath temperature (98°). Establishing a phased sequence of this reaction with the selection of all the conditions and explanation of one of the theoretical versions of the reaction mechanism of nucleophilic conjugationis given in the papers [13] [14].

In general, chemical reactions can be traced to the relationship between the yield of the final product and the indices of reaction ability of the reactants (atomic charges, bond orders, energy of frontier molecular orbitals (MO), the squares of the coefficients of the expansion of frontier MOs by the basis of atomic orbitals (AO) etc.) [15]-[18]. Charges on the atoms and the parameters of frontier MOs are mostly often used. The yield of the final reaction product is determined by the free energy activation, and the indices of reaction ability characterize the energy of the intermolecular interactions of the reagents. However, for certain reactions, these parameters are linked to each other. For example, with an increase in the interaction energy between the reagents its free energy of activation decreases.

Intermolecular interaction energy during the approach of reagents can be divided into three types of contributions: coulon, orbital and steric. Coulon interaction energy depends on the electron density distribution, or the charges on the atoms of the reagents. Therefore, some reactions can succeed finding in a relationship between the charges on the atoms and yield of the final reaction products. Thus, nucleophilic reagents (attacking center is negatively charged) joined mainly to atoms, which are localized by large positive charges and electrophilic (attacking center is positively charged), on the contrary-to atoms, which are localized by large negative charges.

Orbit interaction of any MO pairs is inversely proportional to difference of their energies, *i.e.* the further apart are the orbitals on scale of energies, the less they interact. Therefore, in practice, we usually use the approach of the frontier orbitals. In this approximation, the energy of the orbital interaction depends on the energy of frontier

MO and the coefficients of these MO decomposition by the AO basis. Any of these values can be used as an index of reactivity, but the energy difference of frontier MO is most lyused. Furthermore, the shape of frontier MO allows to make conclusions about the mechanisms of organic reactions as electrophilic attack often comes in the highest values of the highest occupied MO (HOMO) and nucleophilic—in the place of the greatest values of the lowest unoccupied MO (LUMO).

As a result, of molecular modeling of substances molecules entering into the reaction, and the molecules of the cyanine dye with using quantum-chemical method AM1 there were obtained data, confirming that the reaction can go by the mechanism of nucleophilic addition. More precisely, due to the distribution of electronic density in the molecules of DGD and penicillin acid, it can be said that a negatively charged nitrogen atom in the molecule penicillin acid may attack the positively charged carbon atom in the strongly polarized carbonyl group of DGD. In addition, the LUMO energy of penicillin acid is positive, therefore, this molecule is nucleophilic.

The electron density distribution in the molecule of penicillin acid is shown in **Figure 2**.

The electronic density distribution in it is shown in **Figure 3**.

The optimized geometry of the cyanine dye molecule is shown in **Figure 4**.

Geometric and electronic parameters of the objects under study are presented in **Tables 1-3**.

The analysis of the results in **Tables 1-3** follows that the coupling reaction of penicillin acids with DGD with forming cyanine dyes occurs by the interaction of the positively charged carbon of DGD with a free electron pair of a nucleophilic center, *i.e.* negatively charged nitrogen atoms in the molecule of penicillin acid.

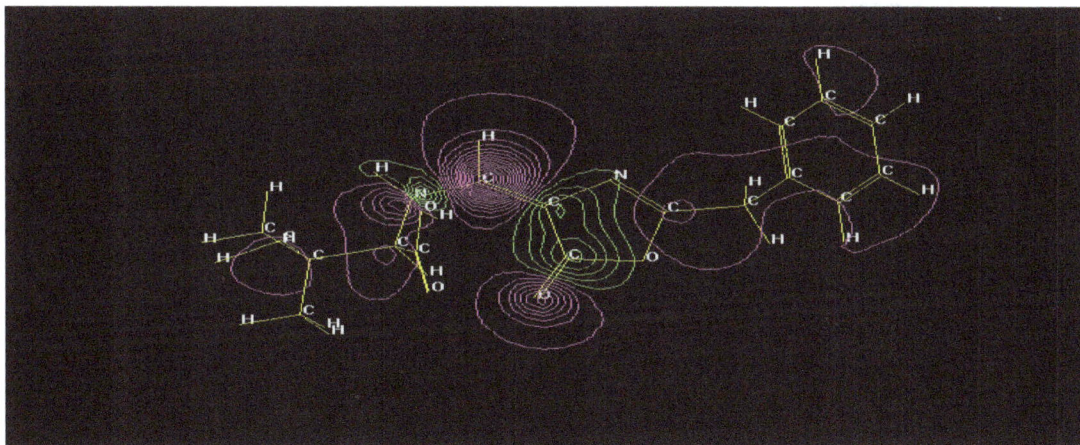

Figure 2. Distribution of the electronic density in the optimized molecule of penicillin acid (ENSMO = 1.69 eV).

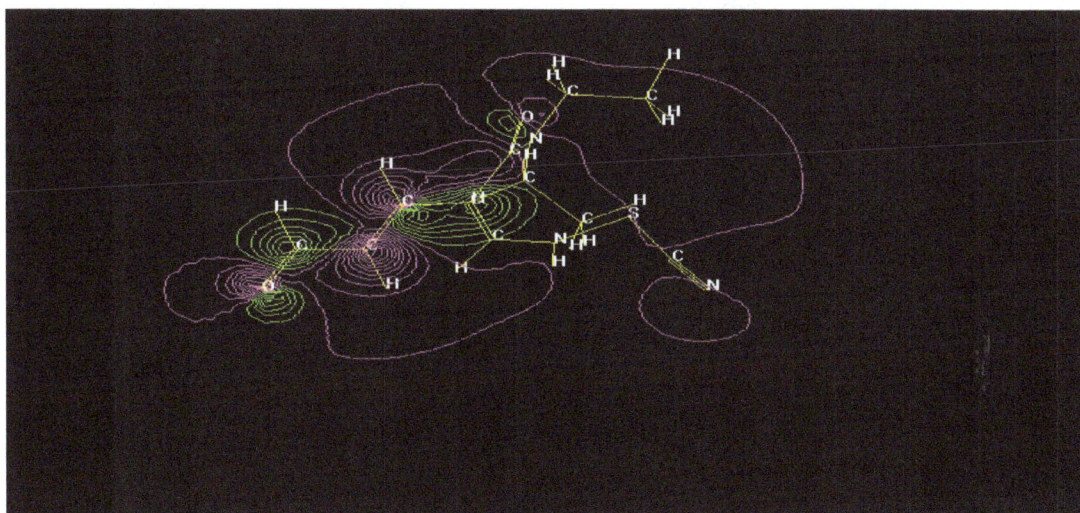

Figure 3. The distribution of electronic density in the optimized molecule of DGD (ENSMO = −1.07 eV).

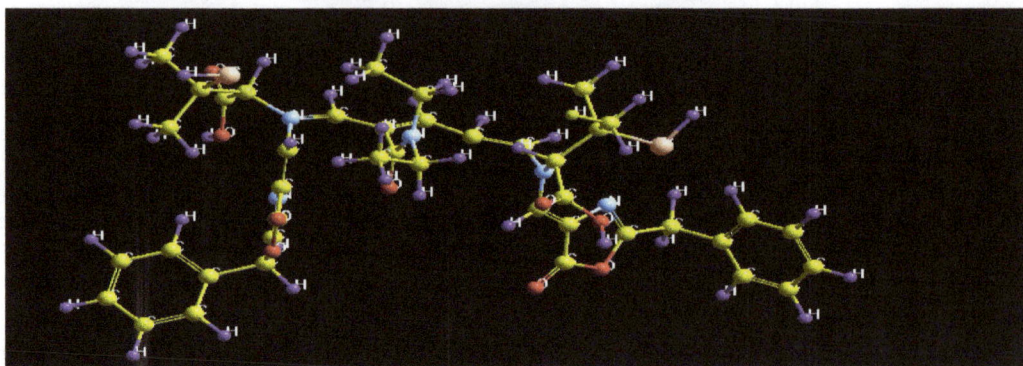

Figure 4. Geometric model of cyanine dye molecule.

Table 1. Geometric baseline parameters, reacting agents and final reaction products.

Geometric parameters		Penicillin acid	DGD	Cyanine dye
Enthalpy of formation ($\Delta N0obr$)		−84.2679 kcal/mol (−352.8 KJ/mol)	−7.8852276 kcal/mol (−3.01 KJ/mol)	−129.631 kcal/mol (−542 KJ/mol)
Dipole moment (μ)		3.571 D	0.953 D	5.145 D
Energy of frontier orbitals	Energy of LUMO (lowest unoccupied molecular orbital)	1.699999	−1.070426	--
	Energy of the HOMO (highest occupied molecular orbital)	−9.739995	−9.370866	--

Table 2. Electronic parameters of the molecules of penicillin acid and DGD.

Parameter	Penicillin acid	Parameter	DGD
μ, D	3.571	μ, D	0.953
q C_1	−0.121483	q C_1	0.210
q C_2	−0.131220	**q O_2**	**−0.284**
q C_3	−0.130831	q C_3	−0.270
q C_4	−0.112560	q C_4	−0.030
q C_5	−0.086901	**q C_5**	**−0.199**
q C_6	−0.110834	q C_6	0.05
q C_{11}	−0.057496	q N7	−0.469
q C_{14}	0.082292	q S13	0.620
q N_{16}	−0.153666	q C14	−0.429
q O_{17}	−0.226165	q N15	0.005
q C_{18}	−0.198795	q C_{16}	0.351
q C_{19}	0.341160	q O_{17}	−0.361
q O_{20}	−0.293519	q N_{18}	−0.322
q C_{21}	0.083508	q C_{19}	−0.028
q N_{23}	**−0.277703**	q C_{20}	−0.234
q C_{25}	0.027240	q C_{21}	−0.022
q C_{27}	0.322425	q C_{22}	−0.237
q C_{28}	−0.145945		
q O_{29}	−0.356745		
q C_{30}	−0.290921		
q C_{32}	−0.213261		
q C_{33}	−0.234911		
q S_{40}	−0.065233		

Table 3. Geometric parameters of penicillin acid and DGD.

Parameter	Penicillin acid	Parameter	DGD
r (C$_1$-C$_2$), nm	0.139	r (C$_1$-O$_2$), nm	0.121
r (C$_1$-H$_8$), nm	0.109	r (C$_1$-H$_8$), nm	0.110
r (C$_5$-C$_{11}$), nm	0.149	r (C$_1$-C$_3$), nm	0.147
r (C$_{11}$-C$_{14}$), nm	0.149	r (C$_3$-C$_4$), nm	0.134
r (C$_{14}$-N$_{16}$), nm	0.131	r (C$_4$-C$_5$), nm	0.145
r (N$_{16}$-C$_{18}$), nm	0.144	r (C$_5$-C$_6$), nm	0.136
r (C$_{14}$-O$_{17}$), nm	0.139	r (C$_6$-N$_7$), nm	0.139
r (O$_{17}$-C$_{19}$), nm	0,139	r (N$_7$-S$_{13}$), nm	0.178
r (C$_{19}$-O$_{20}$), nm	0.121	r (S$_{13}$-C$_{14}$), nm	0.139
r (C$_{19}$-C$_{18}$), nm	0.146	r (C$_{14}$-N$_{15}$), nm	0.116
r (C$_{18}$-C$_{21}$), nm	0.136	r (C$_5$-C$_{16}$), nm	0.147
r (C$_{21}$-H$_{22}$), nm	0.110	r (C$_{16}$-O$_{17}$), nm	0.125
<u>r (C$_{21}$-N$_{23}$), nm</u>	<u>0.139</u>	r (C$_{16}$-N$_{18}$), nm	0.141
<u>r (N$_{23}$-H$_{24}$), nm</u>	<u>0.099</u>	r (N$_{18}$-C$_{21}$), nm	0.149
r (N$_{23}$-C$_{25}$), nm	0.148	r (C$_{21}$-C$_{22}$), nm	0.151
r (C$_{25}$-C$_{27}$), nm	0.153	r (N$_{18}$-C$_{19}$), nm	0.149
r (C$_{27}$-O$_{29}$), nm	0.121	r (C$_{19}$-C$_{20}$), nm	0.151
r (C$_{27}$-O$_{30}$), nm	0.135		
r (O30-H31), nm	0.095		
r (C25-C28), nm	0.155		
r (C28-S40), nm	0.186		
r (S40-H41), nm	0.131		
r (C28-C32), nm	0.152		
r (C28-C33), nm	0.152		

4. Conclusions

The obtained semi-empirical computer modelling calculations can be used to explain the mechanism of the formation reaction of the cyanine dye, and this fact is a novelty of this scientific work. Besides, the defined geometric and electronic parameters of cyanine dye molecules confirm the chemical structure of a given molecule.

Effect of substituents on the electronic density distribution in molecules of penicillin acid and DGD, evaluated by electronic effects-inductive and mesomeric and the electronic density distribution in the molecules of these compounds, explained by using semiempirical AM1 calculations, proves once again that a negatively charged nitrogen atom of penicillin acid molecule can attack the positively charged carbon atom in a strongly polarized carbonyl group of DGD.

Thus, the explanation of the mechanism of this reaction by theoretical possibilities of quantum-chemical method AM1 was confirmed by interpretation of experimental data by methods of spectral analysis (IR, NMR, NMR-spectroscopy) [19], to identify features of the chemistry of the reaction formation of cyanine dyes.

This scientific workout in perspective is a further study of the reaction by other methods in quantum chemistry.

References

[1]   Korenman, I.M. (1975) Photometric Analysis. Methods for the Determination of Organic Compounds. 2nd Edition, Rev. and Add.-Moscow: Chemistry, 359 p.

[2]   Polyudek Fabini R. and Beirikch T. (1981) Organic Analysis: A Guide to the Analysis of Organic Compounds, including Drugs. Substances / Tr. from German, Leningrad: Chemistry, 624 p.

[3]   Beisenbekov, A.S. and Aliyev, A.M. (1986) Application of Formation Reactions of Polymethine Dyes in Pharmaceutical Analysis. Alma-Ata, 95 p.

[4]   Kramarenko, V.F. and Popova, V.I. (1972) Photometry in Pharmaceutical Analysis. Health, Kiev, 191 p.

[5]   Traven, V.F. (2004) Organic Chemistry: A Textbook for High Schools. 2 Volumes/VF Traven.-M.: ICC "Akademkniga", 582 p.

[6]   Sykes, P. (1991) Reaction Mechanisms in Organic Chemistry. 4th Edition, Tr. from English.-M.: Chemistry, 447 p.

[7]   Beisenbekov, A.S., Zhubayeva, R.A. and Boshkayeva, A.K. (1993) Photometric Determination of Anabazin with Fenilmethilpirazolonom Hydrochloride by Reaction Formation of Polymethine Dyes: Method. Instructions/S.D. Asfendyairov ASMI-Almaty, 12.

[8]   Arzamastsev, A.P. (2004) Pharmaceutical Chemistry: Textbook. 3rd Edition, Rev.-M.: GEOTAR Media, 640 p.

[9]   Belikov, B.G. (2007) Pharmaceutical Chemistry. In 2 Parts: Textbook/V. G. Belikov, 4th Edition, Rev. and Add.-M.: MED Press Inform., 624 p.

[10]   Arzamastsev, A.P. (2008) Pharmaceutical Chemistry: Textbook. 3rd Edition, Rev.- M.: GEOTAR Media, 640 p.

[11]   Belikov, V.G. (2008) Pharmaceutical Chemistry: Textbook. 2nd Edition, MED Press Inform, M., 616 p.

[12]   Aksenova, E.N., Andrianova, O.P., Arzamastsev, A.P., *et al.* (2001) Guide to the Laboratory Lessons in Pharmaceutical Chemistry. M.: Medicine, 384 p.

[13]   Boshkayeva, A.K. and Kamilov, Kh.M. (2009) Application in the Analysis of the Reaction Formation of Cyanine Dyes. *Pharmaceutical Journal*, No. 2, 55-59.

[14]   Boshkayeva, A.K., Omarova, R.A. and Beisenbekov, A.S. (2009) Assessment of the Quality of the Cyanine Dye of Betalaktamid by NMR- and H-NMR Spectroscopy. *Bulletin of KazNMU*, No. 2, 55-59.

[15]   Burstein, K.Ya. and Shorygin, P.P. (1989) Quantum Chemical Calculationsin Organic Chemistry and Molecular Spectroscopy. Nauka, Moscow, 104 p.

[16]   Clark, T. (1990) Computer Chemistry. Mir, Moscow, 381 p.

[17]   Stepanov, N.F. (2001) Quantum Mechanics and Quantum Chemistry. Theoretical Foundations of Chemistry. Mir, Moscow, 519 p.

[18]   Boshkayeva, A.K. (2009) The Study by IR Spectroscopy, NMR and 13 C NMR Spectroscopy in Cyanine Dye-Lactamide. *Pharmaceutical Journal*, No. 1, 50-53.

[19]   Boshkayeva, A.K. (2009) The Chemistry of the Formation of Cyanine Dyes. *Scientific Review*, No. 3, 4-7.

18

^{11}C-Labeling of the C(1)-C(10) Dihydroxy Acid Moiety for the Study on the Synthesis of Kulokekahilide-2 PET Tracer

Chunguang Han[1], Hisashi Doi[2], Junji Kimura[3], Yoichi Nakao[4], Masaaki Suzuki[5]*

[1]Research Center for Materials Science, Nagoya University, Nagoya, Japan
[2]RIKEN Center for Life Science Technologies, Kobe, Japan
[3]Department of Chemistry and Biological Science, College of Science and Engineering, Aoyama Gakuin University, Sagamihara, Japan
[4]Department of Chemistry and Biochemistry, School of Advanced Science and Engineering, Waseda University, Tokyo, Japan
[5]National Center for Geriatrics and Gerontology, Obu City, Japan
Email: *suzukims@ncgg.go.jp

Abstract

^{11}C-labeled C1-C10 partial structure of kulokekahilide-2 (**1**) was successfully synthesized based on Pd^0-mediated rapid C-[^{11}C]methylation using [^{11}C]methyl iodide and pinacol alkenylboronate. The preparation of organoboron intermediate via olefin cross-metathesis is also a crucial procedure for the synthesis of ^{11}C-labeling C1-C10 dihydroxy acid moiety of **1**.

Keywords

Kulokekahilide-2, Carbon-11, Boron, C-C Coupling, Isotopic Labeling

1. Introduction

Kulokekahilide-2 (**1**) was isolated from the Hawaiian marine mollusk *Philinopsisspeciosa*, as an aurilide-type metabolite to show potent cytotoxicity against several cell lines in nanomolar concentrations [1]. The 26-membered form of **1** consisting of five amino acids (43-D-Ala, 37-L-Ile, 34-MeGly, 24-D-MePhe, 21-L-Ala) and two hydroxy acids {15-D-2-hydroxyisocaproic acid (15-D-Hica) and (5S,6S,7S,2E,8E)-2,6,8-trimethyl-5,7-dihydroxy-2,8-decadienoic acid (Dtda), **Figure 1**}, possesses almost the same components as the aurilides [2] [3]

---
*Corresponding author.

and lagunamides [4] isolated from the Japanese sea hare *Dolabellaauricularia* and marine cyanobacteria respectively [5]. The aurilide also shows high-level cytotoxicity against renal and prostate cancer cell lines and its molecular target for inducing apoptosis in human cell lines was reported by Sato *et al.* [6] [7]. Recently, Kimura *et al.* revealed that kulokekahilide-2 shows potent cytotoxicity against 39 human cancer cell lines, suggesting a mechanism of action could be different from that of standard anticancer drugs, aurilides, palau'amide [8] [9], and lagunamides [5].

Positron emission tomography (PET), which uses specific probes radiolabeled with short-lived positron-emitting radionuclides (^{11}C, ^{13}N, ^{15}O, ^{18}F etc.), is a powerful non-invasive molecular imaging technique usable for highly accurate diagnoses and investigation of the *in vivo* biochemistry of bioactive compounds. In addition, it is strongly hoped that PET would be applied as a human microdosing study to an early-stage of drug development [10]. Carbon-11 (half-life = 20.4 min) is one of the most meritorious isotopes for PET research because carbon is included in all organic molecules. In recent years, efficient labeling methods of the ^{11}C radioisotope into organic frameworks have continuously been developed by Suzuki *et al.* using palladium(0)-mediated rapid *C*-[^{11}C]methylation (5 min reactions) consisting of the cross-coupling reactions of [^{11}C]methyl iodide and the stannyl or boron substrates [11] [12]. Actually, the rapid cross-coupling reactions have successfully been applied for the syntheses of various disease-oriented PET tracers [13]-[16]. Our interest has been intrigued by extending the ^{11}C labeling reactions to complex natural product. Described herein in the synthesis of ^{11}C-incorporated C(1)-C(10) partial structure in [^{11}C]kulokekahilide-2 focused on the ^{11}C labeling at C10 carbon of **1** using a combination of olefin cross metathesis (CM) and rapid *C*-[^{11}C]methylation.

2. Results and Discussion

Kulokekahilide-2 has several possible positions for ^{11}C radiolabeling. Prior to actual synthesis of ^{11}C-incorporated **1**, we investigated a model study using a partial structure. Here, we are particularly interested in introducing ^{11}C onto the methylene group of **2** as a ^{11}CH$_3$ by our rapid cross-coupling reaction [12], [17] between sp^2$_{vinyl}$- and sp^3-carbon atoms using an organostannyl or boron precursor **3** (**Scheme 1**) where key step of synthetic strategy involves the preparation of precursor **3** derived from methyl ester **2**.

Before synthesizing ^{11}C-labeled **4a**, we prepared the nonradioactive molecule **4b**. Thus, the intermediary compound **5** was synthesized from (*S*)-3-propionyl-4-isopropyl-2-oxazolidinone according to the reported procedure [1]. Protection of the secondary alcohol **5** gave PMB ether **6** in 49% yield. Subsequent deprotection of the TBS group at C7 in **6** afforded the desired compound **4b** in 78% yield after column chromatography on silica gel (**Scheme 2**).

The synthesis of the organostannyl or boron compound **3** started with an acylation of commercially available ox-azolidinone **7** with the treatment of LDA and EtCOCl to give aldol coupling precursor [18]. Subsequent *an-*

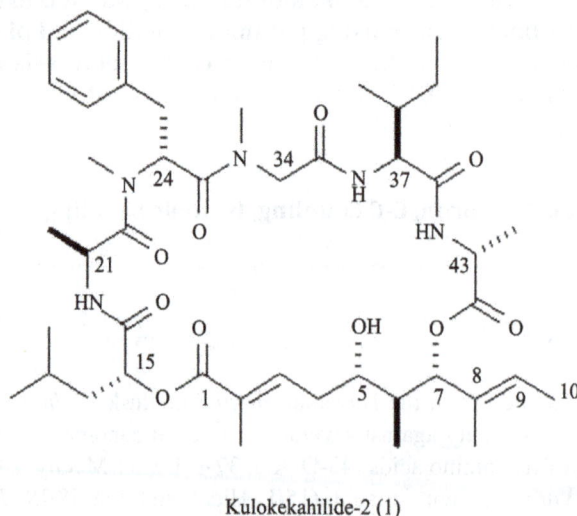

Kulokekahilide-2 (1)

Figure 1. Chemical structure of kulokekahilide-2 (1).

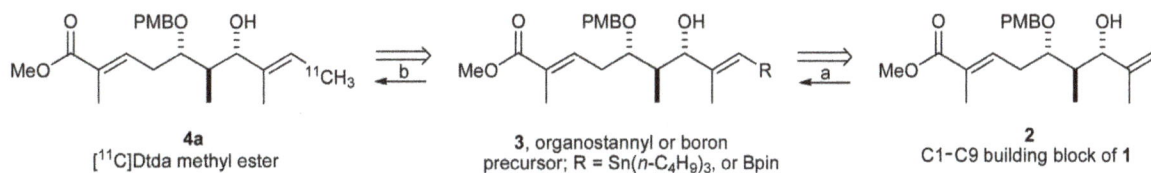

Scheme 1. Synthetic plan: Transformation from terminal alkene **2** to ^{11}C-labeled **4a** by (a) olefin cross metathesis and (b) rapid C-^{11}C]methylation.

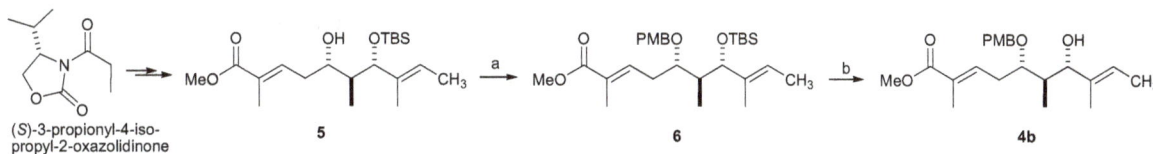

Scheme 2. Synthesis of **4b**. Reagents and Conditions: (a) PMBTCA, Et$_2$O, −78°C added to CF$_3$SO$_3$H, then RT, 2 h, 49%; (b) HF·Py, pyridine, THF, RT, 40°C, overnight, 78%.

ti-selective aldol reaction with methacrylaldehyde to provide the aldol product **8**, which was converted into aldehyde **9** in three steps. Then the BF$_3$·OEt$_2$ mediated Mukaiyamaaldol reaction of **9** with silyl ketene acetal **10** afforded methyl ester (5R)-**11** as a single diastereomer. Successful Moffatt oxidation of **11** to give ketone **12**, and subsequent reduction of the resulting keto group in **12** with NaBH$_4$ stereoselectively (S/R = 22/1) proceeded to give the desired alcohol (5S)-**13** in 69% yield. PMB protection of the secondary hydroxy substituent at C5 in **14**, followed by cleavage of the TBS protecting group at C7 with HF·Py, accomplishing the synthesis of 1,1-disubstituted alkene **2** (**Scheme 3**).

As the key step for synthesis of organostannane or organoboron precursor of **2**, cross metathesis was chosen because it has become a powerful and convenient synthetic technique for the preparation of functionalized alkenes in organic chemistry [19]. With this concern, hydroxy 1,1-disubstituted alkene **15** as a model compound for screening the most effective Ru complexes and cross partners in metathesis. Cross metathesis using **15** prepared by Grignard addition reaction [20] was investigated under various reaction conditions: Grubbs second-generation (G-II) [21] [22] or the Hoveyda-Grubbs second-generation complex (HG-II) [23] in CH$_2$Cl$_2$, benzene, and toluene at reflux or microware irradiation, and the use of an excess amount of cross partners such as vinylstannane **16a**, or vinyl boronates (**16b** and **16c**). These results are summarized in **Table 1**. Cross metathesis of **15** with vinylstannane **16a** using HG-II (1.0 equiv) in CH$_2$Cl$_2$ or toluene at reflux did not give the desired organostannane precursor **17a** presumably due to highly sterically hindered Sn(n-C$_4$H$_9$)$_3$ group in **16a** (**Table 1**, entries 1-2). The lower sterically hinderedvinyl dioxaborolane **16b** compared with tetramethyl vinyl boronate **16c** also did not afford the corresponding organoboron compound **17b** with the use of HG-II or G-II catalysts in thermal or microware heating conditions [24]-[31] (entries 3-5). Grubbs II-catalyzed cross metathesis of **15** with vinyl pinacol boronate **16c** (4.0 equiv) in benzene at refluxdid not afford desired **17c** (entry 6). By contrast, when **15** was treated with more robust and powerful HG-II (25 mol%) in CH$_2$Cl$_2$ at reflux for one day, we observed a small amount of organoboron precursor **17c** (entry 7). The increase of the catalyst to a stoichiometric amount under the same reaction conditions to give **17c** in 35% yield (entry 8) as a single E-isomer as judged by the NOE observation.

We envisioned here that the E-selective olefin cross metathesis using HG-II catalyst in the reaction of **15** and vinyl pinacol boronate **16c** could be applied for synthesis of the organoboron derivative of 1,1-disubstituted alkene **2**, which is crucial precursor for the synthesis of the ^{11}C-labeling dihydroxy acid moiety of **1**. However, contrary to our expectation, cross metathesis using **2** did not proceed under above reaction conditions (**Table 1**, entry 8) with the notice of complete recovery of **2**.The reaction was further conducted under more forcing the reaction conditions, giving the desired (E)-organoboron derivative **3** in 14% yield along with recovered **2** in 32% yield (**Scheme 4**). The geometry of the newly formed double bond was decided by NOE observation as shown below.

By using the precursor **3**, we examined the Pd^0-mediate rapid C-^{11}C]methylations protocol [12] for preparing the ^{11}C-labeled partial structure of **1** under cold conditions (**Scheme 4**). The methylation of **3** was conducted

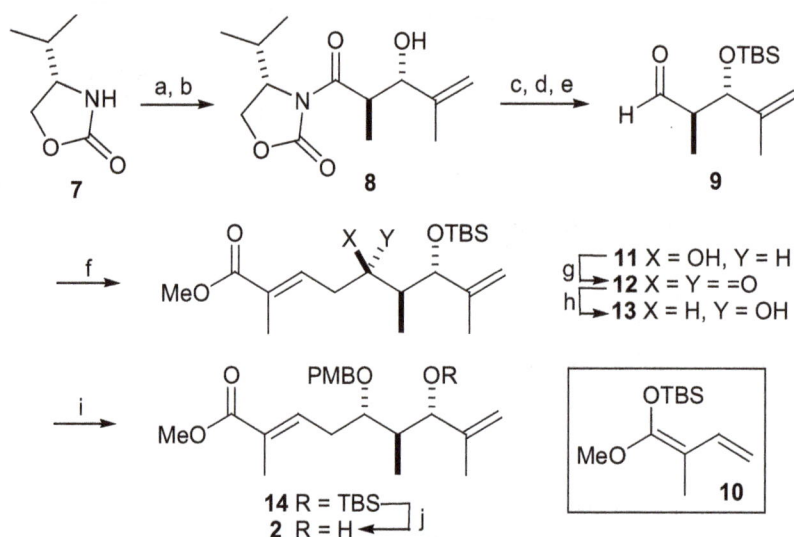

Scheme 3. Synthesis of **2**. Reagents and Conditions: (a) EtCOCl, LDA, THF, −78°C, 1 h, 95%; (b) methacylaldehyde, nBu$_2$BOTf, DIPEA, CH$_2$Cl$_2$, −78°C, 2 h, 47%; (c) MeNH(OMe)HCl, Me$_3$Al, 0°C-RT, 30 min, then 0°C added to **8**, warmed up to RT for overnight; (d) TBSCl, Imidazole, DMF, 94% for two steps; (e) DIBAL-H, THF, −78°C, 1.5 h, 89%; (f) **10**, BF$_3$·OEt$_2$, −78°C, 1 h, 69%; (g) DCC, pyridine trifluoroacetate, DMSO/Et$_2$O (1:1), RT, 1.5 h, 75%; (h) NaBH$_4$, MeOH, −40°C, 2 h, 69%; (i) PMBTCA, Et$_2$O, −78°C added to CF$_3$SO$_3$H, then RT, 2.5 h, 49%; (j) HF·Py, pyridine, THF, RT, 40°C, overnight, 74%.

Table 1. Screening the reactions of Ru complexes and cross metathesis partners[a].

Entry	Cross partner	Catalyst (equiv.)	Product[b]	Yield[f] (%)
1	**16a**	HG-II (1.0)	**17a**	N.R.
2	**16a**	HG-II (1.0)	**17a**[c]	N.R.
3	**16b**	HG-II (1.0)	**17b**	N.R.
4	**16b**	HG-II (1.0)	**17b**[d]	N.R.
5	**16b**	G-II (1.0)	**17b**	N.R.
6	**16c**	G-II (0.1)	**17c**[e]	N.R.
7	**16c**	HG-II (0.25)	**17c**	5
8	**16c**	HG-II (1.0)	**17c**	35

[a]Reaction conditions: **15** (1 equiv), cross partner (4 equiv), in refluxing CH$_2$Cl$_2$ (0.02 M) for 24 h. [b]Only the E isomer was observed in all cases; [c]In toluene (0.02 M), reflux, 24 h; [d] In CH$_2$Cl$_2$ (0.1 M), microware heating, 60°C, 30 min; [e]In benzene (0.02 M), reflux, 24 h. [f]Isolated yield; HG-II: Hoveyda-Grubbs second-generation catalyst; G-II: Grubbs second-generation catalyst.

following the standard procedure: CH$_3$I dissolved in DMF was added to a solution of **3**, Pd$_2$(dba)$_3$, P(o-tolyl)$_3$, and K$_2$CO$_3$ in DMF, and the resulting mixture was heated at 70°C for 5 min, then purified on ODS to give **4b** in 78% yield.

Based on the above-mentioned protocol, we preformed the synthetic of ^{11}C-labeled Dtda methyl ester **4a**, Thus, **3**/Pd$_2$(dba)$_3$/P(o-tolyl)$_3$/K$_2$CO$_3$ (2:1:4:10) dissolved in DMF under argon was mixed with [^{11}C]CH$_3$I, prepared as previously described [32], the solution was heated at 70°C for 5 min (**Scheme 5**). After the mixture was

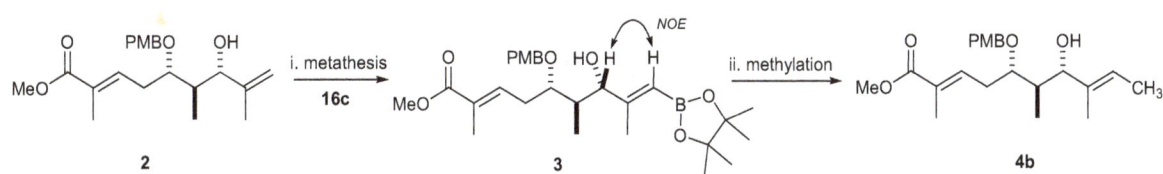

Scheme 4. Synthesis of 3 and 4b. Reagents and Conditions: (i) HG-II (1.3 equiv), toluene (0.01M), 80°C, 24 h, 14%; (ii) MeI, $Pd_2(dba)_3$, P(o-tolyl)$_3$, and K_2CO_3, DMF, 70°C, 5 min, 78%.

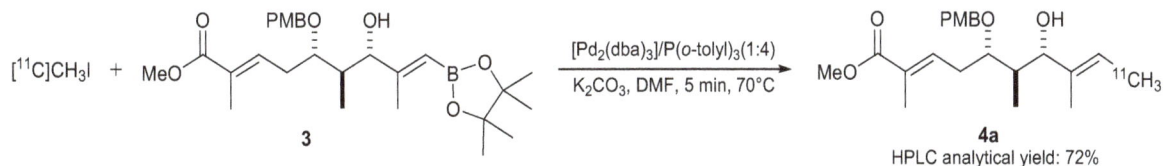

Scheme 5. Synthesis of ^{11}C-labeled Dtda methyl ester 4a.

poured into a separate vial containing a solution of ascorbic acid in acetonitrile, the resulting reaction mixture was submitted to HPLC, and then purified by reverse phase semi-preparative HPLC to give desired [^{11}C]Dtda-methyl ester **4a** in 72% as reverse phase HPLC analytical yield (**Figure 2**). Total synthesis time including HPLC purification was 33 min. The radioactivity of isolated **4a** was 315 MBq and the radiochemical purity was >99%. The chemical identity of **4a** was confirmed by co-injection with the authentic sample of **4b** by analytical HPLC.

3. Conclusion

^{11}C-labeled Dtda methyl ester **4a** as C1-C10 building block of kulokekahilide-2 (**1**) has been successfully synthesized using a combination of olefin cross-metathesis/rapid C-[^{11}C]methylation. Pinacol alkenyl boronate precursor **3** prepared via cross-metathesis is crucial for subsequent Pd^0-mediated rapid C-[^{11}C]methylation using [^{11}C]CH$_3$I. The ^{11}C-labeling would be applied at later stage for the synthesis of ^{11}C-incorporated **1**.

4. Acknowledgements

This work was supported in part by a consignment expense for the Molecular Imaging Program on "Research Base for Exploring New Drugs" from the Ministry of Education, Culture, Sports, Science and Technology (MEXT) of Japan. We thank Dr. Masakatsu Kanazawa (Central Research Lab. Hamamatsu Photonics K. K.), Ms. Mawatari Aya (RIKEN Center for Life Science Technologies) for experimental assistance and Mr. Masahiro Kurahashi (Sumitomo Heavy Industry Accelerator Service Ltd.) for operating the cyclotron.

5. Experimental

5.1. General

Nuclear magnetic resonance (NMR) spectra were recorded on a JEOL JNM-ECX400P spectrometer (400 MHz for ^{1}H), and the chemical shifts in δ (parts per million) were referenced to the solvent peaks of δ_H 7.26 for CHCl$_3$. HR ESI-TOF-MS spectra were measured on an Applied Biosystems Mariner Biospectrometry Workstation using ABN as a calibration standard in the positive mode. Microwave irradiation was carried out in a Biotage Initiator™ (Tokyo, Japan) using a sealed vessel. [^{11}C]Carbon dioxide was produced by a ^{14}N(p, α)^{11}C reaction by using a Sumitomo CYPRIS HM-12S cyclotron (Sumitomo Heavy Industries, Tokyo, Japan), and then converted to [^{11}C]methyl iodide by treatment with lithium aluminum hydride followed by hydriodic acid using an automated synthesis system (Cupid, Sumitomo Heavy Industries). The obtained [^{11}C]methyl iodide was used for palladium(0)-mediated rapid [^{11}C]methylation shown in **Scheme 5**. The synthesis of ^{11}C-labeled Dtda methyl ester **4a** in **Scheme 5** was conducted in a lead-shielded hot-cell with remote control of all operations in RIKEN CLST. Purification with HPLC was performed on a GL Science system (Tokyo, Japan). Radioactivity was quantified with an ATOMLAB™ 300 dose calibrator (Aloka, Tokyo, Japan). Analytical HPLC was performed on a Shimadzu system (Kyoto, Japan), and effluent radioactivity was measured with an RLC700 radio analyser

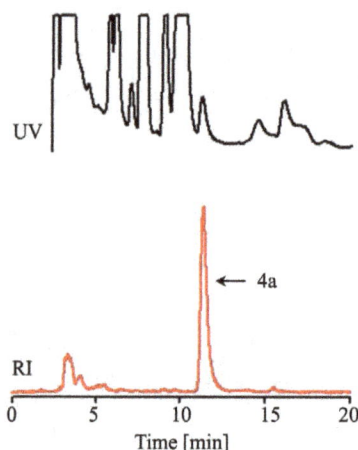

Figure 2. HPLC for purification of the reaction mixture. UV absorbance: 223 nm.

(Aloka). The column used for analytical and semi-preparative HPLC was Develosil ODS-HG-5 (Nomura Chemical, Japan).

5.2. Synthesis of Authentic Sample 4b

p-**Methoxybenzyl ether (6)**: To a stirred solution of alcohol **5** (19 mg, 0.05 mmol) and *p*-methoxybenzyl-2,2,2-trichloroacetimidate (72 mg, 0.26 mmol) in ether (1.6 mL) cooled at −78°C was added trifluoromethane-sulfonic acid (1.5 μL, 0.018 mmol). After the addition was completed, the ice bath was removed and the reaction mixture was warmed to room temperature. After being stirred at rt for 2 h, the reaction mixture was quenched by the addition of saturated aqueous $NaHCO_3$. The separated aqueous layer was extracted with Et_2O (8 mL × 3). The combined organic layer was concentrated under reduced pressure. The residue oil was purified by column chromatography on silica gel (9:1 hexane/ethyl acetate) to give **6** (12 mg, 49%) as a colorless oil. ^{1}H NMR (400 MHz, $CDCl_3$): δ 7.31 (d, J = 6.8 Hz, 2H), 6.92 (d, J = 6.8 Hz, 2H), 6.88 (d, J = 8.4 Hz, 1H), 5.43 (q, J = 6.8 Hz, 1H), 4.57 (d, J = 11.2 Hz, 1H), 4.42 (d, J = 11.2 Hz, 1H), 3.90 (m, 1H), 3.87 (s, 3H), 3.81 (s, 3H), 3.76 (d, J = 9.2 Hz, 1H), 2.33 (m, 2H), 2.21 (m, 1H), 1.90 (s, 3H), 1.67 (d, J = 6.8 Hz, 3H), 1.61 (s, 3H), 0.92 (s, 9H), 0.80 (d, J = 6.8 Hz, 3H), 0.03 (s, 3H), 0.01 (s, 3H).

Authentic sample (4b): *p*-methoxybenzyl ether **6** (12 mg, 0.02 mmol) was dissolved in a 5:3:12 mixture of HF·Py, Pyridine and THF (0.6 mL). The solution was stirred at 40°C for 12 h, diluted with EtOAc (2 mL), and poured into saturated aqueous $NaHCO_3$ (6 mL) cooled at 0°C. The mixture was extracted with EtOAc (4 mL × 3). The combined extracts were washed with brine dried (Na_2SO_4), and concentrated. The residue oil was puri-fied by column chromatography on silica gel (3:1 hexane/ethyl acetate) to give desired **4b** (7 mg, 78%) as a co-lorless oil. ^{1}H NMR (400 MHz, $CDCl_3$): δ 7.26 (d, J = 6.4 Hz, 2H), 6.91 (t, J = 6.4 Hz, 1H), 6.87 (d, J = 6.4 Hz, 2H), 5.43 (q, J = 6.8 Hz, 1H), 4.57 (d, J = 11.2 Hz, 1H), 4.45 (d, J = 11.2 Hz, 1H), 3.83 (m, 1H), 3.80 (s, 3H), 3.72 (s, 3H), 3.69 (m, 1H), 2.59 (m, 1H), 2.43 (m, 1H), 1.96 (m, 1H), 1.87 (s, 3H), 1.61 (d, J = 6.8 Hz, 3H), 1.59 (s, 3H), 0.66 (d, J = 6.8 Hz, 3H). HRMS (ESI-TOF) *m/z*: $[M + H]^+$ Calcd for $C_{22}H_{33}O_5$ 377.2322; Found 377.2298.

5.3. Synthesis of Organoboron Precursor 3

Alcohol **2** was prepared from **14** by deprotection of the second TBS ether using HF·Py as a colorless oil. ^{1}H NMR (400 MHz, $CDCl_3$): δ 7.26 (d, J = 8.8 Hz, 2H), 6.91 (t, J = 6.4 Hz, 1H), 6.87 (d, J = 8.8 Hz, 2H), 4.89 (s, 1H), 4.87 (s, 1H), 4.54 (d, J = 11.2 Hz, 1H), 4.45 (d, J = 11.2 Hz, 1H), 3.92 (d, J = 10 Hz, 1H), 3.80 (s, 3H), 3.75 (s, 3H), 3.72 (m, 1H), 2.57 (m, 1H), 2.43 (m, 1H), 1.92 (m, 1H), 1.87 (s, 3H), 1.71 (s, 3H), 0.73 (d, J = 6.8 Hz, 3H).

Organoboron precursor (3): To a solution of HG-II catalyst (63 mg, 0.1 mmol) and dry toluene (6.5 mL) in a round-bottomed flask equipped with a reflux condenser was added alcohol **2** (28 mg, 0.08 mmol) and vinyl boronate **16c** (66 μL, 0.38 mmol). The solution was refluxed for roughly overnight. The mixture was then concentrated, and the products were purified by SiliaMetS® DMT. The filtrate was concentrated, and then the residue was separated by HPLC [Develosil ODS-HG-5 ($\varnothing$ 10 × 250 mm), flow rate 4 mL/min, 60% - 80% aqMeCN, 40 min] to give **3** (5.3 mg, 14%; t_R = 19.5 min,). ^{1}H NMR (400 MHz, CDCl$_3$): δ 7.25 (d, J = 8.4 Hz, 2H), 6.88 (t, J = 7.6 Hz, 1H), 6.86 (d, J = 8.4 Hz, 2H), 5.28 (s, 1H), 4.52 (d, J = 10.8 Hz, 1H), 4.44 (d, J = 10.8 Hz, 1H), 3.88 (d, J = 7.2 Hz, 1H), 3.80 (s, 3H), 3.75 (s, 3H), 3.71 (m, 1H), 2.57 (m, 1H), 2.43 (m, 1H), 1.97 (m, 1H), 1.95 (s, 3H), 1.86 (s, 3H), 1.26 (s, 12H), 0.74 (d, J = 7.2 Hz, 3H). HRMS (ESI-TOF) m/z: [M + Na]$^+$ Calcd for C$_{27}$H$_{41}$BO$_7$Na 511.2837; Found 511.2814.

5.4. Synthesis of 4b by the Rapid C-Methylation (Cold Conditions)

Iodomethane (0.25 µL, 4 mmol) dissolved in N,N-dimethylformamide (19.75 µL, 0.2 M) was added to a solution of Pd$_2$(dba)$_3$ (1.8 mg, 2.0 µmol), tri(o-tolyl)phosphine (2.4 mg, 7.9 µmol), potassium carbonate (2.8 mg, 20 µmol), and **3** (2.0 mg, 4.0 µmol) in anhydrous N,N-dimethylformamide (200 µL). The resulting mixture was heated at 70°C for 5 min. The reaction solution was evaporated in vacuo and purified on ODS, eluting with MeCN-H$_2$O (7:3) to afford **4b** (1.2 mg) in78% yield.

5.5. Synthesis of ¹¹C-Labeled Dtda Methyl Ester 4a

[^{11}C]CH$_3$I was transported into a vial where the organoboronprecuesor **3** (2.0 mg), Pd$_2$(dba)$_3$ (1.8 mg), P(o-tolyl)$_3$ (2.4 mg), K$_2$CO$_3$ (2.8 mg) were dissolved in anhydrous DMF (200 µL) at room temperature. After the solution was heated at 70°C for 5 min, the mixture was poured into a separated vial. Then the reaction vial was washed with MeCN/water (70:30, 800 µL) and the washing was added to the above mixture solution. The resulting mixture was purified to reverse-phase HPLC [Develosil ODS-HG-5 ($\varnothing$ 10 × 250 mm), flow rate 5.0 mL/min, 70% aqMeCN, detection at 223 nm, **3a**: t_R = 11.4 min]. The desired fractions were collected in a flask containing 25% ascorbic acid (50 µL), and the organic solvent was removed under reduced pressure.

References

[1]   Nakao, Y., Yoshida, W.Y., Takada, Y., Kimura, J., Yang, L., Susan, L.M. and Scheuer, P.J. (2004) Kulokekahilide-2, a Cytotoxic Depsipeptide from a Cephalaspidean Mollusk *Philinopsisspeciosa. Journal of Natural Products*, **67**, 1332-1340. http://dx.doi.org/10.1021/np049949f

[2]   Suenaga, K., Mutou, T., Shibata, T., Itoh, T., Kigoshi, H. and Yamada, K. (1996) Isolation and Stereostructure of Aurilide, a Novel Cyclodepsipeptide from the Japanese Sea Hare *Dolabellaauricularia. Tetrahedron Letters*, **37**, 6771-6774. http://dx.doi.org/10.1016/S0040-4039(96)01464-5

[3]   Suenaga, K., Mutou, T., Shibata, T., Itoh, T., Fujita, T., Takada, N., Hayamizu, K., Takagi, M., Irifune, T., Kigoshi, H. and Yamada, K. (2004) Aurilide, a Cytotoxic Depsipeptide from the Sea Hare *Dolabellaauricularia*: Isolation, Structure Determination, Synthesis, and Biological Activity. *Tetrahedron*, **60**, 8509-8527. http://dx.doi.org/10.1016/j.tet.2004.06.125

[4]   Tripathi, A., Puddick, J., Prinsep, M.R., Rottmann, M. and Tan, L.T. (2010) Lagunamides A and B: Cytotoxic and Antimalarial Cyclodepsipeptides from the Marine Cyanobacterium *Lyngbyamajuscule. Journal of Natural Products*, **73**, 1810-1814. http://dx.doi.org/10.1021/np100442x

[5]   Umehara, M., Negishi, T., Maehara, Y., Nakao, Y. and Kimura, J. (2013) Stereochemical Analysis and Cytotoxicity of Kulokekahilide-2 and Its Analogues. *Tetrahedron*, **69**, 3045-3053. http://dx.doi.org/10.1016/j.tet.2013.01.089

[6]   Nagarajan, M., Maruthanayagam, V. and Sundararaman, M. (2012) A Review of Pharmacological and Toxicological Potentials of Marine Syanobacterial Metabolites. *Journal of Applied Toxicology*, **32**, 153-185. http://dx.doi.org/10.1002/jat.1717

[7]   Sato, S., Murata, M., Orihara, T., Shirakawa, T., Suenaga, K., Kigoshi, H. and Uesugi, M. (2011) Marine Natural Product Aurilide Activates the OPA1-Mediated Apoptosis by Binding to Prohibitin. *Chemistry & Biology*, **18**, 131-139. http://dx.doi.org/10.1016/j.chembiol.2010.10.017

[8]   Williams, P.G., Yoshida, W.Y., Quon, M.K., Moore, R.E. and Paul, V.J. (2003) The Structure of Palau'amide, a Potent Cytotoxin from a Species of the Marine Cyanobacterium *Lyngbya. Journal of Natural Products*, **66**, 1545-1549. http://dx.doi.org/10.1021/np034001r

[9]   Sugiyama, H., Watanabe, A., Teruya, T. and Suenaga, K. (2009) Synthesis of Palau'amide and Its Diastereomers: Confirmation of Its Stereostructure. *Tetrahedron Letters*, **50**, 7343-7345. http://dx.doi.org/10.1016/j.tetlet.2009.10.059

[10] Rowland, M. (2012) Microdosing: A Critical Assessment of Human Data. *Journal of Pharmaceutical Sciences*, **101**,

4067-4074. http://dx.doi.org/10.1002/jps.23290

[11] Suzuki, M., Doi, H., Björkman, M., Andersson, Y., Långström, B., Watanabe, Y. and Noyori, R. (1997) Rapid Coupling of Methyl Iodide with Aryltributylstannanes Mediated by Palladium(0) Complexes: A General Protocol for the Synthesis of [11]CH$_3$-Labeled PET Tracers. *Chemistry—A European Journal*, **3**, 2039-2042. http://dx.doi.org/10.1002/chem.19970031219

[12] Doi, H., Ban, I., Nonoyama, A., Sumi, K., Kuang, C., Hosoya, T., Tsukada, H. and Suzuki, M. (2009) Palladium(0)-Mediated Rapid Methylation and Fluoromethylation on Carbon Frameworks by Reacting Methyl and Fluoromethyl Iodide with Aryl and AlkenylBoronic Acid Esters: Useful for the Synthesis of ^{11}C]CH$_3$-C- and ^{18}F]FCH$_2$-C-Containing PET Tracers (PET=Positron Emission Tomography). *Chemistry—A European Journal*, **15**, 4165-4171. http://dx.doi.org/10.1002/chem.200801974

[13] Kanazawa, M., Furuta, K., Doi, H., Mori, T., Minami, T., Ito, S. and Suzuki, M. (2011) Synthesis of an Acromelic Acid A Analog-Based ^{11}C-Labeled PET Tracer for Exploration of the Site of Action of Acromelic Acid A in Allodynia Induction. *Bioorganic & Medicinal Chemistry Letters*, **21**, 2017-2020. http://dx.doi.org/10.1016/j.bmcl.2011.02.018

[14] Takashima-Hirano, M., Takashima, T., Katayama, Y., Wada, Y., Sugiyama, Y., Watanabe, Y., Doi, H. and Suzuki, M. (2011) Efficient Sequential Synthesis of PET Probes of the COX-2 Inhibitor ^{11}C]Celecoxib and Its Major Metabolite ^{11}C]SC-62807 and *in Vivo* PET Evaluation. *Bioorganic & Medicinal Chemistry*, **19**, 2997-3004. http://dx.doi.org/10.1016/j.bmc.2011.03.020

[15] Suzuki, M., Takashima-Hirano, M., Koyama, H., Yamaoka, T., Sumi, K., Nagata, H., Hidaka, H. and Doi, H. (2012) Efficient Synthesis of ^{11}C]H-1152, a PET Probe Specific for Rho-Kinases, Highly Potential Targets in Diagnostic Medicine and Drug Development. *Tetrahedron*, **68**, 2336-2341. http://dx.doi.org/10.1016/j.tet.2012.01.033

[16] Suzuki, M., Takashima-Hirano, M., Ishii, H., Watanabe, C., Sumi, K., Koyama, H. and Doi, H. (2014) Synthesis of ^{11}C-Labeled Retinoic Acid, ^{11}C]ATRA, via an Alkenylboron Precursor by Pd(0)-Mediated Rapid C-^{11}C]Methylation. *Bioorganic & Medicinal Chemistry Letters*, **24**, 3622-3625. http://dx.doi.org/10.1016/j.bmcl.2014.05.041

[17] Suzuki, M., Doi, H., Koyama, H., Zhang, Z., Hosoya, T., Onoe, H. and Watanabe, Y. (2014) Pd^0-Mediated Rapid Cross-Coupling Reactions, the Rapid C-^{11}C]Methylations, Revolutionarily Advancing the Syntheses of Short-Lived PET Molecular Probes. *The Chemical Record*, **14**, 516-541. http://dx.doi.org/10.1002/tcr.201400002

[18] Evans, D.A., Dow, R.L., Shih, T.L., Takacs, J.M. and Zahler, R. (1990) Total Synthesis of the Polyether Antibiotic Ionomycin. *Journal of the American Chemical Society*, **112**, 5290-5313. http://dx.doi.org/10.1021/ja00169a042

[19] Chatterjee, A.K., Choi, L., Sanders, D.P. and Grubbs, R.H. (2003) A General Model for Selectivity in Olefin Cross Metathesis. *Journal American Chemical Society*, **125**, 11360-11370. http://dx.doi.org/10.1021/ja0214882

[20] Dubowchik, G.M., Vrudhula, V.M., Dasgupta, B., Ditta, J., Chen, T., Sheriff, S., Sipman, K., Witmer, M., Tredup, J., Vyas, D.M., Verdoorn, T.A., Bollini, S. and Vinitsky, A. (2001) 2-Aryl-2,2-difluoroacetamide FKBP12 Ligands: Synthesis and X-Ray Structural Studies. *Organic Letters*, **3**, 3987-3990. http://dx.doi.org/10.1021/ol0166909

[21] Trnka, T.M. and Grubbs, R.H. (2001) The Development of L$_2$X$_2$Ru=CHR Olefin Metathesis Catalysts: An Organometallic Success Story. *Accounts of Chemical Research*, **34**, 18-29. http://dx.doi.org/10.1021/ar000114f

[22] Scholl, M., Ding, S., Lee, C.W. and Grubbs, R.H. (1999) Synthesis and Activity of a New Generation of Ruthenium-Based Olefin Metathesis Catalysts Coordinated with 1,3-Dimesityl-4,5-dihydroimidazol-2-ylidene Ligands. *Organic Letters*, **1**, 953-956. http://dx.doi.org/10.1021/ol990909q

[23] Garber, S.B., Kingsbury, J.S., Gray, B.L. and Hoveyda, A.H. (2000) Efficient and Recyclable Monomeric and Dendritic Ru-Based Metathesis Catalysts. *Journal American Chemical Society*, **122**, 8168-8179. http://dx.doi.org/10.1021/ja001179g

[24] Morrill, C. and Grubbs, R.H. (2003) Synthesis of Functionalized Vinyl Boronates via Ruthenium-Catalyzed Olefin Cross-Metathesis and Subsequent Conversion to Vinyl Halides. *Journal of Organic Chemistry*, **68**, 6031-6034. http://dx.doi.org/10.1021/jo0345345

[25] Morrill, C., Funk, W. and Grubbs, R.H. (2004) Synthesis of Tri-Substituted Vinyl Boronates via Ruthenium-Catalyzed Olefin Cross-Metathesis. *Tetrahedron Letters*, **45**, 7733-7736. http://dx.doi.org/10.1016/j.tetlet.2004.08.069

[26] Funk, T., Efskind, J. and Grubbs, R.H. (2005) Chemoselective Construction of Substituted Conjugated Dienes Using an Olefin Cross-Metathesis Protocol. *Organic Letters*, **7**, 187-190. http://dx.doi.org/10.1021/ol047929z

[27] Fuwa, H., Saito, A. and Sasaki, M. (2010) A Concise Total Synthesis of (+)-Neopeltolide. *Angewandte Chemie International Edition*, **49**, 3041-3044. http://dx.doi.org/10.1002/anie.201000624

[28] Fuwa, H., Suzuki, T., Kubo, H., Yamori, T. and Sasaki, M. (2011) Total Synthesis and Biological Assessment of (–)-Exiguolide and Analogues. *Chemistry—A European Journal*, **17**, 2678-2688. http://dx.doi.org/10.1002/chem.201003135

[29] Coquerel, Y. and Rodriguez, J. (2008) Microwave-Assisted Olefin Metathesis. *European Journal of Organic Chemi-

stry, **7**, 1125-1132. http://dx.doi.org/10.1002/ejoc.200700696

[30] Michaut, A., Boddaert, T., Coquerel, Y. and Rodriguez, J. (2007) Reluctant Cross-Metathesis Reactions: The Highly Beneficial Effect of Microwave Irradiation. *Synthesis*, **18**, 2867-2871. http://dx.doi.org/10.1055/s-2007-983825

[31] Fuwa, H., Noto, K. and Sasaki, M. (2010) Stereoselective Synthesis of Substituted Tetrahydropyrans via Domino Olefin Cross-Metathesis/Intramolecular Oxa-Conjugate Cyclization. *Organic Letters*, **12**, 1636-1639. http://dx.doi.org/10.1021/ol100431m

[32] Suzuki, M., Sumi, K., Koyama, H., Siqin, Hosoya, M., Takashima-Hirano, M. and Doi, H. (2009) Pd^0-Mediated Rapid Coupling between Methyl Iodide and Heteroarylstannanes: An Efficient and General Method for the Incorporation of a Positron-Emitting ^{11}C Radionuclide into Heteroaromatic Frameworks. *Chemistry A European Journal*, **15**, 12489-12495. http://dx.doi.org/10.1002/chem.200901145

19

Simple Reduction of Hydantoins with Sodium Borohydride

Jun-Ichi Yamaguchi*, Emiko Shibuta, Yoshie Oishi

Department of Applied Chemistry, Kanagawa Institute of Technology, Atsugi, Japan
Email: *yamagu@chem.kanagawa-it.ac.jp

Abstract

The reduction of various hydantoins with sodium borohydride gave the corresponding 4-hydroxy-2-imidazolidinones in high yields. In contrast, reduction employing a boron trifluoride etherate-sodium borohydride system generated 2-imidazolidinones. In both reductions, the reactivity of the hydantoin was dependent on its substituents. The Lewis acid-promoted reactions of a 4-hydroxy-2-imidazolidinone with nucleophiles were also investigated.

Keywords

Hydantoin, Reduction, Imidazolidinone

1. Introduction

The hydantoins, five-membered heterocycles containing two nitrogen atoms, have a structural resemblance to 2-imidazolidinones and 2-oxazolidinones (**Figure 1**), both of which are known to act as chiral auxiliaries [1]. However, few studies of these compounds have been reported, with the exception of their utilization as precursors for natural products such as (+)-biotin (vitamin H) [2] [3].

Hydantoin and its derivatives may be useful as important precursors for bioactive compounds, and thus additional information concerning the reactivity of the hydantoins is still required. Herein, we report the reduction of hydantoins (**1**) with sodium borohydride, both with and without boron trifluoride etherate, resulting in the formation of 4-hydroxy-2-imidazolidinones (**2**) and 2-imidazolidinones (**3**), respectively, in high yields (**Scheme 1**). The first reported reduction of a hydantoin was the reaction of a 5-monosubstituted hydantoin with lithium aluminum hydride (LAH) in diethyl ether under reflux or using Red-Al®, which generated **4** rather than **2** except when employing a 5,5-disubstituted hydantoin [4] [5], in which case 4-hydroxy-2-imidazolidinone was obtained since dehydration could not proceed [6] [7]. Reduction of 5-monosubstituted hydantoins with LAH or diisobu-

---

*Corresponding author.

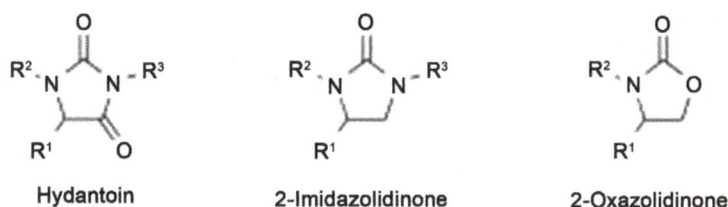

Figure 1. Five-membered heterocycles containing nitrogen atoms.

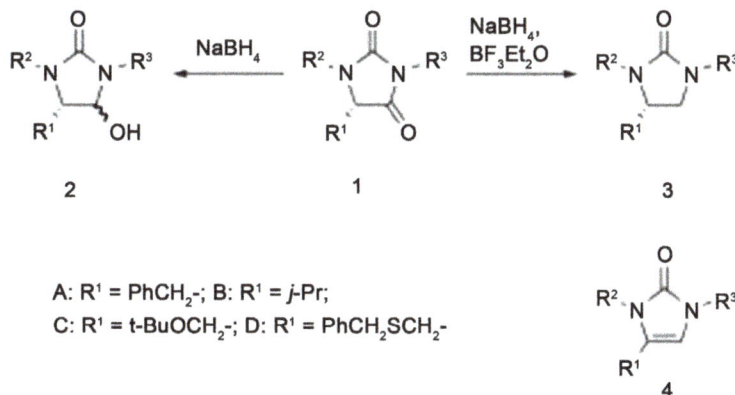

Scheme 1. Reactions applied in the present work.

tylaluminum hydride (DIBAL) below room temperature also resulted in the formation of **4** rather than **2** [8]-[10]. Only one example of the reduction of a 1-alkylhydantoin has been described, in which a compound derived from cysteine was reduced using sodium borohydride by Chavan and co-workers in the synthesis of biotin [3].

2. Results and Discussion

Initial trials involved the treatment of **1** with excess amounts of sodium borohydride in methanol at room temperature, with the results shown in **Table 1**. The reactivity of **1** and the yield of **2** were evidently dependent on the substituents of **1**. In cases in which **1** was derived from phenylalanine, the reduction of **1** bearing a phenyl group at the 3-position proceeded, generating **2** in high yield (Entry 1). However, when the substituent at the 3-position was changed to a phenethyl group, the reduction rate was very slow and only a low yield of 2-imidazolone (**4**), formed by dehydration of **2**, was obtained (Entry 2). We suspected that the lack of a substituent at the 1-position of **1** (that is, R^2 = H) decreased its reactivity, and so the t-butoxycarbonyl (Boc) derivative of **1** was prepared by t-butoxycarbonylation with Boc$_2$O. During reduction of the Boc derivative, the dehydration of **2** was suppressed and thus the yield of **2** was dramatically improved (Entry 3). The reduction of other hydantoins derived from various amino acid amides gave similar results. Based on the above results, the Boc group at the 1-position of the hydantoin ring was effective in promoting the present reduction (Entries 4-9).

Following the initial trials, the Lewis acid-promoted reactions of **2** with an allylsilane or an enol silyl ether to give the coupling products **5** were assessed, with the results presented in **Table 2**. The reaction of **2** with the allylsilane at 0°C resulted in the formation of **5-H** as a single isomer via the removal of the Boc group, with some **4** generated as a by-product (Entry 1). When the reaction was performed at a lower temperature (−78°C), the yield of **5** was increased and formation of **4** was suppressed (Entry 2). In contrast, when using the enol silyl ether as the nucleophile, the reaction gave better results at 0°C than at −78°C and the resulting products, **5-Boc** and **5-H**, were produced as single isomers (Entries 3 and 4). Chavan and co-workers have reported a similar reaction system using 4-hydroxy-2-imidazolidinones, in which the products exhibit exclusively *trans* stereochemistry. Although experimental and spectral data regarding the stereochemistries of **5-Boc** and **5-H** were not obtained in this study, based on the similarity of the present reaction system to that reported by Chavan, as well as the structure of intermediate **6**, we believe that the stereochemistries of both compounds were likely *trans* in the case of the present reactions.

Table 1. Reduction of **1** with sodium borohydride to **2**.

Entry	R^1	R^2	R^3	Time/h	Yield/%	
					2	**4**
1	A	H	Ph	1	77	
2	A	H	Ph(CH$_2$)$_2$	2.5	0	14
3	A	Boc[a]	Ph(CH$_2$)$_2$	0.5	97	
4	B	H	Ph(CH$_2$)$_2$	0.5		NI[b]
5	B	Boc	Ph(CH$_2$)$_2$	0.5	79	
6	C	H	Ph(CH$_2$)$_2$	0.5		NI
7	C	Boc	Ph(CH$_2$)$_2$	0.5	94	
8	D	H	Ph	0.5	35	
9	D	Boc	Ph	1	87	

[a]t-Butoxycarbonyl. [b]Neither **2** nor **4** were isolated.

Table 2. Lewis acid-promoted reaction of **2** with nucleophiles.

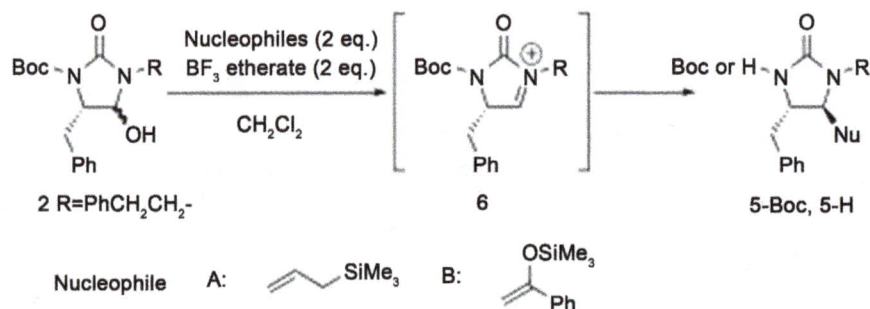

Entry	Nu	Conditions	Yield/%		
			5-Boc	**5-H**	**4**
1	A	0°C-RT, ON		58	38
2	A	−78°C, ON	27	60	
3	B	0°C, 1.5 h		97	
4	B	−78°C, 2 h	24	9	15

ON = overnight.

Although the chiral compounds 2-imidazolidinone (**3**) and 2-oxazolidinone are both well known as chiral auxiliaries [11]-[15], few methods for the preparation of **3** have been reported [16]-[18]. We expected that **1** would be converted to **3** when using a relatively strong reducing agent. Among the many possible reduction methods, a candidate for the conversion to **3** was a sodium borohydride-boron trifluoride etherate system, typically employed to transform amino acids to amino alcohols [19]. When using the hydantoins derived from phenylalanine, the reduction proceeded at room temperature to generate **3** in high yields (**Table 3**, Entries 1-3). Conversely, the reduction of **1** (R^1 = A, R^2 = H, R^3 = Ph) using 4.0 equimolar amounts of a commercial borane-tetrahydrofuran complex gave the corresponding version of **3** in 43% yield, meaning that the present reduction system represents a useful means of preparing many different 2-imidazolidinones. With regard to the present reduction, it was determined that the reactivity of **1** bearing an isopropyl group at the 5-position was affected by the substituent at the 1-position. The conversion of **1** bearing a Boc group to **3** proceeded successfully and generated high yields, with removal of the Boc group (Entry 6). We believe that the carbamic acid-like species **7** or **8**, which is more reactive than **1** (R^2 = H), might be generated in the reaction mixture, and that **3** is ob-

Table 3. Reduction of **1** to **3** in the sodium borohydride-boron trifluoride etherate system.

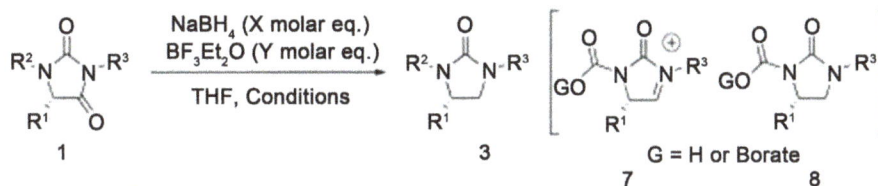

Entry	R^1	R^2	R^3	X	Y	Conditions	Yield
				/eq.	/eq.	/%	
1	A	H	Ph	4	2	RT, 3 h	96
2	A	H	$Ph(CH_2)_2$	4	2	RT, 4.5 h	83
3	A	H	Bu^t	4	2	RT, 22 h	81
4	B	H	Ph	4	2	reflux, 4 h	62
5	B	H	$Ph(CH_2)_2$	4	2	reflux, 3 h	29
6	B	Boc	$Ph(CH_2)_2$	4	4	reflux, 5 h	71[a]
7	C	H	Ph	4	2	reflux, 2 h	61[b]
8	C	H	$Ph(CH_2)_2$	4	2	reflux, 2 h	62[b]
9	D	H	Ph	4	2	reflux, 3 h	78

[a]The Boc group was removed. [b]No removal of the *t*-butyl group occurred.

tained as the final product after a formation of iminium salt 7 and decarboxylation of **8**. The reductions of other hydantoins without a Boc group proceeded, forming the corresponding **3** compounds (Entries 7-9). It was also observed that the *t*-butyl ether bond was stable under these reaction conditions (Entries 7 and 8).

3. Conclusion

In conclusion, the reduction of hydantoins with sodium borohydride in the absence or presence of boron trifluoride etherate gave 4-hydroxy-2-imidazolidinones or 2-imidazolidinones, respectively. Additional novel methods for achieving the conversions of compounds **2** and **3** are now being studied.

4. Spectral Data

4-benzyl-5-hydroxy-1-phenyl-1,3-imidazolidine-2-one: the mixture of diastereomers, Entry 1 in **Table 1**.
   ^{1}H NMR (300 MHz, $CDCl_3$) δ = 2.62 (1H, s), 2.75 - 2.90 (2H, m), 3.35 - 385 (1H, m), 4.00 - 4.10 (1H, m), 6.95 - 7.60 (10H, m).
   4-benzyl-5-hydroxy-1-phenethyl-1,3-imidazolidine-2-one: the mixture of diastereomers, Entry 3 in **Table 1**.
   ^{1}H NMR (300 MHz, $CDCl_3$) δ = 1.54 (9H, s), 2.75 - 2.85 (2H, m), 3.15 - 3.65 (4H, m), 4.05 - 4.20 (2H, m), 4.75 - 4.85 (1H, m), 7.10 - 7.40 (5H, m).
   5-hydroxy-4-isopropyl-1-phenethyl-1,3-imidazolidine-2-one: the mixture of diastereomers, Entry 5 in **Table 1**.
   ^{1}H NMR (300 MHz, $CDCl_3$) δ = 0.65, 0.95, 1.00, and1.02 (6H, 4d, J = 7.1 Hz), 1.50 and 1.51 (9H, 2s), 2.15 and 2.42 (1H, 2 octet, J = 7.1 Hz), 2.85 - 3.00 (2H, m), 3.40 - 4.70 (3H, m), 4.73 and 5.15 (1H, d and t, J = 7.1 Hz), 7.15 - 7.30 (5H, m).
   3-*tert*-butoxycarbonyl-4-*tert*-butoxymethyl-5-hydroxy-1-phenethyl-1,3-imidazolidine-2-one: the mixture of diastereomers, Entry 7 in **Table 1**.
   ^{1}H NMR (300 MHz, $CDCl_3$) δ = 1.16 (9H, s), 1.53 (9H, s), 2.75 - 3.00 (2H, m), 3.45 - 3.55 (2H, m), 3.70 - 3.80 (1H, m), 4.05 - 4.15 (2H, m), 4.60 - 4.66 (1H, m), 5.00 - 5.10 (1H, m), 7.15 - 7.35 (5H, m).
   4-benzylthiomethyl-5-hydroxy-1-phenyl-1,3-imidazolidine-2-one: the mixture of diastereomers, Entry 8 in **Table 1**.
   ^{1}H NMR (300 MHz, $CDCl_3$) δ = 2.75 - 3.10 (2H, m), 3.60-3.80 and 4.20 - 4.40 (2H, m), 4.70 - 4.80 (2H, m) 7.05 - 7.60 (10H, m).
   3-*tert*-butoxycarbonyl-4-benzylthiomethyl-5-hydroxy-1-phenethyl-1,3-imidazolidine-2-one: the mixture of

diastereomers, Entry 9 in **Table 1**.

^{1}H NMR (300 MHz, CDCl$_3$) δ = 1.50 and 1.51 (9H, 2s), 2.50 - 2.60 (1H, m), 2.90 - 3.00 (1H, m), 3.80 and 3.85 (2H, 2s), 4.05 - 4.10 (1H, m), 5.16 - 5.25 (1H, m), 7.15 - 7.70 (10H, m).

4-benzyl-5-phenacyl-1-phenethyl-1,3-imidazolidine-2-one: **5-H**, Entries 1 and 2 in **Table 2**.

^{1}H NMR (300 MHz, CDCl$_3$) δ = 2.50 - 2.55 (1H, m), 2.75 - 3.15 (6H, m), 3.15 - 3.30 (1H, m), 3.40 - 3.45 (1H, m), 3.75 - 3.80 (1H, m), 3.95 - 4.00 (1H, m), 4.44 (1H, brs), 6.95 - 7.35 (10H, m), 7.48 (2H, t, J = 6.6 Hz), 7.58 (1H, t, J = 6.6 Hz), 7.88 (2H, d, J = 6.6 Hz).

4-benzyl-3-*tert*-butoxycarbonyl-5-phenacyl-1-phenethyl-1,3-imidazolidine-2-one: **5-Boc**, Entry 2 in **Table 2**.

^{1}H NMR (300 MHz, CDCl$_3$) δ = 1.56 (9H, s), 2.45 - 2.55 (2H, m), 2.80 - 3.10 (5H, m), 3.65-3.90 (2H, m), 4.00 - 4.05 (1H, m), 7.05 - 7.35 (1H, m), 7.45 (2H, t, J = 7.8 Hz), 7.56 (1H, t, J = 7.8 Hz), 7.78 (2H, d, J = 7.8 Hz).

5-allyl-4-benzyl-1-phenethyl-1,3-imidazolidine-2-one: **5-H**, Entries 3 and 4 in **Table 2**.

^{1}H NMR (300 MHz, CDCl$_3$) δ = 2.22 (2H, t, J = 5.9 Hz), 2.53 (1H, dd, J = 13.4 and 8.3 Hz), 2.63 (1H, dd, J = 13.4 and 5.6 Hz), 3.05 - 3.30 (2H, m), 3.48 (1H, dt, J = 7.2 and 5.9 Hz), 3.79 (1H, ddd, J = 8.3, 7.2, and 5.6 Hz), 4.53 (1H, brs), 5.00 - 5.10 (2H, m), 5.50 - 5.65 (1H, m), 7.05 - 7.40 (10H, m).

5-allyl-4-benzyl-3-*tert*-butoxycarbonyl-1-phenethyl-1,3-imidazolidine-2-one: 5-Boc, Entry 4 in **Table 2**.

^{1}H NMR (300 MHz, CDCl$_3$) δ = 1.58 (9H, s), 2.00 - 2.10 (2H, m), 2.30 - 2.45 (1H, m), 2.70 - 2.90 (2H, m), 3.00 - 3.15 (3H, m), 3.84 - 4.00 (2H, m), 4.90 - 5.00 (2H, m), 5.15 - 5.30 (1H, m), 6.96 (d, J = 6.6 Hz), 7.15 - 7.40 (8H, m).

4-benzyl-1-phenyl-1,3-imidazolidine-2-one: Entry 1 in **Table 3**.

^{1}H NMR (300 MHz, CDCl$_3$) δ = 2.80 - 2.95 (2H, m), 3.60 - 3.70 (1H, m), 3.95 - 4.05 (2H, m), 4.95 (1H, s), 7.05 (1H, t, J = 7.6 Hz), 7.20 - 7.40 (7H, m), 7.53 (2H, d, J = 7.6 Hz).

4-benzyl-1-phenethyl-1,3-imidazolidine-2-one: Entry 2 in **Table 3**.

^{1}H NMR (300 MHz, CDCl$_3$) δ = 2.65 - 2.80 (2H, m), 2.80 (2H, t, J = 7.6 Hz), 3.02 (1H, dd, J = 8.6 and 5.6 Hz), 3.25 - 3.50 (3H, m), 3.79 (1H, quint, J = 6.8 Hz), 4.96 (1H, s), 7.05 - 7.35 (10H, m).

4-benzyl-1-*tert*-butyl-1,3-imidazolidine-2-one: Entry 3 in **Table 3**.

^{1}H NMR (300 MHz, CDCl$_3$) δ = 1.34 (9H, s), 2.75 - 2.85 (2H, m), 3.16 (1H, dd, J = 8.6 and 6.1 Hz), 3.50 (1H, t, J = 8.3 Hz), 3.70 - 3.80 (1H, m), 4.46 (1H, s), 7.15 - 7.35 (5H, m).

4-isopropyl-1-phenyl-1,3-imidazolidine-2-one: Entry 4 in **Table 3**.

^{1}H NMR (300 MHz, CDCl$_3$) δ = 0.96 (3H, d, J = 6.9 Hz), 0.99 (3H, d, J= 6.9 Hz), 1.74 (1H, Octet, J = 6.9 Hz), 3.45 - 3.60 (2H, m), 3.85 - 4.00 (1H, m), 5.89 (1H, s), 7.03 (1H, t, J = 7.3 Hz), 7.33 (2H, t, J = 7.3 Hz), 7.55 (2H, d, J = 7.3 Hz).

4-isopropyl-1-phenethyl-1,3-imidazolidine-2-one: Entries 5 and 6 in **Table 3**.

^{1}H NMR (300 MHz, CDCl$_3$) δ = 0.75 (3H, d, J = 6.8 Hz), 0.86 (3H, d, J = 6.1 Hz), 1.95 - 2.05 (1H, m), 2.82 (2H, t, J = 7.7 Hz), 2.91 (1H, t, J = 8.2 Hz), 3.13 (1H, t, J = 9.2 Hz), 3.25 - 3.50 (3H, m), 7.15 - 7.35 (6H, m).

4-*tert*-butoxymethyl-1-phenyl-1,3-imidazolidine-2-one: Entry 7 in **Table 3**.

^{1}H NMR (300 MHz, CDCl$_3$) δ = 1.20 (9H, s), 3.39 (2H, d, J = 8.8 Hz), 3.58 (1H, dd, J = 4.9 and 9.0 Hz), 3.85 - 3.95 (1H, m), 4.00 (1H, t, J = 9.0 Hz), 5.34 (1H, brs), 7.05 (1H, t, J = 9.0 Hz), 7.36 (2H, t, J = 9.0 Hz), 7.55 (2H, d, J = 9.0 Hz).

4-*tert*-butoxymethyl-1-phenethyl-1,3-imidazolidine-2-one: Entry 8 in **Table 3**.

^{1}H NMR (300 MHz, CDCl$_3$) δ = 1.15 (9H, s), 2.84 (2H, t, J = 9.0 Hz), 2.96 (1H, dd, J = 9.0 and 5.4 Hz), 3.18 (2H, d, J = 10.8 Hz), 3.35-3.55 (3H, m), 3.60-3.75 (1H, m), 4.76 (1H, brs), 7.15-7.35 (5H, m).

4-benzylthiomethyl-1-phenyl-1,3-imidazolidine-2-one: Entry 9 in **Table 3**.

^{1}H NMR (300 MHz, CDCl$_3$) δ = 2.61 (2H, d, J = 7.6 Hz), 3.55 (1H, dd, J = 10.1 and 9.1 Hz), 3.76 (2H, s), 3.70 - 3.80 (1H, m), 3.95 (1H, t, J = 9.0 Hz), 5.29 (1H, brs), 7.03 (1H, t, J = 8.2 Hz), 7.10 - 7.35 (7H, m), 7.57 (2H, d, J = 8.2 Hz).

References

[1] Yamaguchi, J., Harada, M., Narushima, T., Saitoh, A., Nozaki, K., and Suyama, T. (2005) Diastereoselective Conjugate Addition of 1-(a,b-Unsaturated acyl)hydantoin with Nucleophiles. *Tetrahedron Letters*, **46**, 6411-6415. http://dx.doi.org/10.1016/j.tetlet.2005.07.116

[2] Chavan, S.P., Tejwani, R.B. and Ravindranathan, T. (2001) A Switch of Reactivity Profile in Ionic Intramolecular Annulation Reactions: A Short and Efficient Synthesis of D-(+)-Biotin. *Journal of the Organic Chemistry*, **66**, 6197-

6201. http://dx.doi.org/10.1021/jo015730j

[3]   Chavan, S.P., Chittiboyina, A.G., Ravindranathan, T., Kamat, S.K. and Kalkote, U.R. (2005) Diastereoselective Amidoalky-
      lation of (3S,7aR)-6-Benzyl-7-hydroxy-3-phenyltetrahydro-5H-imiazo[1,5-c][1,3]thiazol-5-one: A Short and Highly
      Efficient Synthesis of (+)-Biotin. *Journal of the Organic Chemistry*, **70**, 1901-1903.
      http://dx.doi.org/10.1021/jo0488107

[4]   Wilk, I.J. and Close, W.J. (1950) The Action of Lithium Aluminum Hydride on 3-Methyl-5-phenylhydantoin and 5-
      Phenylhydantoin. *Journal of the Organic Chemistry*, **15**, 1020-1022. http://dx.doi.org/10.1021/jo01151a017

[5]   Li, C.-D., Lee, M.H. and Sartorelli, A.C. (1979) Synthesis and Biological Evaluation of Tetramisole Analogues as In-
      hibitors of Alkaline Phosphatase of the 6-Thiopurine-Resistant Tumor Sarcoma 180/TG[1]. *Journal of Medicinal Chemi-
      stry*, **22**, 1030-1033. http://dx.doi.org/10.1021/jm00195a003

[6]   Marshall, F.J. (1956) Lithium Aluminum Hydride Reduction of Some Hydantoins, Barbiturates and Thiouracils. *The
      Journal of American Chemical Society*, **78**, 3696-3697. http://dx.doi.org/10.1021/ja01596a038

[7]   Cortes, S. and Kohn, H. (1983) Selective Reductions of 3-Substituted Hydantoins to 4-Hydroxy-2-imidazolidinones
      and Vicinal Diamines. *Journal of the Organic Chemistry*, **48**, 2246-2254. http://dx.doi.org/10.1021/jo00161a021

[8]   Cortes, S. and Kohn, H. (1983) Selective Reductions of 3-Substituted Hydantoins to 4-Hydroxy-2-imidazolidinones
      and Vicinal Diamines. *Journal of the Organic Chemistry*, **48**, 2246-2254. http://dx.doi.org/10.1021/jo00161a021

[9]   Liao, Z.-K. and Kohn, H. (1984) Synthesis of Substituted 2-Imidazolidinones and Annelated Hydantoin via Amidoal-
      kylation Transformations. *Journal of the Organic Chemistry*, **49**, 4745-4752. http://dx.doi.org/10.1021/jo00199a001

[10] Moolenaar, M.J., Speckamp, W.N., Hiemstra, H., Poetsch, E. and Casutt, M. (1995) Synthesis of D-(+)-Biotin through
      Selective Ring Closure of N-Acyliminium Silyl Enol Ether. *Angewandte Chemie International Edition*, **34**, 2391-2393.
      http://dx.doi.org/10.1002/anie.199523911

[11] Drewes, S.E., Malissar, D.G.S. and Roos, G.H.P. (1993) Ephedrine-Derived Imidazolidin-2-Ones. Broad Utility Chiral
      Auxiliaries in Asymmetric Synthesis. *Chemische Berichte*, **126**, 2663-2673.
      http://dx.doi.org/10.1002/cber.19931261216

[12] Taguchi, T., Shibuya, A., Sasaki, H., Endo, J., Morikawa, T. and Shiro, M. (1994) Asymmetric Synthesis of Difluoro-
      cyclopropanes. *Tetrahedron: Asymmetry*, **5**, 1423-1426. http://dx.doi.org/10.1016/0957-4166(94)80101-0

[13] Bongini, A., Cardillo, G., Gentilucci, L. and Tomasini, C. (1997) Synthesis of Enantiomerically Pure Aziridine-2-Imides
      by Cyclization of Chiral 3'-Benzyloxyamino Imide Enolates. *Journal of Organic Chemistry*, **62**, 9148-9153.
      http://dx.doi.org/10.1021/jo971254e

[14] Guillena, G. and Najera, C. (1998) PTC and Organic Bases-LiCl Assisted Alkylation of Imidazolidinone-Glycine Im-
      inic Derivatives for the Asymmetric Synthesis of α-Amino Acids. *Tetrahedron: Asymmetry*, **9**, 3935-3938.
      http://dx.doi.org/10.1016/S0957-4166(98)00402-9

[15] Guillena, G. and Najera, C. (2000) 1,5-Dimethyl-4-Phenylimidazolidin-2-One-Derived Iminic Glycinimides: Useful
      New Reagents for Practical Asymmetric Synthesis of α-Amino Acids. *Journal of Organic Chemistry*, **65**, 7310-7322.
      http://dx.doi.org/10.1021/jo000321t

[16] Kim, T.H. and Lee, G.-J. (1999) Regiocontrolled Cyclization Reaction of N-(2-Hydroxy)Ureas by Transfer of Activa-
      tion: One-Pot Synthesis of 2-Imidazolidinones. *Journal of Organic Chemistry*, **64**, 2941-2943.
      http://dx.doi.org/10.1021/jo9820061

[17] Kim, T.H. and Lee, G.-J. (2000) L-Valinol and L-Phenylalaninol-Derived 2-Imidazolidinones as Chiral Auxiliaries in
      Asymmetric Aldol Reactions. *Tetrahedron Letters*, **41**, 1505-1508. http://dx.doi.org/10.1016/S0040-4039(99)02325-4

[18] Nadir, U.K., Krishna, R.V. and Singh, A. (2005) A New and Facile Route for the Synthesis of Chiral 1,2-Diamines and
      2,3-Diamino Acids. *Tetrahedron Letters*, **46**, 479-482. http://dx.doi.org/10.1016/j.tetlet.2004.11.088

[19] Tschantz, M.A., Burgess, L.E. and Meyers, A.I. (1996) (S)-(-)-5-Heptyl-2-pyrrolidineone. Chiral Bicyclic Lactams as
      Templates for Pyrrolidines and Pyrrolidinones. *Organic Synthesis*, **73**, 221-230.

20

Optimization of Extraction Conditions of Some Phenolic Compounds from White Horehound (*Marrubium vulgare* L.) Leaves

Karim Bouterfas[1*], Zoheir Mehdadi[1], Djamel Benmansour[2], Meghit Boumedien Khaled[3], Mohamed Bouterfas[4], Ali Latreche[1]

[1]Laboratory of Vegetal Biodiversity: Conservation and Valorization, Faculty of Life and Natural Sciences, Djillali Liabes University, Sidi Bel-Abbes, Algeria
[2]Laboratory of Statistics and Random Model, Faculty of Natural and Universe Sciences, Abou-Bekr Belkaid University, Tlemcen, Algeria
[3]Department of Biology, Faculty of Natural and Life Sciences, Université Djillali Liabes, Sidi Bel Abbes, Algeria
[4]Laboratory of Microscopy, Microanalysis of the Matter and Molecular Spectroscopy, Faculty of Exact Sciences, Djillali Liabes University, Sidi Bel-Abbes, Algeria
Email: [*]bouterfas_karim@yahoo.fr

Abstract

This research was aimed to optimize the extraction conditions of three phenolic compounds: total phenolics, flavonoids and condensed tannins, from White Horehound's leaves (*Marrubium vulgare* L.). Distilled water and different organic solvents such as: methanol, ethanol and acetone, were used, with various concentrations (20% - 80%, v/v), temperatures (20°C - 60°C) and extraction times (30 - 450 min). Results showed that the maximum total phenolics amounts (293.34 ± 14.60 mg gallic acid equivalent/g dry weigh), were obtained with 60% aqueous methanol at 25°C for 180 min; total flavonoids (79.52 ± 0.55 mg catechin equivalent/g dry weigh) with 80% aqueous methanol at 20°C for 450 min, and condensed tannins (28.15 ± 0.80 mg catechin equivalent/g dry weigh) with 60% aqueous acetone at 50°C and for 180 min. ANOVA test showed the significant effect ([***]$P < 0.001$) of the extraction conditions tested on phenolic compounds. The Principal Component Analysis (PCA) exhibited the positive effect of low temperatures on total phenolics and flavonoids extraction, and the effect of high temperatures on the condensed tannins extraction. The Response Surface Methodology (RSM) provided predicted values of extraction conditions and maximum polyphenols amounts similar to those obtained experimentally.

---

[*]Corresponding author.

Keywords

Marrubium vulgare, Phenolic Compounds, Optimization, Extraction Conditions

1. Introduction

Polyphenols constitute one of the most common and widespread groups of substances in flowering plants, occurring in all vegetative organs, as well as in flowers and fruits. These molecules are involved in many physiological processes such as cell growth, root formation, seed germination and fruit ripening [1]. Moreover, these compounds are considered secondary metabolites involved in the chemical defense of plants against predators, pathogens, environmental stresses and in plant-plant interferences [2].

Nowadays, phenolic compounds represent a unique and a functional place, composed of bioactive products, present in plant-derived foods and beverages and included in the formulations of well-marketed cosmetic and parapharmaceutical products [3]. Furthermore, polyphenols exhibit various biological activities such as anticancer [4], antioxidant [5], antimicrobial [6] and anti-inflammatory activities [7]. Therefore, in recent years, the determination of phenolic compounds concentrations in fruits [8] [9], vegetables [10] and some aromatic and medicinal plants [11] [12] has been of increasing interest in the scientific community as well as among health professionals and business partners.

It is well known that the content of phenolic compounds could be influenced by environmental conditions, such as season [13] [14], sampling period and geographic origin [15], precipitations and temperatures [16], and soil type [17]. Additionally, there are several experimental factors that can influence the rate of extraction and the quality of extracted bioactive phenolic compounds. These factors include extraction method, solvent type used and concentration [18], particle size of medicinal plants, temperature and pH of extraction [19], extraction time [20], number of extractions repetition [1] and solvent-to-sample ratio [21].

In order to recover bioactive compounds from plant raw materials, extraction is widely used and it constitutes the first important step [22]. Different solvents and techniques are used for the extraction of polyphenols from plants [23] [24]. However, there is no one standard extraction method used to extract phenolic compounds from plant materials because of their complexity and their interaction with other bioactive compounds [25]. Furthermore, each plant material has its unique properties, in term of phenolic extraction; different plants may require different extraction conditions to achieve the optimum recovery of phenolic compounds [26].

The family Lamiaceae includes aromatic and medicinal plants, which are used in traditional medicine, although Lamiaceae species are well known for their volatile oil content, their therapeutic activities and other properties. This reflects the existence of other chemical components, such as the polyphenols. *Marrubium vulgare* L. commonly known in Europe as "White horehound", and in the Mediterranean region as "Marute" or as "Merriouet" in Algeria, is a perennial herb of Lamiaceae family, naturalized in North and South America, and Western Asia [27]. In Algeria, *M. vulgare* is used in folk medicine to treat several digestive diseases, diarrhea, as well as diabetes, rheumatism, acute or chronic bronchitis, cough, asthma and other respiratory infections [28]. Earlier, phytochemical investigation of *M. vulgare* have led to the characterization of a very complex metabolic pattern, containing, among other secondary metabolites, diterpenes [29], phenyl propanoids esters [30], tannins [31] and flavonoids [32]. Several activities, traditionally attributed to *M. vulgare*, were approved by intensive modern research and clinical trials, such as hypoglycemic [33], vasorelaxant and antihypertensive [29], analgesic [34], antidiabetic [35], anti-inflammatory [36], and antioxidant properties [37].

However, to the best of our knowledge, optimizing the extraction of phenolic compounds from *M. vulgare* leaves, using different extraction conditions and response surface methodology (RSM), has not been reported yet. Hence, the purpose of the current study was to investigate the effects of different extracting conditions (organic solvent type, concentration of organic solvent, temperature and time) on the extraction of phenolic compounds (total phenolic content, TPC; total flavonoid content, TFC; and condensed tannins content, CTC) from *M. vulgare* leaves.

2. Material and Methods

2.1. Plant Material

Leaves of *M. vulgare* were collected in April 2012, from Tessala Mountain (north-western Algeria, semi-arid climate) at the level of a station which latitudinal coordinates 35°16'33"N, and longitudinal 0°46'27"W, altitude 596 m. The identification of plant specimen was done by Professor Z. Mehdadi and put at the Laboratory of Vegetal Biodiversity: Conservation and Valorization (Faculty of Natural and Life Sciences, Djillali Liabes University of Sidi Bel-Abbes, Algeria).

Upon arrival at the laboratory, samples were thoroughly rinsed with distilled water, dried in the dark for three weeks at room temperature and crushed with the cutting mill. The powdered samples were packaged into a linear-low-density polyethylene (LLDPE) film and stored in dark at room temperature for further experiments.

2.2. Extract Preparation

Two grams of leaf powder *M. vulgare* were extracted using 20 ml of extraction solvent with different concentrations, introduced in conical flask of capacity 100 ml sealed with parafilm and wrapped with aluminum foil to prevent solvent loss and exposure to light. Therefore, the mixture was stirred at 150 rpm in bath water with controlling temperature at a constant speed (level 8) for a particular duration. After completing the extraction process, the *M. vulgare* extract was filtered using Whatman No. 1 filter paper into amber bottle for analysis without storage overnight, in order to obtain a clear crude extract solution.

To determine the optimal conditions for phenolic extraction from *M. vulgare* leaves, the extraction conditions were set according to the experimental design described below:

- *Organic solvents type*: three kinds of organic solvents (methanol, ethanol and acetone) were selected. Distilled water was tested as control. The selection of the three extraction solvents is made firstly after several (about one hundred) referenced protocols in terms of quantification of phenolic compounds and secondly from the viewpoint of the availability of the chemicals used;
- *Solvent concentrations*: four concentrations (20%, 40%, 60% and 80%; v/v) were prepared in distilled water;
- *Extraction temperatures*: six temperatures (20°C, 25°C, 30°C, 40°C, 50°C and 60°C) were used;
- *Extraction times*: six different times (30, 90, 180, 270, 360 and 450 min) were chosen.

2.3. Total Phenolic Contents (TPC)

TPC of *M. vulgare* extracts was determined using Folin-Ciocalteu reagent, according to method suggested by Li *et al.* [38] and slightly modified by Chew *et al.* [39]. Crude extracts were diluted 50 times with deionized water prior to analysis. 1 ml of diluted extract was mixed with 1 ml of diluted Folin-Ciocalteu reagent (10 times diluted with deionized water). After incubating the mixture at room temperature for 4 min, 0.8 ml of 7.5% (w/v), sodium carbonate anhydrous solution was added. The mixture was then vortexed for 10 s and incubated in the dark at room temperature during 2 h. The absorbance of the mixture was measured against blank at 765 nm using UVi light spectrophotometer. Gallic acid with different concentrations (0-100-200-300-400-500 mg/l) was used to calibrate the standard curve. The calibration equation for gallic acid was $Y = 0.0042X - 0.0178$ ($R^2 = 0.9992$). Each crude extract was analyzed in triplicate and the results were expressed in milligrams of gallic acid equivalents per gram of dry weight (mg GAE/g DW). Data were expressed as mean ± standard deviation.

2.4. Total Flavonoids Contents (TFC)

TFC was determined using procedures described by Tan *et al.* [21]. The crude extract was diluted 10 times. An amount of 1.25 ml deionised water followed by 75 µl of 5% sodium nitrite (NaNO2), was added to 0.25 ml of diluted crude extract in an aluminium foil-wrapped 15 ml test tube. The mixture was left standing for 6 min before adding 150 µl of 10% (w/v) aluminium chloride (AlCl3). The mixture was left standing for 5 min before adding 0.5 ml of 1 M sodium hydroxide (NaOH) and 275 µl of deionised water. The tip of the test tube was covered with parafilm and then mixed using vortex mixer for approximately 10 s. The absorbance of the mixture was determined at 510 nm versus the prepared blank using Uvi light spectrophotometer. Catechin with different concentration (0 - 100 - 200 - 300 - 400 - 500 mg/l) was used for calibration. The calibration equation for catechin was calculated as follow: $Y = 0.0035X - 0.0062$ ($R^2 = 0.9995$). Each crude extract was analyzed in triplicate and the results were expressed in milligrams of catechin equivalents per gram of dry weight (mg CE/g DW). Data were expressed as mean ± standard deviation.

2.5. Total Flavonoid Contents (TFC)

TFC was determined using procedures described by Tan *et al.* [21]. The crude extract was diluted 10 times. An amount of 1.25 ml deionised water followed by 75 μl of 5% sodium nitrite ($NaNO_2$), was added to 0.25 ml of diluted crude extract in an aluminium foil-wrapped 15 ml test tube. The mixture was left standing for 6 min before adding 150 μl of 10% (w/v) aluminium chloride (AlCl3). The mixture was left standing for 5 min before adding 0.5 ml of 1 M sodium hydroxide (NaOH) and 275 μl of deionised water. The tip of the test tube was covered with parafilm and then mixed using vortex mixer for approximately 10 s. The absorbance of the mixture was determined at 510 nm versus the prepared blank using Uvi light spectrophotometer. Catechin with different concentration (0 - 100 - 200 - 300 - 400 - 500 mg/l) was used for calibration. The calibration equation for catechin was calculated as follow: $Y = 0.0035X - 0.0062$ ($R^2 = 0.9995$). Each crude extract was analyzed in triplicate and the results were expressed in milligrams of catechin equivalents per gram of dry weight (mg CE/g DW). Data were expressed as mean ± standard deviation.

2.6. Condensed Tannins Contents (TFC)

CTC assay was performed according to the method described by Chew *et al.* [39]. 0.5 ml undiluted crude extract was firstly mixed with 3 ml of vanillin reagent (4%, w/v, in absolute methanol), followed by addition of 1.5 ml of concentrated HCl (37%). After that, the mixture was stored in the dark at room temperature for 15 min. The absorbance of mixture was measured at 500 nm against blank using Uv light spectrometer. Each undiluted crude extract was measured in triplicate. Catechin with different concentrations (0 - 100 - 200 - 300 - 400 - 500 mg/l) was used for calibration of standard curve. The calibration equation for catechin was calculated using the formula $Y = 0.0021X - 0.0143$ ($R^2 = 0.997$). Each crude extract was analyzed in triplicate and the results were expressed in milligrams of catechin equivalents per gram of dry weight (mg CE/g DW). Data were expressed as mean ± standard deviation.

2.7. Statistical Analysis

In order to control the influence of extraction conditions (solvent type, solvent concentration, extraction time, and extraction temperature) on the mean concentrations of each phenolic compound, ANOVA, with more classification criteria, using Fisher's least significant difference test and the significant differences at the 5% level, were calculated. The difference was considered as not significant when $P > 0.05$, significant when $^*P \leq 0.05$, and highly significant for $^{**}P \leq 0.01$ and extremely significant for $^{***}P \leq 0.001$. These analyzes were performed using Minitab 16. Tukey's test was also performed for pair-wise comparisons at the 5% level.

To determine possible correlation among polyphenols concentrations and extraction conditions, PCA was used. This statistical tool is dedicated for data exploration, which allows the reduction of the number of quantitative variables to a small number of components. Moreover, using PCA, interrelationships between different variables could be seen, and detected and sample patterns, groupings, similarities or differences could be interpreted [40]. A matrix is prepared using XLSTAT 2012 software, by taking polyphenols concentrations as observations and extraction conditions as variables.

Optimal conditions for the extraction of phenolic compounds from *M. vulgare* leaves were obtained using response surface methodology (RSM) [41] [42]. This method is adopted for each type of phenolic compound with the solvent giving the highest experimental concentration. The independent variables studies were solvent concentration (X_1), extraction temperature (X_2) and extraction time (X_3); while the dependent variable (Y: response variable) measured was the contents of phenolic compound. To obtain the best combination of independent variables giving a maximum percentage of adjusted R^2, a best subsets regression is performed using Minitab 16 software. Thus, the regression equation of the model is obtained by an analysis of general regression. Finally, the optimum polyphenols contents and extraction conditions are obtained by introducing the regression equation in Maple 6 software. The experimental and predicted values of polyphenols contents and extraction conditions were compared, in order to determine the validity of the model.

3. Results and Discussion

Results of total phenolic, total flavonoids, and condensed tannins amounts are shown in **Figure 1** and **Figure 2**. The influence of extraction's conditions used on phenolic compounds amounts obtained using ANOVA, with more classification criteria, is presented in **Table 1**.

Table 1. Analysis of variance (ANOVA) with more classification criteria of quantified phenolic compounds.

Source	Degree of freedom	Sum of Squares	Mean squares	F-value	P-value
{1}Solvent type	2	40326.1^{TPC}	20163.1^{TPC}	262.15^{TPC}	<0.0001***
		2326.12^{TFC}	1163.06^{TFC}	274.61^{TFC}	<0.0001***
		43.14^{CTC}	21.57^{CTC}	UD^{CTC}	<0.0001***
{2}Solvent concentration	4	107832.8^{TPC}	26958.2^{TPC}	350.49^{TPC}	<0.0001***
		5132.83^{TFC}	1283.21^{TFC}	302.98^{TFC}	<0.0001***
		257.24^{CTC}	64.31^{CTC}	UD^{CTC}	<0.0001***
{3}Extraction time	5	45983.2^{TPC}	9196.6^{TPC}	119.57^{TPC}	<0.0001***
		111.09^{TFC}	22.22^{TFC}	5.25^{TFC}	<0.0001***
		592.16^{CTC}	118.43^{CTC}	UD^{CTC}	<0.0001***
{4}Extraction temperature	5	553938.7^{TPC}	110787.7^{TPC}	1440.39^{TPC}	<0.0001***
		39821.11^{TFC}	7964.22^{TFC}	1880.44^{TFC}	<0.0001***
		10589.43^{CTC}	2117.89^{CTC}	UD^{CTC}	<0.0001***
{1}*{2}	8	31097.1^{TPC}	3887.1^{TPC}	50.54^{TPC}	<0.0001***
		1654.30^{TFC}	206.79^{TFC}	48.82^{TFC}	<0.0001***
		10.91^{CTC}	1.36^{CTC}	UD^{CTC}	<0.0001***
{1}*{3}	10	7922.1^{TPC}	792.2^{TPC}	10.30^{TPC}	<0.0001***
		54.86^{TFC}	5.49^{TFC}	1.30^{TFC}	=0.235^{NS}
		2.59^{CTC}	0.26^{CTC}	UD^{CTC}	<0.0001***
{1}*{4}	10	14801.5^{TPC}	1480.1^{TPC}	19.24^{TPC}	<0.0001***
		2422.48^{TFC}	242.25^{TFC}	57.20^{TFC}	<0.0001***
		0.00^{CTC}	0.00^{CTC}	UD^{CTC}	<0.0001***
{2}*{3}	20	35664.5^{TPC}	1783.2^{TPC}	23.18^{TPC}	<0.0001***
		599.22^{TFC}	29.96^{TFC}	7.07TFC	<0.0001***
		100.97^{CTC}	5.05^{CTC}	UD^{CTC}	<0.0001***
{2}*{4}	20	48446.9^{TPC}	2422.3^{TPC}	31.49^{TPC}	<0.0001***
		7603.54^{TFC}	380.18^{TFC}	89.76^{TFC}	<0.0001***
		233.26^{CTC}	$11 .66^{CTC}$	UD^{CTC}	<0.0001***
{3}*{4}	25	18241.9^{TPC}	729.7^{TPC}	9.49^{TPC}	<0.0001***
		767.42^{TFC}	30.70^{TFC}	7.25^{TFC}	<0.0001***
		92.06^{CTC}	3.68^{CTC}	UD^{CTC}	<0.0001***
{1}*{2}*{3}	40	8775.9^{TPC}	219.4^{TPC}	2.58^{TPC}	<0.0001***
		314.37^{TFC}	7.86^{TFC}	1.86^{TFC}	=0.003**
		1.26^{CTC}	0.03^{CTC}	UD^{CTC}	<0.0001***
{1}*{2}*{4}	40	18002.2^{TPC}	450.1^{TPC}	5.85^{TPC}	<0.0001***
		3061.91^{TFC}	76.55^{TFC}	18.07^{TFC}	<0.0001***
		0.00^{CTC}	0.00^{CTC}	0.00^{CTC}	<0.0001***
{1}*{3}*{4}	50	5440.4^{TPC}	108.8^{TPC}	1.41^{TPC}	=0.050*
		272.27^{TFC}	5.45^{TFC}	1.29^{TFC}	=0.116^{NS}
		0.00^{CTC}	0.00^{CTC}	UD^{CTC}	<0.0001***
{2}*{3}*{4}	100	46098.9^{TPC}	461.0^{TPC}	5.99^{TPC}	<0.0001***
		1176.72^{TFC}	11.77^{TFC}	2.78^{TFC}	<0.0001***
		176.67^{CTC}	1.77^{CTC}	UD^{CTC}	<0.0001***
{1}*{2}*{3}*{4}	200	15383.0	76.09	UD^{TPC}	<0.0001***
		847.06	4.24	UD^{TFC}	<0.0001***
		0.00	0.00	UD^{CTC}	<0.0001***

UD: undefined (the denominator of Fisher's test is null or undefined).

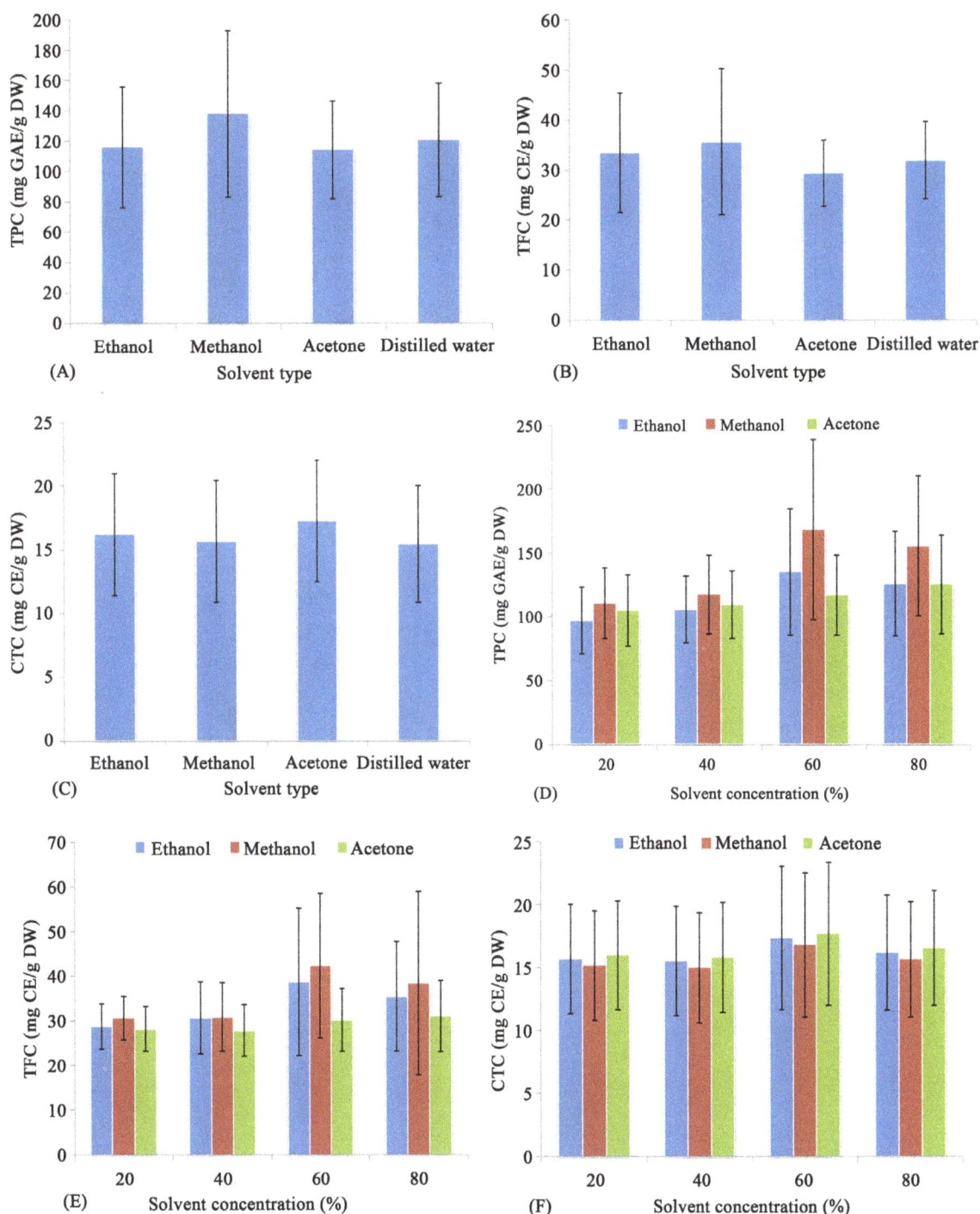

Figure 1. Effects of solvent type ((A), (B), (C)) and solvent concentration ((D), (E), (F)) on TPC, TFC and CTC extracted from *M. vulgare* leaves.

3.1. Effect of Solvent Type on Extraction of Phenolic Compounds

The choice of extraction solvents is important for complex food samples. They allow determining the amount and the type of phenolic compounds to extract. Organic solvents, particularly acetone, ethanol and methanol are the most commonly used in polyphenols extraction from botanical materials [43].

Our results showed that methanol (137.79 ± 54.62 mg GAE/g) was significantly high effective ([***]$P < 0.001$)

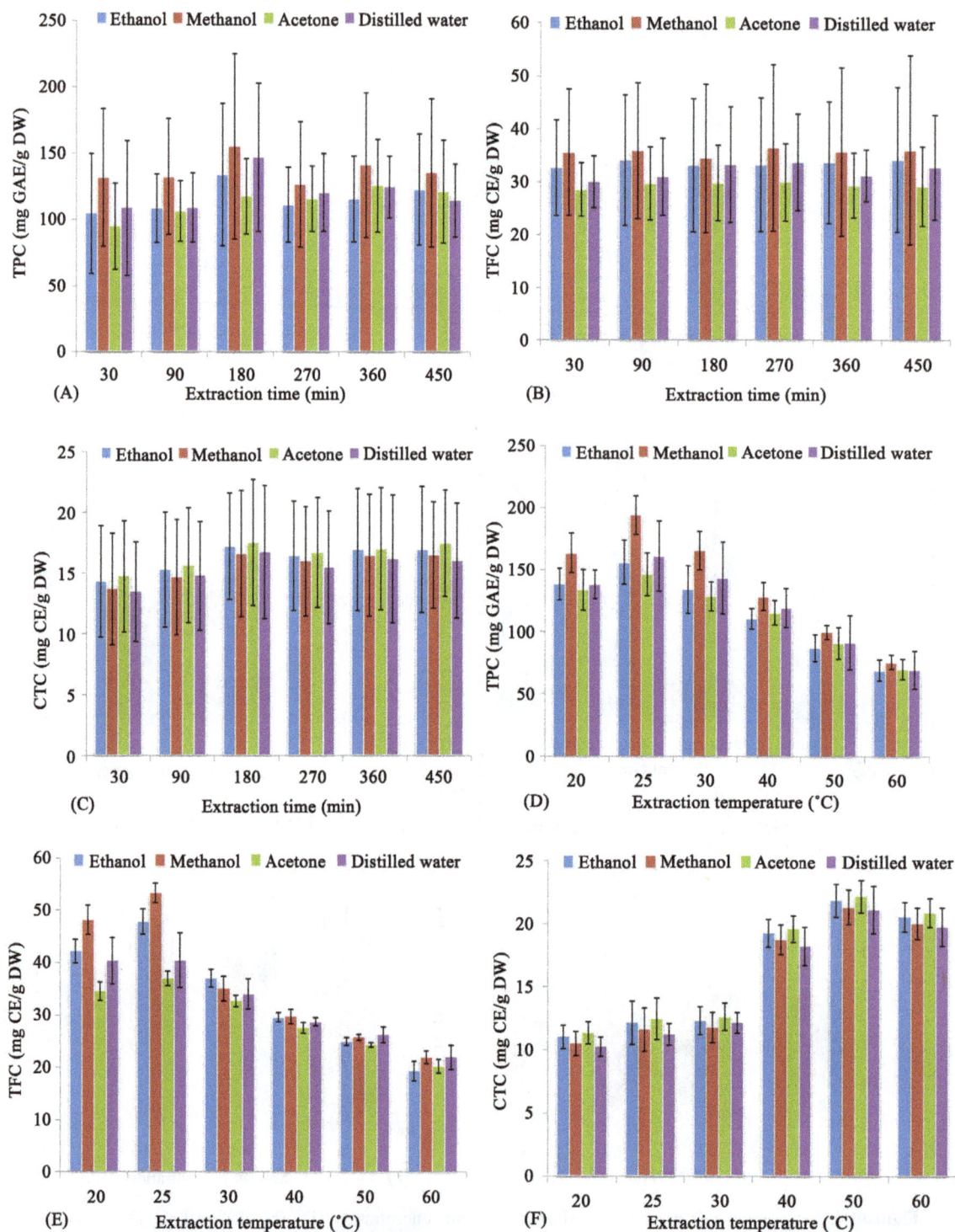

Figure 2. Effects of extraction time ((A), (B), (C)) and extraction temperature ((D), (E), (F)) on TPC, TFC and CTC extracted from *M. vulgare* leaves.

when compared with all other solvent systems used in extracting TPC from *M. vulgare* leaves, as shown in **Figure 1(A)**. Methanol has been considered as an ideal solvent for the TPC extraction from green walnut fruits [44], *Zingiber officinalis* leaves [45], *Tamarix aphylia* leaves [46] and *Artemisia annua* leaves [47]. Several studies confirmed that TPC depends on organic solvent polarity and methanol is one of the most suitable solvents for the extraction of TPC from plants [48] [49]. Mohammedi and Atik [46] reported that the high polarity of methanol, estimated at 6.6, compared to other organic solvents, plays a key role in increasing phenolic solubility. However, Kong et al. [50] found that TPC in methanolic extracts, obtained from *Pouteria campechiana* fruit parts, was about 23 % to 45 % lower than the ethanolic extracts.

Methanol permitted to obtain the highest extraction levels of TFC (35.64 ± 14.61 mg CE/g) from *M. vulgare* leaves (**Figure 1(B)**). Furthermore, a significant difference (***$P < 0.001$) was found between the type of organic solvents used. Similar results were observed for *Moringa oleifera* and *Aloe barbadensis* leaves [51], *Euphorbia helioscopia* leaves, steams and flowers [52]. Our results agree with previous studies, which found that ethanol was less effective than methanol for extracting TFC [53] [54]. In contrast, Settharaksa et al. [55] have found that using water to extract solvent from some Thai medicinal plants, was better than using ethanolic and methanolic solvents to extract TFC. Ghasemzadeh et al. [45] reported that extraction of TFC from *Zingiber officinale* leaves, using methanol, was about 3 times higher when using acetone and 4 times higher when using hexane. Spigno et al. [56] suggested that high level of TPC obtained using methanol, could be explained by the fact that this solvent allows a good solubility of flavonoids' hydroxyl groups.

A maximum level of CTC (17.24 ± 4.77 mg CE/g) was reached using acetone, as shown in **Figure 1(C)**. There was a high significant difference between the four tested solvents (***$P < 0.001$). Our results agree with those of Trabelsi et al. [18] who found that maximum CTC level, from *Limoniastrum monopetalum* leaves, was obtained when acetone solvent was used. Similarly, in other studies performed by Wina et al. [57] and Mazandarani et al. [58], acetone was found to be the most effective solvent extracting CTC from *Acacia mangium* barks and *Onosma dichroanthum* roots, respectively. According to Antwi-Boasiako and Animapauh [59], the four solvents tested in the current study, especially acetone and methanol were the best extraction solvents of CTC from the barks of three tropical hardwoods. Additionally, Uma et al. [60] showed that acetone was the most effective solvent for extraction of condensed tannins as tannins have a relatively high molecular weight.

3.2. Effect of Solvent Concentration on Extraction of Phenolic Compounds

Mixtures of alcohols with different proportions of water have shown to be more effective in extracting phenolic compounds compared to mono-component solvent system [25]. Addition of small quantity of water to organic solvent usually leads to a more polar medium, which facilitates the polyphenols extraction [56].

We observed that maximum rate of TPC (168.41 ± 70.60 mg GAE/g), extracted from *M. vulgare* leaves, was obtained when aqueous methanol was used at 60% (**Figure 1(D)**). However, there were high significant differences (***$P < 0.001$) in TPC among the various concentrations. Our results are in line with those found by Chan et al. [61] and Chew et al. [39] who found that 60% aqueous methanol gave the best effectiveness extracting TPC from *Citrus hystrix* peels and *Centella asiatica* leaves, respectively. Yilmaz and Toledo [62] reported that aqueous mixtures of methanol, ethanol or acetone were better than a mono-component solvent for the extraction of TPC from Muscadine seeds.

The highest value of TFC (42.40 ± 16.22 mg CE/g), as shown in **Figure 1(E)**, was obtained, when we used 60% of aqueous methanol and was significantly different (***$P < 0.001$) compared to the other studied concentrations. In contrast, aqueous methanol at 80% was the best concentration for *Calendula officinalis* flowers [63] and aqueous ethanol at 80% from the leaves of *Limoniastrum monopetalum* [18], *Bauhinia monandra* [64] and *Callicarpa nudiflora* [41]. Musa et al. [65] confirmed that the mixture of an aqueous solvent (distilled water) with an organic solvent (methanol and ethanol) improves the flavonoids yield comparing to water or organic solvent used separately.

Sixty percent (60%) acetone in water gave the highest value of CTC, which was 17.68 ± 5.69 mg CE/g (**Figure 1(F)**). Moreover, there were high significant (***$P < 0.001$) differences in CTC values among the different concentrations used. Our data are in agreement with those reported by Downey and Hanlin [66] on grape skin. In other studies, it has been noticed that 80% aqueous acetone was the best concentration for CTC extraction from *Limonium densiflorum* shoots [67], 60% aqueous methanol from *Cichorium intybus* L. roots, leaves, stems and seeds [68], and 80% aqueous ethanol from the Chinese chestnut [69].

3.3. Effect of Extraction Time on Extraction of Phenolic Compounds

Extraction time represents another key parameter in optimizing the phenolic compounds extraction. In the literature, this parameter might be as short as few minutes or long up to 24 hours, depending on the phenolic compounds present in samples [70].

In this study, extraction time showed a significant effect ([***]$P < 0.001$) on the extraction of TPC from *M. vulgare* leaves. As shown in **Figure 2(A)**, the highest value of TPC (155.16 ± 69.83 GAE/g) was recorded using methanol for 180 min. Chan *et al.* [61] demonstrated that 180 min represented the best extraction time for TPC from the peels of *Citrus hystrix*. In other studies, the best extraction time of TPC was estimated to 18 hours for black tea, *Camellia sinensis* [49], 90 min from grape pomace extracts [71], 45 min from *Areca catechu* seeds [72] and *Azadirachta indica* leaves [70]. Increased time of extraction beyond 180 min (270, 360 and 450 min) induced a loss in TPC. Dent *et al.* [73] highly recommended that the extraction time, should not exceed 3 h for the extraction of TPC from *Salvia officinalis* leaves. Several authors stated that more the extraction time is long, less the content of polyphenols is obtained. This could be the result of loss of phenolic compounds, via oxidation, which might polymerize into insoluble compounds [26] [74]. Therefore, extraction time of TPC depends, not only, on maceration or agitation times, but also on several factors such as filtration time or the time spent during the evaporation of solvents [75].

The maximum of TFC concentration (39.95 ± 17.90 mg CE/g) was obtained using methanol for 450 min as illustrated in **Figure 2(B)**. Extraction time showed significant effect ([***]$P < 0.001$) in TFC. However, 3 hours was considered to be the best extraction time for TFC from *Callicarpa nudiflora* leaves [46] and from some thyme varieties [76]. Additionally, 30 min has been shown to be the most favorable extraction time for TFC from *Gynura medica* leaves [77].

180 min was the longest extraction time, using acetone, for CTC (**Figure 2(C)**) with a maximum rate of 17.51 ± 5.18 CE/g. A high significant difference ([***]$P < 0.001$) was obtained in CTC among different extraction times. Our results agree with those obtained by Zhekova and Pavlov [76] on some thyme varieties. In other works, the best extraction time was of 120 min for mangosteen fruits [78], 150 min for *Parkia clappertoniana* husks [79], 80 min for *Cichorium intybus* different organs [67], and 20 min for *Punica granatum* peels [80].

3.4. Effect of Extraction Temperature on Extraction of Phenolic Compounds

The effectiveness of extraction process of phenolic compounds is largely regulated by different experimental parameters particularly by the extraction temperature [81]. An increase of temperature is mainly due to an increase of the diffusion rate and solubility of the extracted substances. On the other hand, it should be taken into account, that some important biological active substances, such as TPC are damaged at high temperatures [82].

The extraction of TPC, as shown in **Figure 2(D)**, was optimal when methanol was used at 25°C. In this case, 194.16 ± 15.45 mg GAE/g DW were obtained with high significant difference ([***]$P < 0.001$) and between different temperatures. Similar to our finding, 25°C was the most optimal extraction temperature for TPC from *Lawsonia inermis* leaves [60] and *Azadirachta indica* leaves [70]. Extension of extraction temperature beyond 25°C led to an important decrease in TPC. However, it should be noticed that increasing the extraction temperature, beyond certain values, might promote possible concurrent decomposition of phenolic compounds, which were already mobilized at lower temperature. Furthermore, this elevation of temperature can even lead to the breakdown of phenolic compounds that remained in the plant matrix [83] [84]. Hence, heating may affect the polyphenolic composition in many cases; therefore, high-temperature drying should be avoided as much as possible [2]. Usually, TPC extraction is used at room temperature (≈25°C) to avoid the degradation of phenolic compounds [25]. However, in other experiments, the optimal extraction temperature for TPC was found more elevated and estimated to 40°C for *Citrus hystrix* peels [61], 100°C for *Areca catechu* seeds [72], 65°C for *Centella asiatica* leaves [39] and *Orthosiphon stamineus* stems and leaves [85], and 90°C for *Moringa oleifera* leaves [86].

A maximum of TFC rate (53.31 ± 1.83 CE/g) was obtained at 25°C using methanol as solvent (**Figure 2(E)**). In the current investigation we found that extraction temperature significantly ([***]$P < 0.001$) affected the TFC extraction. However, in other researches, the highest extraction temperature was fixed at 90 °C for *Callicarpa nudiflora* leaves [41] and 60°C from some thyme varieties [76]. Like TPC, the extension of extraction temperature higher than 25°C led to an important decrease in TFC. However, the temperature conditions during the extraction procedures of flavonoids have to be carefully adjusted because of the possibility of thermal degradation of flavonoid derivatives, especially hydroxyl groups [87] [88]. In addition, mild heating was also found to have the ability to soften the plant tissues, to weaken the cell wall integrity, and thus to favor the release of bound phenolic compounds [56] [89].

As shown in **Figure 2(F)**, the maximum of CTC rate (22.18 ± 1.27 CE/g) was reached when acetone was used as a solvent and when the extraction temperature was about 50°C. We observed a significant difference ([***]$P < 0.001$) in CTC between the extraction temperatures. Our results agree with those of Zam *et al.* [80] on *Punica granatum* peels. In other studies, the best extraction temperature was slightly elevated compared to those we obtained: 80°C for mangosteen fruits [78], 70°C *Parkia clappertoniana* husks [79] and 60°C for some thyme varieties [73]. Unlike the obtained results for TPC and TFC, the extension of extraction temperature beyond 25°C led to a significant increase in CTC. Al-Farsi and Lee [90] reported that elevated temperature could stimulate the CTC extraction by increasing both diffusion coefficient and solubility of condensed tannins in extraction solvent. Besides that, intense heat from solvent allowed the release of cell wall phenolics and bounded phenolics by breaking down cellular constituents [91] and consequently, increasing the phenolic yield in extract. Moreover, Juntachote *et al.* [89] reported that elevated extraction temperature would increase, on one hand, the mass transfer of condensed tannins, and on the other hand, it would reduce the solvent viscosity and surface tension and, promotes the extraction of phenolic compounds.

3.5. Principal Component Analysis (PCA)

PCA aimed to diminish the size of data collected into a reduced number of components to examine the possible grouping of phenolic compounds according to the different extraction conditions. The first factor, PC1 presented 85.55% of variance accounted for, whereas, the second one, PC2 presented 10.86%. With the two first PCs, the explained variance accumulated was of 96.41%. This great value means that nearly all the variance contained in the original data was explained by just using the first new coordinates. By layering variables projection circle on the observations scatter plot, two groups were obtained as shown in the PCA plot (**Figure 3**):

- Group 1 (Gr1) on the positive side of PC1, formed by TPC and TFC with 0.932 and 0.958 as contributions, respectively. These phenolic compounds are related with ethanolic, methanolic and aqueous extracts at low temperatures 20°C and 25°C; ET1, ET2, MT1, MT2, DWT1, DWT2 with respective contributions of 2.285; 3.040; 3.578; 4.596; 2.253 and 2.618;

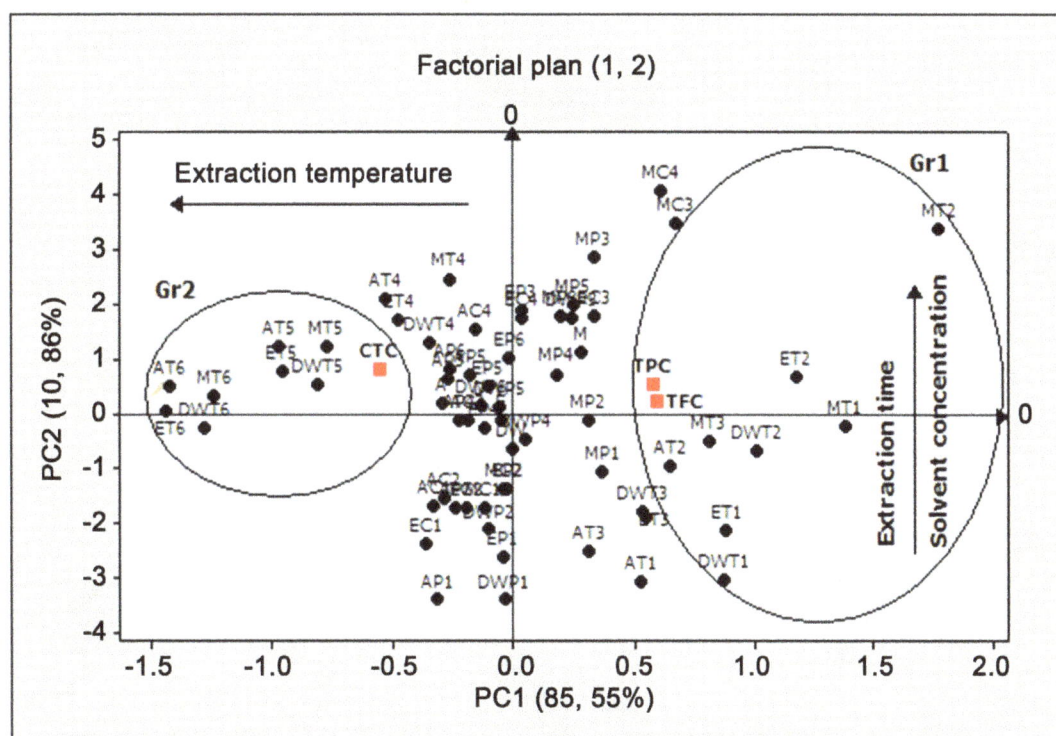

Figure 3. Principal component plot of PC1 and PC2. Gr: group; PC1: first principal component; PC2: second principal component; M: methanol; E: ethanol; DW: distilled water; A: acetone, C: concentration in % (1: 20; 2: 40; 3: 60; 4: 80); T: temperature in °C (1: 20; 2: 25; 3: 30; 4: 40; 5: 50; 6: 60); *P*: time in min (1: 30; 2: 90; 3: 180; 4: 270; 5: 360; 6: 450).

- Group 2 (Gr2) on the negative side of PC1, constituted by CTC with a contribution of −0.883. This phenolic compound was associated to ethanolic, acetonic and aqueous extracts at high temperatures 50°C and 60°C; ET5, ET6, AT5, AT6, DWT5, and DWT6 with contributions of 2.473; 3.705; −2.506; −3.661; −2.089 and −3.290, respectively.
- Moreover, considering the contributions of observations and variables on PC1, we defined two gradients:
- A horizontal gradient, moving from right to left of PC1, formed by the extraction temperature used, which explains the position of TPC and TFC right of PC1, correlated with low temperatures and the location of CTC to the left of PC1 in conjunction with high temperatures;
- A vertical gradient from the bottom to the top of PC1, constituted by the organic solvent concentration and the extraction time. This gradient explains the position of the three phenolic compounds measured at the top of PC1, correlated with high solvent concentrations (60% for TPC and CTC, 80% for TFC) and high times (180 min for TPC and CTC, 270 min for TFC).

3.6. Response Surface Methodology (RSM)

The Response Surface Methodology (RSM) was plotted to study the effects of extraction conditions on phenolic compounds extraction from *M. vulgare* leaves. 3 D response surface plots are illustrated in **Figure 4**. The regression equation of the model selected, for each phenolic compound measured and their adjusted R^2 was as follow:

- TPC = 323.205 − 827.094 X1 − 4.2516 X2 + 0.015029 X3 + 843.115 X12 + 17.1341 X1X2 − 13.5293 X2X12 − 0.0615286 X1X22, with adjusted R2 = 82.03%,
- TFC = 32.4239 + 31.9683 X1 + 0.42576 X2 + 0.0308808 X3 + 98.4907 X12 − 0.0108914 X22 − 2.96617 X1X2 − 0.00163763 X3X2 − 0.00122396 X1X3X2 + 0.0702683 X3X12 − 2.22016 X2X12 + 2.10557e-5 X3X22 + 0.0492972 X1X32, with adjusted R2 = 77.61%,

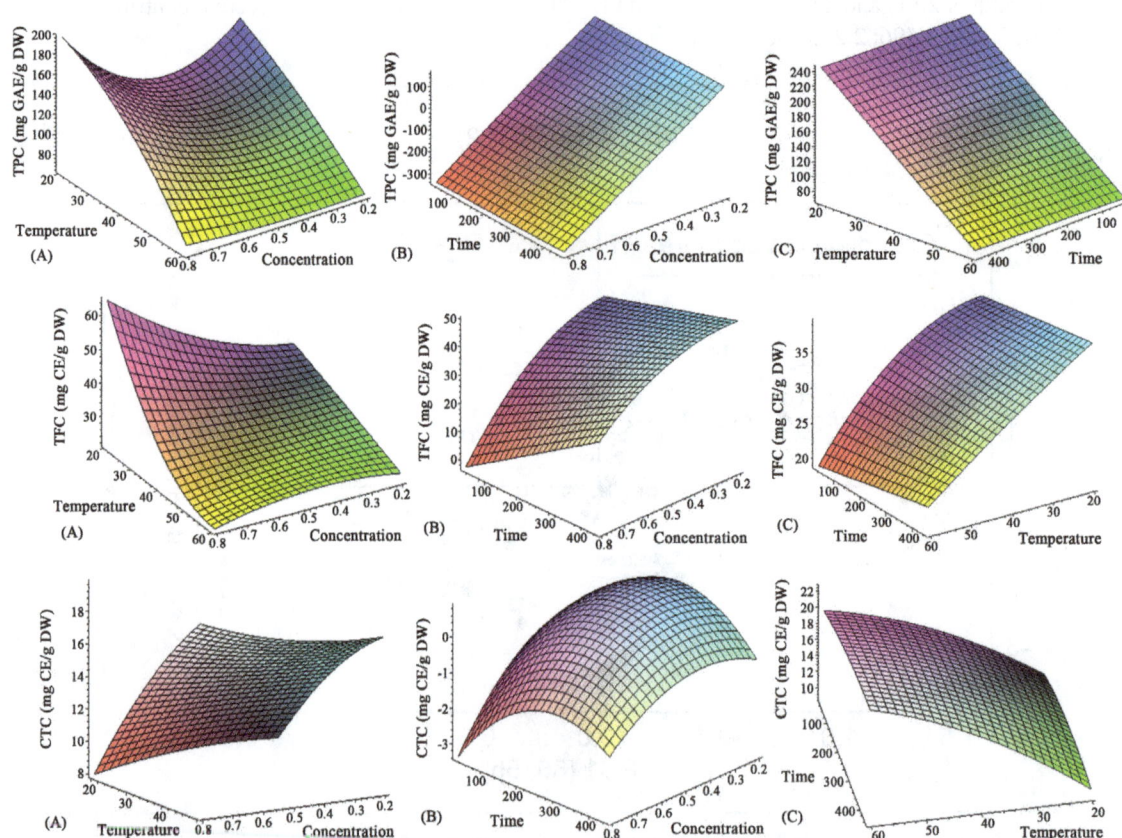

Figure 4. The 3D response surface plots representing the effects of (A) solvent concentration and extraction temperature; (B) solvent concentration and time extraction and (C) extraction temperature and extraction time on the extraction of phenolic compounds from *M. vulgare* leaves.

- CTC = $-2.19999 + 7.81089$ X1 $+ 0.603279$ X2$+ 0.01403$ X3 $- 12.395$ X12 $- 0.00405521$ X22 $- 2.96468e - 5$ X32 $- 0.220734$ X1X2 $+ 0.00680008$ X3X12 $+ 0.388604$ X2X12 $+ 2.36161e - 7$ X2X32, with adjusted R2 $= 90.62\%$.

The comparison between the maximum experimental values of phenolic compounds and extraction conditions, and the predicted ones (**Table 2**), showed that the two sets of values were close. Indeed, the maximum value of TFC (experimental: 79.52 mg CE/g; predicted: 79.46 mg CE/g) and CTC (experimental: 28.15 mg CE/g, predicted: 20.81 mg CE/g) were obtained by the same experimental and predicted extraction conditions (TFC: aqueous methanol 80% at 20°C and for 450 min; CTC: 60% at 50°C and for 180 min). Concerning TPC, the maximum values of experimental and predicted contents and extraction conditions were different. The experimental values were of 293.34 mg GAE/g, obtained with aqueous methanol 60% at 25°C and for 180 min. However, the predicted ones were of 204.75 mg GAE/g, obtained with aqueous methanol 80% at 20°C during 167 min. Both, accounting experimental and predicted results indicated that the experimental model was valid. This implied that there was a high fit degree between the values observed in experiment and those predicted from the regression model. Hence, the response surface modeling could be applied effectively to predict extraction of phenolic compounds from *M. vulgare* leaves.

4. Conclusions

The results obtained in our study indicate that the experimental conditions tested (organic solvent type, solvent concentration, extraction time and extraction temperature) influence notably the values of TPC, TFC and CTC extracted from *M. vulgare* leaves, with extremely significant differences ($^{***}P < 0.001$).

We were able to define the optimum extraction conditions too, for obtaining higher values of some phenolic compounds from *M. vulgare* leaves. Indeed, the optimum level of TPC (293 ± 14.60 mg GAE/g) was obtained using 60% aqueous methanol at 25°C for 180 min, TFC (79.52 ± 0.55 mg CE/g) using 80% aqueous methanol at 20°C for 450 min, and CTC (28.15 ± 0.80 mg CE/g) with 80% aqueous acetone at 50°C for 450 min. These levels highlight the richness of *M. vulgare* in these secondary metabolites. We conclude that this species remains poorly studied compared to other species belonging to the same taxonomic family (Lamiaceae) as thyme and sage.

Our results provide some confirmations on the effect of the variability of extraction procedures and assay on the amount of phenolic compounds recorded and reinforce previous works in this context.

This work is far from exhaustive as other experimental conditions can be tested, as the solid-to-solvent ratio, the number of extractions and the nature of the extraction method. Also, other extraction solvents (ethylene glycol, acetic acid and ethyl acetate), fractional extraction for many times or multistage extraction can be taken into consideration in our future works.

The impact of environmental conditions as the geographical region, altitude, exposure and the harvest season on the levels of polyphenols in *M. vulgare* is a promising perspective to this study and is still under investigation in our laboratory.

Acknowledgements

This research is a part of the Algerian National Project Research (entitled "Flora of Tessala Mountains: inventory, valorization and conservation", code 1/u22/397), financially supported by the Algerian Ministry of High

Table 2. Optimum values of experimental and predicted polyphenols contents and extraction conditions.

	Polyphenols concentrations		Solvent type		Solvent concentration (X_1)		Extraction temperature (X_2)		Extraction time (X_3)	
	EV	PV	EV	PV	EV	PV	EV	PV	EV	PV
TPC	293.34	204.75	Methanol		60	80	25	20	180	167
TFC	79.52	79.46	Methanol		80	80	20	20	450	450
CTC	28.15	20.81	Acetone		60	60	50	50	180	180

EV: experimental values; PV: predicted values.

Education and Scientific Research. Also, the authors are grateful to Dr. Amel Latifi, Bacterial Chemistry Laboratory (Marseille, France), for valuable editorial assistance during the drafting of this manuscript.

References

[1]   Khoddami, A., Wilkes, M.A. and Roberts, T.H. (2013) Techniques for Analysis of Plant Phenolic Compounds. *Molecules*, **18**, 2328-2375. http://dx.doi.org/10.3390/molecules18022328

[2]   Tsao, R. (2010) Chemistry and Biochemistry of Dietary Polyphenols. *Nutrients*, **2**, 1231-1246. http://dx.doi.org/10.3390/nu2121231

[3]   Ferrazzano, G.F., Amato, I., Ingenito, A., Zarrelli, A., Pinto, G. and Pollio, A. (2011) Plant Polyphenols and Their Anti-Cariogenic Properties: A Review. *Molecules*, **16**, 1486-1507. http://dx.doi.org/10.3390/molecules16021486

[4]   Berghe, W.V. (2012) Epigenetic Impact of Dietary Polyphenols in Cancer Chemoprevention: Lifelong Remodeling of Our Epigenomes. *Pharmacological Research*, **65**, 565-576. http://dx.doi.org/10.1016/j.phrs.2012.03.007

[5]   Kazeem, M.I., Akanji, M.A., Hafizur, R.M. and Choudhary, M.I. (2012) Antiglycation, Antioxidant and Toxicological Potential of Polyphenol Extracts of Alligator Pepper, Ginger and Nutmeg from Nigeria. *Asian Pacific Journal of Tropical Biomedicine*, **2**, 727-732. http://dx.doi.org/10.1016/S2221-1691(12)60218-4

[6]   Konaté, K., Hilou, A., Mavoungou, J.F., Lepengué, A.N., Souza, A., Barro, N., Datté, J.Y., M'Batchi, B. and Nacoulma, O.G. (2012) Antimicrobial Activity of Polyphenol-Rich Fractions from *Sida alba* L. (Malvaceae) against Cotrimoxazol-Resistant Bacteria Strains. *Annals of Clinical Microbiology and Antimicrobials*, **11**, 1-6. http://dx.doi.org/10.1186/1476-0711-11-5

[7]   Lolayekar, N. and Shanbhag, C. (2012) Polyphenols and Oral Health. *RSBO*, **9**, 74-84.

[8]   Dragovic-Uzelac, V., Levaj, B., Mrkic, V., Bursac, D. and Boras, M. (2007) The Content of Polyphenols and Carotenoids in Three Apricot Cultivars Depending on Stage of Maturity and Geographical Region. *Food Chemistry*, **102**, 966-975. http://dx.doi.org/10.1016/j.foodchem.2006.04.001

[9]   Ignat, I., Volf, I. and Popa, V.I. (2011) A Critical Review of Methods for Characterisation of Polyphenolic Compounds in Fruits and Vegetables. *Food Chemistry*, **126**, 1821-1835. http://dx.doi.org/10.1016/j.foodchem.2010.12.026

[10]   D'Archivio, M., Filesi, C., Di Benedetto, R., Gargiulo, R., Giovannini, C. and Masella, R. (2007) Polyphenols, Dietary Sources and Bioavailability. *Annali dell'Istituto Superiore di Sanità*, **43**, 348-361.

[11]   N'Guessan, A.H.O., Déliko, C.E.D., Mamyrbékova-Békro, J.A. and Békro, Y.A. (2011) Teneurs en composés phénoliques de 10 plantes médicinales employées dans la tradithérapie de l'hypertension artérielle, une pathologie émergente en Côte d'Ivoire. *Revue de Génie Industriel*, **6**, 55-61.

[12]   Okuda, T. and Ito, H. (2011) Tannins of Constant Structure in Medicinal and Food Plants—Hydrolyzable Tannins and Polyphenols Related to Tannins. *Molecules*, **16**, 2191-2217. http://dx.doi.org/10.3390/molecules16032191

[13]   Cosmulescu, S. and Trandafir, I. (2011) Seasonal Variation of Total Phenols in Leaves of Walnut (*Juglans regia* L.). *Journal of Medicinal Plants Research*, **5**, 4938-4942.

[14]   Generalic, I., Skroza, D., Surjaka, J., Mozinab, S.S., Ljubenkovc, I., Katalinic, A., Simate, V. and Katalinic, V. (2012) Seasonal Variations of Phenolic Compounds and Biological Properties in Sage (*Salvia officinalis* L.). *Chemistry & Biodiversity*, **9**, 441-456. http://dx.doi.org/10.1002/cbdv.201100219

[15]   Raal, A., Orav, A., Pussa, T., Valner, C., Malmiste, B. and Arak, E. (2012) Content of Essential Oil, Terpenoids and Polyphenols in Commercial Chamomile (*Chamomilla recutita* L. Rauschert) Teas from Different Countries. *Food Chemistry*, **131**, 632-638. http://dx.doi.org/10.1016/j.foodchem.2011.09.042

[16]   Ghasemi, K., Ghasemi, Y., Ehteshamnia, A., Nabavi, S.M., Nabavi, S.F., Ebrahimzadeh, M.A. and Pourmorad, F. (2011) Influence of Environmental Factors on Antioxidant Activity, Phenol and Flavonoids Contents of Walnut (*Juglans regia* L.) Green husks. *Journal of Medicinal Plants Research*, **5**, 1128-1133.

[17]   Bruni, R. and Sacchetti, G. (2009) Factors Affecting Polyphenol Biosynthesis in Wild and Field Grown St. John's Wort (*Hypericum perforatum* L. Hypericaceae/Guttiferae). *Molecules*, **14**, 682-725. http://dx.doi.org/10.3390/molecules14020682

[18]   Trabelsi, N., Megdiche, W., Ksouri, R., Falleh, H., Oueslati, S., Soumaya, B., Hajlaoui, H. and Abdelly, C. (2010) Solvent Effects on Phenolic Contents and Biological Activities of the Halophyte *Limoniastrum monopetalum* Leaves. *LWT-Food Science and Technology*, **43**, 632-639.

[19]   Gironi, F. and Piemonte, V. (2011) Temperature and Solvent Effects on Polyphenol Extraction Process from Chestnut Tree Wood. *Chemical Engineering Research and Design*, **89**, 857-862. http://dx.doi.org/10.1016/j.cherd.2010.11.003

[20]   Falleh, H., Ksouri, R., Lucchessi, M.E., Abdelly, C. and Magné, C. (2012) Ultrasound-Assisted Extraction: Effect of Extraction Time and Solvent Power on the Levels of Polyphenols and Antioxidant Activity of *Mesembryanthemum edule* L. Aizoaceae Shoots. *Tropical Journal of Pharmaceutical Research*, **11**, 243-249.

http://dx.doi.org/10.4314/tjpr.v11i2.10

[21] Tan, P.W., Tan, C.P. and Ho, C.W. (2011) Antioxidant Properties: Effects of Solid-to-Solvent Ratio on Antioxidant Compounds and Capacities of Pegaga (*Centella asiatica*). *International Food Research Journal*, **18**, 557-562.

[22] Santana, C.M., Ferrera, Z.S., Padrón, M.E.T. and Rodríguez, J.J.S. (2009) Methodologies for the Extraction of Phenolic Compounds from Environmental Samples: New Approaches. *Molecules*, **14**, 298-320. http://dx.doi.org/10.3390/molecules14010298

[23] Stalikas, C.D. (2010) Phenolic Acids and Flavonoids: Occurrence and Analytical Methods. *Methods in Molecular Biology*, **610**, 65-90. http://dx.doi.org/10.1007/978-1-60327-029-8_5

[24] Jace, D.E. and Shahidul, I. (2012) Effects of Extraction Procedures, Genotypes and Screening Methods to Measure the Antioxidant Potential and Phenolic Content of Orange-Fleshed Sweetpotatoes (*Ipomoea batatas* L.). *American Journal of Food Technology*, **7**, 50-61. http://dx.doi.org/10.3923/ajft.2012.50.61

[25] Garcia-Salas, P., Morales-Soto, A., Segura-Carretero, A. and Fernández-Gutiérrez, A. (2010) Phenolic-Compound-Extraction Systems for Fruit and Vegetable Samples. *Molecules*, **15**, 8813-8826. http://dx.doi.org/10.3390/molecules15128813

[26] Chirinos, R., Rogez, H., Campos, D., Pedreschi, R. and Larondelle, Y. (2007) Optimisation of Extraction Conditions of Antioxidant Phenolic Compounds from Mashua (*Tropaeolum tuberosum* Ruíz & Pavón) Tubers. *Journal of Separation and Purification Technology*, **55**, 217-225. http://dx.doi.org/10.1016/j.seppur.2006.12.005

[27] Kanyonga, P.M., Faouzi, M.A., Meddah, B., Mpona, M., Essassi, E.M. and Cherrah, Y. (2011) Assessment of Methanolic Extract of *Marrubium vulgare* for Antiinflammatory, Analgesic and Anti-Microbiologic Activities. *Journal of Chemical and Pharmaceutical Research*, **3**, 199-204.

[28] Belhattab, R. and Larous, L. (2006) Essential oil Composition and Glandular Trichomes of *Marrubium vulgare* L. Growing Wild in Algeria. *Journal of Essential Oil Research*, **18**, 369-373. http://dx.doi.org/10.1080/10412905.2006.9699116

[29] El Bardai, S., Morel, N., Wibo, M., Fabre, N., Liabres, G., Lyoussi, B. and Quetin-Leclercq, J. (2003) The Vasorelaxant Activity of Marrubenol and Marrubiin from *Marrubium vulgare*. *Planta Medica*, **69**, 75-77. http://dx.doi.org/10.1055/s-2003-37042

[30] Sahpaz, S., Garbacki, N., Tits, M. and Bailleul, F. (2012) Isolation and Pharmacological Activity of Phenylpropanoid Esters from *Marrubium vulgare*. *Journal of Ethnopharmacology*, **79**, 389-392. http://dx.doi.org/10.1016/S0378-8741(01)00415-9

[31] Kurbatova, N.V., Muzychkina R.A., Mukhitdinov, N.M. and Parshina, G.N. (2013) Comparative Phytochemical Investigation of the Composition and Content of Biologically Active Substances in *Marrubium vulgare* and *Marrubium alternidens*. *Chemistry of Natural Compounds*, **39**, 501-502. http://dx.doi.org/10.1023/B:CONC.0000011128.64886.f4

[32] Nawwar Mahmoud, A.M., El-Mousallamy, A.M.D., Barakat, H.H., Buddrus, J. and Linscheid, M. (1989) Flavonoid Lactates from Leaves of *Marrubium vulgare*. *Phytochemistry*, **28**, 3201-3206. http://dx.doi.org/10.1016/0031-9422(89)80307-3

[33] Vergara-Galicia, J., Aguirre-Crespo, F., Tun-Suarez, A., Crespo, A.A., Estrada-Carrillo, M., Jaimes-Huerta, I., Flores-Flores, A., Estrada-Soto, S. and Ortiz-Andrade, R. (2012) Acute Hypoglycemic Effect of Ethanolic Extracts from *Marrubium vulgare*. *Phytopharmacology*, **3**, 54-60.

[34] Meyre-Silva, C., Yunes, R.A., Schlemper, V., Campos-Buzzi, F. and Cechinel-Filho, V. (2005) Analgesic Potential of Marrubiin Derivatives, a Bioactive Diterpene Present in *Marrubium vulgare* L. (Lamiaceae). *Il Farmaco*, **60**, 312-326. http://dx.doi.org/10.1016/j.farmac.2005.01.003

[35] Boudjelal, A., Henchiri, C., Siracusa, L., Sari, M. and Ruberto, G. (2012) Compositional Analysis and *in Vivo* Anti-Diabetic Activity of Wild Algerian *Marrubium vulgare* L. Infusion. *Fitoteparia*, **2**, 286-292.

[36] De Jesus, R.A. and Cechinel Filho, V. (2000) Analysis of the Antinociceptive Properties of Marrubiin Isolated from *Marrubium vulgare* L. *Phytomedicine*, **7**, 111-115. http://dx.doi.org/10.1016/S0944-7113(00)80082-3

[37] Pukalskas, A., Rimantas Venskutonis, P., Salido, S., De Waard, P. and Van Beek., T.A. (2012) Isolation, Identification and Activity of Natural Antioxidants from Horehound (*Marrubium vulgare* L.) Cultivated in Lithuania. *Food Chemistry*, **130**, 695-701. http://dx.doi.org/10.1016/j.foodchem.2011.07.112

[38] Li, H., Wong, C., Cheng, K. and Chen, F. (2008) Antioxidant Properties *in Vitro* and Total Phenolic Contents in Methanol Extracts from Medicinal Plants. *Lebensmittel-Wissenschaft und -Technologie*, **41**, 385-390.

[39] Chew, K.K., Ng, S.Y., Thoo, Y.Y., Khoo, M.Z., Wan Aida, W.M. and Ho, C.W. (2011) Effect of Ethanol Concentration, Extraction Time and Extraction Temperature on the Recovery of Phenolic Compounds and Antioxidant Capacity of *Centella asiatica* Extracts. *International Food Research Journal*, **18**, 571-578.

[40] Ceto, X., Gutiérrez, J.M., Gutiérrez, M., Céspedes, F., Capdevila, J., Manguez, S., Jiménez-Jorquera, C. and Del Valle,

M. (2012) Determination of Total Polyphenol Index in Wines Employing a Voltammetric Electronic Tongue. *Analytica Chimica Acta*, **732**, 172-179. http://dx.doi.org/10.1016/j.aca.2012.02.026

[41] Liao, L., Yin, X. and Wang, Z. (2012) Optimization of Total Flavonoid Extraction in *Callicarpa nudiflora* Hook. et Arn. Using Response Surface Methodology. *Journal of Medicinal Plants Research*, **6**, 5038-5047.

[42] Zheng, N., Wang, Z., Shi, Y. and Lin, J. (2012) Evaluation of the Antifungal Activity of Total Flavonoids Extract from *Patrinia Villosa* Juss and Optimization by Response Surface Methodology. *African Journal of Microbiology Research*, **6**, 586-593. http://dx.doi.org/10.5897/AJMR11.1393

[43] Naczk, M. and Shahidi, F. (2004) Extraction and Analysis of Phenolics in Food. *Journal of Chromatography*, **1054**, 95-111. http://dx.doi.org/10.1016/j.chroma.2004.08.059

[44] Jakopic, J., Veberic, R. and Stampar, F. (2009) Extraction of Phenolic Compounds from Green Walnut Fruits in Different Solvents. *Acta Agriculturae Slovenica*, **93**, 11-15. http://dx.doi.org/10.2478/v10014-009-0002-4

[45] Ghasemzadeh, A., Jaafar, H.Z.E. and Rahmat, A. (2011) Effects of Solvent Type on Phenolics and Flavonoids Content and Antioxidant Activities in Two Varieties of Young Ginger (*Zingiber officinale* Roscoe) Extracts. *Journal of Medicinal Plants Research*, **5**, 1147-1154.

[46] Mohammedi, Z. and Atik, F. (2011) Impact of Solvent Extraction Type on Total Polyphenols Content and Biological Activity from *Tamarix aphylla* (L.) Karst. *International Journal of Pharma and Bio Sciences*, **2**, 609-615.

[47] Iqbal, S., Younas, U., Chan, K.W., Zia-Ul-Haq, M. and Ismail, M. (2012) Chemical Composition of *Artemisia annua* L. Leaves and Antioxidant Potential of Extracts as a Function of Extraction Solvents. *Molecules*, **17**, 6020-6032. http://dx.doi.org/10.3390/molecules17056020

[48] Lapornik, B., Prosek, M. and Wondra, A.G. (2005) Comparison of Extracts Prepared from Plant By-Products Using Different Solvents and Extraction Time. *Journal of Food Engineering*, **71**, 214-222. http://dx.doi.org/10.1016/j.jfoodeng.2004.10.036

[49] Turkmen, N., Sari, F. and Velioglu, S. (2006) Effect of Extraction Solvents on Concentration and Antioxidant Activity of Black and Black Mate Polyphenols Determined by Ferrous Tartrate and Folin-Ciocalteu Methods. *Food Chemistry*, **99**, 838-841. http://dx.doi.org/10.1016/j.foodchem.2005.08.034

[50] Kong, K.W., Khoo, H.E., Prasad, N.K., Chew, L.Y. and Amin, I. (2013) Total Phenolics and Antioxidant Activities of *Pouteria campechiana* Fruit Parts. *Sains Malaysiana*, **42**, 123-127.

[51] Sultana, B., Anwar, F. and Ashraf, M. (2009) Effect of Extraction Solvent/Technique on the Antioxidant Activity of Selected Medicinal Plant Extracts. *Molecules*, **14**, 2167-2180. http://dx.doi.org/10.3390/molecules14062167

[52] Ben Mohamed Maoulainine, L., Jelassi, A., Hassen, I. and Ould Mohamed Salem Ould Boukhari, A. (2012) Antioxidant Proprieties of Methanolic and Ethanolic Extracts of *Euphorbia helioscopia* (L.) Aerial Parts. *International Food Research Journal*, **19**, 1125-1130.

[53] Pérez, M.B., Calderón, N.L. and Croci, C.A. (2007) Radiation-Induced Enhancement of Antioxidant Activity in Extracts of Rosemary (*Rosmarinus officinalis* L.). *Food Chemistry*, **104**, 585-592. http://dx.doi.org/10.1016/j.foodchem.2006.12.009

[54] Karimi, E., Oskoueian, E., Hendra, R. and Jaafar, H.Z.E. (2010) Evaluation of *Crocus sativus* L. Stigma Phenolic and Flavonoid Compounds and Its Antioxidant Activity. *Molecules*, **15**, 6244-6256. http://dx.doi.org/10.3390/molecules15096244

[55] Settharaksa, S., Jongjareonrak, A., Hmadhlu, P., Chansuwan, W. and Siripongvutikorn, S. (2012) Flavonoid, Phenolic Contents and Antioxidant Properties of Thai Hot Curry Paste Extract and Its Ingredients as Affected of pH, Solvent Types and High Temperature. *International Food Research Journal*, **19**, 1581-1587

[56] Spigno, G., Tramelli, L. and De Faveri, D.M. (2007) Effects of Extraction Time, Temperature and Solvent on Concentration and Antioxidant Activity of Grape Marc Phenolics. *Journal of Food Engineering*, **81**, 200-208.

[57] Wina, E., Susana, I.W.R. and Tangendjaja, B. (2010) Biological Activity of Tannins from *Acacia mangium* Bark Extracted by Different Solvents. *Media Peternakan*, **33**, 103-107. http://dx.doi.org/10.5398/medpet.2010.33.2.103

[58] Mazandarani, M., Zarghami Moghaddam, P., Zolfaghari, M.R., Ghaemi, E.A. and Bayat, H. (2012) Effects of Solvent Type on Phenolics and Flavonoids Content and Antioxidant Activities in *Onosma dichroanthum* Boiss. *Journal of Medicinal Plants Research*, **6**, 4481-4488. http://dx.doi.org/10.5897/JMPR11.1460

[59] Antwi-Boasiako, C. and Animapauh, S.O. (2012) Tannin Extraction from the Barks of Three Tropical Hardwoods for the Production of Adhesives. *Journal of Applied Sciences Research*, **8(6)**, 2959-2965.

[60] Uma, D.B., Ho, C.W. and Aida, W.M.W. (2010) Optimization of Extraction Parameters of Total Phenolic Compounds from Henna (*Lawsonia inermis*) Leaves. *Sains Malaysiana*, **39**, 119-128.

[61] Chan, S.W., Lee, C.Y., Yap, C.F., Wan Aida, W.M. and Ho, C.W. (2009) Optimisation of Extraction Conditions for Phenolic Compounds from Limau Purut (*Citrus hystrix*) Peels. *International Food Research Journal*, **16**, 203-213.

[62] Yilmaz, Y. and Toledo, R.T. (2005) Oxygen Radical Absorbance Capacities of Grape/Wine Industry Byproducts and Effect of Solvent Type on Extraction of Grape Seed Polyphenols. *Journal of Food Composition and Analysis*, **19**, 41-48. http://dx.doi.org/10.1016/j.jfca.2004.10.009

[63] Butnariu, M. and Coradini, C.Z. (2012) Evaluation of Biologically Active Compounds from *Calendula officinalis* Flowers Using Spectrophotometry. *Chemistry Central Journal*, **6**, 35-42. http://dx.doi.org/10.1186/1752-153X-6-35

[64] Fernandes, A.J.D., Ferreira, M.R.A., Randau, K.P., De Souza, T.P. and Soares, L.A.L. (2012) Total Flavonoids Content in the Raw Material and Aqueous Extractives from *Bauhinia monandra* Kurz (Caesalpiniaceae). *The Scientific World Journal*, **2012**, Article ID: 923462.

[65] Musa, K.H., Abdullah, A., Jusoh, K. and Subramaniam, V. (2011) Antioxidant Activity of Pink-Flesh Guava (*Psidium guajava* L.): Effect of Extraction Techniques and Solvents. *Food Analytical Methods*, **4**, 100-107. http://dx.doi.org/10.1007/s12161-010-9139-3

[66] Downey, M.O. and Hanlin, R.L. (2010) Comparison of Ethanol and Acetone Mixtures for Extraction of Condensed Tannin from Grape Skin. *South African Journal of Enology and Viticulture*, **31**, 154-159.

[67] Medini, F., Ksouri, R., Falleh, H., Megdiche, W., Trabelsi, N. and Abdelly, C. (2011) Effects of Physiological Stage and Solvent on Polyphenol Composition, Antioxidant and Antimicrobial Activities of *Limonium densiflorum*. *Journal of Medicinal Plants Research*, **5**, 6719-6730.

[68] Shad, M.A., Nawaz, H., Rehman, T., Ahmad, H.B. and Hussain, M. (2012) Optimization of Extraction Efficiency of Tannins from *Cichorium intybus* L.: Application of Response Surface Methodology. *Journal of Medicinal Plants Research*, **6**, 4467-4474.

[69] Zhao, S., Liu, J.Y., Chen, S.Y., Shi, L.L., Liu, Y.J. and Ma, C. (2011) Antioxidant Potential of Polyphenols and Tannins from Burs of *Castanea mollissima* Blume. *Molecules*, **16**, 8590-8600. http://dx.doi.org/10.3390/molecules16108590

[70] Hismath, I., Wan Aida, W.M. and Ho, C.W. (2011) Optimization of Extraction Conditions for Phenolic Compounds from Neem (*Azadirachta indica*) Leaves. *International Food Research Journal*, **18**, 931-939.

[71] Franco, D., Sineiro, J., Rubilar, M., Sanchez, M., Jerez, M., Pinelo, M., Costoya, N. and José Nunez, M. (2008) Polyphenols from Plant Materials: Extraction and Antioxidant Power. *Electronic Journal of Environmental, Agricultural and Food Chemistry*, **7**, 3210-3216.

[72] Sardsaengjun, C. and Jutiviboonsuk, A. (2010) Effect of Temperature and Duration Time on Polyphenols Extract of *Areca catechu* Linn. Seeds. *Thai Pharmaceutical and Health Science Journal*, **5**, 14-17.

[73] Dent, M., Dragovi-Uzelac, V., Peni, M., Brncic, M., Bosiljkov, T. and Levaj, B. (2013) The Effect of Extraction Solvents, Temperature and Time on the Composition and Mass Fraction of Polyphenols in Dalmatian Wild Sage (*Salvia officinalis* L.) Extracts. *Food Technology and Biotechnology*, **51**, 84-91.

[74] Naczk, M. and Shahidi, F. (2006) Phenolics in Cereals, Fruits and Vegetables: Occurrence, Extraction and Analysis. *Journal of Pharmaceutical and Biomedical Analysis*, **41**, 1523-1542. http://dx.doi.org/10.1016/j.jpba.2006.04.002

[75] Garcia-Marquez, E., Roman-Guerrero, A., Perez-Alonso, C. and Cruz-Sosa, F. (2012) Effect of Solvent-Temperature Extraction Conditions on the Initial Antioxidant Activity and Total Phenolic Content of Muitle Extracts and Their Decay upon Storage at Different pH. *Revista Mexicana de Ingeniería Química*, **11**, 1-10.

[76] Zhekova, G. and Pavlov, D. (2012) Influence of Different Factors on Tannins and Flavonoids Extraction of Some Thyme Varieties Representatives of Thymol, Geraniol and Citral Chemotype. *Agricultural Science and Technology*, **4**, 148-153.

[77] Liu, W., Yu, Y., Yang, R., Wan, C., Xu, B. and Cao, S. (2010) Optimization of Total Flavonoid Compound Extraction from *Gynura medica* Leaf Using Response Surface Methodology and Chemical Composition Analysis. *International Journal of Molecular Sciences*, **11**, 4750-4763. http://dx.doi.org/10.3390/ijms11114750

[78] Moosophin, K., Wetthaisong, T., Seeratchakot, L. and Kokluecha, W. (2010) Tannin Extraction from Mangosteen Peel for Protein Precipitation in Wine. *KKU Research Journal*, **15**, 377-385.

[79] Mustapha, M.B., Adefisan, H.A. and Olawale, A.S. (2012) Studies of Tannin Extract Yield from *Parkia clappertoniana*'s Husk. *Journal of Applied Sciences Research*, **8**, 65-68.

[80] Zam, W., Bashour, G., Abdelwahed, W. and Khayata, W. (2012) Effective Extraction of Polyphenols and Proanthocyanidins from Pomegranate's Peel. *International Journal of Pharmacy and Pharmaceutical Sciences*, **4**, 675-682.

[81] Druzynska, B., Stepniewska, A. and Wolosiak, R. (2007) The Influence of Time and Type of Solvent on Efficiency of the Extraction of Polyphenols from Green Tea and Antioxidant Properties Obtained Extracts. *Acta Scientiarum Polonorum. Technologia Alimentaria*, **6**, 27-36.

[82] Jokic, S., Velic, D., Bilic, M., Bucic-Kojic, A., Planinic, M. and Tomas, S. (2010) Modelling of the Process of Solid-Liquid Extraction of Total Polyphenols from Soybeans. *Czech Journal of Food Sciences*, **28**, 206-212.

[83] Perva-Uzunalic, A., Skerget, M., Knez, Z., Weinreich, B., Otto, F. and Grucher, S. (2006) Extraction of Active Ingredients from Green Tea (*Camellia sinensis*), Extraction Efficiency of Major Catechins and Caffeine. *Food Chemistry*, **96**, 597-605. http://dx.doi.org/10.1016/j.foodchem.2005.03.015

[84] Akowuah, G.A., Mariam A. and Chin, J. H. (2009) The Effect of Extraction Temperature on Total Phenols and Antioxidant Activity of *Gynura procumbens* Leaf. *Pharmacognosy Magazine*, **5**, 81-85.

[85] Chew, K.K., Thoo, Y.Y., Khoo, M.Z., Wan Aida, W.M. and Ho, C.W. (2011) Effect of Ethanol Concentration, Extraction Time and Extraction Temperature on the Recovery of Phenolic Compounds and Antioxidant Capacity of *Orthosiphon stamineus* Extracts. *International Food Research Journal*, **18**, 427-1435.

[86] Naeem, S., Ali, M. and Mahmood, A. (2012) Optimization of Extraction Conditions for the Extraction of Phenolic Compounds from *Moringa oleifera* Leaves. *Pakistan Journal of Pharmaceutical Sciences*, **25**, 535-541.

[87] Davidov-Pardo, G., Arozarena, M.R.I. and Marin-Arroyo, M.R. (2011) Stability of Polyphenolic Extracts from Grape Seeds after Thermal Treatments. *European Food Research and Technology*, **232**, 211-220. http://dx.doi.org/10.1007/s00217-010-1377-5

[88] Biesaga, M. and Pyrzynska, K. (2013) Stability of Bioactive Polyphenols from Honey during Different Extraction Methods. *Food Chemistry*, **136**, 46-54. http://dx.doi.org/10.1016/j.foodchem.2012.07.095

[89] Juntachote, T., Berghofer, E., Bauer, F. and Siebenhandl, S. (2006) The Application of Response Surface Methodology to the Production of Phenolic Extracts of Lemon Grass, Galangal, Holy Basil and Rosemary. *International Journal of Food Science and Technology*, **41**, 121-133. http://dx.doi.org/10.1111/j.1365-2621.2005.00987.x

[90] Al-Farsi, M.A. and Lee, C.Y. (2007) Optimization of Phenolics and Dietary Fibre Extraction from Date Seeds. *Food Chemistry*, **108**, 977-985. http://dx.doi.org/10.1016/j.foodchem.2007.12.009

[91] Wang, J., Sun, B., Cao, Y., Tian, Y. and Li, X. (2008) Optimisation of Ultrasound-Assisted Extraction of Phenolic Compounds from Wheat Bran. *Food Chemistry*, **106**, 804-810. http://dx.doi.org/10.1016/j.foodchem.2007.06.062

21

Studies of the Volatile Compounds Present in Leaves, Stems and Flowers of *Vernonanthura patens* (Kunth) H. Rob

Patricia Manzano[1,2*], Migdalia Miranda[1], Tulio Orellana[1], Maria Quijano[1]

[1]Superior Polytechnic School of Litoral, Biotechnology Research Center of Ecuador (ESPOL-CIBE), Guayaquil, Ecuador
[2]Faculty of Natural Sciences and Mathematics of Superior Polytechnic School of Litoral, Guayaquil, Ecuador
Email: [*]manzanopatricial@hotmail.com, [*]pmanzano@espol.edu.ec

Abstract

The study of the volatile components of the leaves, stems and flowers of *Vernonanthura patens* is discussed. A micro solid-phase extraction at constant temperature with a dimethylsiloxane fiber of 100 μm was performed. The compounds extracted were analyzed by gas chromatography coupled to mass spectrometry (GC-MS). 7 monoterpenes structures were assigned to leaves, and three to stems, these compounds were not detected in the flowers with the configuration of the system used. 17 sesquiterpenes were identified in the leaves; 6 in stems and 2 in flowers, finding coincidence in some of them. The major components were α-humulene in leaves, bergamotene in stems and caryophyllene in flowers.

Keywords

V. patens, Volatile, Bergamotene, Caryophyllene, α-Humulene

1. Introduction

Asteraceae (Asteraceae), gather more than 23,500 species spread over about 1600 genera, so the family of Angiosperms is rich and biological diversity [1] [2]. Members of this family are distributed from Polar Regions to the tropics, conquering all available habitats, from dry deserts to swamps and from forests to mountain peaks. In many regions, this family reaches up to 10% of vernacular flora and contains some genera with a large number of species, as *Vernonia* (*Vernonanthura*), with more than 1000 species [3]. Many species have latex and essen-

---

[*]Corresponding author.

tial oils and may or may not be resinous.

Vernonanthura patens is an Asteraceae that grows wild in Ecuador and is employed by people of the Ecuadorian coast to cure various conditions. Manzano *et al.* have reported that alcoholic extracts of the leaves show "*in vitro*" antileishmanial activity against *Leihsmania amazonensis* [4].

Studies conducted on the chemical composition of the species of the Ecuadorian coast have reported the presence of some terpene compounds (mainly pentacyclic triterpenoids), diterpenoids and sesquiterpenoids [5]. Those last are a part of the volatile fraction of the species and not essential oils which are presented in appreciable conditions so this work is carried out to study the volatile components obtained by solid phase micro extraction (SPME) at constant temperature (50°C) with a fiber of 100 microns dimethylsiloxane.

2. Materials and Methods

Leaves, flowers and stems of the species in phenological stage of flowering, collected around the Biotechnology Research Center of Ecuador located at Km 30.5 via Perimeter province of Guayas-Ecuador, were used. A sample of the plant material was taken for botanical identification which was botanized at the National Herbarium of Ecuador (QCNE), Quito, with CIBE37a code.

The volatile compounds were obtained by solid phase microextraction (SPME) at constant temperature of 50°C with a fiber of 100 microns dimethylsiloxane.

The chemical composition of the volatile compounds was analyzed in a gas chromatograph connected to a mass spectrometer (GC-MS) Agilent Technologies, equipped with a J & W capillary column GC of 30 × 250 microns × 0.25. The conditions were as follows: initial temperature in the oven 60°C for 1 minute to 260°C with an increment of 5°C per minute; followed by an increase of 15°C per min to 300°C for 1 minute. Temperatures for injector and detector were 150°C and 280°C, respectively in split mode. The volatiles compounds were assigned by comparison of their spectra with reference compounds existing in the Wiley and Nist library ninth edition 2011 installed on the computer.

3. Results and Discussion

Chromatograms of the volatile compounds from leaves, stems and flowers of *V. patens* is shown in **Figure 1**; the chromatogram of the volatile compounds from the leaves was the most complex and the chromatogram of the flowers was less complex. It is noteworthy that most of the major components are in the range of retention times between 15 and 20 min. Between 20 and 25 min., only the leaves had components with relative abundance. Moreover, the presence of a compound with high relative abundance correspond to the sesquiterpene caryophyllene with a retention time of 18.2 min approximately. This compound had been identified for the leaves of the Ecuadorian species [5] and Vernonia ssp [6].

Table 1 shows the monoterpenes isolated from different plant organs.

It is appreciated that the leaves had a higher number of monoterpene compounds, with a total of seven while in the stems were identified three. In the flowers these compounds could not be detected, or were absent.

Monoterpene compounds eluted from the column between the minutes 5.5 and 8.5 of the run. Retention time

Table 1. Monoterpenes identified in the fraction of volatile compounds of *V. patens*.

	leaves			steams	
tr min	compounds	% abund	tr	compounds	% abund
5.66	α-pinene	0.27	-	-	-
6.50	sabinene	0.23	-	-	-
6.59	β-pinene	3.32	-	-	-
6.86	β-myrcene	0.39	-	-	-
7.70	p-cymene	0.66	7.70	p-cymene	3.01
7.81	D-limonene	0.85	7.81	D-limonene	7.41
8.25	β-ocimeno	0.76	8.25	β-ocimeno	6.46

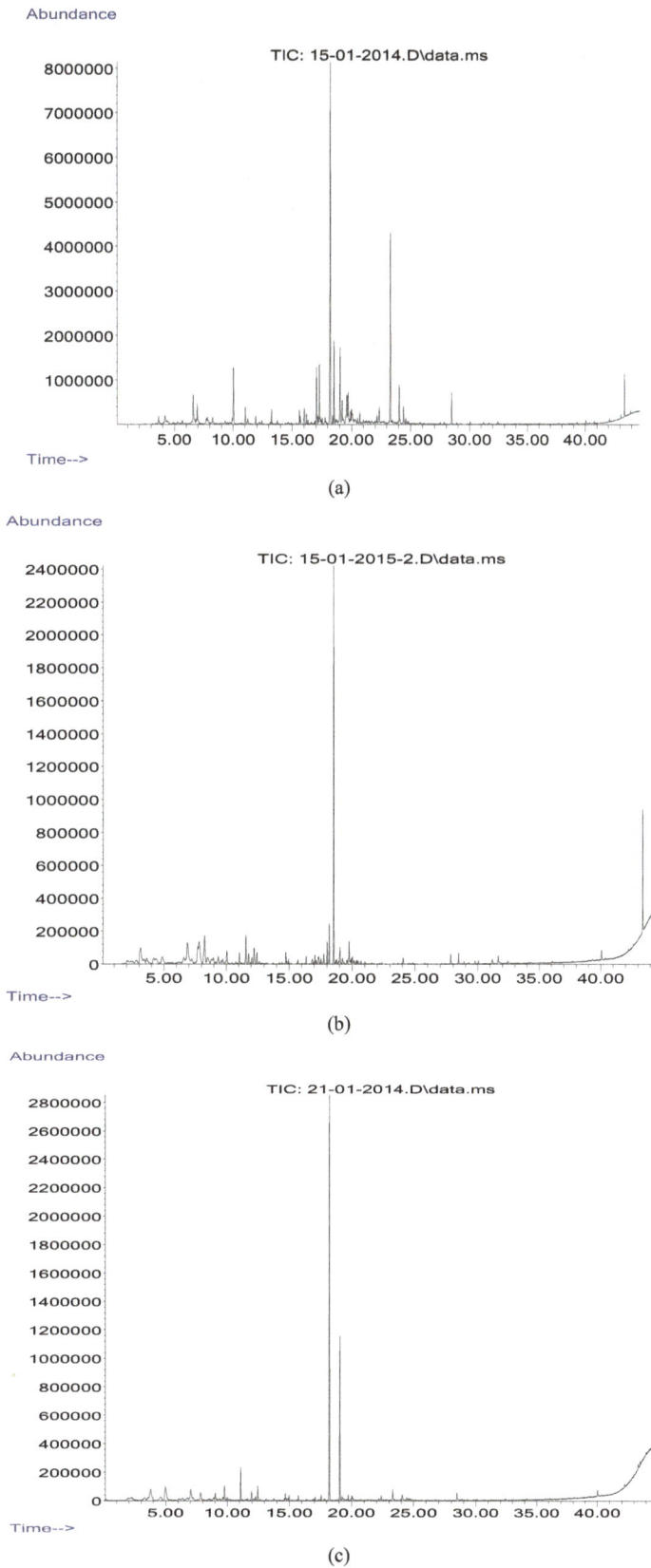

Figure 1. Analytic gas chromatograms of volatile from leaves (a), stem (b) and flowers (c) of *V. patens*.

of the compounds found in leaves and steams was the same, but relative abundances in steam was higher. For leaves the majority monoterpene was α-pinene, whereas in the stems was D-limonene.

The sesquiterpene compounds eluted from the column between 16.3 to 22.2 minutes. 19 compounds were identified in the leaves, 6 in the stems and 3 in the flowers. The results are shown in **Table 2**.

Sesquiterpenes present in the stems and flowers, are in the leaves as well, with the exception of zingiberene that is only present in stems. For those sesquiterpenes, variability is high and the most abundant is the Caryophillene with a relative abundance of 23.42%. However, in the stems the most abundant sesquiterpene was the α-bergamotene with 34.51% of relative abundance and in the flowers was the α-humulene with 19.57% of relative abundance.

Caryophyllene compound and its oxide report properties of pharmacological interest as anti-inflammatory, antitumor, antibacterial, spasmolytic, anti-septic, anti-parasite against *Trypanosoma cruzi* and *Leishmania brasi-*

Table 2. Sesquiterpenes identified in volatile fraction of *V. patens*.

rt min	leaves compound	% abund	steams compound	% abund	flowers compound	% abund
16.01	δ-elemene	1.38				
16.33	α-cubebene	0.25				
17.04	1,4-cadinene	3.46				
17.04			α-copaene	0.78		
17.29	β-bourbonene	3.47				
17.43	β-elemene	0.31				
17.75			zingiberene	1.09		
18.17	caryophyllene		caryophyllene	3.72		
18.20		23.42			caryophyllene	8.30
18.40	germacrene-D	0.47				
18.52	α-bergamotene	4.55	α-bergamotene	34.51		
18.66	aromadendrene	0.42				
18.78	γ-muurolene	0.27				
19.02			α-humulene	1.94		
19.03	α-humulene	4.96				
19.57					α-humulene	19.57
19.58	γ-curcumene	2.23				
19.69	β-cubebene	2.73				
19.74			β-farnesene	1.77		
19.84	cis-muurola-3,5-diene	0.34				
20.06	bicyclogermacrene	0.52				
20.24	α-farnesene	0.23				
20.48	α-amorphene	0.31				
20.68	δ-cadinene	0.71				
22.14	caryophyllene oxide	0.49				

liensis [7] [8]; *β* Ocimene as a powerful agent of tick [9]; *α*-bergamotene compound as antibacterial and anti-oxidant [10]; insecticidal activity is reported for linalool [11]; and *β*-caryophyllene and *α*-humulene present anti-fumigant activity [12].

Unfortunately there is no study on the volatiles of *V. patens* reported in the literature for comparison.

These results confirm the wealth in terpene compounds of *V. patens* from the Ecuadorian coast, for which Manzano *et al.*, [13] [14] identified pentacyclic triterpenoids as major constituents of the species. The results are reported for first time in the species growing in Ecuador.

4. Conclusion

The fraction of volatile compounds from leaves, stems and flowers of *V. patens* is rich in sesquiterpene compounds, which are much more abundant in the leaves of the species together with monoterpene. A total of 20 sesquiterpenoids and 7 monoterpenoides in the samples studied are identified.

Acknowledgements

The authors thank SENESCYT for partial financial support of this work.

References

[1] Thorne, R. (2007) An Updated Classification of the Class Magnoliopsida ("Angiospermae"). The New York Botanical Garden, Bronx.

[2] Jeffrey, C. (2007) Compositae: Introduction with Key to Tribes. In: Kadereit, J.W. and Jeffrey, C., Eds., *Families and Genera of Vascular Plants*, Vol. VIII, *Flowering Plants, Eudicots, Asterales*, Springer-Verlag, Berlin, 61-87.

[3] Bremer, K. (1994) Asteraceae: Cladistics and Classification. Timber Press, Portland.

[4] Manzano Santana, P.I., García, M., Mendiola J., Fernández-Calienes, A., Orellana, T., Miranda, M., Peralta, E. and Monzote, L. (2014) *In Vitro* Anti-Protozoal Assessment of *Vernonanthura patens* Extracts. *Pharmacologyonline*, **1**, 1-6.

[5] Manzano Santana, P.I., Miranda, M., Paz Robles, C., Abreu Payrol, J., Silva, M. and Hernández, V. (2012) Isolation and Characterization of the Hexane Fraction of the Leaves of *Vernonanthura patens* (Kunth) H. Rob. with Antifungal Activity. *Cuban Journal of Pharmacy*, **46**, 352-358.

[6] Bohlmann, F., Jakupovic, J., Gupta, R., King, R. and Robinson, H. (1981) Allenic Germacranolides, Bourbonene Derived Lactones and Other Constituents from *Vernonia* Species. *Phytochemistry*, **20**, 473-480. http://dx.doi.org/10.1016/S0031-9422(00)84169-2

[7] Carneiro, F.B., Júnior, I.D., Lopes, P.Q. and Macêdo, R.O. (2010) Varying the Amount of *β*-Caryophyllene in the Essential Oil of *Plectranthus amboinicus* (Lour.) Spreng., Lamiaceae under Different Growing Conditions. *Brazilian Journal of Pharmacognosy*, **20**, 600-606.

[8] Leite, N., Sobral-Souza, C., Albuquerque, R., Brito, D., Lavor, A., Alencar, L., Tintino, S., Ferreira, J. Figueredo, J., Lima, L., Cunha, F., Pinho, A. and Coutinho, H. (2013) *In Vitro* Cytotoxic Antiparasitic Activity of Caryophyllene and Eugenol against *Trypanosoma cruzi* and *Leishmania brasiliensis*. *Cuban Journal of Medicinal Plants*, **18**, 522-528.

[9] Nchu, F., Magano, S. and Eloff, J. (2012) *In Vitro* Anti-Tick Properties of the Essential Oil of *Tagetes minuta* L. (Asteraceae) on *Hyalomma rufipes* (Acari: Ixodidae). *Onderstepoort Journal of Veterinary Research*, **79**, E1-E5.

[10] Teixeira, B., Marques, A., Ramos, C., Nuno, R., Neng, J., Nogueira, J., Saraiva, A. and Nunes, M. (2013) Chemical Composition and Antibacterial and Antioxidant Properties of *Comercial essential* Oils. *Industrial Crops and Products*, **43**, 587-595.

[11] Kessler, A. and Baldwin, I. (2001) Defensive Function of Herbivore-Induced Plant Volatile Emissions. *Nature Science*, **291**, 2141-2144.

[12] Zoubiri, S. and Baaliouamer, A. (2012) GC and GC/MS Analyses of the Algerian *Lantana Camara* Leaf Essential Oil: Effect against Sitophilus Granaries Adults. *Journal of Saudi Chemical Society*, **16**, 291-297. http://dx.doi.org/10.1016/j.jscs.2011.01.013

[13] Manzano Santana, P.I., Miranda, M., Payrol, J.A., Silva, M., Hernández, V. and Peralta, E. (2013) Gas Chromatography-Mass Spectrometry Study from the Leaves Fractions Obtained of *Vernonanthura patens* (Kunth) H. Rob. *International Journal of Organic Chemistry*, **3**, 105-109. http://dx.doi.org/10.4236/ijoc.2013.32011

[14] Manzano, P.I., Miranda, M., Abreu-Payrol, J., Silva, M., Sterner, O. and Peralta, E.L. (2013) *Pentacyclic triterpenoids with Antimicrobial Activity from the Leaves of Vernonanthura patens* (Asteraceae). *Emirates Journal of Food and Agriculture*, **25**, 539-543

22

Preparation of Polyfunctionally Substituted Pyridine-2(1*H*)-Thione Derivatives as Precursors to Bicycles and Polycycles

Fathi A. Abu-Shanab[1,2]*, Sayed A. S. Mousa[1], Sherif M. Sherif[3], Mohamed I. Hassan[1]

[1]Department of Chemistry, Faculty of Science, Al-Azhar University, Assiut, Egypt
[2]Department of Chemistry, Faculty of Science, Gazan University, Gazan, KSA
[3]Department of Chemistry, Faculty of Science, Cairo University, Giza, Egypt
Email: *fathiabushanab@yahoo.com

Abstract

Reaction of acetylacetone with 1 mole of dimethylformamide dimethyl acetal (DMFDMA) affords enamine 2a which reacts with cyanothioacetamide to give pyridinethione 3a. Pyridinethione 3a reacts with methyl iodide, halogenated compounds, aromatic aldehyde and malononitrile/elemental sulfur to yiled compounds 7-10 respectively. Reactions of thioether 7 in ethanolic K_2CO_3, 1 mole DMFDMA and 4-(dimethylamino)benzaldehyde give compounds 11, 13, 14 respectively. Enaminone 12 can be prepared by reaction of compound 11 with DMFDMA. We have demonstrated some reactions in order to show the potential usefulness of the prepared compounds for the preparation of new bipyridyl compounds 15, 16, 18, bicyclic compounds 17 and uncommon tricyclic compounds 20, 21, 22 and 23 respectively using DMFDMA.

Keywords

Acetyl Acetone, DMFDMA, Malononitrile Dimmer, Bipyridyl, 5-Acetylpyridinethione

1. Introduction

Formamide acetals are useful reagents in organic synthesis; [1] [2] their main application has been used for functional group transformations [3], but they may also be regarded as one-carbon synthons in the construction of carbon skeletons. One type of reaction, which is potentially valuable for the future purpose, is the reaction of *N,N'*-dimethylformamide dimethyl acetal (DMFDMA) with 1,3-dicarbonyl compounds **1** to give enamines **2** [2] [4] (**Figure 1**).

---
*Corresponding author.

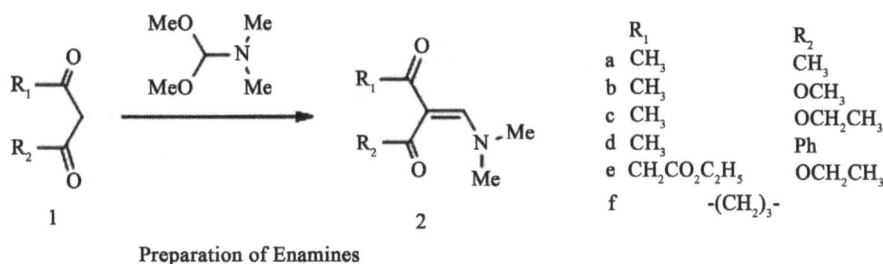

Preparation of Enamines

Figure 1. Preparation of Enamines.

We have reported that enamines **2** were used as precursors in the synthesis of pentasubstituted pyridines **3-6** [5]-[8] (**Figure 2**).

The treatment of acetyacetone (**1a**) with dimethyl formamide dimethylacetal (DMFDMA) in dry DMF under nitrogen and stirring over night afforded the corresponding enamine 2a which on treatment with cyanothioacetamide and sodium hydride in dry DMF (*in situ*) afforded pyridine-2(1H)-thione (**3a**) [6], when the emamine 2a was treated with cyanothioacetamide in ethanol and pepridine as a base afforded the pyridine-2(1H)-thione (**5a**) [7] [12] (**Figure 3**).

2. Results and Discussion

In conjunction of this work, we report here the reaction of acetylacetone **1a** with one mole of *N,N'*-dimetylformamide dimethyl acetal (DMFDMA) in dry dioxane gave the corresponding enamine **2a**. The treatment of this enamine (*in situ*) with cyanothioacetamide in ethanol in the presence of sodium ethoxide under reflux gave 5-acetyl-6-methyl-2-thioxo-1,2-dihydropyridine-3-carbonitrile **3a** with a very good yied [7], **Scheme 1**.

We have found that the prepared compound **3a** included three functional groups which are thioamido group, cyano group and acetyl group. These functional groups can be used for the preparation of bicyclic or polycyclic compounds of biological interest. Thus, some illustrative reactions designed to demonstrate the potential usefulness of 5-acetyl-6-methyl-2-thioxo-1,2-dihydropyridine-3-carbonitrile **3a** for further heterocyclic synthesis. Therefore, the reaction of 5-acetyl-6-methyl-2-thioxo-1,2-dihydropyridine-3-carbonitrile **3a** with methyl iodide in alcoholic sodium hydroxide afforded the corresponding thioether derivative **7**, which in turn is a good intermediate for the preparation of further heterocyclic compounds of biological interest. The structure of the isolated compound **7** is confirmed by spectral analysis. The IR spectrum shows the disappearance of (NH) group. Also, the ^{1}H NMR spectrum shows the disappearance of the thioamide proton and the appearance of a singlet signal corresponding to (SCH$_3$) at δ_H = 2.63 ppm. Also, the mass spectrum shows the molecular ion peak at m/e 206 which corresponding to the molecular formula (C$_{10}$H$_{10}$N$_2$OS). The reaction of 5-acetyl-6-methyl-2-thiox-o-1,2-dihydropyridine-3-carbonitrile **3a** with ethyl chloroacetate or chloroacetamides in ethanolic sodium ethoxide afforded the corresponding 5-acetyl-3-amino-6-methylthieno[2,3-b]pyridine derivatives **8a-c** in a good yield. The structure of the isolated compounds is confirmed by elemental and spectral analysis. The IR spectrum shows the disappearance of cyano group and appearance of amino group at v_{max} = 3427, 3328 cm^{-1} in compound **8a** as example beside the other functional groups. Also, the mass spectra show the molecular ion peaks fit to all compounds **8a-c**. Also, the ^{1}H NMR spectra show signals fit to the structure of all compounds **8a-c**. The presence of acetyl group in 5-acetyl-6-methyl-2-thioxo-1,2-dihydropyridine-3-carbonitrile **3a** is useful for the preparation of fused heterocyclic compounds. So that the reaction of compound **3a** with aldehydes like 4-(dimethylamino) benzaldehyde and 4-methylbenzaldehyde in ethanolic sodium hydroxide afforded the corresponding chalcones **9a,b**. The structure of the isolated chalcones is confirmed by elemental analysis as well as spectral analysis. The mass spectra show the molecular ion peak fit to all compounds **9a,b**. As an example compound **9a** shows the molecular ion peak at m/e 323 which corresponding to the molecular formula (C$_{18}$H$_{17}$N$_3$OS). Also, the ^{1}H NMR spectra of these compounds **9a,b** show the disappearance of the signal corresponding to the methyl of acetyl group and the appearance of two doublets signals corresponding to the two protons of double bond of chalcone. Finally, 5-acetyl-6-methyl-2-thioxo-1,2-dihydropyridine-3-carbonitrile **3a** was treated with malononitrile and sulfur element (Gewald's reaction) in ethanol in the presence of triethylamine as a base to afford 5-(5-amino-4-cyanothiophen-3-yl)-6-methyl-2-thioxo-1,2-dihydropyridine-3-carbonitrile **10** in a good yield, **Scheme 1**. The IR spectrum of compound **10** shows the appearance of amino group at v_{max} = 3435, 3350 cm^{-1} beside the other

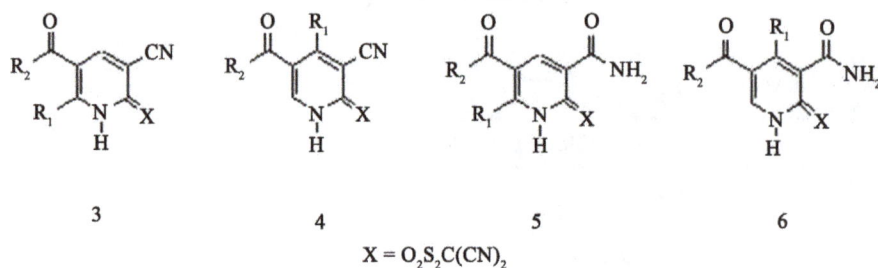

X = O,S,C(CN),

Defferent polysubstituted pyridines have been prepared

Figure 2. Different polysubstituted pridines have been prepared.

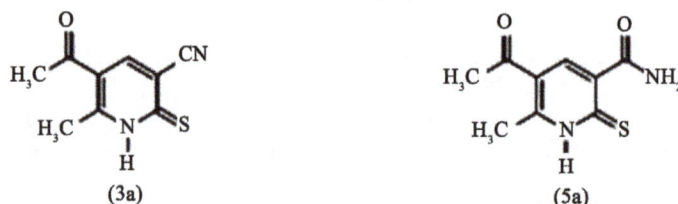

Figure 3. Tetrasubstituted pyridinethione have been prepared.

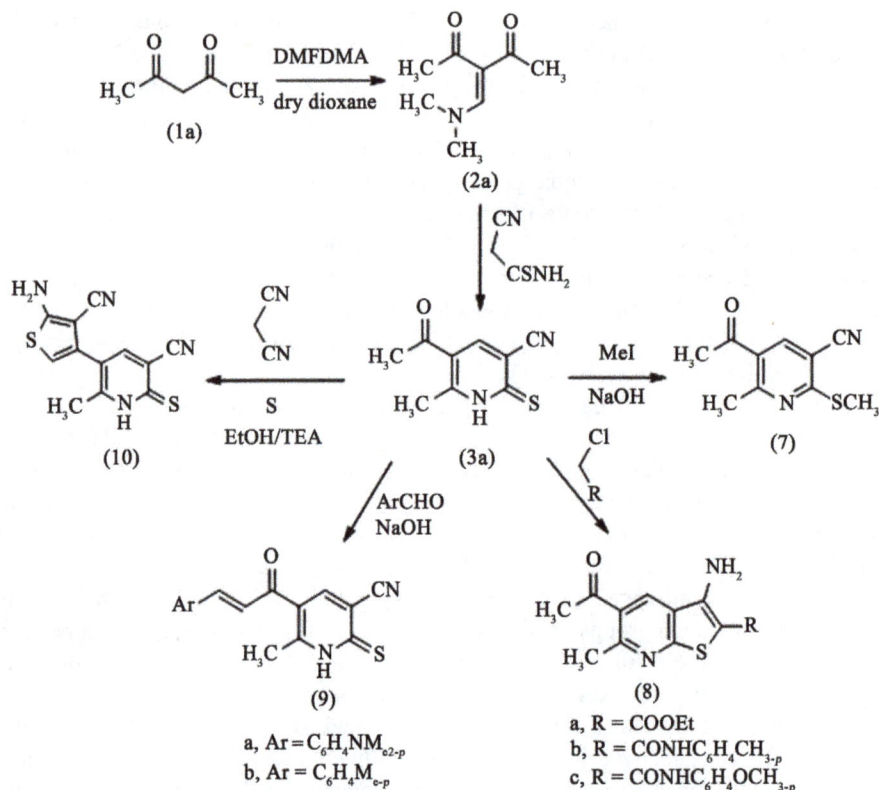

a, Ar = C₆H₄NMₑ₂₋ₚ
b, Ar = C₆H₄Mₑ₋ₚ

a, R = COOEt
b, R = CONHC₆H₄CH₃₋ₚ
c, R = CONHC₆H₄OCH₃₋ₚ

Scheme 1. Synthesis of pyridine (1H)-thione derivative 3a and its reactions with MeI, α-chloroketones, aldehydes and malononitrile.

functional groups. Also, ^{1}H NMR spectrum of compound **10** shows singlet signal at $\delta_H = 2.45$ ppm corresponding to methyl group and singlet signal at $\delta_H = 6.95$ ppm corresponding to amino group and singlet signal at $\delta_H = 7.07$ ppm corresponding to CH thiophene ring and singlet signal at $\delta_H = 7.2$ ppm corresponding to CH pyridine ring.

5-Acetyl-6-methyl-2-(methylthio)nicotinonitrile **7** can be used as intermediate for further preparation of heterocyclic compounds. So that compound **7** was treated with potassium carbonate in ethanol to afford 5-acetyl-2-ethoxy-6-methylnicotinonitrile **11**. This compound was formed by nucleophilic substitution of SMe by OEt group. The structure of the isolated compound is confirmed by elemental and spectral analyses. The mass spectrum shows the molecular ion peak at m/e 204 corresponding to the molecular formula ($C_{11}H_{12}N_2O_2$). Also, the ^1H NMR spectrum shows the disappearance of SMe signal and appearance of two signals; a triplet at $\delta_H = 1.43$ ppm and a quartet at $\delta_H = 4.54$ ppm corresponding to the OEt moiety, in addition to the rest of signals corresponding to the other protons in the molecule. Compound **11** was reacted with N,N'-dimetylformamide dimethyl acetal (DMFDMA) in dry xylene to give the corresponding enamine **12** in a good yield. The mass spectrum of compound **12** shows the molecular ion peak at m/e 259 which corresponding to the molecular formula ($C_{14}H_{17}N_3O_2$). Also, the ^1H NMR spectrum of compound **12** shows the disappearance of the singlet signal which is related to the methyl of acetyl group and the appearance of two singlet signals at $\delta_H = 2.68$ and 3.04 ppm corresponding to the two methyl groups of NMe$_2$ moiety. Consequently the ^1H NMR spectrum shows the appearance of two doublets at $\delta_H = 6.25$ ppm and 7.87 ppm corresponding to the two protons of the enamine double bond.

Enamine **13** can be prepared in a good yield by reaction of 5-acetyl-6-methyl-2-thioxo-1,2-dihydropyri-dine-3-carbonitrile **3a** with two moles of N,N'-dimetylformamide dimethylacetal (DMFDMA) in dry xylene or by the reaction of 5-acetyl-6-methyl-2-(methylthio)nicotinonitrile **7** with one mole of N,N'-dimetylformamide dimethylacetal (DMFDMA) in dry xylene. The structure of the isolated compound is confirmed by elemental and spectral analysis. Whereas the mass spectrum shows the molecular ion peak at m/e 261 which corresponding to the molecular formula ($C_{13}H_{15}N_3OS$). Also, the ^1H NMR spectrum of it shows the disappearance of the singlet signal which is related to the methyl of acetyl group and appearance of two singlet signals at $\delta_H = 2.62$ and 2.64 ppm corresponding to the two methyl groups of NMe$_2$ moiety. Consequently, the ^1H NMR spectrum shows the appearance of two doublets at $\delta_H = 5.28$ ppm and 7.75 ppm corresponding to the two protons of double bond of enamine.

Chalcone **14** can either be prepared by the reaction of compound **7** with (4-(dimethylamino)benzaldehyde) in ethanolic sodium hydroxide or by treatment of compound **9a** with methyl iodide in ethanolic sodium hydroxide. The mass spectrum of compound **14** shows the molecular ion peak at m/e 337 corresponding to the molecular formula ($C_{19}H_{19}N_3OS$). Also, the ^1H NMR spectrum of compound **14** shows singlet signal at $\delta_H = 2.62$ ppm corresponding to methyl group and singlet signal at $\delta_H = 2.66$ ppm corresponding to SCH$_3$ and two singlet signal at $\delta_H = 2.9$, 3.04 ppm corresponding to NMe$_2$ moiety and appearance of some signals of other protons in molecule.

For preparation of bipyridyl derivatives, we have carried out the reaction of chalcones 5-(3-(4-(dimethyla-mino)phenyl)acryloyl)-6-methyl-2-thioxo-1,2-dihydropyridine-3-carbonitrile **9a** and 5-(3-(4-(dimethylamino)phenyl)acryloyl)-6-methyl-2-(methylthio)nicotinonitrile **14** with malononitrile dimmer [9] in acetic acid and ammonium acetate afforded the corresponding bipyridyl derivatives 6-(dicyanomethylene)-4-(4-(dimethylamino)phenyl)-2'-methyl-6'-thioxo-1,1',6,6'-tetrahydro-[2,3'-bipyridine]-5,5'-dicarbonitrile **15** and 6-(dicyanomethylene)-4-(4-(dimethylamino)phenyl)-2'-methyl-6'-(methylthio)-1,6-dihydro-[2,3'-bipyridine]-5,5'-dicarbonitrile **16** respectively. The reaction proceeds by Michael addition followed by cyclization through condensation as shown in **Scheme 2**.

The compound **16** can also be obtained by the reaction of 6-(dicyanomethylene)-4-(4-(dimethylami-no)phenyl)-2'-methyl-6'-thioxo-1,1',6,6'-tetrahydro-[2,3'-bipyridine]-5,5'-dicarbonitrile **15** with methyl iodide in alcoholic sodium hydroxide **Scheme 2**.

The structure of the isolated compounds **15** and **16** is established by elemental and spectral analysis. Whereas the mass spectra of these compounds show the molecular ion peaks at m/e 435 corresponding to the molecular formula ($C_{24}H_{17}N_7S$), and at m/e 449 corresponding to the molecular formula ($C_{25}H_{19}N_7S$) for **15** and **16** respectively. The IR spectra of both compounds **15** and **16** show the disappearance of the carbonyl group and the appearance of NH group. Also, the ^1H NMR spectra of these compounds show signals fit to structures **15** and **16**.

For further preparation of heterocyclic compounds [10], we carried out the following reactions. The reaction of enamine **13** with excess hydrazine hydrate in ethanol afforded 6-methyl-5-(1H-pyrazol-3-yl)-1H-pyrazolo[3,4-b]pyridin-3-amine **17** in a good yield as shown in **Scheme 3**. The IR spectrum of compound **17** shows the disappearance of the cyano group and the appearance of NH$_2$ and NH groups at υ_{max} at 3405 cm^{-1}, 3329 cm^{-1} and 3136 cm^{-1} respectively. Also, the mass spectrum of compound **17** shows the molecular ion peak at m/e 214 corresponding to the molecular formula ($C_{10}H_{10}N_6$). Also, the ^1H NMR spectrum of compound **17** shows signals fit to the structure.

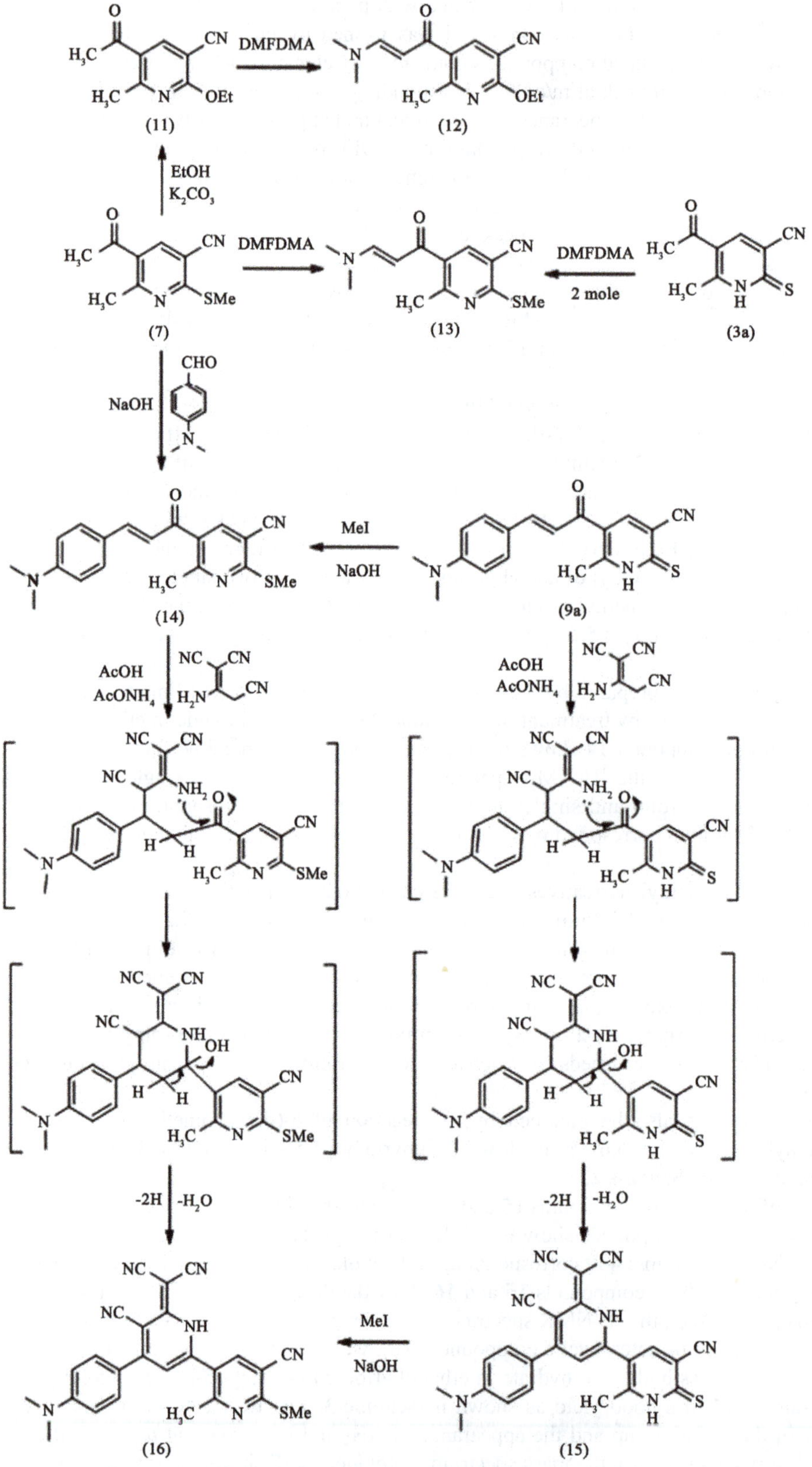

Scheme 2. Reactions of tetrasubstitutedpridine 7 with DMFDMA, alcholic K₂CO₃ and p-N, N-dimethylaminobezaldehyde.

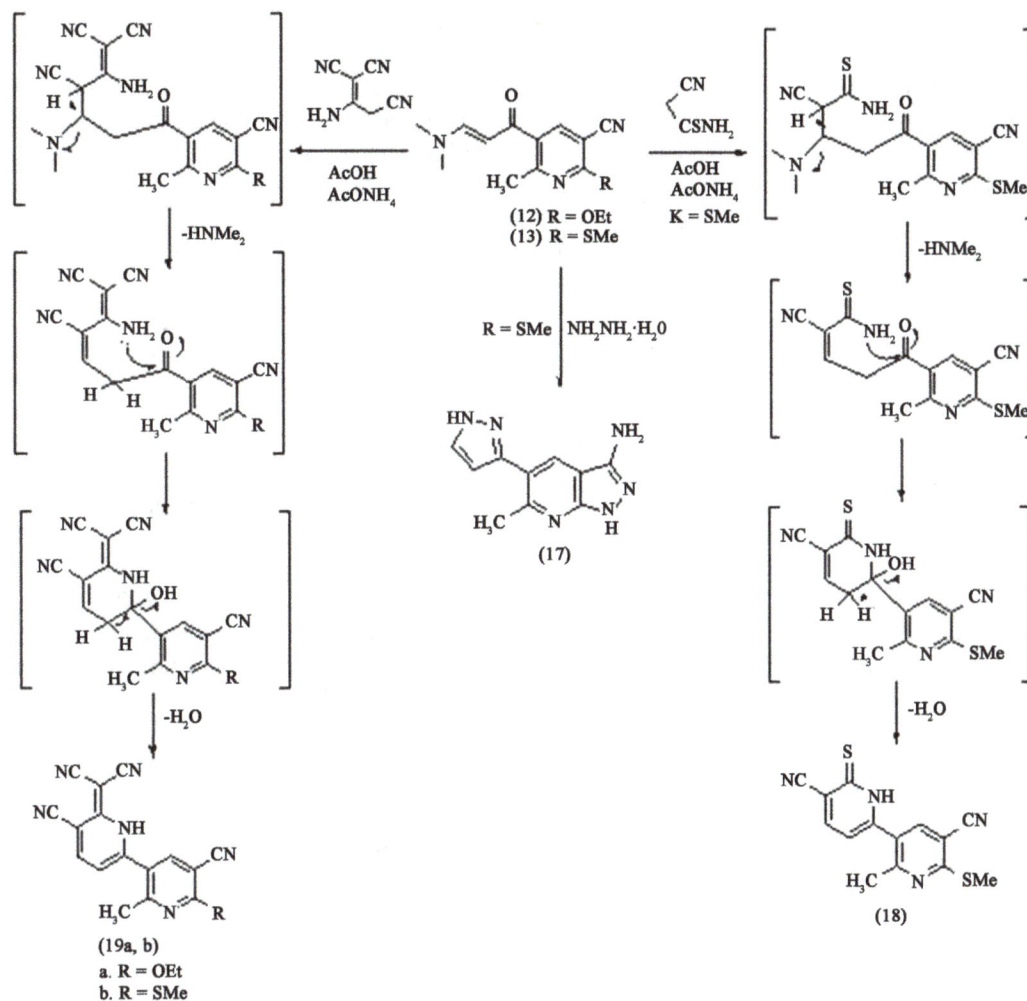

Scheme 3. Reactions of tetrasubstitutedpridine (12,13) with cyanothioacetamide, hydrazine hydrate and malononitrile dimer.

Also, the enamine **13** is treated with cyanothioacetamide in acetic acid and ammonium acetate afforded 2'-methyl-6'-(methylthio)-6-thioxo-1,6-dihydro-[2,3'-bipyridine]-5,5'-dicarbonitrile **18**. The reaction is started by Micheal addition of cyanothioacatamide on the double bond followed by elimination of dimethylamine (HNMe$_2$) and cyclization with the carbonyl group. The structure of the isolated compound **18** is confirmed by elemental and spectral analysis. The IR spectrum of compound **18** shows the disappearance of carbonyl group and appearance of NH group at v_{max} at 3428 cm^{-1}. The mass spectrum of compound **18** shows the molecular ion peak at m/e 298 corresponding to the molecular formula (C$_{14}$H$_{10}$N$_4$S$_2$). Also, the ^{1}H NMR spectrum of compound **18** shows the disappearance of protons of NMe$_2$ moiety and appearance of NH proton beside the other protons.

Another type of bipyridyl derivatives **19a,b** can be prepared by the reaction of the enamines **12** and **13** with malononitrile dimmer in acetic acid and ammonium acetate. This reaction proceeds by Michael addition of malononitrile dimmer, followed by elimination of dimethylamine (HNMe$_2$) and cyclization through condensation of amino group with carbonyl group as shown in **Scheme 3**. The mass spectrum of compound **19a** shows the molecular ion peak at m/e 328 corresponding to the molecular formula (C$_{18}$H$_{12}$N$_6$O), and compound **19b** shows the molecular ion peak at m/e 330 corresponding to the molecular formula (C$_{17}$H$_{10}$N$_6$S). The IR spectra of the compounds **19a,b** shows the disappearance of the carbonyl group and the appearance of NH group beside the other groups. Also, the ^{1}H NMR spectra of compounds **19a,b** show the disappearance of protons of NMe$_2$ moiety and appearance of NH proton beside the other protons.

The tricyclic heterocyclic compounds are biologically interest compounds. They are examples of uncommon ring system [11] [12]. Therefore we are interested for the preparation of this type of heterocyclic compound.

Thus 5-acetyl-3-amino-6-methyl-N-(p-tolyl)benzo[b]thiophene-2-carboxamide **8b** is reacted with *N,N'*-dime-tylformamide dimethyl acetal (DMFDMA) in dry dioxane afforded 8-acetyl-7-methyl-3-(p-tolyl)pyrido [3',2':4,5] thieno[3,2-*d*]pyrimidin-4(3*H*)-one **20**. The IR spectrum of compound **20** shows the disappearance of (NH$_2$) and (NH) groups. The mass spectrum of compound **20** shows the molecular ion peak at m/e 349 which correspond-ing to the molecular formula (C$_{19}$H$_{15}$N$_3$O$_2$S). Also, the ^{1}H NMR spectrum of compound **20** shows the appear-ance of two singlet signals at δ_H = 8.43 ppm, and 8.52 ppm corresponding to two protons of pyrimidinone and pyridine rings respectively beside other signals for other protons.

For further reaction of 5-acetyl-3-amino-6-methyl-N-substituted[b]thiophene-2-carboxamide **8b,c** it reacted with nitrous acid in acetic acid afforded the tricyclic compounds **21a,b** in a good yield as shown in **Scheme 4**. The structures of the compounds **21a,b** are established by elemental and spectral analysis. Whereas the IR spec-tra of both compounds **21a,b** show the disappearance of the bands corresponding to (NH$_2$) and (NH) groups. The mass spectrum of the compound **21a** as an example shows the molecular ion peak at m/e 350 corresponding to molecular formula (C$_{18}$H$_{14}$N$_4$O$_2$S).

Also, the ^{1}HNMR spectra of compounds **21a,b** show the disappearance of the signals which corresponding to (NH$_2$) and (NH) groups beside the appearance the other signals for other groups.

We have found that the prepared tricyclic compounds **20** and **21a,b** contain acetyl group which is very im-portant for the preparation of new heterocyclic compounds. So that the reaction of **21a** with malononitrile and sulphur element in ethanol and triethylamine (Geweld reaction) afforded 2-amino-4-(7-methyl-4-oxo-3-(p-tolyl)-3,4-dihydropyrido[3',2':4,5]thieno[3,2-d][1,2,3]triazin-8-yl)thiophene-3-carbonitrile **22**. The IR spectrum of compound **22** shows the disappearance of the carbonyl group of acetyl moiety and the appearance of amino and cyano groups at υ_{max} at 3427 cm^{-1} and 2208 cm^{-1} respectively. Also, the mass spectrum of this compound **22** shows the molecular ion peak at m/e 430 which corresponding to the molecular formula (C$_{21}$H$_{14}$N$_6$OS$_2$).

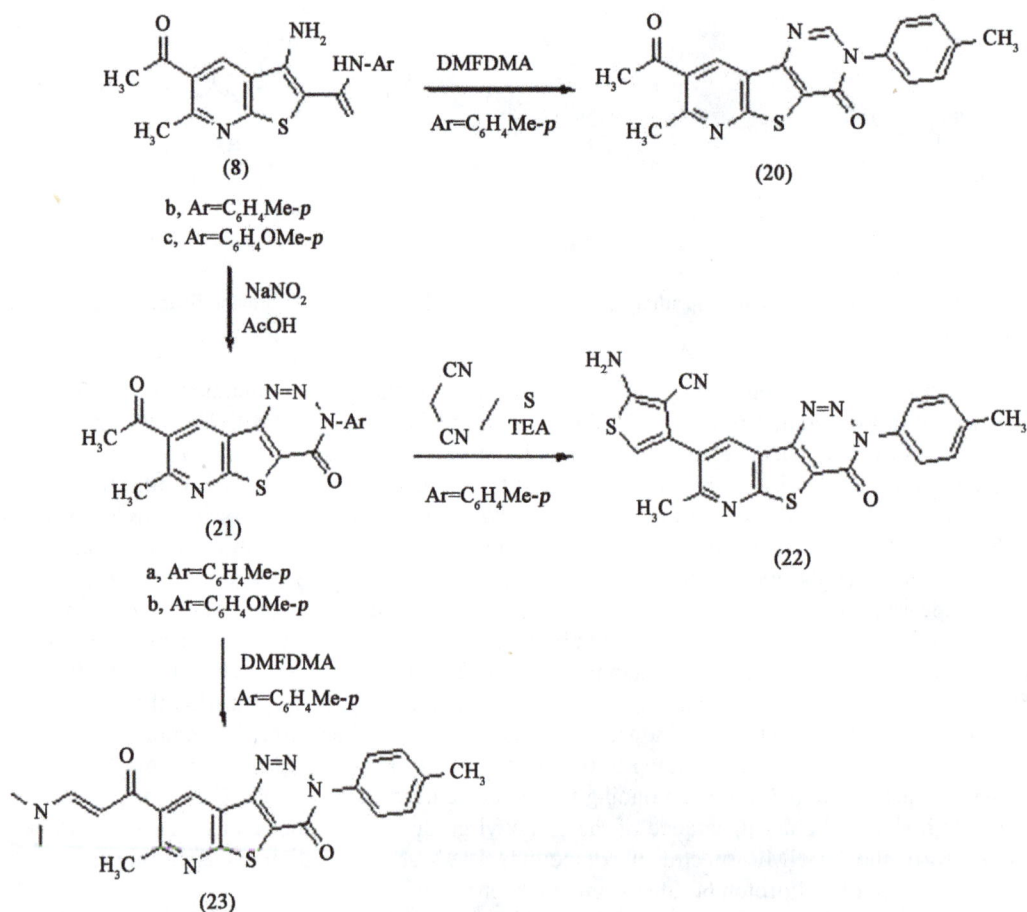

Scheme 4. Reactions of thinopyridine derivatives (8b,c) with DMFDMA and sodium nitrile in acetic.

Also, the compound **21a** is treated with *N,N'*-dimetylformamide dimethyl acetal (DMFDMA) in dry xylene afforded the corresponding enamine 8-(3-(dimethylamino)acryloyl)-7-methyl-3-(p-tolyl)pyrido[3',2':4,5] thieno[3,2-d][1,2,3]triazin-4(3H)-one **23** in a good yield, **Scheme 4**. The mass spectrum of compound **23** shows the molecular ion peak at m/e 405 corresponding to molecular formula ($C_{21}H_{19}N_5O_2S$). Also, the ^{1}H NMR spectrum of compound **23** shows the disappearance of the methyl of acetyl moiety and appearance instead of it two singlet signals at δ_H = 3.63 ppm and 3.67 ppm corresponding to (NMe$_2$) moiety. Also, it shows the appearance of two doublet signals at δ_H = 5.42 ppm and 7.82 ppm respectively corresponding to the double bond protons of ena-minone moiety beside signals for other protons.

3. Experimental

All melting points are uncorrected. IR spectra were recorded on a Perkin-Elmer 17,100 FTIR spectrometer as KBr disks. NMR spectra were recorded on a Varian Gemini (400 MHz) spectrometer for solutions in CDCl$_3$ or DMSO-d$_6$ with tetramethylsilane (TMS) as an internal standard unless otherwise. Mass spectra were obtained on Finnigan 4500 (low resolution) spectrometers using electron impact (EI) at Micro-analytical Center Cairo University Giza Egypt.

Preparation of 5-acetyl-3-cyano-6-methylpyridine-2(1H)-thione (3a):

A mixture of acetylacetone (**1a**) (1.0 g, 10 mmol), dry dioxane (10 mL) and N,N-dimethylformamide dimethyl acetal (1.19 g, 10 mmol) was stirred under dry condition at room for 24 h. In a second flask, a mixture of dry ethanol (15 mL), and sodium metal (0.46 g, 20 mmol) was left under stirring for 10 min. Then cyanothioacetamide (1.0 g, 10 mmol) was added to the mixture. The mixture was left for further 10 min. The contents of the second flask were transferred into the first flask, and the resulting mixture was stirred for further 1 h, fol-lowed by converting stirring into reflux for 4 h. After cooling, the mixture was poured onto acidified ice/cold water. The product was recovered by filtration and recrystallised from ethanol as orange crystals (76%), Mp. 232° - 233°, similar to be published before [7] and mixed Mp.

Preparation of 5-acetyl-6-methyl-2-(methylthio)nicotinonitrile (7)

Mixture of 5-acetyl-6-methyl-2-thioxo-1,2-dihydropyridine-3-carbonitrile **3a** (1.92 g, 10 mmol) in ethanol as solvent and sodium hydroxide (0.4 g, 10 mmol) with stirring for 1 hr., and add methyl iodide (0.62 ml, 10 mmol) with stirring until precipitate formed. The product was recovered by filtration and recrystallised from ethanol as white crystals (1.52 g, 74%), Mp. 140°C - 142°C; ^{1}H-NMR (CDCl$_3$): δ = 2.54 (3H, s, CH$_3$ py.), 2.63 (3H, s, SCH$_3$), 2.77 (3H, s, CH$_3$CO), 8.07 (1H, s, CH py.); IR (KBr) υ 2227 (CN), 1685 cm^{-1} (C=O); MS (EI)$^+$: m/z 206 M$^+$; Anal. Calcd for C$_{10}$H$_{10}$N$_2$OS (206.27): C, 58.23; H, 4.89; N, 13.58. Found: C, 58.03; H, 4.73; N, 13.41.

General procedure for the preparation of compounds **8a-c**

In dry flask, a mixture 5-acetyl-6-methyl-2-thioxo-1,2-dihydropyridine-3-carbonitrile **3a** (1.92 g, 10 mmol) and α-chloro compounds (10 mmol) in ethanol and sodium ethoxide (20 mmol) was left under reflux for two hours. The mixture was left for cooling and poured onto ice cold water. The solid product was recovered by fil-tration and recrystallised from the proper solvent.

Ethyl 5-acetyl-3-amino-6-methylthieno[2,3-b]pyridine-2-carboxylate **(8a):** Obtained using ethyl 2-chloroa-cetate (1.06ml, 10 mmol). The product was recrystallised from acetic acid as yellow crystals (2.16 g, 77.7%), Mp. 220°C - 222°C; ^{1}H-NMR (DMSO): δ = 1.25 (3H, t, CH$_3$ ethyl), 2.6 (3H, s, CH$_3$ py.), 2.66 (3H, s, CH$_3$CO), 4.25 (2H, q, CH$_2$ ethyl), 7.29 (2H, s, NH$_2$), 8.95 (1H, s, CH py.); IR (KBr) υ 3427, 3328 (NH$_2$), 1679 cm^{-1} (C=O); MS (EI)$^+$: m/z 278 M$^+$; Anal. Calcd for C$_{13}$H$_{14}$N$_2$O$_3$S (278.33): C, 56.10; H, 5.07; N, 10.06, Found: C, 55.96; H, 4.94; N, 9.97.

5-Acetyl-3-amino-6-methyl-N-(p-tolyl)thieno[2,3-b]pyridine-2-carboxamide **(8b):** Obtained using 2-chloro-N-(p-tolyl)acetamide (1.83 g, 10 mmol). The product was recrystallised from ethanol as yellow crystals (2.7 g, 79%), Mp. 218°C - 220°C; ^{1}H-NMR (DMSO) δ = 2.26 (3H, s, CH$_3$ Ar), 2.64 (3H, s, CH$_3$ py.), 2.73 (3H, s, CH$_3$CO), 7.12 (2H, d, Ar), 7.47 (2H, s, NH$_2$), 7.55 (2H, d, Ar), 9.04 (1H, s, CH py.), 9.4 (1H, s, NH); IR (KBr) υ 3428 , 3312 (NH$_2$, NH), 1685 cm^{-1} (C=O); MS (EI)$^+$: m/z 339 M$^+$; Anal. Calcd for C$_{18}$H$_{17}$N$_3$O$_2$S (339.42): C, 63.70; H, 5.05; N, 12.38, Found: C, 63.56; H, 4.93; N, 12.15.

5-Acetyl-3-amino-N-(4-methoxyphenyl)-6-methylthieno[2,3-b]pyridine-2-carboxamide **(8c):**

Obtained using 2-chloro-N-(4-methoxyphenyl)acetamide (1.99 g, 10 mmol). The product was recrystallised from ethanol as yellow crystals (2.8 g, 79%), Mp. 240°C - 242°C; ^{1}H-NMR (DMSO) δ = 2.65 (3H, s, CH$_3$ py.), 2.73 (3H, s, CH$_3$CO), 3.76 (3H, s, CH$_3$O), 6.9 (2H, d, Ar), 7.45 (2H, s, NH$_2$), 7.56 (2H, d, Ar), 9.04 (1H, s, CH py.), 9.4 (1H, s, NH); IR (KBr) v 3428, 3310, 3251 (NH$_2$, NH), 1680 cm^{-1} (C=O); MS (EI)$^+$: m/z 355 M$^+$; Anal. Calcd for C$_{18}$H$_{17}$N$_3$O$_3$S (355.42): C, 60.83; H, 4.82; N, 11.82, Found: C, 60.76; H, 4.73; N, 11.69.

General procedure for the preparation of compounds 9a,b

A mixture of 5-acetyl-6-methyl-2-thioxo-1,2-dihydropyridine-3-carbonitrile **3a** (1.92 g, 10 mmol) in ethanol as solvent in presence of sodium hydroxide (0.4 g, 10 mmol) with aromatic aldehydes (10 mmol) with stirring for 2hr. then poured onto ice, cold water and acidified with conc. hydrochloric acid until the precipitate was formed. The solid product was recovered by filtration and recrystallised from ethanol.

5-(3-(4-(Dimethylamino)phenyl)acryloyl)-6-methyl-2-thioxo-1,2-dihydropyridine-3-carbonitrile (9a):

Obtained using 4-(dimethylamino)benzaldehyde (1.49g, 10 mmol). Mp. 140°C - 142°C as yellow crystals (2.45 g, 76%); ^{1}H-NMR (CDCl$_3$) δ = 2.65 (3H, s, CH$_3$), 2.79 (6H, s, NMe$_2$), 6.70 (2H, d, Ar), 7.06 (1H, d, CH chalcone), 7.74 (2H, d, Ar), 7.85 (1H, d, CH chalcone), 8.07 (1H, s, CH py.), 13.2 (1H, br., NH); IR (KBr) v 3437 (NH), 2225 (CN), 1685 cm^{-1} (C=O); MS (EI)$^+$: m/z 323 M$^+$; Anal. Calcd for C$_{18}$H$_{17}$N$_3$OS (323.42): C, 66.85; H, 5.30; N, 12.99, Found: C, 66.69; H, 5.17; N, 12.86.

6-Methyl-2-thioxo-5-(3-(p-tolyl)acryloyl)-1,2-dihydropyridine-3-carbonitrile (9b):

Obtained using 4-methyl-benzaldehyde (1.2 g, 10 mmol). Mp. = 240°C - 242°C as yellow crystals (2.2 g, 74.8%); ^{1}H-NMR (DMSO) δ = 2.31 (3H, s, CH$_3$ Ar), 2.58 (3H, s, CH$_3$ py.), 7.24 (2H, d, Ar), 7.49 (1H, d, CH chalcone), 7.6 (1H, d, CH chalcone), 7.69 (2H, d, Ar), 8.58 (1H, s, CH py.), 13 (1H, br., NH); IR (KBr) v 3434 (NH), 2231 (CN), 1659 cm^{-1} (C=O); MS (EI)$^+$: m/z 294 M$^+$; Anal. Calcd for C$_{17}$H$_{14}$N$_2$OS (294.38): C, 69.36; H, 4.79; N, 9.52, Found: C, 69.19; H, 4.68; N, 9.45.

5-(5-Amino-4-cyanothiophen-3-yl)-6-methyl-2-thioxo-1,2-dihydropyridine-3-carbonitrile (10)

In dry flask, a mixture 5-acetyl-6-methyl-2-thioxo-1,2-dihydropyridine-3-carbonitrile **3a** (1.92 g, 10 mmol), malononitrile (0.66 g, 10 mmol) and sulfur (0.32 g, 10 mmol) in ethanol and few drops of triethylamine as base was left under reflux for three hours. The mixture was left for cooling then poured onto ice cold water. The product obtained was recrystallised from a mixture of ethanol/DMF (3:1) as brown crystals (1.9 g, 69.8%), Mp. > 300°C; ^{1}H-NMR (DMSO) δ = 2.45 (3H, s, CH$_3$), 6.95 (2H, s, NH$_2$), 7.07 (1H, s, CH thiophene), 7.2 (1H, s, CH py.); IR (KBr) v 3435, 3350 (NH$_2$), 3250 (NH), 2210 cm^{-1} (CN); Anal. Calcd for C$_{12}$H$_8$N$_4$S$_2$ (272.35): C, 52.92; H, 2.96; N, 20.57, Found: C, 52.85; H, 2.92; N, 20.40.

5-Acetyl-2-ethoxy-6-methylnicotinonitrile (11)

In dry flask, a mixture of 5-acetyl-6-methyl-2-(methylthio)nicotinonitrile **7** (2.06 g, 10 mmol) in ethanol and potassium carbonate was left under reflux for 3hr. after cooling the mixture was poured onto ice cold water. The product was recovered and recrystallised from EtOH/H$_2$O (1:1) as yellowish crystals (1.6 g, 78%), Mp. 78°C - 80°C; ^{1}H-NMR (CDCl$_3$) δ = 1.43 (3H, t, CH$_3$ ethyl), 2.66 (3H, s, CH$_3$), 3.04 (3H, s, CH$_3$CO), 4.54 (2H, q, CH$_2$ ethyl), 7.85 (1H, s, CH py.); IR (KBr) v 2228 (CN), 1688 cm^{-1} (C=O); MS (EI)$^+$: m/z 204 M$^+$; Anal. Calcd for C$_{11}$H$_{12}$N$_2$O$_2$ (204.23): C, 64.69; H, 5.92; N, 13.72, Found: C, 64.51; H, 5.83; N, 13.54.

5-(3-(Dimethylamino)acryloyl)-2-ethoxy-6-methylnicotinonitrile (12)

In dry flask a mixture of 5-acetyl-2-ethoxy-6-methylnicotinonitrile **11** (2.04 g, 10 mmol) in dry xylene as solvent and N,N'-dimethylformamide dimethyl acetal (DMFDMA) (1.32 ml, 10 mmol) was left under reflux for 2 hr., cool and the solvent was evaporated. The product was recovered and recrystallised from EtOH/H$_2$O (1:1) as yellow crystals (1.9 g, 73.3%), Mp. 68°C - 70°C; ^{1}H-NMR (CDCl$_3$) δ = 1.3 (3H, t, CH$_3$ ethyl), 2.62 (3H, s, CH$_3$), 2.68, 3.04 (6H, 2s, NMe$_2$), 4.58 (2H, q, CH$_2$ ethyl), 6.25 (1H, d, CH), 7.87 (1H, d, CH), 8.2 (1H, s, CH py.); IR (KBr) v 2230 (CN), 1684 cm^{-1} (C=O); MS (EI)$^+$: m/z 259 M$^+$; Anal. Calcd for C$_{14}$H$_{17}$N$_3$O$_2$ (259.31): C, 64.85; H, 6.61; N, 16.20, Found: C, 64.56; H, 6.47; N, 16.11.

5-(3-(Dimethylamino)acryloyl)-6-methyl-2-(methylthio)nicotinonitrile (13)

(A) In dry flask, a mixture of 5-acetyl-6-methyl-2-(methylthio)nicotinonitrile **7** (2.06 g, 10 mmol) in dry xylene as solvent and N,N'-dimethylformamide dimethyl acetal (DMFDMA) (1.32 ml, 10 mmol) was left under reflux for 2 hr., cool and poured in dry backer and the solvent was evaporated. The product was recovered and recrystallised from EtOH/H$_2$O (1:1) as yellow crystals (2 g, 76.6%), Mp. 100°C - 102°C; **(B)** In dry flask a mixture of 5-acetyl-6-methyl-2-thioxo-1,2-dihydropyridine-3-carbonitrile **3a** (1.92 g, 10 mmol) in dry xylene as solvent and N,N'-dimethylformamide dimethyl acetal (DMFDMA) (2.64 ml, 20 mmol) was left under reflux for 2 hr., cool and poured in dry backer and the solvent was evaporated. The product was recovered and recrystallised from EtOH/H$_2$O (1:1) as yellow crystals (2.1 g, 80.4%), Mp. and mixed Mp. 100°C - 102°C; ^{1}H-NMR (CDCl$_3$) δ = 2.62, 2.64 (6H, 2s, NMe$_2$), 2.9 (3H, s, CH$_3$ py.), 3.15 (3H, s, SCH$_3$), 5.28 (1H, d, trans CH), 6.28 (1H, d, cis CH), 7.75 (1H, d, trans CH), 8.07 (1H, s, CH py.), 10.15 (1H, d, cis CH); IR (KBr) v 2227 (CN),

1685 cm^{-1} (C=O); MS (EI)$^+$: m/z 261 M$^+$; Anal. Calcd for C$_{13}$H$_{15}$N$_3$OS (261.35): C, 59.74; H, 5.79; N, 16.08, Found: C, 59.63; H, 5.45; N, 15.8.

5-(3-(4-(*Dimethylamino*)*phenyl*)*acryloyl*)-6-*methyl*-2-(*methylthio*)*nicotinonitrile* (14)

(A) mixture of 5-(3-(4-(dimethylamino)phenyl)acryloyl)-6-methyl-2-thioxo-1,2-dihydropyridine-3-carbonitrile **9a** (3.23 g, 10 mmol) in ethanol as solvent and sodium hydroxide (0.4g, 10mmol) with stirring for 1hr., and add methyl iodide (0.62 ml, 10 mmol) with stirring until precipitate was formed. The product was recovered by filtration and was purified by recrystallised from ethanol as yellow crystals (2.5 g, 74%), Mp. 160˚C - 162˚C; **(B)** mixture of 5-acetyl-6-methyl-2-(methylthio)nicotinonitrile **7** (2.06 g, 10 mmol) in ethanol as solvent in presence of sodium hydroxide (0.4 g, 10 mmol) with 4-(dimethylamino)benzaldehyde (1.49 g, 10 mmol) with stirring for 2 hr., until precipitate formed and dilute with water. The product was recovered by filtration and purified by recrystallised from ethanol as yellow crystals (2.4 g, 71%), Mp. and mixed Mp. 160˚C - 162˚C; ^{1}H-NMR (CDCl$_3$) δ = 2.62 (3H, s, CH$_3$), 2.66 (3H, s, SCH$_3$), 2.9, 3.04 (6H, 2s, NMe$_2$), 6.83 (2H, d, Ar), 7.46 (2H, d, Ar), 6.67 (1H, d, CH), 7.38 (1H, d, CH), 7.85 (1H, s, CH py.); IR (KBr) v 2217 (CN), 1648 cm^{-1} (C=O); MS (EI)$^+$: m/z 337 M$^+$; Anal. Calcd for C$_{19}$H$_{19}$N$_3$OS (337.45): C, 67.63; H, 5.68; N, 12.45, Found: C, 67.49; H, 5.62; N, 12.48.

6-(*Dicyanomethylene*)-4-(4-(*dimethylamino*)*phenyl*)-2'-*methyl*-6'-*thioxo*-1,1',6,6'-*tetrahydro*-[2,3'-*bipyridine*]-5,5'-*dicarbonitrile* (15)

In dry flask, a mixture 5-(3-(4-(dimethylamino)phenyl)acryloyl)-6-methyl-2-thioxo-1,2-dihydropyridine-3-carbonitrile **9a** (3.23 g, 10 mmol) and malononitrile dimmer (1.32 g, 10 mmol) in acetic acid and presence of ammonium acetate was left under reflux for three hours. The mixture was left for cooling and poured onto ice, cold water. The product was recovered by filtration and recrystallisation from ethanol as brown crystals (3.25 g, 74.7%), Mp. 260˚C - 262˚C; ^{1}H-NMR (DMSO) δ = 2.38 (3H, s, CH$_3$), 3.06 (6H, s, NMe$_2$), 6.83 (2H, d, Ar), 7.5 (1H, s, CH py.), 7.93 (2H, d, Ar), 8.21 (1H, s, CH py.), 11.93 (1H, br, NH), 12.4 (1H, br, NH); IR (KBr) v 3334, 3207 (2NH), 2206 cm^{-1} (CN); MS (EI)$^+$: m/z 435 M$^+$; Anal. Calcd for C$_{24}$H$_{17}$N$_7$S (435.51): C, 66.19; H, 3.93; N, 22.51, Found: C, 66.06; H, 3.78; N, 22.35.

6-(*Dicyanomethylene*)-4-(4-(*dimethylamino*)*phenyl*)-2'-*methyl*-6'-(*methylthio*)-1,6-*dihydro*-[2,3'-*bipyridine*]-5,5'-*dicarbonitrile* (16)

(A) In dry flask a mixture of 5-(3-(4-(dimethylamino)phenyl)acryloyl)-6-methyl-2-(methylthio)nicotinonitrile **14** (3.37 g, 10 mmol) and malononitrile dimmer (1.32 g, 10 mmol) in acetic acid acid and ammonium acetate was left under reflux for four hours, cool. The solid product was recovered by filtration and recrystallised from acetic acid as brown crystals (3.4 g, 76%), Mp. 220˚C - 222˚C; **(B)** mixture of 6-(dicyanomethylene)-4-(4-(dimethylamino)phenyl)-2'-methyl-6'-thioxo-1,1',6,6'-tetrahydro-[2,3'-bipyridine]-5,5'-dicarbonitrile **15** (4.35 g, 10 mmol) in ethanol as solvent in presence of sodium hydroxide (0.4 g, 10mmol) and methyl iodide (0.62 ml, 10 mmol) with stirring until precipitate formed. The product was recovered by filtration and recrystallised from acetic acid as brown crystals (3.2 g, 71.5%), Mp. and mixed Mp. 220˚C - 222˚C; ^{1}H-NMR (DMSO) δ = 2.61 (3H, s, CH$_3$), 2.65 (3H, s, SCH$_3$), 2.99, 3.01 (6H, 2s, NMe$_2$), 6.82 (2H, d, Ar), 7.09 (1H, s, CH py.), 7.73 (2H, d, Ar), 8.66 (1H, s, CH py.), 10.3 (1H, br, NH); IR (KBr) v 3345 (NH), 2213 cm^{-1} (CN); MS (EI)$^+$: m/z 449 M$^+$; Anal. Calcd for C$_{25}$H$_{19}$N$_7$S (449.54): C, 66.80; H, 4.26; N, 21.81, Found: C, 66.69; H, 4.18; N, 21.66.

6-*Methyl*-5-(1*H*-*pyrazol*-3-*yl*)-1*H*-*pyrazolo*[3,4-*b*]*pyridin*-3-*amine* (17)

In flask a mixture of 5-(3-(dimethylamino)acryloyl)-6-methyl-2-(methylthio)nicotinonitrile **13** (2.61 g, 10 mmol) and excess of hydrazine hydrate was left under reflux for four hours, cool. The solid product was recovered by filtration and recrystallised from ethanol as yellowish crystals (1.6 g, 75%), Mp. 260˚C - 262˚C; ^{1}H-NMR (DMSO) δ = 2.49 (3H, s, CH$_3$), 5.54 (2H, s, NH$_2$), 6.47 (1H, d, CH pyrazole), 7.8 (1H, s, CH py.), 8.3 (1H, d, CH pyrazole), 11.75 (1H, s, NH), 12.91 (1H, s, NH); IR (KBr) v at 3405, 3329, 3136 cm^{-1} (NH$_2$, NH); MS (EI)$^+$: m/z 214 M$^+$; Anal. Calcd for C$_{10}$H$_{10}$N$_6$ (214.23): C, 56.07; H, 4.71; N, 39.23, Found: C, 55.85; H, 4.56; N, 39.16.

2'-*Methyl*-6'-(*methylthio*)-6-*thioxo*-1,6-*dihydro*-[2,3'-*bipyridine*]-5,5'-*dicarbonitrile* (18)

In dry flask a mixture of 5-(3-(dimethylamino)acryloyl)-6-methyl-2-(methylthio)nicotinonitrile **13** (2.61 g, 10 mmol) and cyanothioacetamide (1 g, 10 mmol) in acetic acid and ammonium acetate was left under reflux for four hours. Cool and poured the mixture onto ice cold water. The product was recovered by filtration and recrystallised from ethanol as brown crystals (2.3 g, 77.1%), Mp. 170˚C - 172˚C; ^{1}H-NMR (DMSO) δ = 2.63 (3H, s, CH$_3$), 2.65 (3H, s, SCH$_3$), 7.7 (1H, d, CH py.), 8 (1H, d, CH py.), 8.14 (1H, s, CH py.), 12.25 (1H, br, NH); IR (KBr) v = 3428 (NH), 2221 cm^{-1} (CN); MS (EI)$^+$: m/z 298 M$^+$; Anal. Calcd for C$_{14}$H$_{10}$N$_4$S$_2$ (298.39): C, 56.35; H, 3.38; N, 18.78, Found: C, 56.12; H, 3.24; N, 18.58.

General procedure for the preparation of compounds **19a,b**

In dry flask, a mixture of 5-(3-(dimethylamino)acryloyl)-2-ethoxy-6-methylnicotinonitrile **12** (2.59 g, 10 mmol) or 5-(3-(dimethylamino)acryloyl)-6-methyl-2-(methylthio)nicotinonitrile **13** (2.61 g, 10 mmol) and malononitrile dimmer (1.32 g, 10 mmol) in acetic acid and ammonium acetate was heated under reflux for four hours, cool. The solid product was recovered by filtration and recrystallised from ethanol

6-(Dicyanomethylene)-6'-ethoxy-2'-methyl-1,6-dihydro-[2,3'-bipyridine]-5,5'-dicarbonitrile (19a):

Obtained using 5-(3-(dimethylamino)acryloyl)-2-ethoxy-6-methylnicotinonitrile **12**. Mp. 200°C - 202°C as brown crystals (2.4 g, 73.1%); ^{1}H-NMR (DMSO) δ = 1.39 (3H, t, CH$_3$), 2.62 (3H, s, CH$_3$), 4.50 (2H, q, CH$_2$), 7.58 (1H, d, CH py.), 8.48 (1H, d, CH py.), 8.7 (1H, s, CH py.), 11.3 (1H, br, NH); IR (KBr) v 3330 (NH), 2218 cm^{-1} (CN); MS (EI)$^+$: m/z 328 M$^+$; Anal. Calcd for C$_{18}$H$_{12}$N$_6$O (328.34): C, 65.85; H, 3.68; N, 25.60, Found: C, 65.71; H, 3.52; N, 25.43.

6-(Dicyanomethylene)-2'-methyl-6'-(methylthio)-1,6-dihydro-[2,3'-bipyridine]-5,5'-dicarbonitrile

(19b): Obtained using 5-(3-(dimethylamino)acryloyl)-6-methyl-2-(methylthio)nicotinonitrile **13**. Mp. 190°C - 192°C as brown crystals (2.3 g, 69.7%); ^{1}H-NMR (DMSO) δ = 2.58 (3H, s, CH$_3$), 2.64 (3H, s, SCH$_3$), 6.5 (1H, d, CH py.), 8.2 (1H, d, CH py.), 8.69 (1H, s, CH py.), 11.31 (1H, br, NH); IR (KBr) v 3340 (NH), 2212 cm^{-1} (CN); MS (EI)$^+$: m/z 330 M$^+$; Anal. Calcd for C$_{17}$H$_{10}$N$_6$S (330.37): C, 61.80; H, 3.05; N, 25.44, Found: C, 61.63; H, 2.89; N, 25.27.

8-Acetyl-7-methyl-3-(p-tolyl)pyrido[3',2':4,5]thieno[3,2-d]pyrimidin-4(3H)-one (20)

A mixture of 5-acetyl-3-amino-6-methyl-N-(p-tolyl)thieno[2,3-b]pyridine-2-carboxamide **8b** (3.39 g, 10 mmol) in dry dioxane and DMFDMA (1.32 ml, 10 mmol) with stirring for 12 hrs. The product was recovered by filtration and recrystallised from acetic acid as gray crystals (2.6 g, 74.5%), Mp. 200°C - 202°C; ^{1}H-NMR (DMSO) δ 2.26 (3H, s, CH$_3$ Ar), 2.68 (3H, s, CH$_3$ py.), 2.69 (3H, s, CH$_3$CO), 7.16 (2H, d, Ar), 7.52 (2H, d, Ar), 8.43 (1H, s, CH pyrimidinone), 8.52 (1H, s, CH py.); IR (KBr) v at 1691, 1649 cm^{-1} (2C=O); MS (EI)$^+$: m/z 349 M$^+$; Anal. Calcd for C$_{19}$H$_{15}$N$_3$O$_2$S (349.41): C, 65.31; H, 4.33; N, 12.03, Found: C, 65.19; H, 4.26; N, 11.95.

General procedure for the preparation of compounds **21a,b**

A mixture of N-substituted-5-acetyl-3-amino-6-methylthieno[2,3-b]pyridine-2-carboxamide **8b,c** (10 mmol) in acetic acid and sodium nitrite (1.38 g, 20mmol) with stirring for 1 hr. the precipitate was formed and dilute with water. The product was recovered by filtration and recrystallised from ethanol.

8-Acetyl-7-methyl-3-(p-tolyl)pyrido[3',2':4,5]thieno[3,2-d][1,2,3]triazin-4(3H)-one (21a): Obtained using 5-acetyl-3-amino-6-methyl-N-(p-tolyl)thieno[2,3-b]pyridine-2-carboxamide **8b** (3.39g, 10 mmol). Mp. 170°C - 172°C as gray crystals (3 g, 85.7%); ^{1}H-NMR (DMSO) δ = 2.4 (3H, s, CH$_3$ Ar), 2.74 (3H, s, CH$_3$ py.), 2.77 (3H, s, CH$_3$CO), 7.4 (2H, d, Ar), 7.54 (2H, d, Ar), 9.17 (1H, s, CH py.); IR (KBr) v 1700, 1687 cm^{-1} (2C=O); MS (EI)$^+$: m/z 350 M$^+$; Anal. Calcd for C$_{18}$H$_{14}$N$_4$O$_2$S (350.40): C, 61.70; H, 4.03; N, 15.99, Found: C, 61.56; H, 3.94; N, 15.78.

8-Acetyl-3-(4-methoxyphenyl)-7-methylpyrido[3',2':4,5]thieno[3,2-d][1,2,3]triazin-4(3H)-one (21b): Obtained using 5-acetyl-3-amino-N-(4-methoxyphenyl)-6-methylthieno[2,3-b]pyridine-2-carboxamide **8c** (3.55 g, 10 mmol). Mp. = 220°C - 222°C as gray crystals (2.9 g, 79.4%); ^{1}H-NMR (DMSO) δ = 2.74 (3H, s, CH$_3$ py.), 2.81 (3H, s, CH$_3$CO), 3.85 (3H, s, CH$_3$O), 7.15 (2H, d, Ar), 7.61 (2H, d, Ar), 9.26 (1H, s, CH py.); IR (KBr) v 1687 cm^{-1} (2C=O); Anal. Calcd for C$_{18}$H$_{14}$N$_4$O$_3$S (366.40): C, 59.01; H, 3.85; N, 15.29, Found: C, 58.90; H, 3.76; N, 15.17.

2-Amino-4-(7-methyl-4-oxo-3-(p-tolyl)-3,4-dihydropyrido[3',2':4,5]thieno[3,2-d][1,2,3]triazin-8-yl)thiophene-3-carbonitrile (22):

In dry flask a mixture 8-acetyl-7-methyl-3-(p-tolyl)pyrido[3',2':4,5]thieno[3,2-d][1,2,3]triazin-4(3H)-one **21a** (3.5g, 10 mmol), malononitrile (0.66 g, 10 mmol) and elemental sulfer (0.32 g, 10mmol) in ethanol and few drops of triethylamine as base was heated under reflux for three hours. The mixture was left for cooling and poured onto ice cold water. The product was recovered by filtration and recrystallised from a mixture of ethanol/DMF (3:1) as brown crystals (3 g, 69.7%), M.p 260°C - 262°C; IR (KBr) v 3427 (NH$_2$), 2208 (CN), 1683 cm^{-1} (C=O); MS (EI)$^+$: m/z 430 M$^+$; Anal. Calcd for C$_{21}$H$_{14}$N$_6$OS$_2$ (430.51): C, 58.59; H, 3.28; N, 19.52, Found: C, 58.43; H, 3.14; N, 19.36.

8-(3-(Dimethylamino)acryloyl)-7-methyl-3-(p-tolyl)pyrido[3',2':4,5]thieno[3,2-d][1,2,3]triazin-4(3H)-one (23):

In dry flask a mixture 8-acetyl-7-methyl-3-(p-tolyl)pyrido[3',2':4,5]thieno[3,2-d][1,2,3]triazin-4(3H)-one **21a** (3.5 g, 10 mmol) and DMFDMA (1.32 ml, 10 mmol) in dry dioxane was left under reflux for two hours. The mixture was left for cooling and evaporates the solvent. The product was recovered by filtration and recrystallised from ethanol as brown crystals (2.9 g, 71.6%), Mp. 210°C - 212°C; ^{1}H-NMR (DMSO) δ = 2.39 (3H, s, CH$_3$ Ar), 2.66 (3H, s, CH$_3$ py.), 3.63, 3.67 (6H, 2s, NMe$_2$), 5.42 (1H, d, CH), 7.41 (2H, d, Ar), 7.54 (2H, d, Ar), 7.82 (1H, d, CH), 9.12 (1H, s, CH py.); IR (KBr) υ 1693, 16.44 cm^{-1} (2C=O); MS (EI)$^+$: m/z 405 M$^+$; Anal. Calcd for C$_{21}$H$_{19}$N$_5$O$_2$S (405.48): C, 62.21; H, 4.72; N, 17.27, Found: C, 62.08; H, 4.59; N, 17.11.

4. Conclusion

From the biological importance of pyridine-2(1H)-thione derivatives, we have used it in order for the preparation of biologically important bipyridyles, bi- and uncommon tricyclic compounds.

References

[1]   Granik, V.G., Zhidkova, A.M. and Glushkov, R.G. (1977) Advances in the Chemistry of the Acetals of Acid Amides and Lactams. *Russian Chemical Reviews*, **46**, 361. http://dx.doi.org/10.1070/RC1977v046n04ABEH002137

[2]   Abdulla, R.F. and Brinkmeyer, R.S. (1979) The Chemistry of Formamide Acetals. *Tetrahedron*, **35**, 1675-1735.

[3]   Anelli, P.L., Brocchetta, M., Palano, D. and Visigalli, M. (1997) Mild Conversion of Primary Carboxamides into Carboxylic Esters. *Tetrahedron Letters*, **38**, 2367-2368. http://dx.doi.org/10.1016/S0040-4039(97)00350-X

[4]   Malesic, M., Krbavcic, A., Golobic, A., Golic, L. and Stanovenik, B. (1997) The Synthesis and Transformation of Ethyl 2-(2-acetyl-2-benzoyl-1-ethenyl)amino-3-dimethylaminopropenoate. A New Synthesis of 2,3,4-Trisubstituted Pyrroles. *Journal of Heterocyclic Chemistry*, **34**, 1757-1762.

[5]   Abu-Shanab, F.A., Elnagdi, M.H., Aly, F.M. and Wakefield, B.J. (1994) α,α-Dioxoketene Dithioacetals as Starting Materials for the Synthesis of Polysubstituted Pyridines. *Journal of the Chemical Society, Perkin*, **1**, 1449-1452. http://dx.doi.org/10.1039/p19940001449

[6]   Abu-Shanab, F.A., Redhouse, A.D., Thompson, J.R. and Wakefield, B.J. (1995) Synthesis of 2,3,5,6-Tetrasubstituted Pyridines from Enamines Derived from N,N-Dimethylformamide Dimethyl Acetal. *Synthesis*, **5**, 557-560. http://dx.doi.org/10.1055/s-1995-3954

[7]   Abu-Shanab, F.A., Aly, F.M. and Wakefield, B.J. (1995) Synthesis of Substituted Nicotinamides from Enamines Derived from N,N-Dimethylformamide Dimethyl Acetal. *Synthesis*, **8**, 923-925.

[8]   Abu-Shanab, F.A., Hessen, A.M. and Mousa, S.A.S. (2007) Dimethylformamide Dimethyl Acetal in Heterocyclic Synthesis: Synthesis of Polyfunctionally Substituted Pyridine Derivatives as Precursors to Bicycles and Polycycles. *Journal of Heterocyclic Chemistry*, **44**, 787-791. http://dx.doi.org/10.1002/jhet.5570440406

[9]   Carboni, R.A., Conffman, D.D. and Howard, E.G. (1958) Cyanocarbon Chemistry. XI.[1] Malononitrile Dimer. *Journal of the American Chemical Society*, **80**, 2838-2840. http://dx.doi.org/10.1021/ja01544a061

[10]   Abu-Shanab, F.A., Sherif, S.M. and Mousa, S.A.S. (2009) Dimethylformamide Dimethyl Acetal as a Building Block in Heterocyclic Synthesis. *Journal of Heterocyclic Chemistry*, **46**, 801-827. http://dx.doi.org/10.1002/jhet.69

[11]   Melani, F., Cecchi, L., Colotta, V., Filacchini, G., Martini, C., Giannicini, G. and Lucacchini, A. (1989) Dipyrazolo[5,4-b:3',4'-d]pyridines. Synthesis, Inhibition of Benzodiazepine Receptor Binding and Structure-Activity Relationships. *Farmaco*, **44**, 585-594.

[12]   Abu-Shanab, F.A.M., Mousa, S.A.S., Eshak, E.A., Sayed, A.Z. and Al-Harrasi, A. (**2011**) Dimethylformamide Dimethyl Acetal (DMFDMA) in Heterocyclic Synthesis: Synthesis of Polysubstituted Pyridines, Pyrimidines, Pyridazine and Their Fused Derivatives. *International Journal of Chemistry*, **1**, 207-214.

23

The Research Progress of Hexafluorobutadiene Synthesis

Jing Zhu, Shuang Chen, Baohe Wang, Xiaorong Zhang

Research and Development Center of Petrochemical Technology, Tianjin University, Tianjin, China
Email: cj_zhu1975@tju.edu.cn

Abstract

Hexafluorobutadiene is a new plasma etching gas for semiconductor molectron which has perfect properties and also is a preceding monomer that can be used for synthesizing many fluorinated compounds. This paper described the different synthesis methods of perflurobutadiene from different materials, and contrasted the characteristic of each synthetic method. The route from tetrafluoroethylene has more industrialization prospects.

Keywords

Perfluorobutadiene, Synthesis, Diiodoperfluoroalkanes, Trifluoromonochloroethylene, Tetrafluoroethylene, 1,2-Difluoro-1',2'-dichloroethylene

1. Introduction

Hexafluorobutadiene, which is a fully fluorinated compound with double bonds, has the boiling point of 5.6°C and the density of 1.4 g/ml (15°C). It shows a good foreground as monomer in the synthesis process of fluororesins, fluoroplastics and fluororubbers. In addition, it can be copolymerized with other monomers to synthesis high performance fluoroelastomer and resins with excellent electric properties [1]-[3]. However, the present researches on hexafluorobutadiene application focus mainly on dry etching of VLSI, and studies suggest that it is a fine electrical etching gas with high selectivity and accuracy [4]-[7]. So far, only a few companies produce hexafluorobutadiene all over the world because of the difficulties in its preparation. The preparation methods of hexafluorobutadiene are introduced in this article in order to provide some ideas for the industrialized development.

2. Hexafluorobutadiene Application

2.1. Novel Cyanine Dye

Cyanine dye first appeared in 1856, and it is widely used in synthetic organic chemistry and physical chemistry due to its structural variability. Recently, researchers have found emerging applications of cyanine dye in solar energy utilization, for instance, Rensmo [8] indicated that nanocrystalline ZnO electrode, which was sensitized by organic dye, can be used as photo-anodes in photoelectrochemical solar cell, and photoelectric conversion efficiency of the photoelectrochemical solar cell was 2%. Yagupolskii [9] suggested that a fluorine atom introduced in the connecting bridge could effectively increase the absorbance range of the dye, and Yagupolskii [9] successfully introduced the flooring system into the cyanine dyes via reaction of perfluorobutadiene and benzothiazole, so that the maximum absorption wavelength of the dye was increased from 453 nm to 578 nm. The synthesis process is mainly divided into three steps: synthesized 2-perfluorobutadiene benzothiazole by the reaction of benzothiazole basic magnesium chloride and perfluorobutadiene in tetrahydrofuran solvent at −40°C, and then, prepared N-methyl-2-perfluorobutadiene benzothiazoles boron tetrafluoride salt(I) by the alkylation reaction of 2-perfluorobutadiene benzothiazole with iodomethane under the catalysis of $AgBF_4$. Lastly, the dehydrofluorination reaction of the salt(I) and N-methyl-2-fluorine methylene benzothiazole was conducted to get the cyanine dyes. The specific procedures were as follows (**Scheme 1**).

2.2. Plasma Atching Agent

Recently, many researches had focused on dry etching process of VLSI [10] using hexafluorobutadiene. In the manufacturing process of semiconductor devices, etching process is to use as chemical solvent, corrosive gas or plasma to remove unwanted parts in the wafer or wafer surface layer. Wet etching, which always happened in a chemical solution, conducted isotropic etching reaction under the action of strong acids and can also etch the covered parts. In contrast, dry etching with corrosive gas or plasma ion could realize anisotropic etching in vertical direction on the wafer. Thus, dry etching is applicable to high-precision fine craft, such as large scale integrated circuit (VLSI) etching process.

Hexafluorobutadiene, which is an environmental friendly laser etching gas and just appeared on the market in 2004, etches the line width of 90 nm or less [11]. Compared with octafluorocyclobutane (C_4F_8) which is widely used at present and etches the line width of 130 nm, hexafluorobutadiene has several distinguished features: 1) faster degradation speed in atmosphere. C_4F_6 can be degraded in two days while C_4F_8 requires 3200 years; 2) lower greenhouse effect. Greenhouse effect produced by C_4F_6 is only equivalent to 1/1,000,000 of carbon dioxide and 1/870,000 of octafluorocyclobutane; 3) higher aspect ratio. Hexafluorobutadiene is suitable for the extremely narrow line width process, and its aspect ratio can be up to 10 while the aspect ratio of octafluorocyclobutane is only 3; 4) higher selectivity. It only etches silicon oxide membrane without affecting the photoresist, silicon membrane or nitride membrane. Hexafluorobutadiene (C_4F_6) is 4.5 times the selectivity of carbon tetrafluoride (CF_4), while octafluorocyclobutane (C_4F_8) is 4 times the selectivity of carbon tetrafluoride (CF_4).

Recently, there are several companies focus on the research and development of hexafluorobutadiene, such as Ausimont of Italy, Kanto electrification of Japan, Asahi Glass, Dakin, Russia NITs Medkhim, United States PCBU SERVICES INC. However, only Ausimont and Kanto electrification have the capacity of tons productive scale. With the increasing demand of very large scale integrated circuit and the attention to greenhouse gases, it

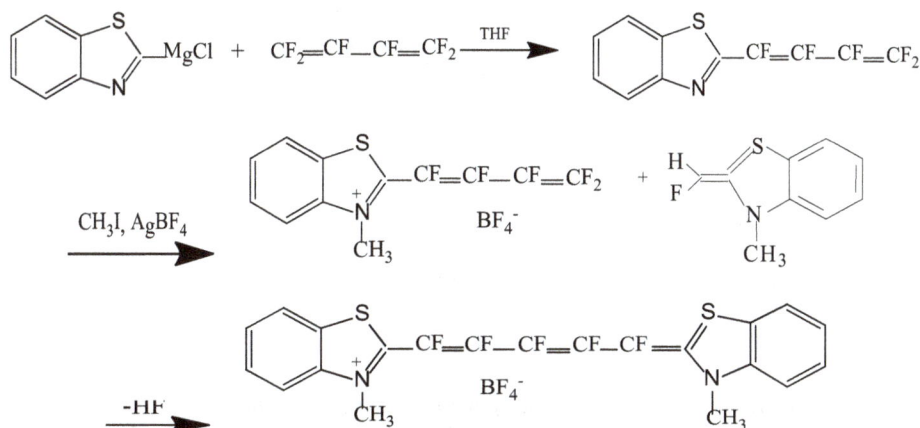

Scheme 1. The synthesis process of the flooring system into the cyanine dyes via reaction of perfluorobutadiene and benzothiazole.

is foreseeable that hexafluorobutadiene, which has the characteristics of the best etching effect and environmental friendly, will become the leading product and will be widely used in the laser etching agent market.

3. The Synthesis of Hexafluorobutadiene

3.1. Using 1,2-Difluoro-Dichloroethylene (CFCl=CFCl) as Raw Material

In 1956, Ruh [12] firstly prepared hexafluorobutadiene using 1,2-difluoro-dichloroethylene as raw materials. The autoclave was charged with 1,2-difluoro-dichloroethylene, and then heated to 275°C. The reaction was conducted in presence of mercury salts catalyst for 6 h, then the reaction mixture was distilled to obtain 1,3,4,4-tetrafluorotetrachloro-1-butene, and the yield was 87.4% (mol). Thereafter, addition reaction between the above product and chlorine was conducted in presence of photocatalyst to get 1,1,2,3,4,4-hexachloro-tetrafluorobutane, and the yield was 99.4%. After that the 1,1,2,3,4,4-hexachloro-tetrafluorobutane and SbF_3Cl_2 were added to the autoclave and stirred at 250°C for 5 h, then1,2,3,4-tetrachloro-hexafluorobutane was generated and the yield was 89.6%. At last, 1,2,3,4-tetrachloro-hexafluorobutane was dissolved in anhydrous ethanol in a glass bottle with zinc powder to act dechlorination reaction. Hexafluorobutadiene was obtained and the yield was 93.5%. The specific reaction equations were as follows:

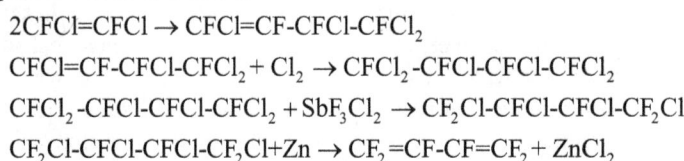

$$2CFCl=CFCl \rightarrow CFCl=CF\text{-}CFCl\text{-}CFCl_2$$

$$CFCl=CF\text{-}CFCl\text{-}CFCl_2 + Cl_2 \rightarrow CFCl_2\text{-}CFCl\text{-}CFCl\text{-}CFCl_2$$

$$CFCl_2\text{-}CFCl\text{-}CFCl\text{-}CFCl_2 + SbF_3Cl_2 \rightarrow CF_2Cl\text{-}CFCl\text{-}CFCl\text{-}CF_2Cl$$

$$CF_2Cl\text{-}CFCl\text{-}CFCl\text{-}CF_2Cl + Zn \rightarrow CF_2=CF\text{-}CF=CF_2 + ZnCl_2$$

William T. Miller [13] reacted 1,2-difluoro-dichloroethylene with fluorine under 0.9 to 1.0 MPa at room temperature for 26 h, then1,2,3,4-tetrachloro-hexafluorobutane was obtained directly and the yield was about 51% (mol). Thereafter, the above product and zinc powder were mixed in the diethylene glycol monobutyl ether, and the 1,2,3,4-tetrachloro-hexafluorobutane was dechlorinated directly to prduce hexafluorobutadiene. The equation was as follows:

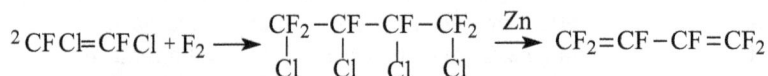

$$2\,CFCl=CFCl + F_2 \longrightarrow \begin{array}{cccc} CF_2 & -CF & -CF & -CF_2 \\ | & | & | & | \\ Cl & Cl & Cl & Cl \end{array} \xrightarrow{Zn} CF_2=CF-CF=CF_2$$

Since this method has the disadvantages of harsh reaction conditions, lower yield and poisonous and highly corrosive materials, there is no application of this method in industry yet.

3.2. With Chlorotrifluoroethylene (CF₂=CFCl) as Raw Material

In 1957, Haszeldine [14] prepared hexafluorobutadiene using chlorotrifluoroethylene as raw material. The reaction firstly proceeded at 35°C to 40°C to give 1,2-dichloro-1,2,2-trifluoro-1-iodoethane and 2,2-dichloro-1,1,2-trifluoro-1-iodoethane, and the yield was 97%. Then 1,2-dichloro-1,2,2-trifluoro-1-iodoethane was purified and mixed with the same amount of mercury. Under UV irradiation for 48h, 1,2,3,4-tetrachloro-hexafluorobutane was obtained, and the yield was 95%. After that, hexafluorobutadiene was got through the dechlorination reaction between the above product and zinc powder in ethanol solution, the yield was 98%. The detailed reaction equations were as follows:

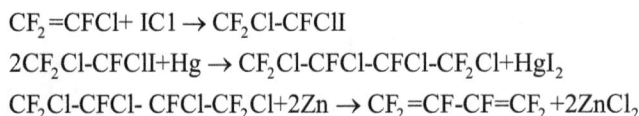

$$CF_2=CFCl + ICl \rightarrow CF_2Cl\text{-}CFClI$$

$$2CF_2Cl\text{-}CFClI + Hg \rightarrow CF_2Cl\text{-}CFCl\text{-}CFCl\text{-}CF_2Cl + HgI_2$$

$$CF_2Cl\text{-}CFCl\text{-}CFCl\text{-}CF_2Cl + 2Zn \rightarrow CF_2=CF\text{-}CF=CF_2 + 2ZnCl_2$$

William. T. Miller [13] decomposed trifluorochloroethylene in Pyrex tube under normal pressure at 550°C to generate l,2-dichloro-hexafluorocyclobutane and 3,4-dichloro-hexafluoro-1-butene，the conversion was about 36.2%. Because of having similar boiling point, the above products were directly added into a glass tube without separation to react with liquid chlorine. Under the light of 200 w at room temperature for 24 h, 3,4-dichlorohexafluoro-1-butene converted to 1,2,3,4-tetrachloro-perfluorobutane with the yield of 15.16% (mol). The reaction products was separated to obtain 1,2,3,4-tetrachloro-hexafluorobutane and l,2-dichloro-hexafluorocyclobutane. Finally, 1,2,3,4-tetrachlorohexafluorobutane was dechlorinated using zinc powder in ethanol to get hexafluorobutadiene. The reaction process was as follows:

$$2\,CF_2{=}CFCl \longrightarrow \begin{array}{c} CF_2{-}CFCl \\ | \qquad | \\ CF_2{-}CFCl \end{array} \xrightarrow{\cdot Cl} \begin{array}{c} CF{-}CF_2 \\ \| \qquad | \\ CF{-}CF_2 \end{array}$$

$$2\,CF_2{=}CFCl \longrightarrow \begin{array}{c} CF_2{=}CF{-}CF{-}CF_2 \\ \quad\;\; | \quad\; | \\ \quad\;\; Cl \quad Cl \end{array} \longrightarrow \begin{array}{c} CF_2{-}CF{-}CF{-}CF_2 \\ | \quad\; | \quad\; | \quad\; | \\ Cl \quad Cl \quad Cl \quad Cl \end{array} \xrightarrow{Zn} CF_2{=}CF{-}CF{=}CF_2$$

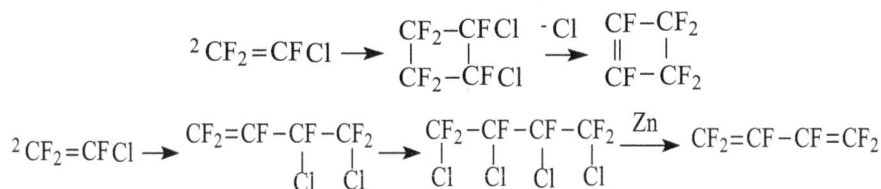

The generated 1,2-dichlorohexafluorocyclobutane is dehydrogenized using zinc powder in ethanol to get hexafluorocyclobutene [15]. D. Young [16] investigated the reaction of 1,2,3,4-tetrahydro-perfluorobutane and hydrogen at 250°C - 300°C for 21.5 h under the pressure of 280 psig in the presence of iodine catalyst, and the yield of product 1,2,3,4-tetrahydroperfluorobutane was more than 90%. Thereafter, the product was distillated under reduced pressure and added into Vycor tube which contained a small amount of H_2O and excessive chlorine, after irradiation of 100 w tungsten filament lamp for 85 h, 1,2,3,4-tetrachloridehexafluorobutane and a handful of impurities were obtained. The 1,2,3,4-tetrachloridehexafluorobutane was separated from the mixtures after washing with alkaline, and then dechlorinated with zinc powder in ethanol. The reaction equations were as follows:

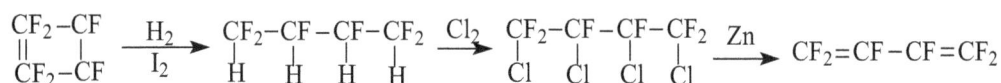

$$\begin{array}{c} CF_2{-}CF \\ \| \qquad | \\ CF_2{-}CF \end{array} \xrightarrow[I_2]{H_2} \begin{array}{c} CF_2{-}CF{-}CF{-}CF_2 \\ | \quad\; | \quad\; | \quad\; | \\ H \quad H \quad H \quad H \end{array} \xrightarrow{Cl_2} \begin{array}{c} CF_2{-}CF{-}CF{-}CF_2 \\ | \quad\; | \quad\; | \quad\; | \\ Cl \quad Cl \quad Cl \quad Cl \end{array} \xrightarrow{Zn} CF_2{=}CF{-}CF{=}CF_2$$

Veeraraghavan [17] developed a one-step synthetic method of hexafluorobutadiene using chlorotrifluoroethylene as raw materials. In the presence of various salt catalysts, reactions of trifluorochloroethylene and trifluorobromoethylene were studied. Reactions using copper salts catalyst and iron salts catalyst were preferable. Specific steps were presented as follows: Three necks' bottle was charged with zinc powder and anhydrous DMF, after cooling of dry ice, trifluorobromoethylenee was added slowly. Then, reaction system was heated up to 70°C, and trifluoroethene zinc-bromine was formed after 1 h. Under the condition of 0°C - 5°C, 100 mmHg vacuum and high purity nitrogen flow, displacement reaction was conducted by slowly adding iron salt ($FeCl_3$ or $FeBr_3$) or copper salt (Cu (OTf)$_2$ or Cu (OAc)$_2$). The generated gas was collected by a refrigerant of $-78°C$. Lastly, reaction system was heated up to 40°C and keep stirring for 2 h in order to discharge all the produced gas. The total yield was between 62% and 70%. Specific process was as follows:

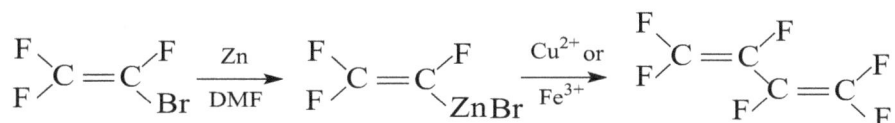

$$\begin{array}{c} F \\ \diagdown \\ \diagup \\ F \end{array}\!\! C{=}C \!\!\begin{array}{c} F \\ \diagup \\ \diagdown \\ Br \end{array} \xrightarrow[DMF]{Zn} \begin{array}{c} F \\ \diagdown \\ \diagup \\ F \end{array}\!\! C{=}C \!\!\begin{array}{c} F \\ \diagup \\ \diagdown \\ ZnBr \end{array} \xrightarrow[Fe^{3+}]{Cu^{2+}\,or} \begin{array}{c} F \\ \diagdown \\ \diagup \\ F \end{array}\!\! C{=}C \!\!\begin{array}{c} F \\ \diagup \\ \diagdown \\ \end{array}\!\! C{=}C \!\!\begin{array}{c} F \\ \diagup \\ \diagdown \\ F \end{array}$$

Robertovich [18] synthesized perfluorobutadiene in the presence of $PdCl_2$ (PhCN)$_2$ and three p-phenyl phosphine p(o-$CH_3C_6H_5$) as catalyst at 50°C - 60°C by trifluorochloroethylene directly dechlorination with excessive zinc powder, and the maximum yield was 34%.

Wang Yi [19] add the 1,1,2-tribromo-trifluoroethane which was synthesized by bromo-trifluoroethylene and bromine slowly into the mixture of excessive zinc powder and DMF at 50°C - 70°C, then the mixtures was heated up to 80°C - 90°C, and maintained for 1h. After the mixture was cooled to 0°C - 5°C, a solution of ferric trichloride and DMF was added. The generated hexafluorobutadiene is cooled at $-25°C$, with total yield of 42.12%.

In the above synthetic routes, many serious environmental pollutions exist due to the using of poisonous and harmful materials during the dehalogenation and coupling reactions of 1,2-dichloro-1,2,2-trifluoro-1-ethyl iodide, such as acetic anhydride and iodine chloride. In addition, utilization rate is low. Meanwhile, dechlorination reaction products at the last step contain a lot of unreactive chlorine compounds, which brought greater difficulty to subsequent purification. However, this synthesis route is one of the earlier developed and mature route and the raw materials is cheap, so that some companies in Russia and Japan established tons scale industrial production devices using this method to product hexafluorobutadiene. Although the one-step method had the advantages of simple operation and wide raw material sources, the condition was rigorous and not easy to control. So far it is confined to the laboratory synthesis, but it has a good industrialization prospect if several key problems will be solved.

3.3. With Trichloroethylene as Raw Material

Vito Tortelli *et al.* [20] synthesized $C_4H_2F_2Cl_6$ by reacting trichloroethylene (TCE) and fluorine diluted with 10 times helium in autoclave until mole ratio of TCE and fluorine was 6.6:1, the flow rate of fluorine was 1 Nl/h, the conversion rate of TCE is 24% and selectivity of $C_4H_2F_2Cl_6$ is 60%. Under the condition of reaction temperature less than 35°C, the reaction of $C_4H_2F_2Cl_6$ and 20% NaOH solution was catalyzed with methyltrioctylammonium chloride by strongly stirring for 8h to obtain 1,2,3,4-tetrachloro-difluoro-butadiene, and the yield was 93%. Then the mixture of the above product, fluorine and helium were added in 50 ml autoclave which was charged of 50.1 gram $CF_3OCFClCF_2Cl$, the flow rates of 1,2,3,4-tetrachloro-difluoro-butadiene, fluorine and helium were 9.4 g/h, 9.4 Nl/h and 0.75 Nl/h respectively. The reaction proceed at 10°C for 1 h, the conversion of 1,2,3,4-tetrachloro-difluoro-butadiene was 97.8%, and selectivity of 1,2,3,4-tetrachloro-perfluorobutane (CFC316) was 64%. At last, hexafluorobutadiene was formed by reaction of 1,2,3,4-tetrachloro-perfluorobutane with zinc powder in 2-propanol, the yield was 95%, while overall yield in the process of TCE was 33.92%. Specific process was as follows:

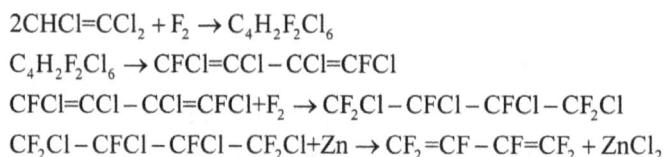

$$2CHCl{=}CCl_2 + F_2 \rightarrow C_4H_2F_2Cl_6$$
$$C_4H_2F_2Cl_6 \rightarrow CFCl{=}CCl{-}CCl{=}CFCl$$
$$CFCl{=}CCl{-}CCl{=}CFCl{+}F_2 \rightarrow CF_2Cl{-}CFCl{-}CFCl{-}CF_2Cl$$
$$CF_2Cl{-}CFCl{-}CFCl{-}CF_2Cl{+}Zn \rightarrow CF_2{=}CF{-}CF{=}CF_2 + ZnCl_2$$

3.4. With Tetrafluoroethylene ($CF_2{=}CF_2$) as Raw Materials

The route of using tetrafluoroethylene as main raw material is divided into two steps: the first step is that tetrafluoroethylene react with iodine or bromine to generate 1,2-dihalo -tetrafluoroethane, and then telomeric reaction between 1,2-dihalo-tetrafluoroethane and tetrafluoroethylene is conducted under certain temperature and pressure to produce a series of α, ω-dihalo-perfluoroalkanes; The second step is that the separated 1,4-dihalo-octafluorobutane dehalogenate to generate hexafluorobutadiene.

3.4.1. Synthesis of 1, 4-Dihaloperfluorobutane

The earliest report of synthesis of diiodoperfluoroalkanes was by Haszeldine in 1951 [21] and after that, large research efforts had been focused on this issue because of the increasing industrial demand. Dindi Hasan and Hagedorn [22] used tetrafluoroethylene and 1,2-diiodotetrafluoroethane to produce diiodoperfluoroalkanes, the total conversion of 1,2-diiodotetrafluoroethane reached 96% after 12 - 20 hours reaction under 245°C and 3.65 - 3.86 Mpa. Catalysis telomerization reaction was also developed [23], using benzoyl peroxide as catalyst; after 26 hours reaction under 80°C, 650 psig, the mixture of $I(C2F4)_nI$ (n = 1, 2, 3, 4)was acquired. Zhang Zongli [24] using Cu as catalyst, synthesized diiodoperfluoroalkanes by telomerization method with tetrafluoroethylene and 1,2-diiodotetrafluoroethane; after 4 hours under 260°C, the conversion reached 57%, the selectivity reached 75%. V. Tortelli and C. Tonelli [25] has developed a one step route in which tetrafluoroethylene and iodine was used to synthesize 1,2-diiodotetrafluoroethane at first, then 1,2-diiodotetrafluoroethane continued to react with superfluous tetrafluoroethylene. Suzuki Kaichirou [26] synthesized diiodoperfluoroalkanes by thermal deiodination reaction of 1,2-diiodotetrafluoroethane under 250°C, the conversion of 1,2-diiodotetrafluoroethane was 70%. Overall, thermal deiodination reaction and telomerization are currently the prevailing methods to synthesize DIPFAs. However, both methods need 1,2-diiodotetrafluoroethane as raw material.

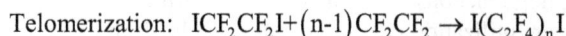

Thermal pyrolysis: $ICF_2CF_2I \rightarrow I(C_2F_4)_n I + (n{-}1)I_2$

Telomerization: $ICF_2CF_2I + (n{-}1)CF_2CF_2 \rightarrow I(C_2F_4)_n I$

3.4.2. Synthesis of Hexafluorobutadiene

1,4-diiodooctafluorobutane separated from homologue of diiodoperfluoroalkanes could be transformed into hexafluorobutadiene by the two methods: one is that 1,4-diiodooctafluorobutane and a certain metal like Zn or Mg deiodinate in protic solvent to produce hexafluorobutadiene; The other is that 1,4-diiodooctafluorobutane react with Grignard reagent in aprotic solvents to get hexafluorobutadiene.

MIKI, Jun and YOSHIMI [27] synthesized hexafluorobutadiene by reacting I-CF_2-CF_2-CF_2-CF_2-I and zinc powder in DMF, the process was as follow: $I-CF_2-CF_2-CF_2-CF_2-I$ was mixed with zinc powder , slowly heated to 120°C for 30 minutes, a mixture of DMF and Zn was droped into the reactants, and the reaction was kept moderate. Then gas mixture containg hexafluorobutadiene was obtained after 30 minutes, the content of hexafluorobutadiene was 65%, and the yield was 53.66%. In contrast, when perfluorochemical (such as FC-30) was added into the reaction as solvent and reaction proceed under 140°C, the content of hexafluorobutadiene in the mixed gases is 88%, and the overall yield of the process is 64.74%. Reaction equations were as follows:

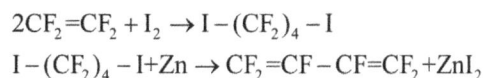

$$2CF_2=CF_2 + I_2 \rightarrow I-(CF_2)_4-I$$
$$I-(CF_2)_4-I+Zn \rightarrow CF_2=CF-CF=CF_2+ZnI_2$$

The above reaction has advantages of simple process and available raw materials, but once it was triggered, the reaction would appear some phenomena, such as releasing a large amount of heat, reacting violently, controlling difficultly, and producing a plenty of by-products, furthermore, the total yield of the products is relatively low. So far, it hasn't been found for industrialization.

Gianangelo synthesized hexafluorobutadiene in the following ways. At first, 1,4-diiodooctafluorobutane [28] or 1,4-dibromooctafluorobutane [29] and tetrahydrofuran (THF) were mixed, heated to boil, and Grignard reagent made from bromoethane and Mg with a concentration of 1 mole was slowly added, the reaction was controlled moderately. The generated gas was only hexafluorobutadiene, the yield was 71.6%. By analysis, the solution still contained a small amount of hexafluorobutadiene, the overall yield was 96%.

Hae-Seok Ji, et al. [30] reported that magnesium powder and a little of Grignard reagent with concentration of 1 mole were added into 1 litre toluene, the mole rate of magnesium powder to 1,4-diiodoperfluorobutane is 1.5:1, under strongly stirring, the mixture was heated to boil, and then slowly droped 400 grams of 1,4-dibro-moperfluorobutane in 3 hours, thereafter, product was collected using cold media with a temperature of −40°C. The weight of product was 178 grams. The content of hexafluorobutadiene was 96%. The yield of hexafluorobutadiene was 94.93%. Reaction equations were as follows:

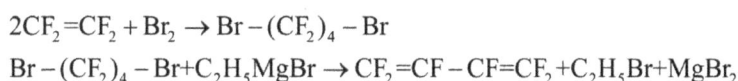

$$2CF_2=CF_2 + Br_2 \rightarrow Br-(CF_2)_4-Br$$
$$Br-(CF_2)_4-Br+C_2H_5MgBr \rightarrow CF_2=CF-CF=CF_2+C_2H_5Br+MgBr_2$$

Comparing to the dehalogenation reactions using metals, the reaction using Grignard reagent to get hexafluorobutadiene has the obvious advantages: moderate reaction conditions, low impurity content, relatively high yield and selectivity. However, products in the reaction using Grignard reagent contain the isomer of perfluorocyclobutane, whose boiling point is only higher 0.8°C than perfluorobutadinene, so that the products are purified with more difficultly. At the same time, using Grignard reagent as raw material had an extremely harsh using condition, that's to say it should be prepared at service times. It increased the difficulty of realizing industrial production to a certain extent, thus, the method hasn't been found in the industrial production. Although these synthesis methods had quite a few shortcomings, α, ω-diiodoperfluoroalkanes, which was an intermediate for many synthesis of useful fluorinated compound, was used extensively, in addition, it could obviously reduce the production cost of hexafluorobutadiene and made this method have extremely strong competitiveness.

3.4.3. With Waste PTF as Raw Materials

Considering the severe reaction conditions and using of the high purity tetrafluoroethylene monomer, a new synthesis method of 1,2-diiodotetrafluoroethane has been developed. The detail process is as follows: firstly, waste tetrafluoroethylene was pyrolyzed to obtain tetrafluoroethylene monomer at 500°C and 2 kPa [31]. The conversion of waste tetrafluoroethylene pyrolysis reaction was 99.6%, and the yield of tetrafluoroethylene monomer was 95.52%. And then the pyrolysis gas, which also contained hexafluoropropylene and octafluorocyclobutane, directly reacted with iodine at 150°C and 1.2 MPa without separation, and the conversion of iodine was 98.05%, while the selectivity was 98.4%. In terms of the conversion and selectivity of iodine, there was no essentially difference between the two means of using pyrolysis gas and high purity tetrafluoroethylene as raw material respectively.

Based on the above green synthesis process, a further study on the one-step synthesis method of α, ω-diiodoperfluoroalkanes [24] has been also developed. Preparation of 1,2-diiodotetrafluoroethane and synthesis of 1,4-

diiodooctafluorobutane were conducted by one step reaction, which not only reduced the loss of materials, but also greatly shortened the reaction time and improved the yield of target product. Detail process is as follows: the autoclave was charged with iodine, solid copper powder and 1,2-diiodotetrafluoroethane as solvent, and then heated to 160°C to synthesize 1,2-diiodotetrafluoroethane. After iodine reacted completely by inletting into pyrolysis gas, C_2F_4 was continuously fed to keep a certain pressure in the autoclave, and then the reaction proceed for 4 hours at 260°C. After that the reaction temperature was reduced to 160°C and C_2F_4 gas was continuously added to react with elemental iodine generated during thermal pyrolysis reaction. 1,2-diiodotetrafluoroethane in the product was separated by distillation to recycle as solvent. According to $I(C_2F_4)I$, the total mole yield of 1,4-diiodooctafluorobutane was 68.15%.

1,4-diiodooctafluorobutane, which was obtained by distillating from the above reaction product, reacted with the Grignard reagent to get hexafluorobutadiene through reactive-distillation process [32]. Specific reaction process was as follows: 400 ml tetrahydrofuran was placed into the tower kettle with a thermocouple, Grignard reagent and tetrahydrofuran solution contained diiodooctafluorobutane were respectively added in the two dropping funnel, Grignard reagent was prepared with bromobenzene and Mg in the diethyl ether, and the addtion amount of Grignard reagent and tetrahydrofuran solution were 670 ml (2.01 mol) and 454 g (1.0 mol) respectively. Under the protection of nitrogen with a flow rate of 0.5l/h, tetrahydrofuran was heated to boil, the cooling temperature of tower top was maintained between 6°C - 7°C, and a certain return flow on the top of the tower was kept. After total reflux, Grignard reagent and diiodooctafluorobutane solution were droped slowly from the middle of distillation tower during 3 - 4 h, and the mass of 164 g of crude product hexafluorobutadiene was got by collecting with a low temperature of −90°C in cold trap. The conversion rate of 1,4-diiodooctafluorobutane was 99.53%, the yield of hexafluorobutadiene was 96.75%, and the selectivity was 97.2%.

4. Conclusion

Hexafluorobutadiene (C_4F_6) is a new plasma etching agent which is used in the manufacture process of the large scale integrated circuit and memory chips with high speed and high capacity. It is also a synthetic monomer of many materials, such as a new type of fluorine resin, fluorine plastic and fluorine rubber. However, only Italy and Japan have the capacity of more than ton scale production of hexafluorobutadiene, which makes its market price high ($500 - $600/kg) at present. So it has extremely good economic prospect to develop industrial production approach of hexafluorobutadiene as soon as possible. Among these several synthetic methods of hexafluorobutadiene, the synthetic method using trifluorochloroethylene as material is mainly route being used in industrial manufacture. This route has advantages of mature technology, raw material easy to get and simple operation, but the chlorine compound generating in the reaction and raw material trifluorochloroethylene all destroys the Ozone layer, and is prohibited to use in many countries. The strict requirement to the content of chlorine compound in hexafluorobutadiene makes the refinement of production become more difficult and the cost of production becomes higher. In the contrast, the route using tetrafluoroethylene as raw material has a lower costing of production because the 1,6-diiodoperfluorohexane and 1,8-diiodoperfluorooctane by-products also have a high price. Furthermore, the intermediates are iodine compounds, so that this route has none environmental problems. Now, the waste polytetrafluoroethylene using as raw material can further reduce the costing of hexafluorobutadiene production. If the restrictions of Grignard reagent application can be reduce, the route will have very good prospects for industrial application.

References

[1]   Toshio, Y. and Takahiro, N. (2008) Coating Composition, Coating Formed Therfrom, Anti-Reflection Coating, Anti-Reflection Film, and Image Display Device. US7371786.

[2]   Hsing-Yeh, P. and Willia, L. (2001) Preparation of Fluorinated Polymers. US6218464.

[3]   Massimo, M. and Dario, S. (1999) Novelties and Prospects in the Synthesis of Perfluoropolyethers by Oxidative Polymerization of Fluoroolefins. *Journal of fluorine Chemistry*, **95**, 19-25.
      http://dx.doi.org/10.1016/S0022-1139(98)00295-4

[4]   Sun, W. and Wang, X. (2010) Soldering-Pan and Forming Method Thereof. CN101645408-A.

[5]   Park, S., Cheong, J. and Park, S.S. (2008) Method for Fabricating Capacitor in Semiconductor Device. US200808-1429-A1.

[6]   Jun, M., Hitoshi, Y. and Hirozaku, A. (2003) Process for Production of Perfluoroalkadienes. US6610896.

[7] Hung, H., Caulfield, J.P and Shan, H. (2003) Process for Etching Oxide Using Heafluorobutadiene or Related Fluoro-carbons and Manifesting a Wide Process Window. US6602434.

[8] Rensmo, H., Keis, K., Lindström, H., et al. (1997) High Light-to-Energy Conversion Efficiencies for Solar Cells Based on Nanostructured ZnO Electrodes. Journal of Physical Chemistry B, 101, 2598-2601. http://dx.doi.org/10.1021/jp962918b

[9] Yagupolskii, I..M., Chernega, O.I., Kondratenko, N.V., et al. (2010) Synthesis of the First Representative of Dicar-bonthiacyanine Dyes with Completely Fluorinated Polymethine Chain. Journal of Fluorine Chemistry, 131, 165-171

[10] Hung, R., Caulfield, J.P., Shan, H., et al. (2003) Highly Selective Oxide Etching Process Using Hexafluorobutadiene. US 6613691.

[11] Hung, H., Caulfield, J.P., Shan, H.Q., et al. (2003) Highly Selective Process for Etching Oxide over Nitride Using Hexafluorobutadiene. US 2003/0000913.

[12] Palmer, R.R. and Ralph, D. (1958) Preparation of Hexafluorobutadiene. GB798407.

[13] Miller, W.T. (1950) Polyunsaturated Fluoroolefins. US2668182.

[14] Haszeldine, R.N. (1962) Coupling of Halogenated Organic Compounds. US3046304.

[15] Harmon, J. (1943) Polyfluorocyclobutenes. US2436142.

[16] Young, D. (1960) Improvements in or Relating to the Preparation of 1,2,3,4-tetrahydroperfluorobutane and Perfluoro-butadiene. GB839756.

[17] Veeraraghavan Ramachandran, P. and Venkat Reddy, G. (2008) Preparative-Scale One-Pot Syntheses of Hexafluo-ro-1,3-Butadiene. Journal of Fluorine Chemistry, 129, 443-446. http://dx.doi.org/10.1016/j.jfluchem.2008.01.015

[18] Robertovich, M.O. (2008) Method of Obtaining Hexafluorobutadiene. RU2340588.

[19] Wang, Y., Wang, Q.M., Shen, D.X., et al. (2012) Method for Preparing Hexafluorobutadiene-1,3. CN102399128.

[20] Tortelli, V., Millefanti, S. and Carella, S. (2010) Process for the Synthesis of Perfluorobutadiene. US2010/0280291.

[21] Haszeldine, R.N. (1951) Synthesis of Fluorocarbons, Perfluoroalkyl Iodides, Bromides and Chlorides, and Perfluo-roalkyl Grignard Reagnts. Nature, 167, 139-140. http://dx.doi.org/10.1038/167139a0

[22] Dindi, H. and Hagedorn, J.J. (2004) Process for Manufacturing Diiodoperfluoroalkanes. US 006825389.

[23] Thiokol Chemical Corporation (US) (1973) Method of Making Alpha-Omega-Diiodoperfluoroalkanes. GB1301617.

[24] Zhang, Z.L., Zhu, J., Jing, X., et al. (2005) Method for Preparing Alpha, Omega Diiodoperfluo-Alkane. CN1686985.

[25] Tortelli, V. and Tonelit, C. (1990) Telomerization of Tetrafluoroethylene and Hexafluoropropene: Synthesis of Dii-odoperfluoroalkanes. Journal of Fluorine Chemistry, 47, 199-217. http://dx.doi.org/10.1016/S0022-1139(00)82373-8

[26] Suzuki, K., Uchijima, Y., Munakata, S., et al. (1978) Preparation of 1, 4-diiodoperfluorobutane. JP53144507 (A).

[27] Jun, M. and Htoshi, Y. (2002) Process for Production of Perfluoroalkadienes. EP1247791.

[28] Bargigia, G. and Tortelli, V. (1987) Process for the Synthesis of Hexafluorobutadiene and of Higher Perfluoronated dienes. US4654448.

[29] Bargigia, G., Tortelli, V. and Tonelli, C. (1987) Process for the Synthesis of Perfluoroalkandiene. EP 0270956.

[30] Ji, H.S., Cho, O.J. and Ahn, Y.H. (2009) Method for Preparing Perfluoroalkandiene. US7504547.

[31] Zhu, J., Wang, B.H. and Liu, D.Z. (2013) Synthesis of 1,2-diiodotetrafluoroethane with Pyrolysis Gas of Waste Poly-terafluoroethylene as Raw Material. Green Chemistry, 15, 1042-1047. http://dx.doi.org/10.1039/c3gc36880g

[32] Zhu, J., Liu, C., Wang, B.H., et al. (2010) Preparation Method of Hexafluorobutadiene. CN101774884.

24

Synthesis, Spectral and Antimicrobial Studies of Bis (Cyclopentadienyl) Titanium (IV) Bis (O,O'-Dialkyl and Alkylenedithiophosphate) Complexes

Adnan A. S. El Khaldy[1]*, Florence Okafor[2], Alaa M. Abu Shanab[3]

[1]Department of Physics, Chemistry, and Mathematics, College of Engineering, Technology and Physical Sciences, Alabama A&M University, Normal, AL, USA
[2]Department of Biological & Environmental Sciences, College of Agricultural, Life and Natural Sciences, Alabama A&M University, Normal, AL, USA
[3]Chemistry Department, Al-Aqsa University, Gaza, Palestine
Email: *adnan.elkhaldy@aamu.edu

Abstract

A new complexes of $Cp_2Ti[S_2P(OR)_2]_2$ (where R = Et, Pr-n, Pr-i, Bu-i and Ph) and $Cp_2Ti[S_2POGO]_2$ (where G = -$CH_2CMe_2CH_2$-, -$CH_2CEt_2CH_2$- and -CMe_2CMe_2-) were prepared by the dropwise addition of the appropriate O,O'-dialkyl or -alkylenedithiophosphoric acid to biscyclopentadienyl titanium dichloride in 1:2 molar ratio and refluxed in benzene solution. These novel deep red colored complexes were characterized by elemental analyses, molecular weight measurements and spectroscopic techniques (IR., NMR ^{1}H, ^{13}C and ^{31}P NMR). These titanium (IV) dithio complexes have also been screened for their antibacterial activities.

Keywords

Titanium (IV) Dialkyl and Alkylenedithiophosphate Complexes

1. Introduction

The synthesis of coordination compounds with sulfur containing ligands has been in the center of interest in chemical research for many years [1]-[6]. A survey of literature on dithiophosphato derivatives of titanium and

---

*Corresponding author.

organotitanium reveals that only simple derivatives (e.g., those containing organic and halo substituents on tita-nium on addition to the dithiophosphato group) have been described [7] [8]. Derivatives containing other mo-nodentate ligands in addition to dithiophosphate have not been isolated. Dialkyl and alkylenedithiophosphates exhibit a variety of coordination modes of bonding [9]-[12] and their metal complexes have important biochem-ical, analytical and industrial applications [13]-[17]. The biocidial importance of organophosphorus compounds is well known. The synthesis of organotitanium (IV) dithiophosphate compounds provide model systems of in-terest because the presence of biologically active organophosphorus and organo titanium moieties in a single molecule could provide new information about the bioactivity of titanium compound. In view of the ready con-version of Ti-Cl bonded into Ti-S bonded compounds, the preparative route chosen for the above compounds was the direct interaction between Biscyclopentadienyl Titanium (IV) dichloride and dialkyl (or alkylene) dio-thiophosphoric acids. Thus, the reactions of Titanocene with dithiophosphoric acids have been carried out in 1:2 molar ratios under mild condition.

Before discussing the results of the above reactions, it may be relevant to mention the structural features of the Cp_2TiCl_2 [18]. Titanocene does not adopt the typical "sandwich" structure like ferrocene due to the 4 ligands around the metal center, but rather takes on a distorted tetrahedral shape [19]. Although its crystal structure has not been determined, its structure has been demonstrated in a number of other organotitanium dithiophosphates. In dilute solution, however, the titanocene structure which is monomeric species containing unidentate dithio-phosphates and 4-coordinated titanium atoms.

2. Experimental

Stringent precautions were taken to exclude moisture. Solvents (benzene, n-hexane) were dried by standard me-thods. Glycols were distilled before use; Titanocene (Merck) was used as received. Dialkyl and alkylenedithio-phosphoric acids were prepared by the reaction of phosphorus pentasulfide and alcohols in a 1:4 ratio, and in a 1:2 ratio with glycols as described in the literature [20]. Sulfur was determined by Messenger's method as ba-rium sulfate. Titanium was determined titanium oxide (cupferron method). Infrared spectra were recorded as Nujol mulls using CsI cells in the region 4000 - 200 cm^{-1} on an FT-IR 8201PC spectrophotometer.

^1H and ^{13}C spectra were recorded on a Jeol-FT NMR spectrometer-LA300 and using TMS as the internal ref-erence. ^{31}P NMR spectra were recorded in $CHCl_3$ using H_3PO_4 as an external reference on the same instrument. The following synthetic details for a specific 1:2 reaction represent the procedure used to synthesize all com-pounds.

2.1. Reaction between Biscyclopentadienyl Titanium Dichloride with Dialkyl (OPr-n) and Alkylenedithiophosphoric Acids in 1:2 Molar Ratios

A benzene (~10 ml) solution of $HS_2P(OPr-n)_2$ (0.818 g; 3.82 mmol) was added to benzene (~15 ml) solution of Cp_2TiCl_2 (0.475 g; 1.91 mmol) dropwise with stirring at room temperature. The reaction mixture was refluxed for ~5 hour, during which the color of the reaction mixture changed from color red to dark red. The excess solvent was removed under reduced pressure and the product washed repeatedly by n-hexane and the desired product was finally dried under reduced pressure.

2.2. Antimicrobial Studies

Bioactivity studies were conducted using three bacterial strains; *Escherichia coli*, *Bacillus cereus*, and *Pseudo-monas aeruginosa* as test microorganisms.

Susceptibility of the microorganisms to the novel chemical compounds was determined using the Agar diffu-sion method and in accordance with the CLSI (formerly NCCLS) guidelines [21]. This method was used as a qualitative method to determine whether each bacterium is resistant, intermediately resistant or susceptible to the synthesized chemicals. The broth cultures of the microorganisms were first grown until they had an optical density (OD) or absorbance of 0.8 - 1.0 at a 600 nm wavelength. Inoculum of approximately 10^6 colony forming unit (CFU) of each isolate was plated on to the Mueller-Hinton Agar plates to form a confluent or lawn growth. The test chemical (400 ug/ml) or control (70% ethanol, v/v) was applied to sterile disks and placed on the inoculated plates and incubated at 37°C for 24 h.

The zones of inhibition/clearance of microbial growth around the disks containing the extracts/control were measured. The zone of inhibition was defined as the shortest distance (in mm) from the outside margin of the initial point of microbial growth. Three replicates were made for each test organism.

3. Results and Discussion

Biscyclopentadienyl titanium bis (dialkyl and alkylenedithiophosphate) have been synthesized by the reaction of biscyclopentadienyl titanium dichloride (Titanocene) with dialkyl and alkylenedithiophosphoric acids in 1:2 molar ratios in refluxing benzene as in Equations (1) and (2).

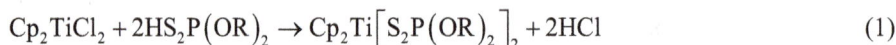

$$Cp_2TiCl_2 + 2HS_2P(OR)_2 \rightarrow Cp_2Ti\left[S_2P(OR)_2\right]_2 + 2HCl \tag{1}$$

where R = Et, Pr-n, Pr-i, Bu-i and Ph

$$Cp_2TiCl_2 + 2HS_2\overline{POGO} \rightarrow Cp_2Ti\left[S_2\overline{POGO}\right]_2 + 2HCl \tag{2}$$

G = -CH$_2$CMe$_2$CH$_2$-, -CH$_2$CEt$_2$CH$_2$- and -CMe$_2$CMe$_2$.

The color of the reaction medium changed from red to deep red color with the progress of reaction. Biscyclopentadienyl titanium bis (dialkyl and alkylenedithiophosphate) derivatives are deep red color solids in open and cyclic chain complexes. All these compounds are soluble in common organic solvents like benzene, dichloromethane and chloroform. The molecular weight of all these products determined by cryoscopic method in benzene indicated the monomeric nature of these products (**Table 1**).

3.1. IR Spectra

IR spectra of biscyclopentadienyl titanium bis (dialkyl and alkylenedithiophosphate), have been recorded in the region 4000 - 400cm^{-1} [22] [23]. The band shown by the parent acids in the region 2544 - 2400 cm^{-1}, due to SH stretching vibration, are absent for biscyclopentadienyl titanium bis(dialkyl and alkylenedithiophosphate) derivatives, indicating the formation of Ti-S bond with the appearance a new band in the regions 428 - 400 cm^{-1} [22] [24]. The bands present in the region 1104.0 - 1014.5 cm^{-1} and 937.3 - 800 cm^{-1} have been assigned to ν (P)-O-C and νP-O-(C) stretching vibrations respectively. Strong bands in the region 995 - 921.9 cm^{-11} are due to dioxaphospholane and dioxaphosphorinane ring vibrations [25]-[27]. The bands observed present in the region 704.0- 638.0 cm^{-11} can be assigned to ν P=S vibrations [28]. The bands in medium intensities in the region 602 - 513.0 cm^{-11} may be attributed to vibration of νP-S asymmetric and symmetric vibrations [29]. Details regarding the individual bands have been included in **Table 2**.

Table 1. Physical properties and analytical data of biscyclopentadienyl titanium bis (O,O-dialkyl and alkylene dithiophosphate) compounds.

SI. No.	Compounds	Physical State	M. P. °C	Mol. Wt. Found/(Calc.)	% H Found/(Calc.)	%C Found/(Calc.)	% S Found/(Calc.)	% Ti Found/(Calc.)
1	Cp$_2$Ti[S$_2$P(OEt)$_2$]$_2$	Deep Red Solid	137°	542.22/(548.50)	5.54/(5.51)	38.99/(39.41)	22.43 / (23.38)	7.10/(8.72)
2	Cp$_2$Ti[S$_2$P(OPr-n)$_2$]$_2$	Deep Red solid	153°	598.54/(604.61)	6.38/(6.33)	43.8/(43.7)	20.88 / (21.21)	7.80/(7.91)
3	Cp$_2$Ti[S$_2$P(OPr-i)$_2$]$_2$	Deep Red Solid	121°	598.69/(604.61)	6.29/(6.33)	43.92/(43.70)	20.98 / (21.21)	7.83/(7.91)
4	Cp$_2$Ti[S$_2$P(OBu-i)$_2$]$_2$	Deep Red Solid	195°	659.32/(660.72)	7.23/(7.01)	47.16/(47.26)	18.94 / (19.41)	6.88/(7.24)
5	Cp$_2$Ti[S$_2$P(OPh)$_2$]$_2$	Deep Red Solid	217°	738.11/(740.67)	4.23/(4.08)	54.89/(55.13)	16.47/ (17.32)	5.91/(6.46)
6	Cp$_2$Ti[S$_2$POCH$_2$CMe$_2$CH$_2$O]$_2$	Deep Red Solid	187°	568.67/(572.52)	5.48/(5.28)	42.16/(41.95)	21.97/ (22.40)	7.95./(8.36)
7	Cp$_2$Ti[S$_2$POCH$_2$CEt$_2$CH$_2$O]$_2$	Deep Red Solid	173°	627.78/(628.63)	5.98/(6.09)	46.22/(45.85)	19.97 / (20.40)	7.81./(7.61)
8	Cp$_2$Ti[S$_2$POCMe$_2$CMe$_2$O]$_2$	Deep Red Solid	203°	599.43/(600.58)	5.98/(5.70)	43.62/(43.99)	21.41/ (21.36)	7.60/(7.97)

Table 2. IR spectral data (cm^{-1}) of biscyclopentadienyl titanium bis (O,O-dialkyl and alkylene dithiophosphate) compounds.

SI. No.	Compounds	ν(P)-O-C	νP-O-(C)	Ring Vibration	ν P=S	νP-S	ν (Ti-S)
1	Cp$_2$Ti[S$_2$P(OEt)$_2$]$_2$	1014.5 s	817.8 s	--	644.2 m	530.0 w	400.0 m
2	Cp$_2$Ti[S$_2$P(OPr-n)$_2$]$_2$	1060.0 m	827.4 m	--	655.8 m	520.0 m	408.0 w
3	Cp$_2$Ti[S$_2$P(OPr-i)$_2$]$_2$	1022.2 m	800.4 m	--	638.0 m	540.0 w	400.0 m
4	Cp$_2$Ti[S$_2$P(OBu-i)$_2$]$_2$	1018.3 s	804.3 s	--	661.5 m	550.0 m	406.0 w
5	Cp$_2$Ti[S$_2$P(OPh)$_2$]$_2$	1104.0 s	820.8 s	--	682.5 s	513.0 w	409.0 w
6	Cp$_2$Ti[S$_2$POCH$_2$CMe$_2$CH$_2$O]$_2$	10415 s	815.8 m	987.5 s	667.3 m	601.7 m	410.0 w
7	Cp$_2$Ti[S$_2$POCH$_2$CEt$_2$CH$_2$O]$_2$	1066.6 s	937.3 m	995.0 s	671.2 m	602.0 m	428 0 m
8	Cp$_2$Ti[S$_2$POCMe$_2$CMe$_2$O]$_2$	1022.5 s	800.4 s	921.9 m	704.0 m	584.4 m	414.0 m

s = strong, m = medium, w = weak and b = broad absorption band.

3.2. ^{1}H NMR Spectra

The ^{1}H NMR spectra Biscyclopentadienyl titanium bis (dialkyl and alkylenedithiophosphate) recorded in CDCl$_3$, show the characteristic resonance due to alkoxy and glycoxy (dithio moiety) protons. These ^{1}H NMR spectral data are given in **Table 3**. The singlet peak at (3.1 - 3.5 ppm) in the parent dithiophosphoric acids and assigned to SH proton, is absent from the spectra of Titanium bis (dithiophosphate) derivatives indicating deprotonation of SH group and forming of Ti -S bond [30].

3.3. ^{13}C NMR Spectra

The ^{13}C NMR spectra of biscyclopentadienyl titanium bis (dialkyl and alkylenedithiophosphate) complexes were recorded in deuterated chloroform at ambient temperature (**Table 4**). The spectra show very small chemical shifts when compared to those obtained for the parent dithiophosphoric acids and indicate no substantial difference in the structure [31].

3.4. ^{31}P NMR Spectra

The proton decoupled ^{31}P NMR spectra of biscyclopentadienyl titanium (IV) bis (dialkyl and alkylenedithiophosphate) derivatives, **Table 3**, show only one signal peak for each complex in the region 77.9 - 93.7 ppm. The observation of only one sharp singlet for all compounds reflects the equivalent nature of phosphorous nuclei and the purity of the compound. However, no notable difference was observed in comparison to the parent acids [32] [33]. According to Glidewell, these small shifts indicate monodentate behavior of the ligand [34].

3.5. Structural Elucidation

Considering the normal mode of bonding of dithiophosphate with the metal as bidentate chelating ligand and based on the above spectral studies of the complexes using IR, NMR (^{1}H, ^{13}C, ^{13}P), molecular weight determination and elemental analyses. We suggest the following structure (**Figure 1** and **Figure 2**).

3.6. Results of Bioactivity Tests

The preliminary results show that the chemical compounds slightly inhibited the growth of *Escherich coli and Bacillus cereus* after a 24 h incubation period, but had little or no effect on *Pseudomonas aeruginosa* under similar conditions. See (**Table 5**) and **Figure 3**. The results are means of three replicate experiments. *Escherichia coli* a Gram-negative bacterium commonly found in the lower intestine of warm-blooded organisms [35] appeared to be the most susceptible to all chemicals tested with zones of inhibition ranging from 5.0 ± 0.76 (Cp$_2$Ti [S$_2$P(OPr-n)$_2$]$_2$) to 11.1 ± 0.54 for (Cp$_2$Ti [S$_2$P(OEt)$_2$]$_2$). The test chemicals had very little inhibitory effect on *Pseudomonas aeruginosa*.

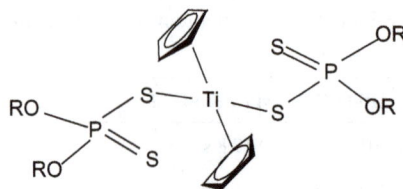

Figure 1. Suggested structure for biscyclopentadienyl titanium bis (dialkyldithiophosphate) derivatives.

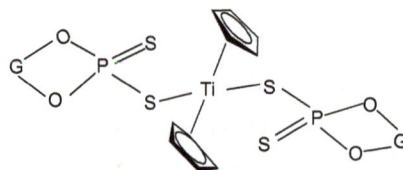

Figure 2. Suggested structure for biscyclopentadienyl titanium bis (alkylenedithiophosphate) derivatives.

Table 3. ^{1}H and ^{31}P NMR spectral data of biscyclopentadienyl titanium bis (O,O-dialkyl and alkylene dithiophosphate) compounds.

Sl. No.	Compounds	^{1}H chemical shift in δ ppm in CDCl$_3$	^{31}P chemical (parent acid)
1	Cp$_2$Ti[S$_2$P(OEt)$_2$]$_2$	1.14, t (J = 6.5 Hz), 12H(CH$_3$), 3.7, q (J = 6.0 Hz), 8H(OCH$_2$) 6.42, s, 10H(C$_5$H$_5$)	86.0. (85.7)
2	Cp$_2$Ti[S$_2$P(OPr-n)$_2$]$_2$	0.76, t (J = 7.5 Hz), 12H(CH$_3$) 1.34, m (J = 6.5 Hz), 8H(CH$_2$) 4.0 - 4.1, t (J = 7.5 Hz, 8H(OCH$_2$) 6.6, s, 10H(C$_5$H$_5$)	86.2 (86.1)
3	Cp$_2$Ti[S$_2$P(OPr-i)$_2$]$_2$	1.21, d (J = 6.6 Hz), 24H(CH$_3$) 4.40 - 4.42, m J (PH) = 12 Hz, 4H(OCH) 6.42, s, 10H(C$_5$H$_5$)	82.3 (82.3)
4	Cp$_2$Ti[S$_2$P(OBu-i)$_2$]$_2$	0.8, d (J = 7 Hz), 24H(CH$_3$) 1.92, m (J = 6.5 Hz), 4H(CH) 3.80, d (J = 7 Hz), 8H(OCH$_2$) 6.60, s, 10H(C$_5$H$_5$)	85.6 (85.7)
5	Cp$_2$Ti[S$_2$P(OPh)$_2$]$_2$	7.2 - 7.4, m, 2OH(OC$_6$H$_5$) 6.4, s, 10H(C$_5$H$_5$)	79.9 (79.9)
6	Cp$_2$Ti[S$_2$POCH$_2$CMe$_2$CH$_2$O]$_2$	0.82, s, 12H(CH$_3$) 4.10, d, 8H(OCH$_2$), J(PH) = 15.6 Hz 6.35, s, 10H(C$_5$H$_5$)	77.4 (77.3)
7	Cp$_2$Ti[S$_2$POCH$_2$CEt$_2$CH$_2$O]$_2$	0.71, t (J = 7.5 Hz), 12H(CH$_3$) 1.11, q (J = 7.5 Hz), 8H(CH$_2$) 4.02, d, 8H(OCH$_2$), J(PH) = 16 Hz 6.28, s, 10H(C$_5$H$_5$)	78.3 (78.5)
8	Cp$_2$Ti[S$_2$POCMe$_2$CMe$_2$O]$_2$	1.06, s, 24H(CH$_3$) 6.50, s, 10H(C$_5$H$_5$)	93.4 (93.1)

Table 4. ^{13}C NMR spectral data of biscyclopentadienyl titanium bis (O,O-dialkyl and alkylene dithiophosphate.

Sl. No.	Compound	^{13}C Chemical shift, in ppm					
		CH$_3$	CH$_2$	CH	C	CO	C$_5$H$_5$
2	Cp$_2$Ti[S$_2$P(OPr-n)$_2$]$_2$	10.0 s	23.0 s			70.2 s	120.4 s
4	Cp$_2$Ti[S$_2$P(OBu-i)$_2$]$_2$	18.8 s	28.5 s			74.2 s	120.4 s
7	Cp$_2$Ti[S$_2$POCH$_2$CEt$_2$CH$_2$O]$_2$	6.8 s	22.2 s		36.9 s	76.2 s	120.7 s

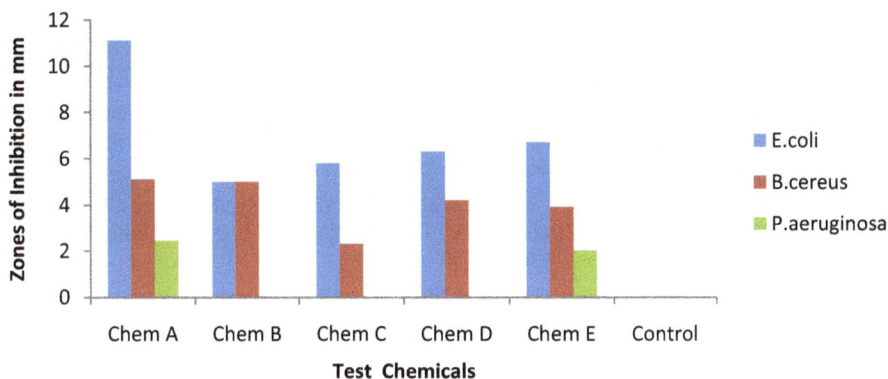

Figure 3. Antimicrobial activity of test chemicals.

Table 5. Zones of inhibition (mm) of test chemicals against *E. coli*, *B. cereus* and *P. aeruginosa*.

TEST chem. compounds	*Escherichia coli*	*Bacillus cereus*	*Pseudomonas aeruginosa*
$Cp_2Ti[S_2P(OEt)_2]_2$	11.1 ± 0.54	5.1 ± 0.32	2.45 ± 1.5
$Cp_2Ti[S_2P(OPr-n)_2]_2$	5.0 ± 0.76	5.0 ± 0.78	0.00
$Cp_2Ti[S_2P(OPr-i)_2]_2$	5.8 ± 0.81	2.3 ± 0.21	0.00
$Cp_2Ti[S_2P(OBu-i)_2]_2$	6.3 ± 0.41	4.2 ± 0.46	0.00
$Cp_2Ti[S_2P(OCH_2C(Et)_2CH_2O)_2]_2$	6.7 ± 0.72	3.9 ± 0.18	2.00 ± 3.0

4. Conclusion

We have successfully synthesized and characterized the new biscyclopentadienyl titanium (IV) bis (O,O dialkyl and alkylenedithiophosphate) compounds. The molecular weight of all these products determined by cryoscopic method in benzene indicated the monomeric nature of these products. The IR, ^{1}H, ^{13}C and ^{31}P NMR spectra and the elemental analysis of all of these titanium complexes are consistent with the proposed tentative structure: (see **Figure 1** and **Figure 2**). These novel chemical compounds used in this study show monodentate bond with the metal center and antimicrobial activity against *E. coli and B. cereus* and weak or no effect on *P. aeruginosa*.

Acknowledgements

The authors are thankful for financial support from Evans-Allen Federal Appropriated Funds. The authors also wish to express their profound appreciation for the support received from Dr. L. Walker, Prof. Matthew Edwards, Dean Chance M. Glenn, Ms. Dianne Kirnes and Shonda Scott.

References

[1]   Ma, N., Li, Y., Xu, H., Wang, Z. and Zhang, X. (2010) Well-Defined, Reversible Boronate Crosslinked Nanocarriers for Targeted Drug Delivery in Response to pH and cis-Diols. *Journal of the American Chemical Society*, **132**, 442-443.

[2]   El khaldy, A.A.S., Abushanab, A.M. and Abu Alkhair, E. (2011) Synthesis and Antimicrobial Studies of Bis (O,O'-Dialkyl and Alkylene Dithiophosphoric Acids) Adducts of Diphenyl Diselenide. *Applied Organometallic Chemistry*, **25**, 487-595.

[3]   Maheshwari, S., Drake, J.E., Kori, K., Light, M.E. and Ratnani, R. (2009) Synthesis and Spectroscopic Characterization of Tris(O,O'-Ditolyldithiophosphato) Arsenic/Antimony/Bismuth(III) Compounds: Crystal Structures of [As{S$_2$P(OC$_6$H$_4$Me-m)(2)}(3)]Center Dot 0.5C(6)H(14), [Sb{S$_2$P(OC$_6$H$_4$Me-m)(2)}(3)] and [Bi{S$_2$P(OC$_6$H$_4$Me-m)(2)}(3)]. *Polyhedron*, **28**, 689-694.

[4]   Bingham, A.L., Drake, J.E., Hursthouse, M.B., Light, M.E., Nirwan, M. and Ratnani, R. (2007) Synthesis, Characterization and Spectral Studies of Nitrogen Base Adducts of bis(O,O'-Ditolyldithiophosphato)Nickel(II). Crystal Structures of Ni[S$_2$P(OC$_6$H$_4$Me-*p*)$_2$]$_2$·C$_{10}$H$_8$N$_2$ and Ni[S$_2$P(OC$_6$H$_4$Me-o)$_2$]$_2$·C$_{14}$H$_{12}$N$_2$·C$_6$H$_6$. *Polyhedron*, **26**, 2672-2678.

[5]   Cotero-Villegas, A.M., Toscano, R.-A., Muñoz-Hernándeza, M., López-Cardosoa, M, García y Garcíaa, P. and Cea-

Olivares, R. (2005) Synthesis, Spectroscopic Characterization of O,O-Alkylene Dithiophosphates of Tellurolane and 1-Oxa-4-Tellurane. Single Crystal Structures of $C_4H_8Te[S_2P(OCH_2)_2CMe-nPr]_2$ and $C_4H_8OTe[S_2P(OCH_2)_2CEt_2]_2$. *Journal of Organometallic Chemistry*, **690**, 2872-2879. http://dx.doi.org/10.1016/j.jorganchem.2005.01.058

[6]   Dave, G.V. and Vyas, P.J. (2012) Synthesis, Structural Elucidation and Antimicrobial Activities of Some Alkylene Dithiophosphate Derivatives of Macrocyclic Complexes of Ni (II) Having N_2S_2 Potential Donors in 18 to 24 Membered Rings. *Journal of Current Chemical & Pharmaceutical Sciences*, **2**, 133-148.

[7]   El khaldy, A.A., Hussien, A.R., Abushanab, A.M. and Wasse, M.A. (2011) Synthesis and Characterization of Chloro-Bis (Cyclopentadienyl) Titanium(IV) and Zirconium(IV) O,O'-Dialkyl and Alkylene Dithiophosphates. *Phosphorus, Sulfur, and Silicon and the Related Elements*, **186**, 589-597.

[8]   Yadav, J.S., Mehrotra, R.K. and Srivastava, G. (1987) Metal and Organometal Complexes of Oxy- and Thio-Phosphorus Acids—I. O,O-Alkylene and Dialkyl Dithiophosphates of Titanium(IV). *Polyhedron*, **6**, 1687-1693. http://dx.doi.org/10.1016/S0277-5387(00)80772-X

[9]   Chauhan, H.P.S., Singh, U.P., Shaik, N.M., Mathur, S. and Huch, V. (2006) Synthetic, Spectroscopic, X-Ray Structural and Antimicrobial Studies of 1,3-Dithia-2-Stibacyclopentane Derivatives of Phosphorus Based Dithiolato Ligands. *Polyhedron*, **25**, 2841-2847. http://dx.doi.org/10.1016/j.poly.2006.04.027

[10]   Drake J.E., Gurnani, G., Hursthouse, M.B., Light, M.E., Nirwan, M. and Ratnani, R. (2007) Synthesis and Spectroscopic Characterization of Dimethyl/di(*n*-butyl)tin(IV)bis(*O,O*'-Ditolyl Dithiophosphate) Complexes. Crystal Structures of $Me_2Sn[S_2P(OC_6H_4Me-o)_2]_2$ and n-$Bu_2Sn[S_2P(OC_6H_4Me-o)_2]_2$. *Applied Organometallic Chemistry*, **2**, 539-544. http://dx.doi.org/10.1002/aoc.1265

[11]   Chauhan, H.P.S. and Singh, U.P. (2007) Synthetic, Spectral, Thermal and Antimicrobial Studies on Some *Bis*(*N,N*'-Dialkyldithiocarbamato) Antimony(III) Alkylenedithiophosphates. *Applied Organometallic Chemistry*, **21**, 880-889. http://dx.doi.org/10.1002/aoc.1290

[12]   Sharma, P.K., Rehwani, H., Gupta, R.S. and Singh, Y.P. (2007) The Antispermatogenic Activity of Some Phenylbismuth(III) *O,O*'-Dialkyldithiophosphates. *Applied Organometallic Chemistry*, **21**, 701-710. http://dx.doi.org/10.1002/aoc.1238

[13]   Shah, F.U., Glavatskih, S., Höglund, E., Lindberg, M. and Antzutkin, O.N. (2011) Interfacial Antiwear and Physicochemical Properties of Alkylborate-Dithiophosphates. *ACS Applied Materials & Interfaces*, **3**, 956-968. http://dx.doi.org/10.1021/am101203t

[14]   Juliano, R.C., Curtius, A.G. and Borges, D.L.G. (2012) Diethyldithiophosphate (DDTP): A Review on Properties, General Applications, and Use in Analytical. *Applied Spectroscopy Reviews*, **47**, 583-619. http://dx.doi.org/10.1080/05704928.2012.682286

[15]   Lee, Y.A., McGarrah, J.F., Lachicotte, R.J. and Eisenberg. R. (2002) Multiple Emissions and Brilliant White Luminescence from Gold(I) *O,O*'-Di(alkyl)Dithiophosphate Dimers. *Journal of the American Chemical Society*, **124**, 10662-10663. http://dx.doi.org/10.1021/ja0267876

[16]   Margielewski, L. (2010) The Effect of Zinc Dithiophosphates on Friction and Wear of Partially Stabilized Zirconia. Part I, Zinc Di-n-alkyldithiophosphates Tribological Properties. *Tribologia*, **230**, 87-104.

[17]   Hernandez-Molina, R., Gonzalez-Platas, J., Kovalenko, K.A., Sokolov, M.N., Virovets, A.V., Llusar, R. and Vicent, C. (2011) Cuboidal Mo_3S_4 and Mo_3NiS_4 Complexes Bearing Dithiophosphates and Chiral Carboxylate Ligands: Synthesis, Crystal Structure and Fluxionality. *European Journal of Inorganic Chemistry*, **5**, 683-693. http://dx.doi.org/10.1002/ejic.201000795

[18]   Klapoetke, T.M., Koepf, H., Tornieporth-Oetting, I.C. and White, P.S. (1994) Synthesis, Characterization, and Structural Investigation of the First Bioinorganic Titanocene(IV).alpha.-Amino Acid Complexes Prepared from the Antitumor Agent Titanocene Dichloride. *Organometallics*, **13**, 3628-3633. http://dx.doi.org/10.1021/om00021a040

[19]   Clearfield, A., Warner, D.K., Saldarriaga-Molina, C.H., Ropal, R. and Bernal, I. (1975) Structural Studies of $(\pi\text{-}C_5H_5)_2MX_2$ Complexes and their Derivatives. The Structure of Bis(π-cyclopentadienyl)titanium Dichloride. *Canadian Journal of Chemistry*, **53**, 1622-1629. http://dx.doi.org/10.1139/v75-228

[20]   Chauhan, H.P.S., Bhasin, C.P.G., Srivastava, G. and Mehrotra, R.C. (1983) Synthesis and Characterization of 2-Mercapto-2-Thiono-1,3,2-Dioxaphospholanes and Dioxaphosphorinanes. *Phosphorus and Sulfur and the Related Elements*, **15**, 99-104. http://dx.doi.org/10.1080/03086648308073283

[21]   Clinical and Laboratory Standards Institute (2012) Approved Standard CLSI Document M07-A9. 9th Edition, Clinical and Laboratory Standards Institute, Wayne.

[22]   Kato, S., Hori, A., Shiotani, H., Mizuta, M., Hayashi, N. and Takakuwa, T. (1974) Infrared and Raman Spectra of (Thioacetoxythio)Triorgano Derivatives of Silicon, Germanium, Tin and Lead. *Journal of Organometallic Chemistry*, **82**, 223-228. http://dx.doi.org/10.1016/S0022-328X(00)90359-0

[23]   Sowerby, D.B., Haiduc, I., Barbul-Rusu, A. and Salajan. M. (1983) Antimony(III) Diorganophosphoro- and Diorga-

nophospinodithioates: Crystal Structure of Sb[S$_2$P(OR)$_2$]$_3$ (R = Me and i-Pr). *Inorganica Chimica Acta*, **162**, 87-96. http://dx.doi.org/10.1016/S0020-1693(00)88943-4

[24]   Pavia, D.L., Lampman, G.M. and Kris, G.S. (1996) Introduction to Spectroscopy. 2nd Edition, Saunders Golden Sunburst Series, Orlando.

[25]   Corbridge, D.E.C. (1969) Infra-Red Spectra of Phosphorus Compounds. *Topics in Phosphorus Chemistry*, **6**, 235-366.

[26]   Ohkaku, N. and Nakamoto, N. (1973) Metal Isotope Effect on Metal-Liquid Vibrations. X. Far-Infrared Spectra of Trans Adducts of Tin(IV) Tetrahalide with Unidentate Ligands. *Inorganic Chemistry*, **12**, 2440-2446. http://dx.doi.org/10.1021/ic50128a043

[27]   Lockhart, T.P. and Manders, W.P. (1986) Structure Determination by NMR Spectroscopy. Dependence of |2J(119Sn,1H)| on the Me-Sn-Me Angle in Methyltin(IV) Compounds. *Inorganic Chemistry*, **25**, 892-895. http://dx.doi.org/10.1021/ic00227a002

[28]   Drew, M.G.B., Baricalli, P.J., Mitchell, P.C.H. and Read, A.R. (1983) Crevice Co-Ordination: Binding of a Ligand Molecule in a Molecular Crevice. Crystal and Molecular Structures of μ-Oxo-μ-pyridine-μ-sulphido-bis[(*OO'*-di-isopropyl phosphorodithioato)oxo-molybdenum(V)] and μ-Oxo-μ-pyridazine-μ-sulphido-bis[(*OO'*-di-isopropyl phosphorodithioato)oxomolybdenum(V)]. *Journal of the Chemical Society, Dalton Transactions*, 649-655. http://dx.doi.org/10.1039/dt9830000649

[29]   Gupta, R.K., Rai, A.K., Mehrotra, R.C. and Jain, V.K. (1984) Cyclic *O,O*-Alkylenedithiophosphates of Phenyl-Arsenic and -Antimony. *Inorganica Chimica Acta*, **88**, 201-207. http://dx.doi.org/10.1016/S0020-1693(00)83597-5

[30]   Gupta, R.K., Rai, A.K., Mehrotra, R.C., Jain, V.K., Hoskins, P.F. and Tiekink, E.R.T. (1985) Phenylarsenic(III) and Phenylantimony(III) Bis(dialkyl dithiophosphates): Synthesis and Multinuclear (Proton, Carbon-13, Phosphorus-31) NMR and Mass Spectral Studies. Crystal and Molecular Structures of C$_6$H$_5$M[S$_2$P(OCHMe$_2$)$_2$]$_2$ [M = Sb(III) and As(III)]. *Inorganic Chemistry*, **24**, 3280-3284. http://dx.doi.org/10.1021/ic00214a037

[31]   Srivastava, S.K., Tomar, S., Rastogi, R. and Saxena, R. (2010) Substituted Diorganotin(IV) *O,O'*-Alkylene Dithiophosphates: Synthesis and Spectral Aspects. *Phosphorus, Sulfur, and Silicon and the Related Elements*, **185**, 634-640. http://dx.doi.org/10.1080/10426500902894959

[32]   Chauhan, H.P.S. (1998) Chemistry of Diorganodithiophosphate (and Phosphinate) Derivatives with Arsenic, Antimony and Bismuth. *Coordination Chemistry Reviews*, **173**, 1-30. http://dx.doi.org/10.1016/S0010-8545(97)00071-4

[33]   Chauhan, H.P.S. (2003) The Chemistry and Applications of Alkoxy, Aryloxy and Applied Derivatives of Elements. RBSA Publishers, Jaipur.

[34]   Glidewell, C. (1977) Ambident Nucleophiles: VI. Solution Metal-Ligand Binding Modes in Phosphorodithioate Complexes. A Phosphorus-31 N.M.R. Study. *Inorganica Chimica Acta*, **25**, 159-163. http://dx.doi.org/10.1016/S0020-1693(00)95706-2

[35]   Abu-Basha, E.A., Gharaibeh, S.M. and Thabet, A.M. (2012) *In Vitro* Susceptibility of Resistant, *Escherichia coli* Field Isolates to Antimicrobial Combinations. *The Journal of Applied Poultry Research*, **21**, 595-602. http://dx.doi.org/10.3382/japr.2011-00500

25

The First Synthesis of Sessiline

Viktor Ilkei[1]*, Kornél Faragó[1], Zsuzsanna Sánta[2], Miklós Dékány[2], László Hazai[1], Csaba Szántay Jr.[2], Csaba Szántay[1], György Kalaus[1]

[1]Department of Organic Chemistry and Technology, Budapest University of Technology and Economics, Budapest, Hungary
[2]Gedeon Richter Plc, Budapest, Hungary
Email: *viktor.ilkei@gmail.com

Abstract

Sessiline is an alkaloid which was recently isolated from the fruits of *Acanthopanax sessiliflorus*. The molecule contains two five-membered heterocyclic units joined together by an acylaminocarbinol-ether type bond. Here, we describe the first, simple synthesis of sessiline from 5-hydroxypyrrolidin-2-one and 5-hydroxymethylfurfural, which are prepared from succinimide and furfuryl alcohol, respectively. The coupling reaction takes place on moderate heating under neat conditions.

Keywords

Sessiline, Acylaminocarbinol, Iminium Ion, Alkaloid Synthesis

1. Introduction

Sessiline (**1**) (**Figure 1**) was isolated in 2002 from the fruits of *Acanthopanax sessiliflorus*, a herbaceous plant, which is distributed in East Asia [1]. Its structure was elucidated by spectroscopic methods. The molecule consists of two heterocyclic units joined together by an ether-bond. The alkaloid is found in the plant as a racemate.

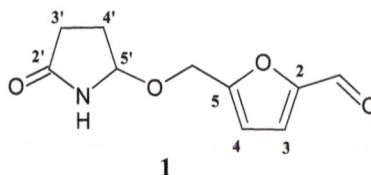

1

Figure 1. The structure of sessiline (**1**).

---

*Corresponding author.

It is known that acylaminocarbinols, as well as the iminium ions that arise from them, are electrophilic reagents, which can be utilized in the synthesis of sessiline (1). Taking the above assumption into consideration, we outlined a simple retrosynthetic scheme for 1 (Scheme 1). By disconnecting the ether bond, we obtain two known compounds: 5-hydroxypyrrolidin-2-one (2) and 5-hydroxymethylfurfural (3).

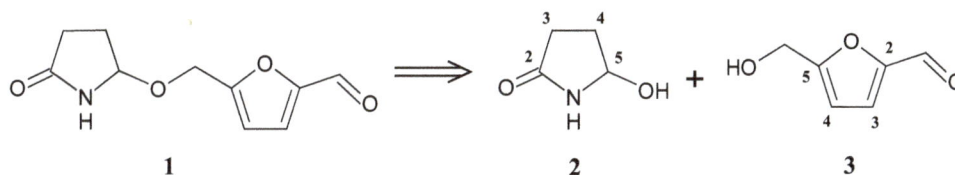

Scheme 1. Retrosynthetic analysis of sessiline (1).

2. Results and Discussion

In order to accomplish our goal, we synthesized 5-hydroxypyrrolidin-2-one (2) and 5-hydroxymethylfurfural (3) (Scheme 2). Compound 2 can be prepared from succinimide (4) in two steps. The partial reduction of 4 yields 5-ethoxypyrrolidin-2-one (5) [2], which can be hydrolysed in boiling water to give the amidocarbinol 2 [3].

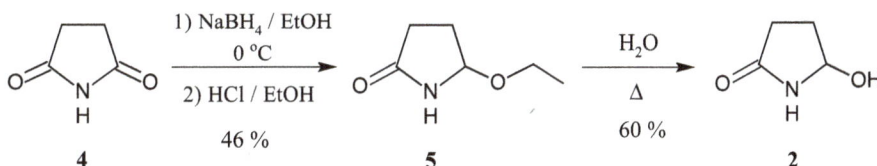

Scheme 2. Synthesis of 5-hydroxypyrrolidin-2-one (2).

Next, we prepared 5-hydroxymethylfurfural (3) from furfuryl alcohol (6) (Scheme 3). First, 6 was protected by acetylation to give furfuryl acetate (7) [4], which was subjected to *Vilsmeier formylation* to yield 5-(formyl)furfuryl acetate (8) [5]. Then, 3 was obtained by deacetylation [6].

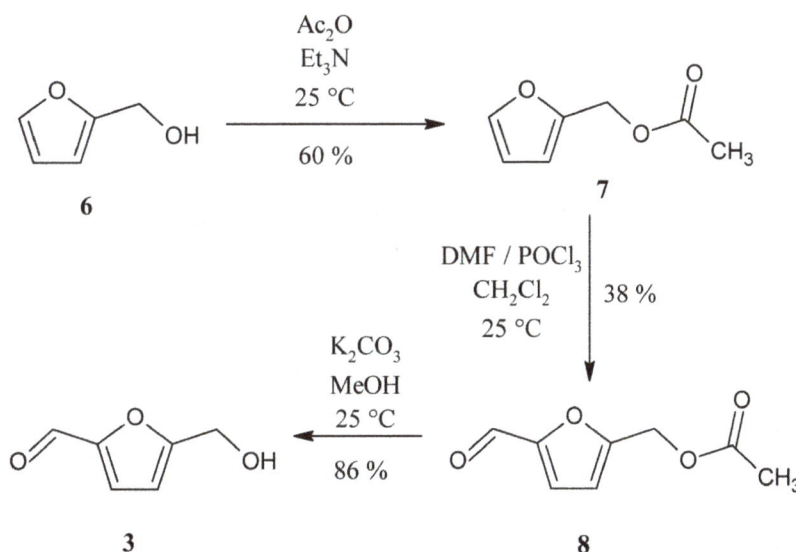

Scheme 3. Synthesis of 5-hydroxymethylfurfural (3).

Next, we needed to synthesize target molecule 1 by coupling 2 and 3. Although the type of the planned reaction is known, relatively few examples can be found in the literature. The majority of these transformations were carried out under mild conditions using acid catalysis [7] [8].

In the present case, **2** was allowed to react with an excess of **3** at 60°C under neat conditions. The reaction gave sessiline (**1**) in an acceptable yield (**Scheme 4**).

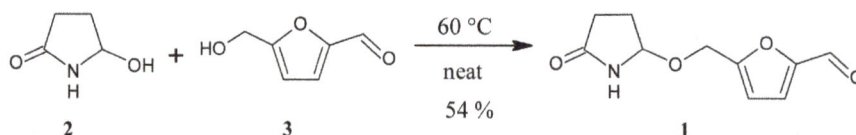

Scheme 4. The synthesis of sessiline (**1**).

Surprisingly, no acid catalysis or solvent was necessary for the reaction to take place, raising of the temperature proved to be sufficient. In light of our successful synthesis, we proposed a plausible reaction mechanism for the formation of sessiline (**1**) (**Scheme 5**). According to our assumption, higher temperatures cause the equilibrium between the amidocarbinol **2** and its ionic form **2/a** to shift towards the latter, which results in a higher concentration of acyliminium ions in the reaction mixture, thus enabling the reaction to take place.

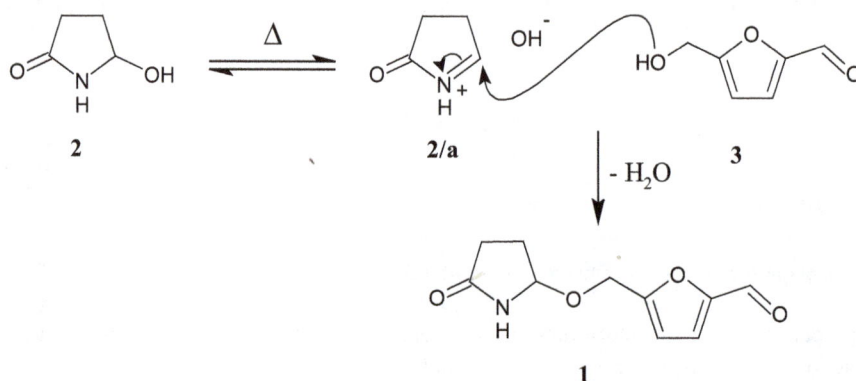

Scheme 5. Proposed reaction mechanism for the formation of sessiline (**1**).

3. Experimental

3.1. General

Melting points were measured on a SANYO Gallenkamp apparatus and are uncorrected. IR spectra were recorded on a Bruker FT-IR instrument. ^{1}H-NMR and ^{13}C-NMR measurements were performed on Varian 400 MHz, Varian 500 MHz and Varian 800 MHz spectrometers. Chemical shifts are given on the delta scale as parts per million (ppm) with tetramethylsilane (TMS) (^{1}H) or dimethylsulfoxide-d_6 (^{13}C) as the internal standard (0.00 ppm and 39.5 ppm, respectively). MS spectra were recorded on VG-Trio-2 and Finnigan MAT 95SQ instruments using EI or ESI techniques. HRMS analyses were performed on an LTQ FT Ultra (Thermo Fischer Scientific, Bremen, Germany) system. TLC was carried out using Kieselgel 60 F_{254} (Merck) coated glass plates. Column chromatography was performed using Geduran Si 60 (Merck) silica.

3.2. Furfuryl Acetate (7)

To a mixture of 56.6 g (577 mmol; 50 ml) furfuryl alcohol (**6**) and 12.1 g (120 mmol; 16.7 ml) triethylamine was added 67 g (656 mmol; 62.4 ml) of acetic anhydride dropwise over 15 minutes. The reaction mixture was stirred at ambient temperature for 21 hours, then it was extracted with 3 × 30 ml water. The organic phase was dried over magnesium sulphate and evaporated under reduced pressure. The crude product was distilled *in vacuo* (bp. 76°C/15 mmHg; lit.: 67°C/8 mmHg [4]). 48 g (60%) pure **7** was obtained as a clear liquid. n_d: 1.4619 (lit.: 1.4603 [4]). IR ν_{max} (film, cm^{-1}): 1738, 1502, 1437, 1231, 1150, 1079, 1015, 918, 885, 817, 744.

3.3. 5-(Formyl)Furfuryl Acetate (8)

A mixture of 97 g (1.33 mol; 103 ml) dimethylformamide and 300 ml dichloromethane was cooled to 0°C. To this mixture 152 g (0.99 mol; 91 ml) phosphoryl chloride was added dropwise at 0°C over 5 minutes. The reaction mixture was stirred at 0°C for 1 hour, then 48 g (0.343 mol; 43 ml) furfuryl acetate (7) was added dropwise over 5 minutes. The reaction mixture was stirred at ambient temperature for 22 hours, after which it was neutralised with 15% sodium carbonate solution. The precipitate was filtered and washed with 100 ml dichloromethane. The aqueous-organic mixture was extracted with 10 × 100 ml dichloromethane, the organic phase was dried over magnesium sulphate and evaporated under reduced pressure. The crude product was crystallized from ether, after which 22.7 g (39%) of pure 8 was obtained as colourless crystals. R_f 0.81 (ethyl acetate:hexane = 2:1). Mp.: 55°C - 56°C (lit.: 55°C - 56°C [5]). IR v_{max} (KBr, cm^{-1}): 3122, 2943, 2833, 1740, 1675, 1588, 1523, 1437, 1403, 1366, 1273, 1221, 1022. ^{1}H NMR (800 MHz, DMSO-d_6) δ_H 2.08 (s, 3H, Ac); 5.15 (s, 2H, CH$_2$O); 6.81 (d, 1H, J = 3.5 Hz, H-4); 7.53 (d, 1H, J = 3.5 Hz, H-3); 9.60 (s, 1H, CH=O). MS(EI): 168 (C$_8$H$_8$O$_4$). EI-MS (rel. int.%): 168(1); 126(100); 109(22); 97(7); 79(30); 53(9); 44(22); 43(28).

3.4. 5-Hydroxymethylfurfural (3)

To 2.60 g (15.5 mmol) 5-(formyl)furfuryl acetate (8) dissolved in 20 ml methanol, 0.330 g (2.39 mmol) potassium carbonate was added. The reaction mixture was stirred at ambient temperature for 1 hour, after which 15 ml water was added and the mixture was extracted with 10 × 15 ml dichloromethane. The organic phase was dried over magnesium sulphate and evaporated under reduced pressure. The crude product was subjected to column chromatography using a mixture of ethyl acetate:hexane = 2:1 as the eluent. 1.671 g (86%) of pure 3 was obtained as a yellow liquid, which crystallized on cooling. R_f 0.55 (ethyl acetate:hexane = 2:1). Mp.: 28°C - 32°C (lit.: 31°C - 32°C [6]). IR v_{max} (KBr, cm^{-1}): 3405, 3123, 2926, 2851, 1675, 1523, 1397, 1370, 1334, 1280, 1192, 1023. ^{1}H NMR (800 MHz, DMSO-d_6) δ_H 4.51 (d, 2H, J = 6.0 Hz, CHO); 5.59 (t, 1H, J = 6.0 Hz, OH); 6.61 (d, 1H, J = 3.5 Hz, H-4); 7.50 (d, 1H, J = 3.5 Hz, H-3); 9.55 (s, 1H, CH=O). MS(EI): 126 (C$_6$H$_6$O$_3$). EI-MS (rel. int.%): 126(96); 109(10); 97(100); 81(4); 69(24); 53(9); 41(50); 39(22).

3.5. 5-Ethoxypyrrolidin-2-One (5)

A solution of 7.156 g (72.22 mmol) succinimide (4) in 300 ml ethanol was cooled to 0°C. To this solution 4.00 g (105.74 mmol) sodium borohydride was added in one portion. The solution was stirred at 0°C for 4 hours, during which time every 15 minutes 5 drops of 2 M ethanolic hydrogen chloride solution were added. Then the reaction mixture was acidified to pH = 3 with 2 M ethanolic hydrogen chloride solution over 30 minutes, after which it was stirred at 5°C for 45 minutes. Then the reaction mixture was neutralised (pH = 7) with 5% ethanolic potassium hydroxide solution and evaporated to dryness under reduced pressure. The remaining syrupy solid was suspended in 80 ml chloroform, filtered, and the precipitate washed with 3 × 20 ml chloroform. The filtrate was evaporated under reduced pressure, and the remaining colourless oil was dissolved in 80 ml dichloromethane and washed with 3 × 10 ml water. The aqueous phase was extracted with 6 × 20 ml dichloromethane, then the organic phases were unified, dried over magnesium sulphate and evaporated under reduced pressure. The remaining colourless oil crystallized on standing, after which 4.327 g (46%) pure 5 was obtained as colourless crystals. R_f 0.67 (acetone). Mp.: 51°C - 53°C (lit.: 48°C - 53°C [2]). IR v_{max} (KBr, cm^{-1}): 3200, 2978, 1707, 1689, 1668, 1457, 1282, 1250, 1067, 986. ^{1}H NMR (400 MHz, DMSO-d_6) δ_H 1.10 (t, 3H, J = 7.0 Hz, CH$_3$); 1.78 - 1.89 (m, 1H, H$_x$-4); 1.95 - 2.05 (m, 1H, H$_x$-3); 2.11 - 2.30 (m, 2H, H$_y$-4, H$_y$-3); 3.27-3.35 (m, 1H, OCH$_{2x}$); 3.44 - 3.54 (m, 1H, OCH$_{2y}$); 4.83 - 4.89 (m, 1H, H-5); 8.62 (s, 1H, NH). ^{13}C NMR (100 MHz, DMSO-d_6) δ_C 15.1 (CH$_3$); 27.7 (C-4); 28.1 (C-3); 61.6 (OCH$_2$); 85.0 (C-5); 177.4 (CON). MS(ESI): 130 (C$_6$H$_{12}$NO$_2$). ESI-MS-MS (cid = 35) (rel. int.%): 84(100).

3.6. 5-Hydroxypyrrolidin-2-One (2)

2.0 g (15.5 mmol) 5-ethoxypyrrolidin-2-one (5) was dissolved in 25 ml water and the solution was refluxed for 3 hours. Then the water was evaporated under reduced pressure. The remaining oil was triturated with ethyl acetate, from which a white solid crystallized on cooling. The solid was filtered and recrystallized from acetone. 0.89 g (60%) pure 2 was obtained as colourless crystals. R_f 0.19 (acetone:hexane = 2:1). Mp.: 94°C - 96°C (lit.: 90°C [3]). IR v_{max} (KBr, cm^{-1}): 3254, 2996, 2962, 1668, 1475, 1415, 1323, 1271, 1166, 1101, 1070, 1016. ^{1}H

NMR (400 MHz, DMSO-d_6) δ_H 1.66 - 1.75 (m, 1H, H_x-4); 1.94 - 2.03 (m, 1H, H_x-3); 2.14 - 2.24 (m, 1H, H_y-4); 2.23 - 2.33 (m, 1H, H_y-3); 5.06 (m, 1H, H-5); 5.70 (d, 1H, J = 6.9 Hz, OH); 8.18 (s, 1H, NH). ^{13}C NMR (100 MHz, DMSO-d_6) δ_C 28.4 (C-3); 30.3 (C-4); 78.5 (C-5); 176.7 (CON). HRMS: 102.05491 ($C_4H_8NO_2$; calc. 102.05496). ESI-MS-MS (cid = 55) (rel. int.%): 85(100).

3.7. Sessiline (1)

200 mg (1.98 mmol) 5-hydroxypyrrolidin-2-one (**2**) and 800 mg (6.34 mmol) 5-hydroxymethylfurfural (**3**) were stirred at 60°C under neat conditions for 1 hour. Then the reaction mixture was diluted with 1 - 2 ml dichloromethane, filtered, the solid washed with 3 × 2 ml dichloromethane, and air-dried with suction. 225 mg (54%) pure **1** was obtained as colourless crystals. R_f 0.47 (acetone:hexane = 2:1). Mp.: 169°C - 172°C (lit.: 171°C - 172°C [1]). IR v_{max} (KBr, cm^{-1}): 3321, 3176, 3117, 2965, 2896, 2863, 2790, 2763, 1772, 1709, 1666, 1531, 1463, 1412, 1389, 1338, 1284, 1276, 1262, 1250, 1208, 1201, 1177, 1096, 1062, 1030, 1007. ^{1}H NMR (400 MHz, DMSO-d_6) δ_H 1.86 - 1.95 (m, 1H, H_x-4'); 2.01 - 2.10 (m, 1H, H_x-3'); 2.16 - 2.34 (m, 2H, H_y-4', H_y-3'); 4.49 + 4.58 (AB, 2 × 1H, J_{gem} = 13.2 Hz, CHO); 5.02 (m, 1H, H-5'); 6.73 (d, 1H, J = 3.4 Hz, H-4); 7.51 (d, 1H, J = 3.4 Hz, H-3); 8.80 (s, 1H, NH); 9.58 (s, 1H, CH=O). ^{13}C NMR (100 MHz, DMSO-d_6) δ_C 27.5 (C-4'); 27.9 (C-3'); 60.3 (CH$_2$O); 85.2 (C-5'); 111.8 (C-4); 123.9 (C-3); 152.2 (C-2); 157.7 (C-5); 177.6 (CON); 178.3 (CH=O). HRMS: 210.07605 ($C_{10}H_{12}NO_4$; calc. 210.07608). ESI-MS-MS (cid=65) (rel. int.%): 192(100); 164(8); 126(2); 109(4).

Acknowledgements

The authors are grateful to Gedeon Richter Plc for the financial support.

References

[1] Lee, S., Ji, J., Shin, K.H. and Kim, B.-K. (2002) Sessiline, A New Nitrogenous Compound from the Fruits of *Acanthopanax sessiliflorus*. *Planta Medica*, **68**, 939-941. http://dx.doi.org/10.1055/s-2002-34925

[2] Hubert, J.C., Wijnberg, J.B.P.A. and Speckamp, W.N. (1975) NaBH$_4$ Reduction of Cyclic Imides. *Tetrahedron*, **31**, 1437-1441. http://dx.doi.org/10.1016/0040-4020(75)87076-1

[3] Cue Jr., B.W. and Chamberlain, N. (1979) An Improved Method for the Preparation of 5-Hydroxy-2-Pyrrolidone. *Organic Preparations and Procedures International*, **11**, 285-286. http://dx.doi.org/10.1080/00304947909355413

[4] Renvall, I. and Mattila, T. (1977) Esterification of Furfuryl Alcohol and Its Derivatives. US Patent No. 4008256.

[5] Mehner, A., Montero, A.L., Martinez, R. and Spange, S. (2007) Synthesis of 5-Acetoxymethyl- and 5-Hydroxymethyl-2-Vinylfuran. *Molecules*, **12**, 634-640. http://dx.doi.org/10.3390/12030634

[6] Schinzer, D., Bourguet, E., Ducki, S. (2004) Synthesis of Furano-Epothilone D. *Chemistry—A European Journal*, **10**, 3217-3224. http://dx.doi.org/10.1002/chem.200400125

[7] Toja, E., Gorini, C., Zirotti, C., Barzaghi, F. and Galliani, G. (1991) Amnesia-Reversal Activity of a Series of 5-Alkoxy-1-Arylsulfonyl-2-Pyrrolidinones. *European Journal of Medicinal Chemistry*, **26**, 403-413. http://dx.doi.org/10.1016/0223-5234(91)90101-R

[8] Toja, E., Gorini, C., Zirotti, C., Barzaghi, F. and Galliani, G. (1991) Amnesia-Reversal Activity of a Series of 5-Alkoxy-1-Arylcarbonyl-2-Pyrrolidinones and 5-Alkoxy-1-Arylmethyl-2-Pyrrolidinones. *European Journal of Medicinal Chemistry*, **26**, 415-422. http://dx.doi.org/10.1016/0223-5234(91)90102-S

26

Synthesis of Novel Fluorine Substituted Isolated and Fused Heterobicyclic Nitrogen Systems Bearing 6-(2'-Phosphorylanilido)-1,2,4-Triazin-5-One Moiety as Potential Inhibitor towards HIV-1 Activity

Reda M. Abdel-Rahman, Mohammed S. T. Makki, Abeer N. Al-Romaizan*

Department of Chemistry, Faculty of Science, King Abdul Aziz University, Jeddah, KSA
Email: *ar-orkied@hotmail.com

Abstract

Novel 6-(5'-fluoro-2'-diphenylphosphorylanilido)-3-hydrazino-1,2,4-trizin-5 (2H) one (3) is achieved from hydrozinolysis of the corresponding 3-thioxo-analoges 2. Compound 2 is also obtained from phosphorylation of 6-(5'-fluoro-2'-aminophenyl)-3-thioxo-1,2,4-triazin-5(2H) one (1). Novel fluorine substituted isolated and/or fused heterobicyclic nitrogen systems bearing and/or containing, 6-phosphoryl anilido-1,2,4-trizin-5 (2H) one moiety (4 - 22) have been synthesized from ring closure reactions of compound 3 with π-acceptors activated carbon compounds in different medium and conditions. Structures of the products are characterized by MS, IR, UV-VIS, CH, N, and ¹H/¹³CNMR spectral data. The new products have been evaluated as potential inhibitors towards HIV-1 activity.

Keywords

Synthesis, Fluorine, Phosphorus, Sulfar 1,2,4-Trizinones HIV-1

1. Introduction

Organophosphorus systems are ubiquitous in nature and exhibit many applications in the field of agriculture medicine and industry [1] [2]. Many multi-ring phosphorus heterocycles are used as pesticide [3], bactericide

---

*Corresponding author.

[4]-[6], antibiotics [4], and acts as HIV protease inhibitors [7]. Thus, synthesis of new phosphorus bearing a heterocycles has attracted the attention of researchers. Phosphorylation of organic compounds often improves their biological activity, especially through a vital energy, because the P-O is the store of energy for metabolism process. For example, phophorylated-N in the nucleocapsid affects the interaction between the N-atom and the genomic RNA. The charge repulsion between the negatively charged phosphoserine and the negatively charged RNA may weaken the interaction between N-atom and RNA, thus enabling the viral polymerase to gain access to genomic RNA and to initiate viral RNA transcription and replication [8]. On the other hand, chemistry of N-phosphorylheterocycles showed that these compounds from two dimensional polymeric chains via intermolecular P-O^{-+}H-N hydrogen bonds [9]. Also, phosphodiester compound had a type of action, especially enzymetic of DNA replication [10] on DNA ligase as (**Figure 1**).

It is interesting that fluorine containing aheterocycles bearing functional groups exhibits highly effective in biological process, pharmaceuticals, agrochemicals, polymers and a wide range of consumer products [11]. It reflects its resistance to metabolic change due to the strength of the C-F bond providing biological stability and the application of its nonstick-interfacial physical characteristics [12]-[16]. Abdel-Rahman *et al.* [17]-[21] reported the synthesis and chemical reactivity of 3-thioxo-1,2,4-triazin-5 one derivatives as a bioactive molecules, especially as anti-cancer, anti-AIDS, Amyllolytic, cellobiase and antimicrobial agents. Based upon these observations, the aim of this work is to study the formation of 6-(5'-fluoro-2'-phosphoryl anilido-3-hydrazino-1,2,4-triazin-5 (2H) one then study their behaviour as electron donor towards different electron acceptors such as carbo-sulfur, oxygen, halogen and nitrogen compounds; finally, a type of isolated and/or fused heterobicyclic systems obtained and evaluation as potential inhibitors towards HIV-1 virus.

2. Experimental

Melting points were determined with an electro-thermal Bibbly Stuart Scientific Melting point SMPI (UK). A perkin Elmer (Lambda EZ-2101) double beam spectrophotometer (190 - 1100 nm) used for recording the electronic spectra. A Perkin Elmer model RXI-FT-IR 55,529 cm^{-1} used for recording the IR spectra (EtOH as solvents). A Brucker advance DPX 400 MHz using TMS as an internal standard for recording the ^{1}H/^{13}C NMR spectra in deuterated DMSO (δ in pp m). AGC- MS-QP 1000 Ex model is used for recording the mass spectra. Hexafluorobenzene was used as external standard for ^{19}FNMR at 84.25 MHz and ^{31}P (in CDCl$_3$, 101.25 MHZ).

Elemental analysis was performed on Micro analytical Center of National Reaches Center-Dokki, Cairo, Egypt. Compound **1** prepared according the reported method [17].

6-(2'-aminos-5'-fluorophenyl-3-thioxo-1,2,4-triazin-5(2H)one (1)

A mixture of 5-fluoroisatin (0.01 mol) in sodium hydroxide solution (5%, 50 ml) warm for 10 min, then thiosemicarbazide (0.01 mol, in hot water, 10 ml) add and complete the refluxing for 2 h. The reaction mixture cooled then poured onto ice and neutralize with diluted HCl. The solid thus obtained filtered off and crystallized from ethanol to give **1** as yellow crystals, yield 80%; m.p. 265°C. Analytical data; found: C, 44.91; H, 2.90; F, 7.58: N, 23.40; S, 13.29%.Calculated for C$_9$H$_7$FN$_4$OS (238); C, 45.37; H, 2.94; F, 7.98; N, 23.52; S, 13.44%.UV: λ$_{max}$ (EtOH) = 280 nm. IR $v$cm^{-1}: 3424(NH$_2$), 3258, 3169 (NH, NH), 1685 (C=O), 1618 (deform. NH$_2$), 1545 (C=N), 1263 (C-F), 858,818 (aryl CH), 685 (C-F). ^{1}H NMR (DMSO) δ = 14.66, 12.66, 10.90 (each s, 1H, 3NH), 8.68 - 8.06, 7.69 - 7.64, 7.39 - 7.30, (s, 3H, aryl protons). ^{13}C NMR (DMSO): δ 179.47(C=S), 162 (C=O), 159 - 157 (C-F), 138.54 (C=N), 131.82, 121.8, 121.51 (aromatic carbons, 78.14, 77.71 (C$_5$ - C$_6$, 1,2,4-trinzine). M/Z (Int. %); 256 (M + H$_2$O, 5%), 68 (100), 148(21), 138(18), 110 (30), 96 (50), 82(58.0), 70(78).

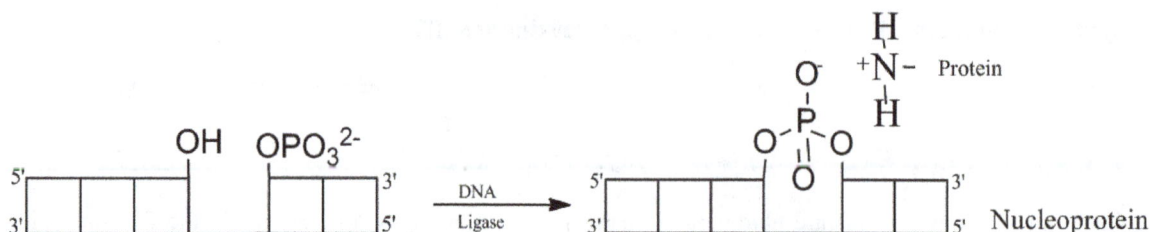

Figure 1. The reaction of DNA ligase phospho diester link.

6-(5'-Fluoro-2'-diphenylphosphorylanilido)-3-thioxo-1,2,4-triazin-5(2H)one (2)

Equimolar mixture of compound **1** and diphenylphosphoryl chloride in DMF (20 ml) warm for 1h, cooled then poured onto ice. The produce solid filter off and crystallized from methanol to give **2** as deep yellow crystals. Yield 70%; m.p. 218-220°C. Analytical data; found: C, 53.39; H, 3.51; F, 3.88: N, 11.69%.Calculated for $C_{21}H_{16}FN_4PSO_4$(470); C, 53.61; H, 3.70; F, 4.04; N, 11.91%. IR νcm^{-1}: 3133.8 (NH), 1688.4 (C=O), 1574 (C=N), 1370 (Cyclic NCSN), 1262 (C-F), 1200 (P=O), 1150 (C-S), 10.96 (Ph-O-P), 858, 801 (substituted phenyls) M/Z: (Int.%); 473 (M + 3, 5.11), 110 (100).

3-Hydrazino-6-(5'-fluoro-2'-diphenylphosphorylanilido)-1,2,4-triazin-5(2H)one (3)

A mixture of compound **2** (1 gm) and hydrazine hydrate (2ml) in ethanol (50ml) reflux for 2h. cooled. The solid thus produce filter off and crystallized from ethanol to give **3** as orange crystals, yield 85%; m.p. 178°C - 180°C. Analytical data; found: C, 53.45; H, 3.59; F, 3.80%.Calculated for $C_{21}H_{18}FN_6PO_4$(468), C, 53.84; H, 3.84; F, 4.02; N, 17.14%. IR ν cm^{-1} = 3381, 3340, 3220 - 3190 (NH, NH, NH$_2$), 1682 (C=O), 1558 (C=N), 1266.9 (C-F), 1210-1156 (P=O). 1050 (Ph-O-P), 860,805 (substituted phenyls).^{1}H NMR(DMSO) δ = 10.7, 10.2, 8.7 (each s, 3H, 3NH, 1,2,4-triazino NH-P=O), 7.511, 7.27-2.24, 7.23-7.22 (each s, 3H, aryl protons); 7.143 - 7.002; 6.860 - 6.796 (each m, 10H, phenyl protons), 2.88 (s, 2H, NH$_2$ of hydrazine), ^{13}C NMR (DMSO) δ = 163.78, 159.44, 157.86, 134.97, 129.07, 127.65, 123.66, 122.87, 120.12, 120.09, 113.47, 113.31, 110.77, 110.71, 107.93, 105.27, 105.10, 77.76, 77.33.

3-(3'-Amino-4'-carboxy-5-(4-chlorophenyl)-4',5'-dihydro-pyrazolin-1'-yl)-6-(5'-fluoro-2'-diphenylphosphorylanilido)-1,2,4-triazin-5(2H)one (4)

Equimolar amounts of **3** and α-(4-chlorophenylidene) cyano acetic acid in ethanol (100 ml) and a few drops piperidine (0.5 ml) reflux for 8h, cooled, then added off and crystallized from dioxan to give **4** as yellowish crystals, yield 65% m.p., 158°C - 160°C. Analytical data; found: C, 54.81; H, 3.11; F, 2.55; Cl, 5.09; N, 14.29%. Calculated for $C_{31}H_{24}FClN_7PO_6$(675). C, 55.11; H, 3.55; F, 2.81; Cl, 5.18; N, 14.51%; M/S: 675 (1.11), 95(100). IR νcm^{-1}: 3420 (OH), 3381, 3155 (NH, NH$_2$), 1680 (C=O) 1558(C=N), 1478 (deform aliphatic CH), 1269 (C-F), 1200 (P=O), 1043 (Ph-O-P), 993.861, 804 (Aryl CH).^{1}H NMR (DMSO) δ = 10.67, 10.14, 8.72 (each δ, 3H, 3NH, 1,2,4-triazinare NH-P=O), 9.77 (δ, 1H, OH), 8.15 - 7.52, 7.36 - 7.29, 7.13 - 7.00, 6.991 - 6.811 (each m, 19H, 7H aryl, 10H phenyls pyrazole protons); 3.46 (δ, 2H, NH$_2$).^{13}C NMR (DMSO) δ: 179.64; 163.76; 163.05; 159.42; 157.84; 138.58; 134.94; 128.65; 127.59; 123.72; 123.66; 121.13; 121.07; 117.67; 117.51; 113.45; 113.28; 112.06; 112.01; 110.75; 110.70; 109.66; 108.09; 107.92; 105.25; 105.08; 77.78; 77.35; 23.64.

3-(3'-Methyl-5'-arylamino-pyrazolin-1'-yl)-6-(5'-fluoro-2'-diphenylphosphorylanilido)-1,2,4-trinzin-5(2 H) one (5)

A mixture of **3** (0.01 mol) and acetyl acetanilide derivative (0.01 mol) in DMF (50 ml) reflux for 2h, cooled then poured onto ice. The solid thus yield filtere off and crystallized from dioxan to give **5** as deep-yellowish crystals, yield 70%; m.p. 100°C - 101°C. Analytical data; found C, 56.03; H, 3.41; F, 2.23; N, 16.17; S, 3.91%. Calculated for $C_{36}H_{29}FN_9PSO_6$(765). C, 56.47; H, 3.79; F, 2.48; N, 16.47; S, 4.18%. IR νcm^{-1}: 3424, 3220 - 31.70 (NH, NH), 2932 (CH$_3$), 1694 (C=O), 1620 (C=N), 1358 (SO$_2$-NH), 1268 (C-F), 1220 (P=O), 1054 (Ph-O-P), 952, 909, 810, 758 (aryl CH).^{1}H NMR (DMSO) δ = 12.28, 12.27, 10.36, 10.21 (each s, 4H, 4NH) 9.7 (s, 1H, C$_4$- of pyrazole), 7.979, 7.971, 7.556, 7.545 (m, 4-H of pyridine); 7.34, 7.336, 7.322, 7.091(m, 4H of aryl-P-SO$_2$NH) 6.95 - 6.83; 6.66 - 6.52; 6.41 - 6.16 (each m, 13H of FC$_6$H$_3$, 2C$_6$H$_5$). 0.46 - 0.45 (s, 3H, CH$_3$). ^{13}C NMR(DMSO) δ = 179.64, 163.75, 163.03, 162.41, 158.08, 157.99, 157.47; 151.97; 138.58, 134.94, 130.23, 129.10, 123.72, 123.66, 117.66, 117.50, 115.11, 113.41, 113.25, 112.97, 112.07, 112.02, 110.76, 110.71, 108.09, 107.92, 105.21, 105.04, 77.83-77.40, 36.40, 31.29.

3-(3'-Amino-4',5'-dihydro-5'-oxo-pyrazolin-1'-yl)-6-(5'-fluoro-2'-diphenylphosphorylanilido)-1,2,4-triazin-5(2H)one (6)

Equiomolar of **3** and ethyl cyanoacetate in THF (100ml) reflux for 4H, cooled. The solid thus produce filtered off and crystallized from THF to give **6** as faint yellow crystals, yield 68%; m.p. 200-202°C. Analytical data; found: C, 53.53; H, 3.31; F, 3.29; N, 18.01%. Calculated for $C_{24}H_{19}FN_7PO_5$(535); C, 53.83; H, 3.55; F, 3.55; N, 18.31%. IR νcm^{-1}: 3425 (NH$_2$), 3220-3190 (NH $\rightleftarrows$ OH), 1694 (C=O), 1471 (deformation. CH$_2$), 1310 (N-N), 1250 (C-F), 1210 (P=O), 1058 (Ph-O-P), 952, 915, 808 (aryl CH).^{1}H NMR (DMSO) δ: 12.7, 10.67, 8.8 (each s, 3H, 3NH), 10.24 (s, 1H, OH of pyrazole), 7.66 - 7.400, 7.39-7.00, 6.99-6.79 (each m, 14H, aryl & phenyl protons). ^{13}C NMR(DMSO) δ:179.54, 163.70, 159.31, 157.74, 134.91, 128.98, 127.31, 123.68, 117.57, 113.29, 113.13, 111.95, 110.69, 110.64, 108.23, 105.08, 104.91, 78.0-77.54.

3-(3',5'-Diaminopyrazolin-1'-yl)-6-(5'-fluoro-2'-diphenylphosphorylanilido)-1,2,4-triazin-5(2H)one (7)

A mixture of **3** (0.01mol) and malononitrile (0.01 mol) in ethanol (50ml) and piperidine (0.5 ml) reflux for 8h, cooled. The solid obtained filtered off and crystallized from ethanol to give **7** as deep orange crystals. Yield 70%; m.p. 205°C - 207°C. Analytical data; found: C, 53.55, H, 3.12; F, 3.31; N, 20.71%. Calculated for $C_{24}H_{21}FN_8PO_4$ (535); C, 53.83; H, 3.55; F, 3.55; N, 20.93%. IR: vcm^{-1} 3430 - 3380 (NH$_2$), 1700 (C=O), 1620 (deform. NH$_2$), 1580 (C=N) 1265 (C-F), 1200 (Ph-P=O), 1050 (Ph-O-P), 930, 910, 850, 800(aryl CH).^{1}H NMR (DMSO) δ: 12.7, 10.7, 10.2 (each s, 3H, 3NH), 8.8 (s, 1H, NH-P=O), 7.55 - 7.311, 7.131 - 6.991, 6.87 - 6.76 (each m, 14 H, aryl & phenyl protons), 3.89 (s, 4H, 2NH$_2$). ^{13}C NMR (DMSO)δ: 163.73, 159.38, 157.80, 134.93, 127.49, 123.67, 113.38, 113.22, 110.66, 105.18; 105.01, 77.84-77.41, M/S (Int.%): 534 (536, M + 2, 1.55%), 248 (1.11), 97 (3.18) 96 (5.55), 95 (100), 93 (18.11), 68 (42.00), 67 (3.11), 62 (37.15).

3-(5'-Phenyl-3'-oxo-2,3,4,5-tetrahydro-pyrazolin-1'-yl)-6-(5'-fluoro-2'-diphenyl phosphorylanilido)-1,2,4-triazin-5-(2H)one(8)

Equimolar amounts of **3** and cinnamoyl chloride in DMF (20ml) reflux for 4 h, cooled then poured onto ice. The produce solid filtered off and crystallized from dioxan to give **8** as deep yellowish crystals, yield 60%, m.p. 160°C - 162°C Analytical data; found: C, 59.89; H, 3.55; F, 3.01; N, 13.75%. Calculated for $C_{30}H_{23}FN_6PO_5$ (597); C, 60.30; H, 3.85; F, 3.18; N, 14.07%. IR vcm^{-1}: 3428 (OH), 3180 (NH), 2937 (CH$_2$), 1694 (C=O), 1610 (C=N), 1482 (deform. CH$_2$), 1314 (N-N), 1230 (C-F), 1169 (P=O), 1059 (Ph-O-P), 954, 909, 809 (aryl CH).^{1}H NMR (DMSO) δ: 14.7, 13.5, 12.6 (each s, 3H, 3NH), 10.80 (s, 1H, OH of pyrazole), 8.8 (s, 1H, NH-P=O), 7.99 - 7.41, 7.04 - 7.32, 7.10 - 6.93, 6.88 - 6.40 (each m, 20H, aryl and phenyl protons), 3.23 (s, 2H, NH$_2$). ^{13}C NMR (DMSO) δ: 163.70, 163.58, 162.90, 162.24, 159.34, 157.83, 157.76, 144.07, 134.29, 130.01, 128.77, 127.84, 121.16, 118.94, 118.84, 117.43, 114.23, 113.13, 112.97, 111.94, 111.89, 111.48, 111.43, 110.64, 110.59, 108.30, 108.12, 78.13 - 77.91, 36.26, 31.14.

3-(3'-(4''-Nitrophenyl)-5'-(4''-fluorophenyl)-4'-5'-dihydro-pyrazolin-1'-yl)-6-(5'-fluoro-2'-diphenyl-phos-phorylanilido)-1,2,4-triazin-5(2H)one (9)

A mixture of **3** (0.01 mol) and a chalcone (0.01 mol) in ethanol (50 ml), and piperidine (0.5 ml) reflux for 8 h, cooled then poured on ice-HCl. The solid produce filtered off and crystallized from THF to give **9** as yellow crystals, yield 82%, m.p. 148°C - 150°C. Analytical data; found: C, 59.49; H, 3.51; F, 5.21; N, 13.29%. Calculated for $C_{36}H_{26}F_2N_7PO_6$ (721); C, 59.91; H, 3.60; F, 5.27; N, 13.59%. IR vcm^{-1}: 3180 (NH), 2827 (CH$_2$), 1675 (C=O), 1595 (C=N), 1547 (C=N), 1500, 1423 (deform. CH$_2$), 1291 (C-F), 1225 (Ph-P=O), 1097 (Ph-O-P), 921, 847, 762 (aryl CH), 684 (C-Cl). ^{1}H NMR (DMSO) δ = 10.57, 10.115 (each s, 2H, 2NH), 8.160 - 8.157, (s, 1H, NH-P=O), 8.118 - 8.095, 7.838 - 7.49, 7.23 - 7.010, 6.92 - 6.90, 6.73 - 6.70 (each m, 23H, aryl & phenyl protons) 2.49 - 2.48 (δ, 2H, CH$_2$). ^{13}C NMR (DMSO) δ = 163.72, 159.35, 157.78, 134.92, 130.85, 130.80, 129.46, 127.41, 123.86, 123.73, 123.67, 121.10, 113.34, 113.18, 111.98, 110.72, 110.67, 108.09, 107.92, 105.13, 104.97, 77.92 - 77.50, 39.56.

3-(3',5'-Dioxo-2',3',4',5'-tetrahydro-pyrazolin-1'-yl)-6-(5'-fluoro-2'-diphenyl phosphorylanilido)-1,2,4-triazin-5'(2H)one (10)

Equimolar mixture of **3** and diethyl malonate in THF (100 ml) reflux for 8 h, cooled. The solid thus obtain filtered off and crystallized from dioxan to give **10** as faint yellow crystals, yield 66%, m.p. 189-190°C. Analytical data; found: C, 53.35; H, 3.18; F, 3.35; N, 15.38%. Calculated for $C_{24}H_{18}FN_6PO_6$(536); C, 53.73; H, 3.35; F, 3.54; N, 15.67%. IR vcm^{-1}: 3300 (NH), 3169 (NH), 1694 (C=O), 1626 (C=N), 1579 (C=N), 1471 (deform. CH$_2$), 1304 (N-N), 1265 (C-F), 1196 (P=O), 1040 (Ph-O-P), 903, 863, 809 (aryl CH).^{1}H NMR (DMSO) δ = 12.7, 10.69, 10.35 (each s, 3H, NH, OH, OH) 9.21 (s, 1H, NH-P=O), 8.15 (s, 1H, C$_4$ of pyrazole), 7.79 - 7.38, 7.10 - 7.003, 6.99 - 6.79 (each m, 15H, aryl & phenyl protons), 4.18 - 4.17, (δ, CH$_2$ of pyrazole). ^{13}CNMR (DMSO) δ = 179.39, 163.58, 162.91, 159.18, 157.61, 134.86, 126.95, 123.74, 123.68, 117.44, 117.28, 113.13, 112.97, 111.94, 110.64, 110.58, 108.12, 107.95, 104.89, 104.72, 78.29 - 77.85, 61.09, 13.94. M/S (Int.%): 536 (538, M + 2, 2.28), 99 (5.5), 96 (13.11), 95 (100), 93 (36.11), 69 (21.85), 68(42.35), 62(5.18).

3-(5'-Aryl-3'-thioxo-2',3'-dihydro-1',2',4'-triazol-1'-yl)-6-(5'-fluoro-2'-diphenylphosphorylanilido)-1,2,4-triazin-5(2H) one (11)

A mixture of **3** (0.01 mol) and P-methoxybenzoylisothiocyanate (0.01 mol) in dioxan (20ml) reflux for 4 h, cooled. The solid produce filtered and crystallized from dioxan to give **11** as orange yellowish crystals. Yield 65%, m.p. 141°C - 192°C. Analytical data, found: C, 55.62; H, 3.41; F, 2.74; N, 15.02; S, 4.72%. Calculated for $C_{30}H_{23}FNPSO_5$ (643); C, 55.98; H, 3.57; F, 2.95; N, 15.24; S, 4.97%. IR vcm^{-1} = 3220-3180 (b, NH), 1693 (C=O), 1624, 1537 (C=N), 1477 (deform. MeO), 1305(N-N), 1266(C-F), 1155 (C-S), 1099 (P=O), 1039(P-O), 980, 840, 809 (aryl CH). ^{1}H NMR(DMSO) δ = 12.74, 10.80 - 1066, 8.73 (each s, 3H, 3NH), 8.40 - 7.52, 7.38 -

7.30, 7.29 - 7.00, 6.99 - 6.800 (each m, 17H) aryl & phenyl protons), 3.68 - 3.67 (s, 3H, OMe). ^{13}C NMR(DMSO) δ: 179.64, 163.77, 163.04, 159.40, 158.02, 157.85, 138.58, 134.94, 132.10, 127.60, 123.72, 123.66, 121.13, 121.07, 117.69, 117.53, 113.46, 113.28, 112.02, 110.76, 110.71, 108.10, 107.93, 105.25, 105.07, 77.81 - 77.35, 66.91.

3-(3'-(4'-Methoxyphenyl)-5'-thioxo-4',5'-dihydro-1'2',4'-triazol-1'-yl)-6-(5'-fluoro-2'-diphenylphos p-horylanilido)-1,2,4-triazin-5(2H)one (12)

A mixture of 3 (0.01 mol) and 4-methoxybenzoyl isothiocynate (0.01 mol) is DMF (20 ml) reflux for 4 h, cooled then poured onto ice. The produce solid filtered off and crystallized from ethanol (to give 12 as orange crystals, yield 70%, m.p. 186°C - 188°C. Analytical data: found: C, 55.69; H, 3.55; F, 2.70; N, 14.89; S, 4.79%. Calculated for $C_{30}H_{23}FN_7PSO_5$ (643); C, 55.98; H, 3.57; F, 2.95; N, 15.24; S, 4.97%. IR vcm^{-1}: 3382 (NH), 3318 (NH), 3204 (NH), 1684 (C=O), 1562 (C=N), 1484 (deform. MeO), 1311 (N-N), 1270 (C-F), 1169 (C=S), 1041 (Ph-O-P), 995, 881, 807 (aryl CH).

3-Methyl-1H-7-(5'-fluoro-2'-diphenyl phosphorylanilido)-1,2,4-triazino[4,3-b][1,2,4] triazin-4,8-dione (13)

A mixture of 3 (0.01 mol) and sodium pyruvate (0.01 mol) in sodium hydroxide solution (5%, 100 ml) warm under reflux for 2 h, cooled them poured onto ice HCl. The solid thus obtained filtered off and crystallized 70%, m.p. 220°C - 222°C. Analytical data; found: C, 55.19; H, 3.31; F, 3.38; N, 15.88%. Calculated for $C_{24}H_{18}FN_6PO_5$ (520); C, 55.38; H, 3.46; F, 3.65; N, 15%. IR vcm^{-1}: 3380, 3162 (NH, NH) 1683 (C=O), 1589, 1553 (C=N), 1458 (deform. CH$_3$), 1308 (N-N), 1250 (C-F), 1204 (P=O), 1040 (Ph-O-P), 909, 861, 804 (aryl CH). ^{1}H NMR (DMSO) δ: 14.55, 12.8 (each s, 2H, 2NH), 10.06 (s, 1H, NH), 7.66 - 7.40, 7.35 - 7.26, 6.93 - 6.76 (each m, 13H, aryl & phenyl protons), 5.58 (s, 1H, OH), 1.256 (s, 3H, CH$_3$). ^{13}C NMR(DMSO)δ:177.40, 176.13, 158.14, 139.40, 128.24, 119.11, 118.95, 118.82, 117.83, 114.59, 114.43, 114.35, 114.29, 112.36, 112.19, 112.09, 112.04, 111.84, 111.77, 111.60, 110.87, 109.95, 109.90, 108.90, 108.37, 77.67 - 77.24, 55.38, 17.21.

Indolo[3,4-e][1,2,4] triazino [4,3-b]-1,2,4-triazin-11-one (14)

Equimolar amounts of 3 and isatin in DMF (20 ml) reflux for 2h, cooled then poured onto ice. The yielded solid filtered off and crystallized from dioxan to give 14 as deep-brown crystals, yield 80%, m.p. 198-200°C. Analytical data found: C, 59.89; H, 3.01; F, 2.98; N, 16.59%. Calculated for $C_{29}H_{19}FN_7PO_4$(579); C, 60.10; H, 3.28; F, 3.28, N, 16.92%. IR vcm^{-1}: 3139 (NH), 1694 (C=O), 1626, 1540 (C=N), 1305 (N-N), 1250 (C-F), 1157 (P=O), 1080 (Ph-O-P), 970, 900, 850, 811 (aryl CH).^{1}H NMR(DMSO) δ=10.76-10.61 (s, 1H, NH), 8.011 (s, 1H, NH-P=O), 7.67 - 7.28, 7.27 - 7.12, 7.05 - 6.94, 6.87 - 6.816 (each m, 17H, aryl and phenyl protons). ^{13}CNMR (DMSO) δ: 163.77, 163.68, 162.45, 159.48, 157.32, 141.29, 136.82, 134.95, 95.133.65, 130.07, 122.27, 120.80, 120.30, 117.73, 117.57, 113.54, 113.50, 113.38, 112.10, 112.04, 110.79, 110.73, 108.09, 107.92, 105.34, 105.17, 77.67 - 77.24. M/Z (Int. %) 579 (581, M + 2; 13.15%) 331 (31.88), 275 (75.19), 248 (1.15); 141 (18.20), 95 (100), 93 (20.51), 63 (5.13).

3-Phenyl-1,2,3,4-tetrahydro-7-(5'-fluoro-2'-diphenyl phosphorylanilido)-1,2,4-triazino [4,3-b] [1,2,4] triazin-8-one (15)

A mixture of 3 (0.01 mol) and phenacyl bromide (0.01 mol) in ethanolic KOH (5%, 20 ml) reflux for 2 h, then poured onto ice-HCl. The solid produced filtered off and crystallized from THF to give 15 as brownish crystals. Yield 65% m.p., 98°C - 100°C. Analytical data; Found: C, 59.98; H, 3.55; F, 3.20; N, 14.51%. Calculated for $C_{29}H_{22}FN_6PO_4$ (568); C, 61.26; H, 3.87; F, 3.34; N, 14.78%. IR vcm^{-1}: 3158 (NH), 1655 (C=O), 1580 (C=N), 1474 (deform. CH$_2$), 1308 (N-N), 1230 (C-F), 1180 (P=O), 1092 (Ph-O-P), 980, 950, 900, 850, 801 (aryl CH).^{1}H NMR(DMSO) δ = 12.75 (s, 1H, NH), 10.78 (s, 1H, NH-P=O), 7.84 - 7.55, 7.49 - 7.30, 7.29 - 7.00, 6.99 - 6.79 (each m, 18H, aryl & phenyl protons). 3.4 (s, 2H, CH$_2$). ^{13}C NMR (DMSO) δ:179.66, 163.04, 159.62, 158.02, 138.60, 129.51, 129.14, 128.67, 128.45, 128.18, 127.18, 126.89, 126.67, 125.86, 125.44, 123.31, 121.05, 120.20, 117.72, 117.56, 113.90, 113.74, 112.21, 112.13, 112.07, 110.47, 110.04, 108.10, 107.92, 77.72-77.29, 36.62.

7-(5'-Fluoro-2'-diphenylphophorylanilido)-1,2,3,4-tetrahydro-1,2,4-triazino[4,3-b][1,2,4]triazin-4,8-dione (16)

A mixture of 3 (0.01 mol) and monochloroacetic acid (0.01 mol) in DMF (20 ml) reflux for 2 h, cooled then poured onto ice. The solid thus obtained filtered off and crystallized from THF to give 16 as yellowish crystals. Yield 70%, m.p. 165°C - 167°C. Analytical data; found: C, 54.01; H, 3.25; F, 3.58; N, 16.31%. Calculated for $C_{23}H_{18}FN_6PO_5$ (508); C, 54.33; H, 3.54; F, 3.74; N, 16.53%. IR vcm^{-1}: 3382 (NH), 3224 (NH), 1685 (C=O), 1590 (C=N), 1477 (deform. CH$_2$), 1250 (C-F), 1180 (P=O), 1041 (Ph-O-P), 807, 760 (aryl CH). ^{1}H NMR (DMSO) δ: 12.97, 10.70, 10.42 (each s, 3H, 2NH, OH), 9.19 (s, 1H, NH-P=O), 7.99 - 7.00, 6.99 - 6.86, 6.85 - 6.79 (each m, 13H, aryl & phenyl protons), 4.33 (s, 2H, CH$_2$). ^{13}C NMR (DMSO) δ = 173.41, 163.77, 163.67,

159.42, 157.84, 157.70, 136.69, 136.04, 121.23, 121.06, 113.87, 113.81, 113.24, 113.08, 111.48, 111.43, 110.966, 110.699, 111.644, 109.28, 107.00, 106.83, 105.02, 104.02, 78.08 - 77.65, 31.22.

7-(5'-Fluoro-2'-diphenylphosphorylanilido)-1,2-dihydro-1,2,4-triazino [4,3-b][1,2,4] triazin-3,8-dione (17)

Equimolar mixture of **3** and 1,1-dichloroacetic acid in DMF (20ml) reflux for 2h, cooled then poured onto ice. The yield obtained filtered and crystallized from dioxan to **17** as yellowish crystals. Yield 78%, m.p. 180°C - 182°C. Analytical data: found C, 54.03; H, 2.89; F, 3.55; N, 16.31%. Calculated for $C_{23}H_{16}FN_6PO_5$ (506); C, 54.54; H, 3.16; F, 3.75; N, 16.60%. IR vcm^{-1}: 3100 (NH), 1791 (C=O), 1655 (C=O), 1590 (C=N), 1547 (C=N), 1295 (N-N), 1235 (C-F), 1158 (P=O), 1094 (Ph-O-P), 984, 865, 830 (aryl CH).^{1}H NMR (DMSO) δ: 12.7 (s, 1H, NH), 10.53 (s, 1H, NH-P=O), 9.07 (s, 1H, CH=N), 7.91 - 7.16, 6.92 - 6.81, 6.77 - 6.71 (each m, 13H, aryl & aryl protons). ^{13}C NMR(DMSO) δ:179.58, 163.82, 163.01, 162.39, 159.50, 157.92, 142.32, 138.56, 136.67, 136.03, 121.28, 121.12, 117.45, 115.59, 115.43, 113.90, 112.13, 111.47, 110.67, 109.50, 109.34, 108.08, 107.91, 107.12, 106.95, 77.88 - 77.45, 66.36.

7-(5'-Fluoro-2'-diphenylphosphorylanilido)-1,2,3,4-tetrahydro-1,2,4-triazino[4,3-b][4,34]triazin-3,4,8-trione (18)

A mixture of **3** (0.01 mol) in dry C_6H_6 treat with oxalyl chloride (0.01 mol) added dropwise then, TEA added drop wise (few drops) then heated under reflux for 2h, cooled. The solid produce filtered off and crystallized from C_6H_6 to give **18** as deep-yellowish crystals, yields 60%, m.p. 168°C - 170°C. Analytical data: found: C, 52.49; H, 2.85; F, 3.51; N, 15.88%. Calculated for $C_{23}H_{16}FN_6PO_6$ (522); C, 52.87; H, 3.06; F, 3.63; N, 16.09%. IR vcm^{-1}: 3382, 3319, 3156 (3NH), 1680 (C=O), 1557 (C=N), 1270 (C-F), 1143 (P=O), 1041 (Ph-O-P), 994, 860, 804 (aryl CH).^{1}H NMR (DMSO) δ: 14 (s, 1H, NH), 12.9 (s, 1H, NH), 10.8 (s, 1H, NH-P=O) 9.3, 8.50, 8.0 (each s, aryl protons), 7.45, 7.35, 7.039, 7.035, 7.031, 7.02, 6.886, 6.878, 6.870, 6.86 (m, 10H phenyl protons). ^{13}C NMR (DMSO) δ: 167.31, 163.07, 159.53, 158.95, 129.48, 128.26, 120.21, 118.50, 118.26, 112.38, 112.33, 112.19, 112.14, 77.62 - 77.202. M/Z (Int.%) 522 (525, M + 3, 18.55), 248 (3.15), 96 (23.11), 95 (100), 93 (8.9), 68 (15.0).

The acid hydrazido derivative 20

A mixture of **3** (0.01 mol) and oxazolone**19** (0.01 mol) in ethanol (50 ml) with H_2O (20 ml) reflux for 2 h, cooled then poured onto ice. The produced solid filtered off and crystallized from dioxan to give **20** as yellow crystals. Yield 58%, m.p. 155-157°C. Analytical data; found: C, 59.89, H, 3.55, F, 5.00; N, 13.05%. Calculated for $C_{37}H_{28}F_2N_7PO_6$ (735); C, 60.40; H, 3.80; F, 5.17; N, 13.13%. IR vcm^{-1}: 3600 - 3200 (b, OH, NH, NH), 1592 (CONH), 1480 (deform. CH=C), 1240 (C-F), 1170 (P=O), 1050 (Ph-O-P), 850, 752 (aryl CH).

3-[5'-(4''-Fluorophenylidene)-3'-phenyl-1'H-6'-oxo-1',2',4'-triazin-2'-yl]-6-(5'-fluoro-2'-diphenylphospho-rylanilido)-1,2,4-triazin-5-(2H)one (21)

Compound **20** (1 mg) and glacial acetic acid (10ml) reflux for 2h, cooled then poured onto ice. The solid thus obtained filtered off and crystallized from ethanol to give **21** as deep yellowish crystals. Yield 80%, m.p. 233°C - 235°C, Analytical data; found: C, 61.81; H, 3.49; F, 5.09; N, 13.55%. Calculated for $C_{37}H_{26}F_2N_7PO_5$ (717); C, 61.92; H, 3.62; F, 5.29; N, 13.66%. IR vcm^{-1}: 3192 (NH), 2880, 2810 (aliphatic CH), 1699 (C=O), 1632 (CONH), 1479 (deform. CH=), 1303 (N-N), 1240 (C-F), 1210 (P=O), 1056 (Ph-O-P), 900, 870, 810, 801 (aryl CH).

6-[5'-fluoro-2'-(diphenylphosphato)aminophenyl]-3,3,3-triphenyl-3-λ^5-1,2,4,3-triazaphopholino[4,5-b][1,2,4]triazine-7(8H) one (22)

A mixture of compound **2** (0.01 mol) and triphenyl phosphine (0.01 mol) in acetonitryl (20ml) warm for 30 min, cooled. The solid obtained filtered off and crystallized from dioxan to give **22** as deep yellow crystals, yield 70%, m.p. 283°C - 285°C. Analytical data; found: C, 64.01; H, 4.15; F, 2.39, N, 11.35%. Calculated for $C_{39}H_{31}FN_6P_2O_4$ (728); C, 64.28; H, 4.25; F, 2.60; N, 11.53%. IR vcm^{-1}: 3300 - 3100 (b, NH, NH), 1694 (C=O), 1620 (C=N), 1307(N-N), 1264 (C-F), 1179 (P=O), 1027 (Ph-O-P), 910, 880, 793, 745 (aryl CH).

3. Results and Discussion

α-Aminophosphonic acids continue to elicit study due to interest in their biological properties as herbicides [22], plant growth regulators [23] and most notably those species heaving a direct P-N bond are investigated as transition state analogues of the tetrahedral transition-state involved in peptide hydropysis [24]. Ali *et al.* [25] [26] studied the reactivity of α-amino phosphonates as dipolar ion structure and have type of tautomeric formula due to the higher e-withdrawing of two phenoxy and P=O groups (**Figure 2**). Thus, α-aminophosphonate group had

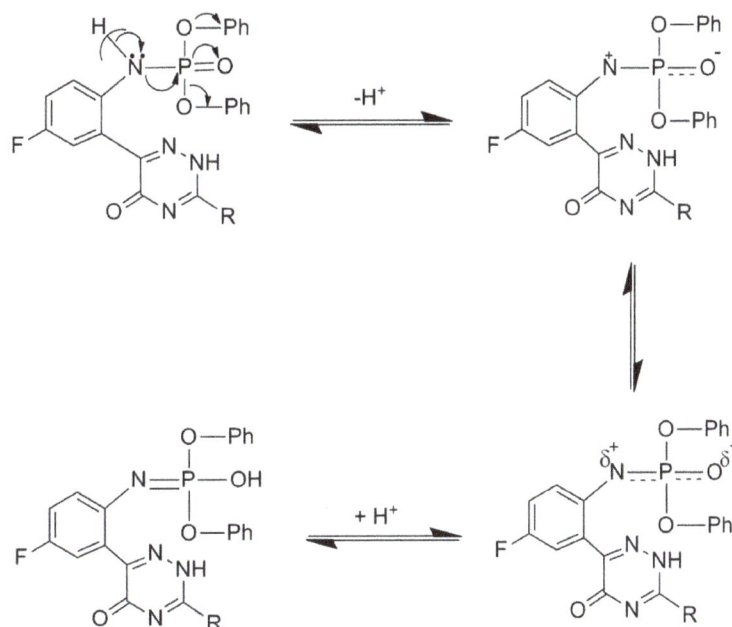

Figure 2. A possible present formula of the new synthesized isolated systems.

a higher degree of stability towards any attack of reagents, which attribute to presence of differ factors of stability [27]. Phosphorus elements in these systems, was determine by using the spectrophotomeric. The method is based on the development of floated complex of molybdophosphonic acid (MPA) and methylene blue (MB) with N, N'-diphenylbenzamide (DPBA) in toluene and its subsequent dissolution in acetone [28].

A series of some new fluorine substituted phosphoryl-amino-1,2,4-trinzines bearing a functionally pyrazole ring have been obtained via cycloaddition and/or cyclocodension of 3-hydrazino-1,2,4-trinzinone**3** with π-acceptor activated carbon atom reagents. In addition, a type of 1,2,4-trinzino [4,3-b] [1,2,4] triazindiones have been also obtained from cyclocondensation of compound **3** with α, β-bifunctional oxygen and halogen compounds in different conditions. The former structures of the new products confirmed from correct elemental analysis and their spectral measurements. Keeping in view the diverse medicinal activities associated with organo-heterocyclilc systems substituted fluorine, phosphorus and 3-thioxo-1,2,4-trinzinone, which intend to construct novel fluorinated substituted phosphoryl amino bearing 1,2,4-triazinone moiety hoping to active additive effects towards their HIV-1 activity.

3.1. Chemistry

The starting material 3-hydrazino-6-(5'-fluoro-2'-diphenylphosphorylanilido)-1,2,4-trizino-5(2H) one (**3**) obtained from treatment of 6-(5'-fluoro-2'-aminophenyl)-3-thioxo-1,2,4-triazin-5(2H)one (**1**) with diphenylphosphoryl chloride in DMF to give 6-(5'-fluoro-2'-diphenylphosphorylanilido)-3-thioxo-1,2,4-triazin-5-one (**2**) followed by hydrozinolysis in boiling ethanol (**Scheme 1**).

Recently, the most reactions of activated nitrites take place in basic medium leading to novel heterocyclic systems [29] [30]. Similarly, amino-pyrazolyl-1,2,4-triazinano derivatives **4-7** were obtained from the interaction between compound **3** and arylidenecyanoacetic acid (EtOH-piperidine), acetyl acetanilide (DMF), ethyl cyanoacetate (TFH) and/or malono nitrile (EtOH-piperidine) (**Scheme 2**).

These reactions are carried out via cycloaddition and/or cyclocondensation reactions [31] (**Figure 3**).

It is interested that 3-perhydropyrazo-1-yl-1,2,4-trizinones **8-10** also obtained from the ring closure reaction of compound **3** with cinnmoyl chloride (DMF), chalcone (EtOH-piperidine) and or diethylmalonate (THF) (**Scheme 3**).

Formation of **8** may be take place via aroylation then cycloadditon reaction [32] (**Figure 4**).

The greater reactivity of the polyfunctional compound as aroylisothiocynate towards the hydrazino group as bi-nucleophile is presumably due to its favourable location between both carbonyl and thiocarbonyl functional

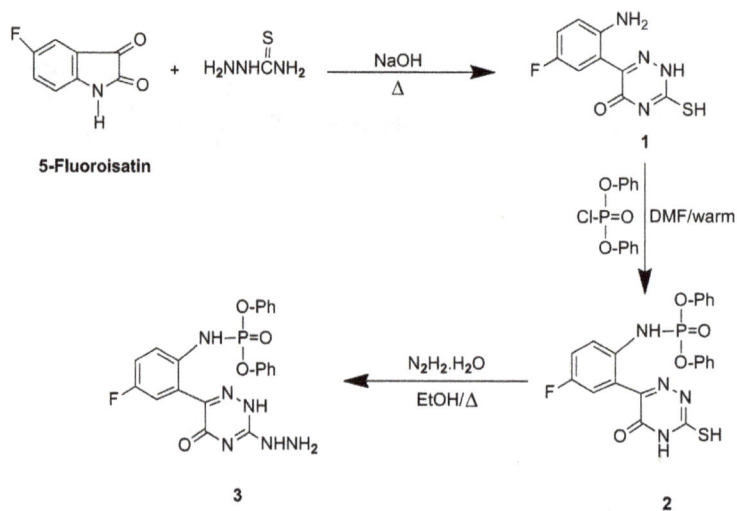

Scheme 1. Formation compounds 1-3.

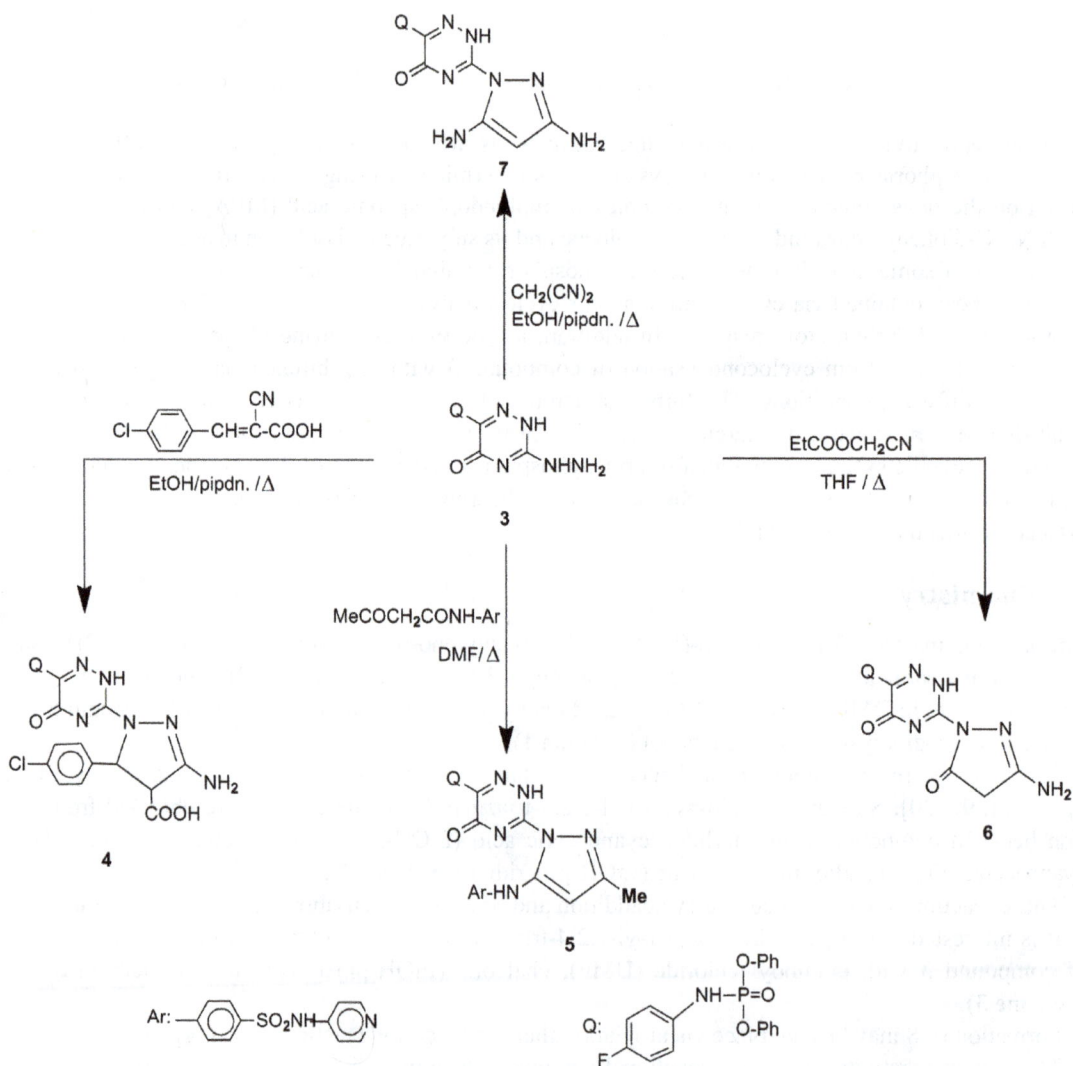

Scheme 2. Formation compounds 4-6.

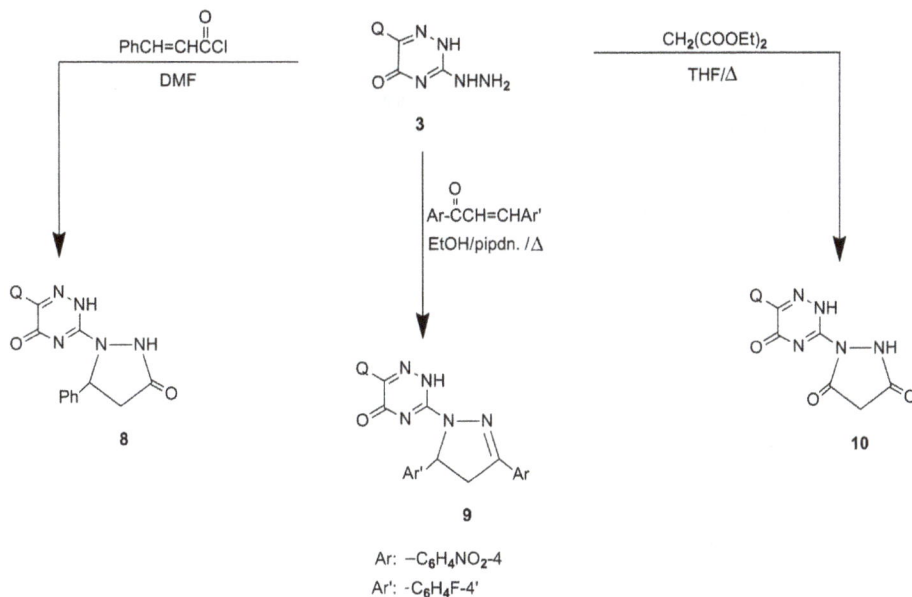

Scheme 3. Formation compounds 8-10.

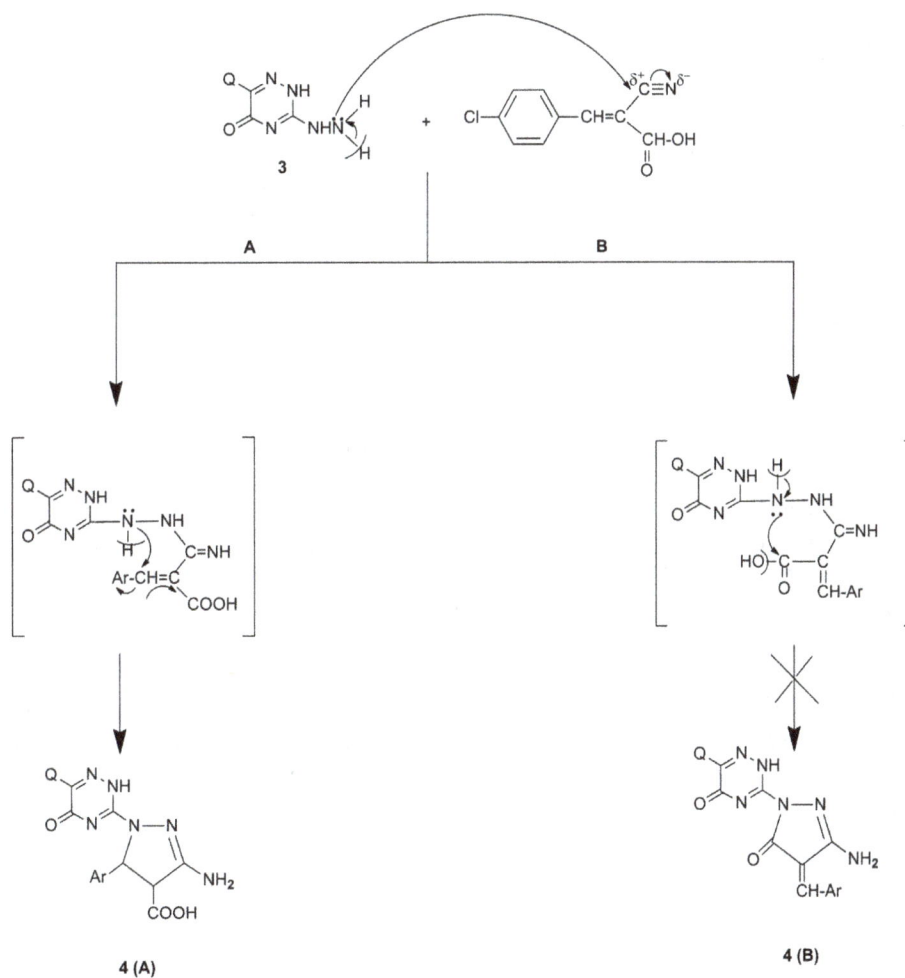

Figure 3. Formation of compound 4 from compound 3.

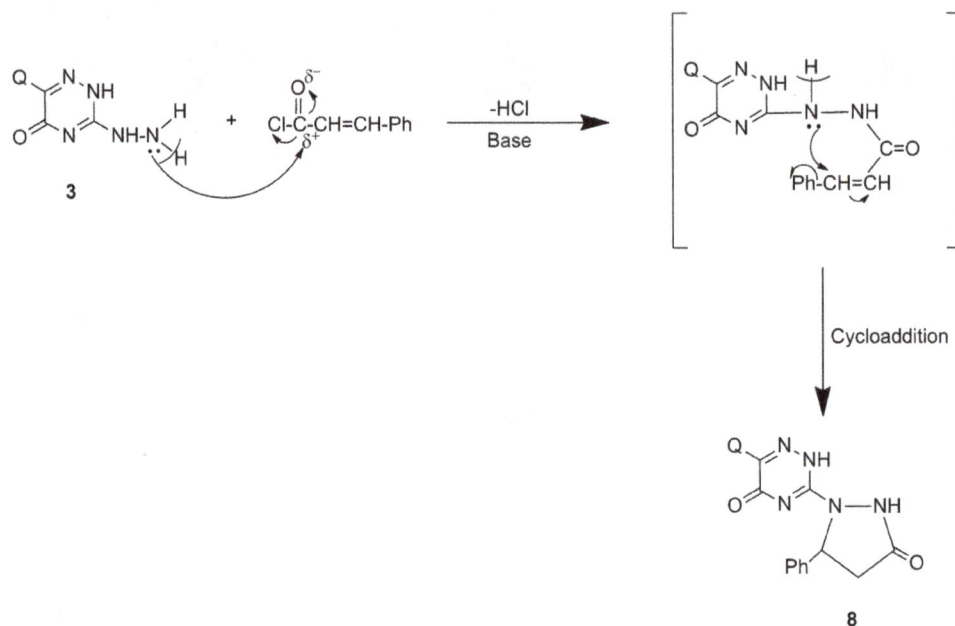

Figure 4. Formation of compound 8 from compound 3.

groups [33]. Thus, treatment of 3-hydrazino-1,2,4-trizinone **3** with aroylisothiocyanate in boiling non-polar solvent as dioxan yield 3-(5'-aryl-3'-mercapto-1',2',4'-triazol-1'-yl)-6-(5'-fluoro-2'-diphenly phosphorylanilido)-1, 2,4-triazin-5(2H)one (**11**), while that reaction when carried out in DMF, 3-(3'-aryl-5'-mercapto-1', 2',4'-tri-azol-1'-yl)-6-(5'-fluoro-2'-di-phenylphosphorylanilido)-1,2,4-trizin-5-(2H) one (**12**) isolated (**Scheme 4**). Formation of compounds **11** and **12** starting from compound **3** were outlined in (**Scheme 4**).

Reactivity of α,β-bifunctional carbonyl compounds towards an hydrazino groups arrived us to synthesize new fused heterobicyclic nitrogen systems. Thus, the interaction between compound **3** and sodium pyruvate in warming sodium hydroxide solution afforded 1H-3-methyl-7-aryl-1,2,4-triazino[4,3-b][1,2,4] traizin-4,8-dione (**13**), while cyclo-condensation of compound **3** with isatinas 1,2-bicarbonyl compound in boiling DMF yield indolo [2,3-e] [1,2,4] trinzino [4,3-b][1,2,4] triazin one (**14**). Refluxing of **3** with phenacyl bromide in ethanolic KOH furnish the tetrahydro-1,2,4-triazino [4,3-b][1,2,4] triazin-8-one (**15**) (**Scheme 5**).

A large degree of the biological activity is attributed of the nature of substituent's and a degree of electronic distribution over the active center of the 1,2,4-triazines [34] [35]. Thus, direct nucleophilic displacement of chlorine atoms by nitrogen or other nucleophilic can easily occur if present α-carbonyl groups. Based on these facts, treatment of compound **3** with activated halogen as chloroacetic acid (DMF), dichloroacetic acid (DMF) and or oxalyl chloride (C₆H₆/TEA), produce perhydro 1,2,3,4-tetrahydro-1,2,4-triazino [4,3-b] [1,2,4] triazin-4,8-dione (**16**); 1,2-dihydro-1,2,4-triazino [4,3-b] [1,2,4] triazin-4,8-dione (**17**) and 1,2,3,4-tetrahydro-1,2, 4-triazino [4,3-b] [1,2,4] triazin-3,4,8-trione (**18**) derivatives (**Scheme 6**).

In view of interesting results obtained from the reaction of 1,3-oxazolium salts and of 1,3-oxazol-2-one derivatives with hydrazine derivatives[36], [37], it was worthwhile to investigate the behavior of oxazolone**19** towards hydrazine-derivative. Similarly, 3-hydrazino-6-aryl-1,2,4-triazin-5 (2H) one (**3**) when react with oxazole derivation **19** in boiling aqueous ethanol, the acid hydrazide derivative **20** isolated. Ring closure reaction of **20** by refluxing with glacial acetic acid afforded 3-(1'H-3-phenyl-5'-arylidene-6'-oxo-1,2,4-triazin-2'-yl)-1-6-(5'-fluoro-2'-diphenylphosphorylanilido)-1,2,4-triazin-5(2H)one(**21**)Finally,6-[5'-fluoro-2'-(diphenylphosphato)aminoph-enyl]-3,3,3-triphenyl-3-λ⁵-1,2,4,3-triazaphopholino[4,5-b][1,2,4]triazine-7(8H) one (**22**) isolated from treat compound **3** with triphenylphosphine in THF (**Scheme 7**).

3.2. Elucidation the Former Structures

3.2.1. UV Spectra Study

UV absorption study of the new compounds, synthesize give us a good indication about electronic distribution

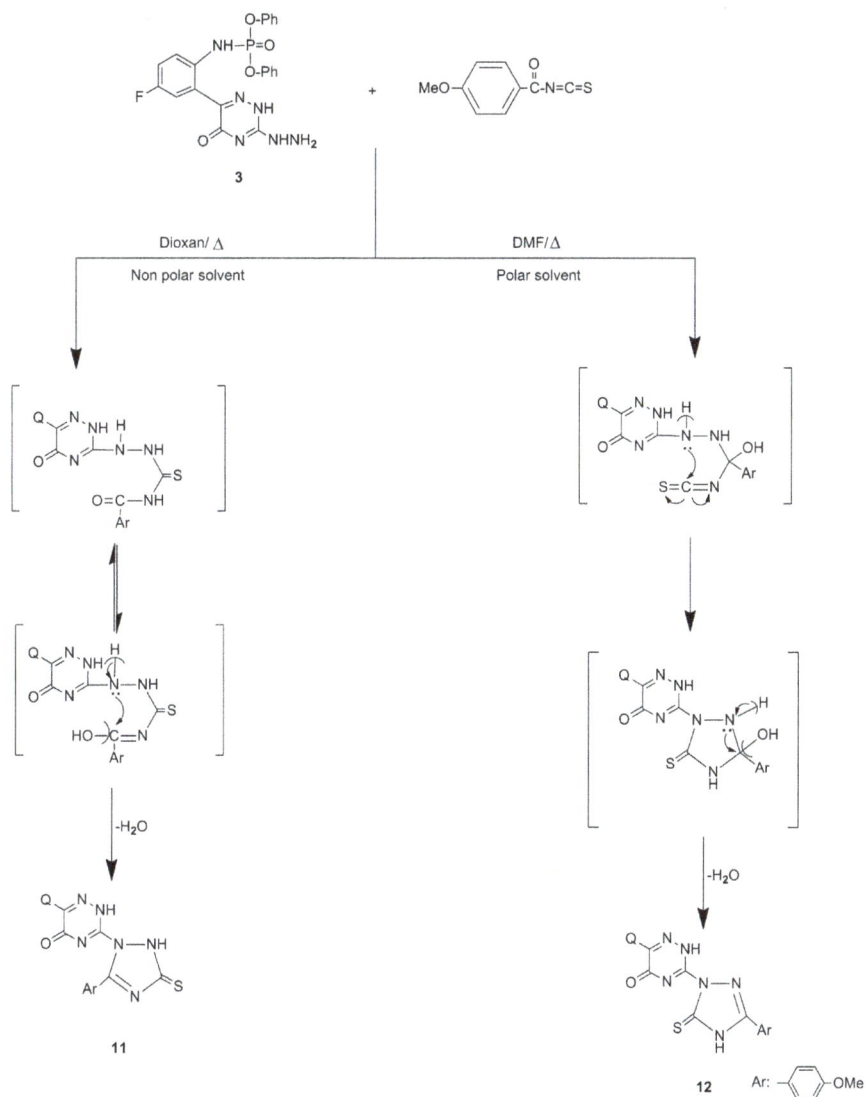

Scheme 4. Formation compounds 11 & 12.

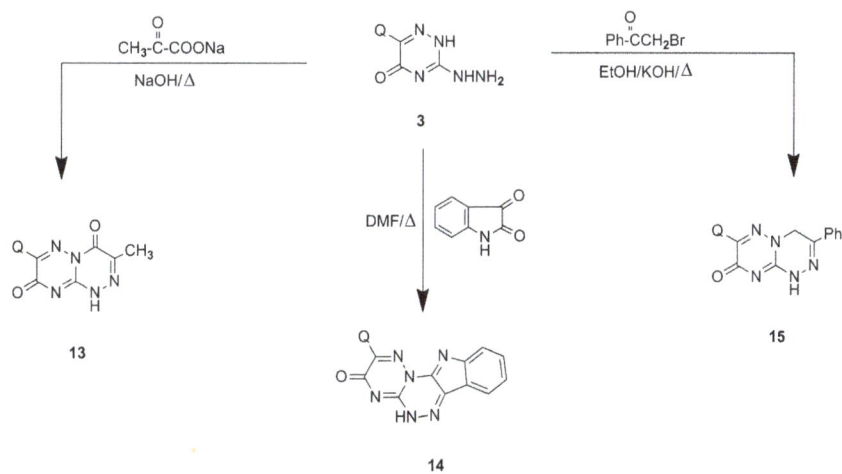

Scheme 5. Formation compounds 13-15.

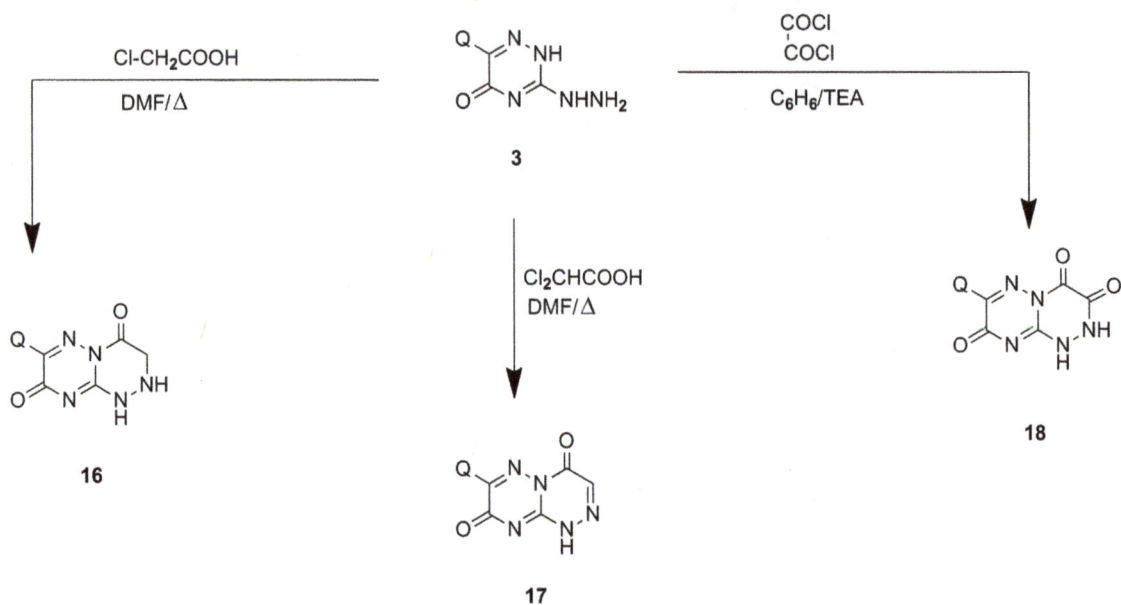

Scheme 6. Formation compounds 16-18.

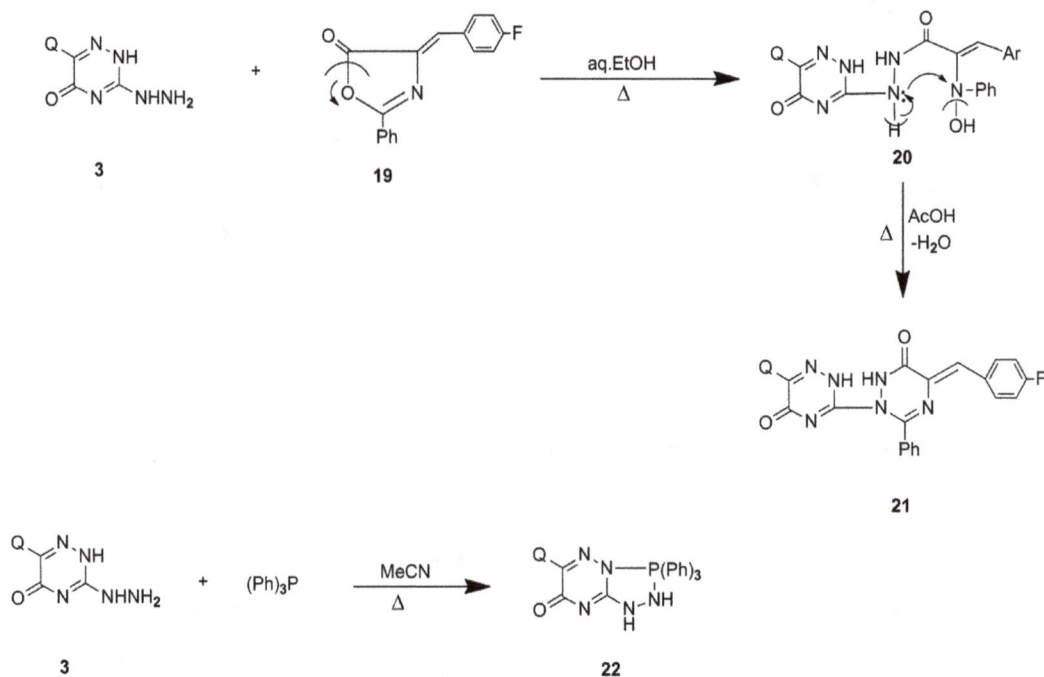

Scheme 7. Formation compounds 19-22.

and molecular configuration as possible. In general, all the new compounds record the presence of n-π^*, n-σ^*, π-π^* and σ-σ^* electronic transition. UV-absorption spectrum of compound 3 as state recorded λ_{max} (EtOH) at 346.48 nm, while that of compounds 14 (361.55) and 12 (361.12). A higher value of λ_{max} for 14 and 12 than 3 is may be a lack's of -OH groups (which generate of H-bending (**Figure 5**).

On the other hand, UV absorption spectra of selected compound record a lower λ_{max} than compound 3. λ_{max} of 10 (330.56) and 18 (317.22) nm. A lower λ_{max} of these compounds than the start is may be due to the presence of O-H group, which generate a type of H-bonding. The Keto-enol forms of 10 and 18 led to the inhibition of heteroconjugative, in addition a type of H-bonding which is closed to 1,2,4-triazine moiety (**Figure 6**).

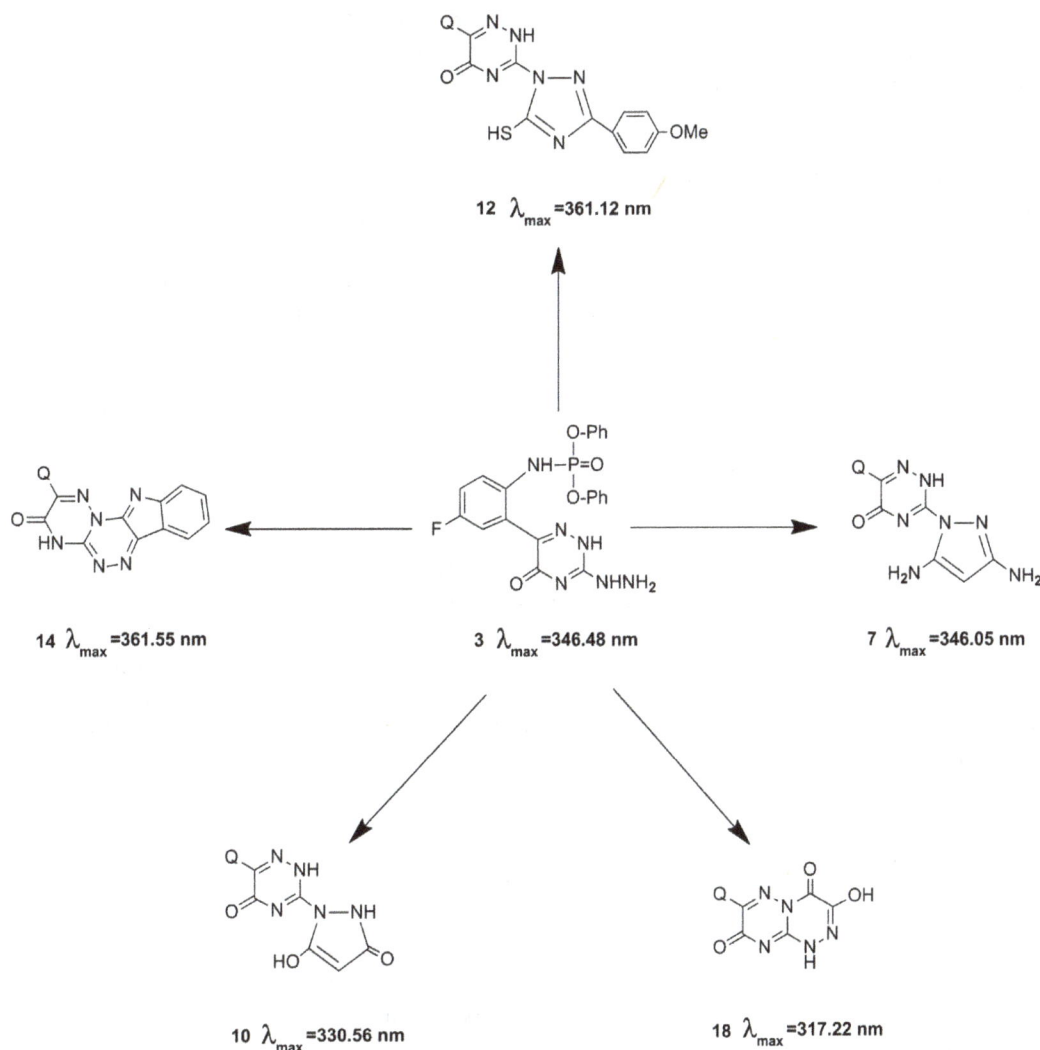

Figure 5. UV-absorbtion data of compound 3 and some prepared compounds.

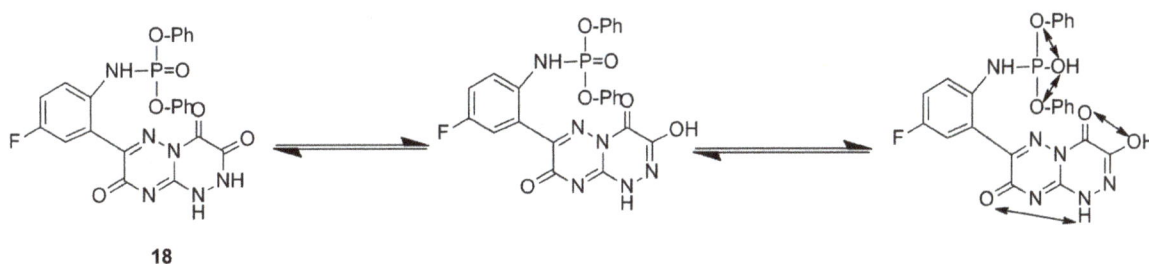

Figure 6. A possible conformer structure of 18 (fused system).

3.2.2. IR-Absorption Study

The IR absorption spectral data show that most of the new compounds lack's a band of NH-P=O, which is due to a type of H-bonding present (**Figure 6**), while that of these compounds record only NH bond of new 1,2, 4-triazinones and/or pyrazoles moiety. In addition, IR absorption spectra of all the synthesized compounds exhibit an absorption bands at 3300 - 3190 (NH), 1690 - 1650 (C=O). Moreover presence of a characteristic bands at v 1390 - 1370 for cyclic NCN, 1250 and 1220 - 1200 cm^{-1} for C-F and P=O functional. Also, all the new compounds record v at 1100 - 1050 cm^{-1} attribute to Ph-O-P group. Only, the compounds **3, 4, 6**, and **7** record a v for

NH_2 at 3390 - 3400 cm^{-1}, while that of the compounds **4, 6, 8, 10, 13, 16-21** showed v for C=O in addition the original C=O of 1,2,4-trinzine. Finally, IR spectra of the compounds **4, 5, 6, 8-13** and **15, 16** record a type of bands characteristic for aliphatic groups (deformation 1480 - 1440 cm^{-1}).

3.2.3. NMR Spectral Study

1) ^{1}H NMR Spectral Study

The NH proton signals in all the new compounds appear as doublets at δ 8.8 - 8.2 ppm (JP-N-H=6 - 5.5 H_z) is due to its coupling with phosphorus. Also, H-bonding with oxygen of P=O and the deshielding effect of phenoxyphosphoryl group (Ph-O-P=O) are obviously the contributing factors for the downfield shift of the NH proton. On the other hand, phenoxy protons resonated at δ 7.50 - 7.1 ppm and their integration corresponds to five protons with no splitting of the signals. This shows that all the protons as magnetically equivalent. Normally, exo and endo NH protons of 1,2,4-triazinone reveal at δ12 - 11 ppm, while what of the 1,2,4-tiazinone addujent of CH_2 or NH protons show as enolic protons at δ11 - 10 ppm (**5,7,17,18,9**).In addition, all the new compounds **4 & 6**, exhibited a resonated signals at δ5 - 4 ppm as NH_2 protons and NH proton at δ11 - 10 ppm. Finally, the perhydro pyrazolyl-1,2,4-triazinones and the 1,2,4-trinzino-1,2,4-trinzinones which containing an aliphatic protons show a resonated signals at δ4 - 3 and 1 - 0.5 ppm for CH_2 and CH_3 protons.

2) ^{13}CNMR Spectral Study

The ^{13}C Chemical shifts of phenoxy moiety are agreeing well with the reported values [38]. But, the coupling constants for ^{2}Jare concurring with those of equationally oriented P-O-Ar groups [39] showing the 1,2,4-trinzine ring has probably half chair conformation with phosphorus atom projecting upwards and the O-Ar group orienting equationally. The carbons C, which are connected to phosphorous through NH, resonated at δ 112.97 with ^{2}J P-N-C (d, J 8.1) and the difference in their chemical shifts may be attributed to the variation of shielding effect of NH [40]. In addition, all the new compounds record the resonated attribute to C=O (170 - 160), C-F (150 - 140), C=N(140 - 130), C-N (111 - 110) of 1,2,4-triazinone, with a differ type of carbons of pyrazolyl as well as carbons of other 1,2,4-triazine formed (**Figure 7** and **Figure 8**).

3.2.4. Mass Spectrometric Measurements

The mass spectral investigations of the isolated heterobicyclic system is differ than the fused heterobicyclic systems (**14 & 18**) for example, M/Z of **7** and/or **10** recorded the molecular ion Peak's at m/z 534 and 336 respectively with a lower abounds percent's, which indicating the fragile nature of these systems. While M/Z of compound **14** showed the highest value peak at 579 with a base peak at m/z 95, which give us a high degree of stability for this Skelton. Moreover, M/Z of **18** exhibits a molecular ion peak at m/z 522 with moderate abounds percent. From these data, we can conclude that fused heterobicyclic nitrogen systems are more stabilized [41]

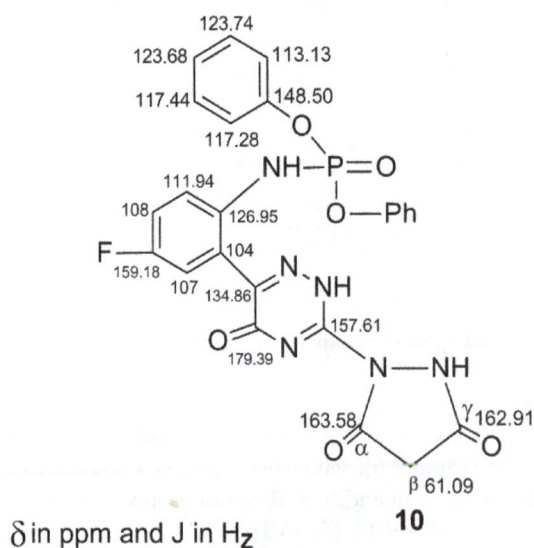

Figure 7. ^{13}C NMR data of compound 10.

than other isolated heterobicyclic systems. It's interest that in all mass fragmentation pattern, 4-fluorophenyl ion is a base peak followed by N-phosphorus oxide ions, while that heterocycle supported that a large fragmentation bath way (**Figures 9-12**).

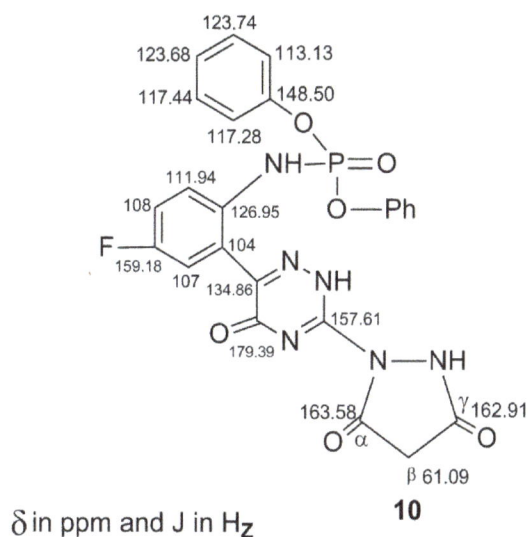

δ in ppm and J in H$_z$

10

Figure 8. ^{13}C NMR data of compound 18.

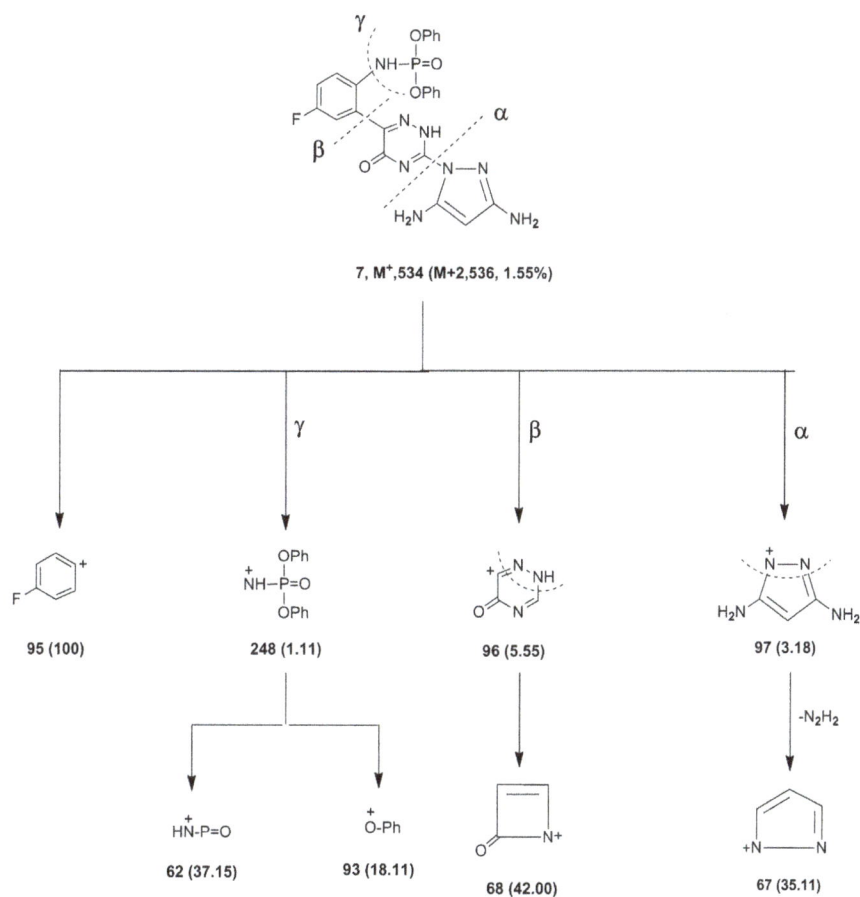

Figure 9. Mass Fragmentation pattern of compound 7.

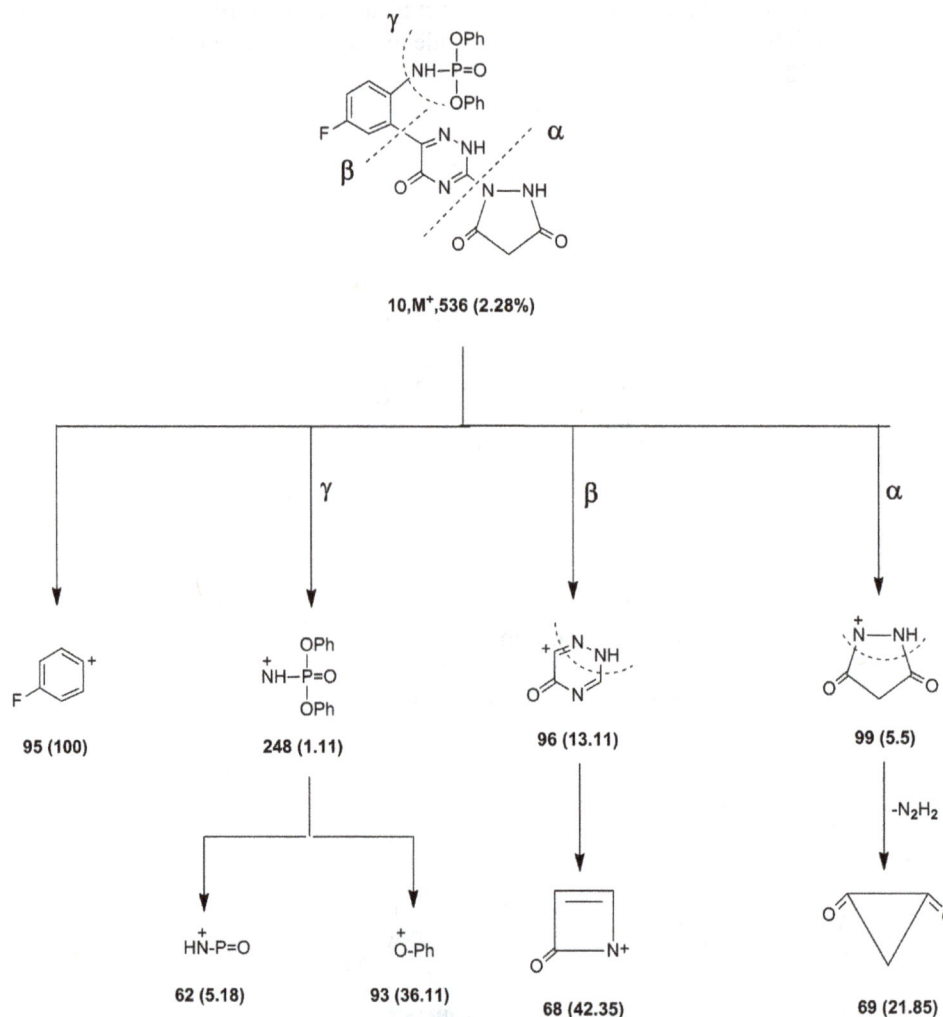

Figure 10. Mass Fragmentation pattern of compound 10.

4. HIV-1 Inhibition (Enzyme Inhibition)

Human immunodeficiency virus type-1 is the causative agent of acquired immunodeficiency syndrome (AIDS) which is one of the most serious health problems [42]. Since reverse transcriptase (RT) is an essential enzyme for the replication [43] of HIV, it is the most favoured target for the antiviral chemotherapy against HIV infection [44]. 3'-Azido-2',3'-dideoxythymidine (AZT) [45] and 2',3'-dideoxyinosine (DDI) [46], 2',3'-dideoxycytydine (DDC) [47] and 2',3'-didehydro-3-deoxythymidine (DT4) [48] are the well-known potent nucleoside reverse transcriptase inhibitors clinical use, but unfortunately they produce serious side effects such as bone marrow suppression. The search for a more effective and less toxic agent has brought into focus potent yet structurally different non-nucleoside HIV-1 reverse transcriptase inhibitors (NNRTIs) [49]. Shakil *et al.*, [50] reported that increase or decrease of electro-negativity and hydrophobicity of the bioactive drugs, cytotoxicity will also increase or decrease accordingly. So less electronegative and less hydrophobic substituents would be preferred to design the less cytotoxic drugs. The large number of research papers published every year indicate that the development of an effective drug for the inhibition of HIV-1 via enzymes inhibitors [51] [52]. HIV PIs for example, prevent the cleavage of the gag and gag-pol precursor polyproteins to the structural proteins and functional proteins, thus arresting maturation and thereby blocking infectivity of the nascent virions. e.g. Tipranavir showed loss cross-resistance to HIV strains that were resistant to the established (peptidomimetic) inhibitors of HIV protease. Also, tipranavir retained marked activity against HIV-1 isolates derived from patients with multidrug resistance to other PIs (**Figure 13**).

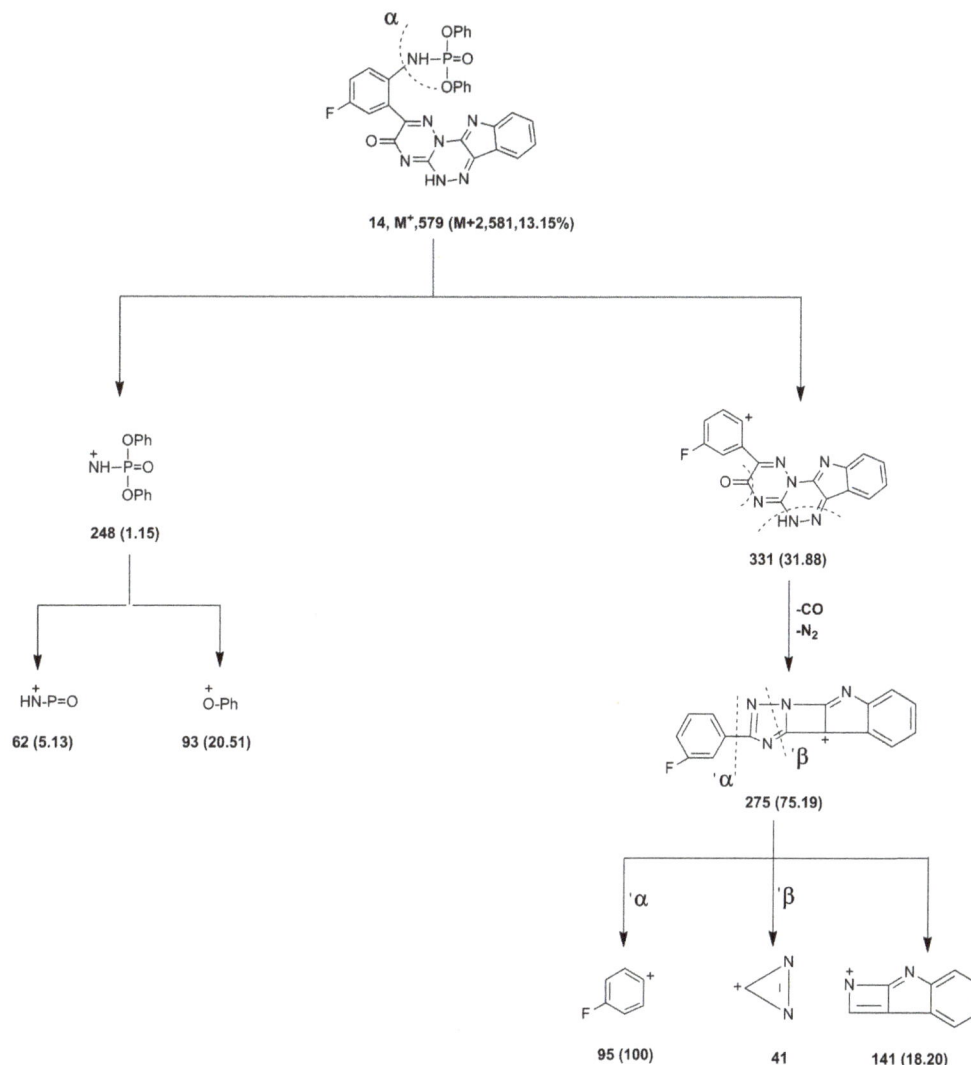

Figure 11. Mass Fragmentation pattern of compound 14.

In search for new poly substituted 1,2,4-triazine bearing a phosphoryl group. The present work is synthesize novel fluorine substituted phosphorylanilido-1,2,4-triazin-ones and evaluate as potential inhibitors for HIV-1. The procedure used in the National Cancer Institute's Test for agents active against HIV is designed to detect agents acting at any stage of the virus reproductive cycle [53]. The assay basically involves the killing of T_4 lymphocytes by HIV. Small amounts of HIV are added to cells and two cycles of virus reproduction are necessary to obtain the required cell killing. Agents that interact with virions, cells or virus gene-products to interfere with viral activities will protect cells from cytolysis. The tetrazolium salt XTT is added to all wells and cultures are incubated to allow formazan color development by viable cells used analyzed spectrophotometrically (**Figure 14**).

Anti-HIV-1 screening results of some of compounds (**Table 1**) shows that these systems found inactive, probably due to their inability to exist in butterfly like conformation as explained in a similar case [54]. Only the compounds **7, 12, 18** and **22** exhibited a higher protection% (concentration required in protect MT-4 cells against the cytopathogenicity of HIV by 50%). The order reactivity increases as **18 > 10 > 7 > 12 > 22**. A higher effects of compound **18** (**Table 2**) is may be due to a higher possibility to form a type of H-bonds with proteins of virus, which led to a moderate degree of inhibition of HIV-1 activity. Also, compound **18** had a differ type of phenolic bonds, which give a variety of differ effects towards HIV-1 activity. Moreover, compound **18** had a less possibility to form a type of intra-molecular H-bond, thus their hydroxyl group become a higher degree of free action

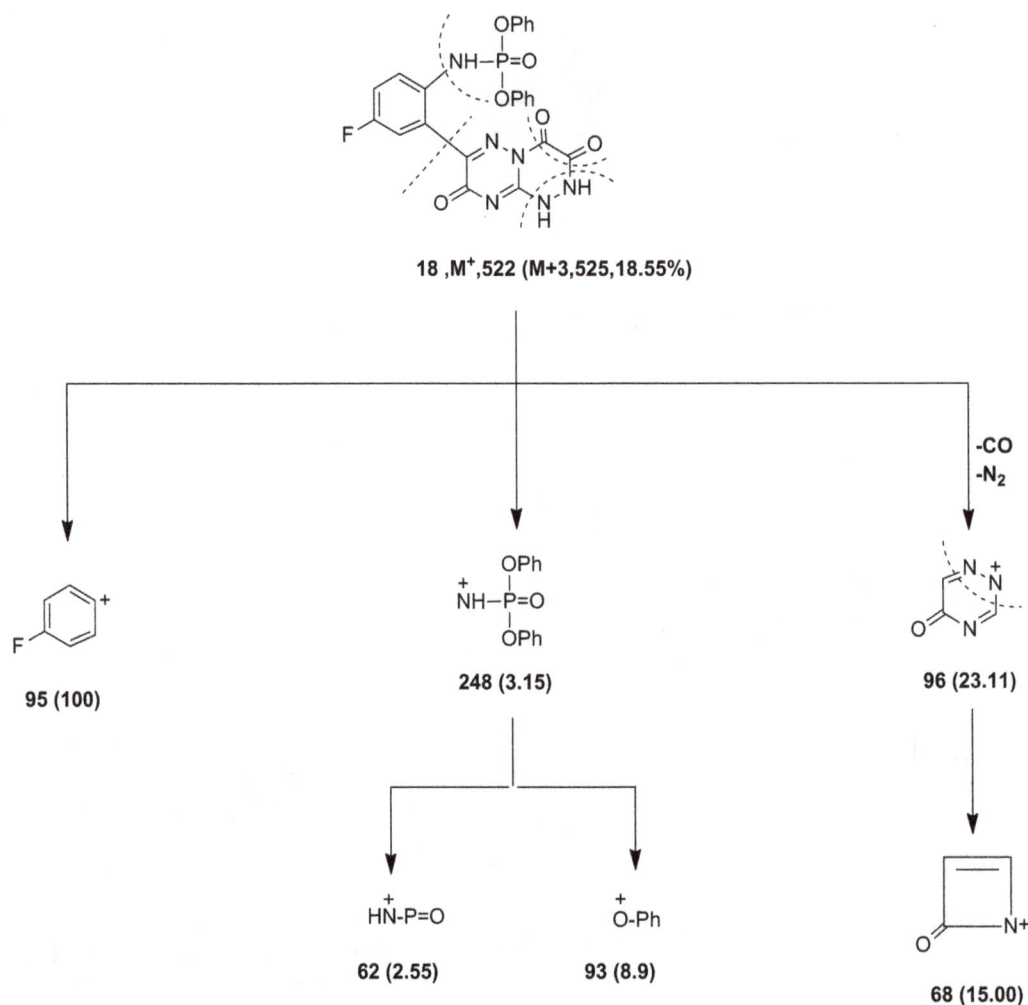

Figure 12. Mass Fragmentation pattern of compound 18.

Table 1. The *in vitro* anti-HIV-1 screening results of new compounds **3-22** in MT-4 cells.

Compd. No.	Dose (Molar)	Anti-HIV-1 Activity			
		IC$_{50}$ (µg/mL)	Max Production %	Percent of Control	
	2.00×10^{-6}			Infected	Uninfected
3		>55.13	13.5	10.11	101.11
6		>20.20	19	17.38	106.32
7		>27.20	26.5	63.51	97.93
10		>19.40	30.5	80.29	96.11
12		>28.57	25.0	35.40	105.77
14		>88.30	9.5	0.04	0.51
16		>3.99	16.00	14.16	101.91
18		>42.60	35.00	83.47	104.97
22		>94.40	23.5	18.19	104.12

Table 2. The *in vitro* anti-HIV-1 screening results of compound **18**.

Index	Concentration	Dose	Percent of Protection	Percent of Control	
				Infected	Uninfected
IC_{50} (Molar)	1.12×10^{-5}	6.33×10^{-9}	7.55	13.99	101.90
EC_{50} (Molar)	4.26×10^{-7}	2.0×10^{-8}	10.10	17.11	106.20
TIC_{50} (IC/E_C)	$2.63 \times 10^{+1}$	6.35×10^{-8}	12.00	18.15	104.21
		2.00×10^{-7}	30.01	32.31	105.30
		6.33×10^{-7}	60.66	62.15	98.10
		2.00×10^{-6}	82.23	83.47	104.99
		6.34×10^{-6}	75.66	80.13	99.10
		2.00×10^{-5}	-7.55	0.040	0.55

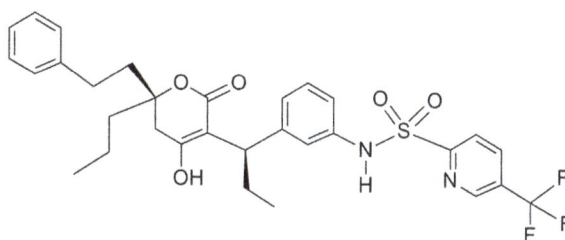

Figure 13. Tipranavir (PNU-140690).

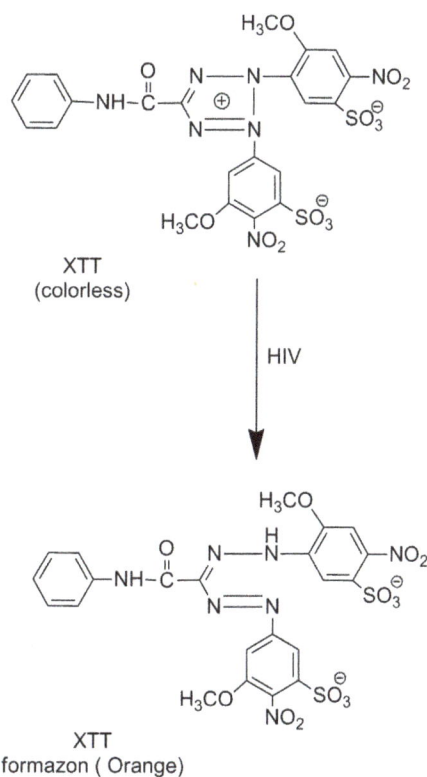

Figure 14. The standard indicator for HIV present (XXT).

with a active center of proteins of HIV-1 virus.

5. Conclusion

In search for new anti-HIV-1, novel fluorine substituted isolated and fused heterobicyclic nitrogen systems bearing 6-(2'-phosphorylanilido)-1,2,4-triazin moiety have been obtained from a ring closure reactions of the corresponding 3-hydrazino-1,2,4-triazinone with π-acceptable reagents. Some compounds exhibited a mark's inhibitors as anti-HIV-1 activity in hope to a possible control on the HIV-1 activity.

References

[1]   Hartley, F.R. and Dahl, O. (1996) The Preparation and Properties of Tervalent Phosphorus Acid Derivatives. In: Brecuer, E. and Heartly, F.R., Eds., *The Chemistry of Organophosphorus Compounds: Ter- and Quinque-Valent Phosphorus Acids and Their Derivatives*, John Wiley & Sons, New York, 653.

[2]   Prakasha, T.K., Day, R.O. and Holmes, R.R. (1994) Pentacoordinated Molecules. 101. New Class of Bicyclic Oxy-phosphoranes with an Oxaphosphorinane Ring: Molecular Structures and Activation Energies for Ligand Exchange. *Journal of the American Chemical Society*, **116**, 8095-8104. http://dx.doi.org/10.1021/ja00097a016

[3]   Fest, C. and Schmidt, K.J. (1982) The Chemistry of Organophosphorus Pesticides. Springer-Verlag, Berlin, 12. http://dx.doi.org/10.1007/978-3-642-68441-8

[4]   Bhatia, M.S. and Jit, P. (1976) Phosphorus-Containing Heterocycles as Fungicides: Synthesis of 2,2'-Diphenylene Chlorophosphonate and 2,2'-diphenylene Chlorothiophosphonate. *Experienia*, **32**, 1111. http://dx.doi.org/10.1007/BF01927572

[5]   Manne, P.N., Deshmukh, S.D., Rao, N.G.N., Dodale, H.G., Tikar, S.N. and Nimbalkar, S.A. (2000) Efficacy of Some Insecticides against *Helicoverpa armigera* (HUB). *Pestology*, **34**, 65.

[6]   Hendlin, D., Stapley, E.O., Jackson, M., Wallick, H., Miller, A.K., Woll, F.J., Miller, T.W., Chaiet, L., Kahan, F.M., Foltz, E.L., Woodruff, H.B., Mata, J.M., Hernandez, S. and Mochales, S. (1969) Phosphonomycin a New Antibiotic Produced by Strains of Streptomyces. *Science*, **166**, 122-123. http://dx.doi.org/10.1126/science.166.3901.122

[7]   Polozov, A.M. and Cremer, S.E. (2002) Synthesis of 2H-1,2-Oxaphosphorin 2-oxides. *Journal of Organometallic Chemistry*, **646**, 153-160. http://dx.doi.org/10.1016/S0022-328X(01)01207-4

[8]   Wu, X., Lei, X. and Fu, Z.F. (2003) Rabies Virus Nucleoprotein Is Phosphorylated by Cellular Casein Kinase II. *Biochemical and Biophysical Research Communication*, **304**, 333-338.

[9]   Gholivand, K., Shariatinia, Z., Mahzouni, H.R. and Amiri, S. (2007) Phosphorus Heterocycles: Synthesis, Spectroscopic Study and X-Ray Crystallography of Some New Diazaphosphorinanes. *Structural Chemistry*, **18**, 653-660. http://dx.doi.org/10.1007/s11224-007-9197-3

[10] Page, D.S. (1981) Principles of Biological Chemistry. 2nd Edition, Willard Grant Press, Boston.

[11] Hugcl, H.M. and Jackson, N. (2012) Special Feature Organo-Fluorine Chemical Science. *Applied Sciences*, **2**, 558-565. http://dx.doi.org/10.3390/app2020558

[12] Ismail, F.M.D. (2002) Important Fluorinated Drugs in Experimental and Clinical Use. *Journal of Fluorine Chemistry*, **118**, 27-33. http://dx.doi.org/10.1016/S0022-1139(02)00201-4

[13] Elliott, A.J. (1995) Chemistry of Organic Fluorine Compounds. In: Hudlicky, M. and Pavlath, A.E., Eds., *Chemistry of Organic Fluorine Compounds II: A Critical Review*, American Chemical Society, Washington DC, 1119-1125.

[14] Dolbier Jr., W.R. (2005) Fluorine Chemistry at the Millennium. *Journal of Fluorine Chemistry*, **126**, 157-163. http://dx.doi.org/10.1016/j.jfluchem.2004.09.033

[15] Smart, B.E. (2001) Fluorine Substituent Effects (on Bioactivity). *Journal of Fluorine Chemistry*, **109**, 3-11. http://dx.doi.org/10.1016/S0022-1139(01)00375-X

[16] Ojima, I., McCarthy, J.R. and Welch, J.T., Eds. (1996) Biomedical Frontiers of Fluorine Chemistry. American Chemical Society, Washington DC. http://dx.doi.org/10.1021/bk-1996-0639

[17] Al-Romaizan, A.N., Abdel-Rahman, R.M. and Makki, M.S.T. (2014) Synthesis of New Fluorine/Phosphorus Substituted 6-(2'-Amino phenyl)-3-thioxo-1,2,4-triazin-5 (2H, 4H) One and Their Related Alkylated Systems as Molluscicidal Agent as against the Snails Responsible for Bilharziasis Diseases. *International Journal of Organic Chemistry*, **4**, 154-168.

[18] Abdel-Rahman, R.M. (2001) Role of Uncondenced1,2,4-triazinecompounds and Related Heterobicyclic Systems as Therapeutic Agents. *pharmazie*, **56**, 18-30.

[19] Abdel-Rahman, R.M. (2001) Role of Uncondensed 1,2,4-Triazinederivatives as Biological Plant Protection Agents.

Pharmazie, **56**, 195-212.

[20] Abdel-Rahman, R.M. (1999) Synthesis and Chemistry of Fluorine Containing 1,2,4-triazines. *Pharmazie*, **54**, 791-804.

[21] Abdel-Rahman, R.M. (2002) Synthesis of New Phosphaheterobicyclic Systems Containing 1,2,4-Triazine Moiety Part IX; Straightforward Synthesis of New Fluorine Bearing 5-Phospha-1,2,4-triazin/1,2,4-triazepine-3-thions-part X. *Trends in Heterocyclic Chemistry*, **8**, 187-195.

[22] Treov, K.D. (2006) Chemistry and Application of H-Phosphonates. *El-Services*, Amsterdam, 256.

[23] Cherkasov, R.A. and Galki, V.I. (1998) The Kabachnik-Fields Reaction: Synthetic Potential and the Problem of the Mechanism. *Russian Chemical Reviews*, **67**, 857-940. http://dx.doi.org/10.1070/RC1998v067n10ABEH000421

[24] Annie Bligh, S.W., Mc Grath, C.M., Failla, S. and Finocchiaro, P. (1996) α-Aminophosphate Monoester in One Step. *Phosphorus, Sulfured Silicon*, **118**, 189-194.

[25] El-Sayed Ali, T. (2008) Synthesis and Characterization of Novel Bis-(α-Aminophosphonates) with Terminal Chromone Moieties. *ARKIVOC*, **2**, 71-79.

[26] El-Sayed Ali, T., Abdul-Ghaffar, S.A., El-Mahdy, K.M. and Abdel-Karim, S.M. (2013) Synthesis, Characterization, and Antimicrobial Activity of Some New Phosphorus Macrocyclic Compounds Containing Pyrazole Rings. *Turkish Journal of Chemistry*, **37**, 160-169.

[27] Abdel-Rahman, R.M. and Ali, T.E. (2013) Synthesis and Biological Evaluation of Some New Polyfluorinated-4-Thiazolidinone and α-Amino Phosphonic Acid Derivatives. *Monatshefte fur Chemie*, **144**, 1243-1252.

[28] Chakravarty, S. and Mishra, R.K. (1994) Spectrophotometric Determination of Phosphorus with N,N'-Diphenylbenzamidine and Methylen-Blue. *Journal of Indian Chemical Society*, **71**, 717-719.

[29] Bukowski, L. (2001) Some Reactions of 2-cyanomethyl-3-methyl-3H-imidazo[4,5-b]pyridine with Isothiocyanates. Antituberculotic Activity of the Obtained Compounds. *Pharmazie*, **56**, 23-27.

[30] Abdel-Rahman, R.M. and Abdel-Monem, W.R. (2007) Chemical Reactivity of 3-Hydrazino-5,6-diphenyl-1,2,4-triazine towards π-Acceptors Activated Carbonitriles. *Indian Journal of Chemistry*, **46B**, 838-846.

[31] Abdel-Rahman, R.M. (1988) Reactions of 3-Hydrazino-5,6-diphenyl-1,2,4-triazine with Unsymmetrical 1,3-bicarbonyl Compounds: Synthesis of Some New 3-(3',5'-Disubstituted pyrazol-1'-yl(-5-6-diphenyl-1,2,4-triazines of Their Antimicrobial Activity. *Indian Journal of Chemistry*, **27B**, 548-553.

[32] Zaher, H.A., Abdel-Rahman, R.M. and Abdel-Halim, A.M. (1987) Reactions of 3-Hydrazino-5,6-diphenyl-1,2,4-triazine with α,β-Bifunctional Compounds. *Indian Journal of Chemistry*, **26B**, 110-115.

[33] Gamal, A.A. (1997) Studies on 3-Amino-1,2,4-triazole. *Journal of the Indian Chemical Society*, **74**, 624-625.

[34] El-Gendy, Z., Morsy, J.M., Allimony, H.A., Abdel-Monem, W.R. and Abdel-Rahman, R.M. (2001) Synthesis of Heterobicyclic Nitrogen Systems Bearing the 1,2,4-triazine Moiety as Anti-HIV and Anticancer Drugs, Part III. *Pharmazie*, **56**, 376-382.

[35] El-Gendy, Z., Morsy, J.M., Allimony, H.A., Abdel-Monem, W.R. and Abdel-Rahman, R.M. (2003) Synthesis of Heterobicyclic Nitrogen Systems Bearing a 1,2,4-Triazine Moiety as Anticancer Drugs, Part IV. *Phosphorus, Sulfur and Silicon*, **179**, 2055-2071. http://dx.doi.org/10.1080/10426500390228738

[36] Haddadin, M.J. and Hassner, A. (1973) Cycloaddition Reactions. XIV. Thermal and Photochemical Reactions of Some Bicyclic Aziridine Enol Ethers. *Journal of Organic Chemistry*, **38**, 3466-3471. http://dx.doi.org/10.1021/jo00960a005

[37] Abdel-Rahman, R.M. and El-Gendy, Z. (1989) Synthesis of Some New1,2,4-Benzotriazine Derivatives from 2-Methyl Benzoxazole. *Indian Journal of Chemistry*, **28B**, 1072-1076.

[38] Buchannan, G.W., Whitman, R.H. and Malaiyandi, M. (1982) A Carbon-13 Nuclear Magnetic Resonance Spectral Investigation of Substituted Triphenyl Phosphates. *Organic Magnetic Resonance*, **19**, 98-101. http://dx.doi.org/10.1002/mrc.1270190211

[39] Al-Ravi, J.M.A., Behnam, G.O., Naceur, A. and Kruemer, R. (1985) Carbon-13 Chemical Shift Assignment of Some Organophosphorus Compounds. IV—2-oxo- and 2-thio-2-phenoxy-1,3,2-Diazaphosphorinanes and Related P(IV) Compounds. *Magnetic Resonance in Chemistry*, **23**,728-731. http://dx.doi.org/10.1002/mrc.1260230910

[40] Gorenstein, D.G. and Kar, D.J. (1977) Effect of Bond Angle Distortion on Torsional Potentials. *Ab Initio* and CNDO/2 Calculations on Dimethoxymethane and Dimethyl Phosphate. *Journal of the American Chemical Society*, **99**, 672-677.

[41] Reddy, C.D., Reddy, G.T. and Reddy, M.S. (1993) ^{1}H,^{13}C and ^{31}P NMR Studies of 2-(Substituted Phenoxy) 2, 3-Dihydro-1H-Naphtho-[1, 8-de]-1, 3,2-Diazaphosphorine-2-Oxides. *Asian Journal of Chemistry*, **5**, 291-295.

[42] Clercq, E.D. (2002) New Developments in Anti-HIV Chemotherapy. *Biochemical et Biophysica Acta*, **1587**, 258-275.

[43] Xie, L., Takeuchi, Y., Mark, L. and Lee, K.H. (1999) Anti-AIDS Agents. 37. Synthesis and Structure-Activity Relationships of (3'R,4'R)-(+)-*cis*-Khellactone Derivatives as Novel Potent Anti-HIV Agents. *Journal of Medicinal Chemistry*, **42**, 2662-20877. http://dx.doi.org/10.1021/jm9900624

[44] Connoly, K.J. and Hammer, S.M. (1992) Comparative Pharmacokinetics, Distributions in Tissue, and Interactions with Blood Proteins of Conventional and Sterically Stabilized Liposomes Containing 2',3'-Dideoxyinosine. *Antimicrobial Agents and Chemotherapy*, **36**, 245-254.

[45] Mitsuya, H., Weinhold, K.J., Fuman, P.A., Clair, M.H., Lehrman, S.N., Gallo, R.C., Bolognesi, D., Barry, D.W. and Broder, S. (1985) 3'-Azido-3'-deoxythymidine (BW A509U): An Antiviral Agent That Inhibits the Infectivity and Cytopathic Effect of Human T-lymphotropic Virus Type III/Lymphadenopathy-Associated Virus *in Vitro. Proceedings of the National Academy of Sciences of the United States of America*, **82**, 709670-707100.

[46] Mitsuya, H. and Broder, S. (1986) Inhibition of the *in Vitro* Infectivity Andcytopathic Effect of Human T-Lymphotrophic Virus Type III/Lymphadenopathy-Associated Virus (HTLV-III/LAV) by 2',3'-dideoxynucleosides. *Proceedings of the National Academy of Sciences of the United States of America*, **83**, 1911-1915. http://dx.doi.org/10.1073/pnas.83.6.1911

[47] Bozzette, S.A. and Richman, D.D. (1990) Salvage Therapy for Zidovudine-Intolerant HIV-Infected Patients with Alternating and Intermittent Regimens of Zidovudine and Dideoxycytidine. *The American Journal of Medicine*, **88**, S24-S26. http://dx.doi.org/10.1016/0002-9343(90)90418-D

[48] Dunkel, L., Cross, A., Martin, R., Brown, M. and Murray, H., (1990) Dose-Escalating Study of Safety and Efficacy of Dideoxydidehydrothymidine (d4T) for HIV Infection. *Antiviral Research*, **13**, 116. http://dx.doi.org/10.1016/0166-3542(90)90217-U

[49] Romero, D.L., Busso, M., Tan, C.K., Reusser, F., Palmer, J.R., Poppe, S.M., Aristoff, P.A., Downey, K.M., So, A.G., Resnick, L. and Tarpley, W.G. (1991) Non-Nucleoside Reverse Transcriptase Inhibitors That Potently and Specifically Block Human Immunodeficiency Virus Type 1 Replication. *Proceedings of the National Academy of Sciences of the United States of America*, **88**, 8806-8810. http://dx.doi.org/10.1073/pnas.88.19.8806

[50] Srivastava, A.K., Khan, A.A. and Shakil, M. (2001) Quantitative Structure Activity Relationship (QSAR) Studies on Anti-HIV-1 and Cytotoxic Arylpyrrolylsulfones. *Journal of the Indian Chemical Society*, **78**, 154-157.

[51] Lu, C. and Li, A.P. (2010) Enzyme Inhibition in Drug Discovery and Development: The Good and the Bad. Wiley, OU143, E 605.

[52] Sergei, V., Gulni, K., Elena, A. and Michael, E. (2010) Enzyme Inhibition in Drug Discovery and Developmental Edited, Lu & Li. Wiley & Sons, Inc., New York, 749.

[53] Weislow, O.W., Kiser, R., Fine, D., Bader, J., Shoemaker, R.N. and Boyd, M.R. (1989) New Soluble Formazan Assay for HIV-1 Cytopathiceffects, Application to High Flux Screening of Synthetic and Natural Products for AIDS-Antiviral Activity. *Journal of the National Cancer Institute*, **81**, 577-586. http://dx.doi.org/10.1093/jnci/81.8.577

[54] Chaouni, B.A., Galtier, C., Allouchi, H., Kherbeche, A., Chavignon, O., Teulade, J.C., Witvrouw, M., Pannecouque, C., Snoeck, R., Andrei, G., Balzarini, J., De Clercq, E., Fauvelle, F., Enguehard, C. and Gueiffier, A. (2001) 3-Benzamido, Ureido and Thioureidimidazo[1,2-a]Pyridine Derivatives as Potential Antiviral Agents. *Chemical and Pharmaceutical Bulletin*, **49**, 1631-1635. http://dx.doi.org/10.1248/cpb.49.1631

Permissions

List of Contributors

Toratane Munegumi
Department Science Education, Naruto University of Education, Naruto, Japan

Shokichi Ohuchi and Kaoru Harada
Department of Chemistry, University of Tsukuba, Tsukuba, Japan

Rammohan Pal and Taradas Sarkar
Department of Chemistry, Acharya Jagadish Chandra Bose College, Kolkata, India

Zaman Ashraf
Department of Chemistry, Allama Iqbal Open University, Islamabad, Pakistan

Aamer Saeed
Department of Chemistry, Quaid-I-Azam University, Islamabad, Pakistan

Pulimamidi Rabindra Reddy, Ravula Chandrashekar, Hussain Shaik and Battu Satyanarayana
Department of Chemistry, Osmania University, Hyderabad, India

Abeer N. Al-Romaizan, Mohammed S. T. Makki and Reda M. Abdel-Rahman
Department of Chemistry, Faculty of Science, King Abdul Aziz University, Jeddah, KSA

Khalid Hussain
Mewat Engineering College (WAKF), Nuh, India

Deepak Wadhwa
Department of chemistry, Kurukshetra University, Kurukshetra, India

Pascal Binda
Department of Chemistry and Forensic Science, Savannah State University, Savannah, GA, USA

Leslie Glover
Science Division, Southern Wesleyan University, Central, SC, USA

Yohei Oe and Tetsuo Ohta
Department of Biomedical Information, Faculty of Life and Medical Sciences, Doshisha University, Kyoto, Japan

Hiroaki Kishimoto, Nahoko Sugioka, Daisuke Harada, Yukio Sato and Isao Furukawa
Department of Molecular Science and Technology, Faculty of Engineering, Doshisha University, Kyoto, Japan

Zilma Escobar and Olov Sterner
Centre for Analysis and Synthesis, Department of Chemistry, Lund University, Lund, Sweden

Martin Johansson, Anders Bjartell and Rebecka Hellsten
Division of Urological Cancers, Department of Clinical Sciences Malmö, Lund University, Malmö, Sweden

Tomoko Mineno, Yu Takebe, Chiaki Tanaka and Sho Mashimo
Laboratory of Medicinal Chemistry, Faculty of Pharmacy, Takasaki University of Health and Welfare, Takasaki, Gunma, Japan

Yuji Ueki, Seiichi Saiki, Takuya Shibata, Hiroyuki Hoshina, Noboru Kasai and Noriaki Seko
Environment and Industrial Materials Research Division, Quantum Beam Science Center, Sector of Nuclear Science Research, Japan Atomic Energy Agency, Takasaki, Japan

Aly A. Hebeish and Amal A. Aly
Textile Research Division, National Research Center, Cairo, Egypt

Qi Lin, Huirong Zheng, Guocai Zheng, Xinzhong Li and Benyong Lou
Department of Chemistry and Chemical Engineering, Minjiang University, Fuzhou, China

Rami Y. Morjan, Basam S. Qeshta, Hussein T. Al-Shayyah, Basam A. Abu-Thaher and Adel M. Awadallah
Department of Chemistry, Islamic University of Gaza, Gaza, Palestine

John M. Gardiner
Manchester Institute of Biotechnology, School of Chemistry and EPS, The University of Manchester, Manchester, UK

Santosh Kumar Yadav
Department of Organic Chemistry & FDW, Andhra University, Visakhapatnam, India

Boshkayeva Assyl, Omarova Roza, Pichkhadze Guram and Shalpykova Nasiba
Asfendiyarov Kazakh National Medical University, Almaty, Kazakhstan

Chunguang Han
Research Center for Materials Science, Nagoya University, Nagoya, Japan

Hisashi Doi
RIKEN Center for Life Science Technologies, Kobe, Japan

Junji Kimura
Department of Chemistry and Biological Science, College of Science and Engineering, Aoyama Gakuin University, Sagamihara, Japan

Yoichi Nakao
Department of Chemistry and Biochemistry, School of Advanced Science and Engineering, Waseda University, Tokyo, Japan

Masaaki Suzuki
National Center for Geriatrics and Gerontology, Obu City, Japan

Jun-Ichi Yamaguchi, Emiko Shibuta and Yoshie Oishi
Department of Applied Chemistry, Kanagawa Institute of Technology, Atsugi, Japan

Karim Bouterfas, Ali Latreche and Zoheir Mehdadi
Laboratory of Vegetal Biodiversity: Conservation and Valorization, Faculty of Life and Natural Sciences, Djillali Liabes University, Sidi Bel-Abbes, Algeria

Djamel Benmansour
Laboratory of Statistics and Random Model, Faculty of Natural and Universe Sciences, Abou-Bekr Belkaid University, Tlemcen, Algeria

Meghit Boumedien Khaled
Department of Biology, Faculty of Natural and Life Sciences, Université Djillali Liabes, Sidi Bel Abbes, Algeria

Mohamed Bouterfas
Laboratory of Microscopy, Microanalysis of the Matter and Molecular Spectroscopy, Faculty of Exact Sciences, Djillali Liabes University, Sidi Bel-Abbes, Algeria

Patricia Manzano
Superior Polytechnic School of Litoral, Biotechnology Research Center of Ecuador (ESPOL-CIBE), Guayaquil, Ecuador
Faculty of Natural Sciences and Mathematics of Superior Polytechnic School of Litoral, Guayaquil, Ecuador

Migdalia Miranda, Tulio Orellana and Maria Quijano
Superior Polytechnic School of Litoral, Biotechnology Research Center of Ecuador (ESPOL-CIBE), Guayaquil, Ecuador

Fathi A. Abu-Shanab
Department of Chemistry, Faculty of Science, Al-Azhar University, Assiut, Egypt
Department of Chemistry, Faculty of Science, Gazan University, Gazan, KSA

Sayed A. S. Mousa and Mohamed I. Hassan
Department of Chemistry, Faculty of Science, Al-Azhar University, Assiut, Egypt

Sherif M. Sherif
Department of Chemistry, Faculty of Science, Cairo University, Giza, Egypt

Jing Zhu, Shuang Chen, Baohe Wang and Xiaorong Zhang
Research and Development Center of Petrochemical Technology, Tianjin University, Tianjin, China

Adnan A. S. El Khaldy
Department of Physics, Chemistry, and Mathematics, College of Engineering, Technology and Physical Sciences, Alabama A&M University, Normal, AL, USA

Florence Okafor
Department of Biological & Environmental Sciences, College of Agricultural, Life and Natural Sciences, Alabama A&M University, Normal, AL, USA

Alaa M. Abu Shanab
Chemistry Department, Al-Aqsa University, Gaza, Palestine

Viktor Ilkei, Kornél Faragó, László Hazai Csaba Szántay and György Kalaus
Department of Organic Chemistry and Technology, Budapest University of Technology and Economics, Budapest, Hungary

Zsuzsanna Sánta, Miklós Dékány and Csaba Szántay Jr.
Gedeon Richter Plc, Budapest, Hungary

Reda M. Abdel-Rahman, Mohammed S. T. Makki and Abeer N. Al-Romaizan
Department of Chemistry, Faculty of Science, King Abdul Aziz University, Jeddah, KSA

www.ingramcontent.com/pod-product-compliance
Lightning Source LLC
Champersburg PA
CBHW080247230326

41458CB00097B/4081